Algorithms and Combinatorics

Volume 28

For further volumes:
http://www.springer.com/series/13

Jaroslav Nešetřil • Patrice Ossona de Mendez

Sparsity

Graphs, Structures, and Algorithms

Springer

Jaroslav Nešetřil
Computer Science Institute of
Charles University (IUUK)
and
Department of Applied Mathematics
Charles University
Prague
Czech Republic

Patrice Ossona de Mendez
CNRS
Centre d'Analyse et de
Mathématique Sociales (UMR 8557)
Paris
France

Algorithms and Combinatorics ISSN 0937-5511
ISBN 978-3-642-42776-3 ISBN 978-3-642-27875-4 (eBook)
DOI 10.1007/978-3-642-27875-4
Springer Heidelberg New York Dordrecht London

Mathematics Subject Classification (2010): 05-02; 05C05, 05C10, 05C15, 05C42, 05C60, 05C62, 05C65, 05C75, 05C76, 05C78, 05C83, 05C85, 05C90, 03C13, 68R10, 90C35

Supported by a grant of the Czech Ministry of Education

Printed on acid-free paper

Springer is part of Springer Science+Business Media (www.springer.com)

To Helena and Thị Mỹ Liên.

The drawings under the parts' title are freely inspired by works of

- *Jan Vredeman de Vries*,
- *Vassily Kandinsky*,
- *Jiří Načeradský* and *Jaroslav Nešetřil*.

The image on the first page of the front-matter is an original drawing by *Jaroslav Nešetřil*, which is freely inspired by the ink on paper "Elegant Rocks and Sparse Trees" by *Zhao Mengfu*.

Preface

This text is aimed at doctoral students and researchers, who are interested in Combinatorics and Graph Theory or who would just like to learn about some active topics and trends. But the book may be also interesting to researchers in mathematics, physics, chemistry, computer science, etc. who would seek for an introduction to the tools available for analysis of the properties of discrete structures, and sparse structures particularly. The dichotomy between sparse and dense objects is one of the main paradigm of the whole mathematics which transcends boundaries of particular disciplines. This is also reflected by our book.

The book is organized in three parts, called *Presentation*, *Theory*, and *Applications*.

The first part, *Presentation*, gives a general overview of the covered material and of its relationships with other domains of contemporary mathematics and computer science. In particular, Chap. 2 is devoted to the exposition of some typical examples illustrating the scope of this book.

The second part, *Theory*, is the largest part of the book and it is divided into eleven chapters. Chapter 3 introduces all the relevant notions and results which will be used in the book: basic notions and standard terminology, as well as more involved concepts and constructions (such as homomorphisms, minors, expanders, Ramsey theory, logic, or complexity classes), or more specific considerations on graph parameters, structures, and homomorphism counting. Chapter 4 introduces the specific notions used to study the density properties, *shallow minors*, *shallow topological minors*, or *shallow immersions* of individual graphs, as well as the related fundamental stability results. These results are applied in Chap. 5, and this leads to the *nowhere dense/somewhere dense* classification and to the notion of *classes with bounded expansion* (which are sparser than general nowhere dense classes). This classification is very robust and it can be characterized by virtually all main combinatorial invariants. Several first characterizations are included in Chap. 5, and more characterizations are given in Chaps. 7, 8, 12, and 11. Chapter 5 ends with a discussion about the connection to model theory and the various approaches to handle general relational structures. Although the study of dense graphs frequently relies on the properties of dense homogeneous core structures (like complete graphs or even random graphs), it will be shown that sparse graph properties are intimately related to the properties of trees, and particularly to the ones of bounded height trees. Fundamental results on bounded height trees and, more generally, on graphs with bounded tree-depth are proved in Chap. 6. They open the way to the main decomposition theorem, which is the subject of Chap. 7. The decomposition scheme introduced there, which we call *low tree-depth coloring*, is a deep generalization of the concept of proper coloring. The low tree-depth colorings also lead to an alternative characterization of the nowhere dense/somewhere dense dichotomy. Yet another characterization of this dichotomy is proved in Chap. 8, that relies on the notion of independence through the notion of *quasi-wideness* (which has been introduced in the context of mathematical logic). Chapters 9 and 11 deal with *homomorphism dualities*. Bounded expansion classes are proved to have the richest spectrum of finite dualities and, in the oriented case, they are actually characterized by this property. Meanwhile, Chap. 10 establishes a connection to model theory and deals particularly with relativizations of the homomorphism preservation theorem of first-order logic. A last characterization of the somewhere dense/nowhere dense dichotomy is proved in Chap. 12 by considering the asymptotic logarithmic density of a fixed pattern in the shallow minors of the graph of a class. In a sense, one can view this last result as a characterization of the dichotomy in probabilistic terms. The *Theory* part ends with Chap. 13 where the results of the previous chapters are gathered and put to service in the study of the characteristics of nowhere dense classes, of classes with bounded expansion, and of classes with bounded tree-depth (which are derived from trees with bounded height). It is pleasing to see how these characterizations are nicely related.

The third part, *Applications*, concerns both theoretical and algorithmic applications of the concepts and results introduced in the second part. This part opens with Chap. 14 which gives several examples of classes with bounded expansion, such as classical classes defined in the context of geometric graphs and graph drawing, as well as classes admitting bounded non-repetitive colorings. It is also the occasion for a connection with the Erdős-Rényi model of random graphs. Some applications are considered in Chap. 15, such as the existence of linear matching (and more generally unions of long disjoint paths), connection with the Burr-Erdős conjecture, with game coloring, and with spectral graph theory. In Chap. 16, the use of a density driven criterion for the existence of sublinear vertex separators links our study to the sparse model of property testing, via the concept of *hyperfiniteness*.

We provide in Chap. 17 core algorithms related to our study. In particular, we detail a fast iterative algorithm to compute a low tree-depth decomposition, the number of colors being controlled by a polynomial dependence on the densities of the shallow minors of the graph. The fact that this algorithm is nearly linear for sparse classes is one of the main advantages of our approaches. In Chap. 18 we consider algorithmic applications, which mainly derive from the fast low tree-depth coloring algorithm. These cover various well-known algorithmic problems, such as *subgraph isomorphism*, decidability of first-order properties, as well as their counting versions.

The title of the last chapter—*Further Directions*—is self-explanatory.

This book contains some previously unpublished results of the authors, as can be expected in a fast developing field. The extensive literature reflects the multiplicity of connections, applications, and similarities to other parts of mathematics and theoretical computer science.

We included exercises at the end of nearly every chapter. These exercises may complement previous material by a small question but often they suggest further study or extension of the main text. Such exercises may also contain hints for solutions. Some hints are also included at the end of the book.

This book is the result of the collaboration of the authors for over a decade in both Paris and Prague (and elsewhere). This was made possible thanks to the generous support of institutions at both ends: École des Hautes Études en Sciences Sociales, École Normale Supérieure, and Université Paris VI in Paris, as well as the Institute of Theoretical Computer Science (ITI) and the Department of Applied Mathematics (KAM) and most recently by Computer Science Institute (IUUK) of Charles University in Prague. We thank our colleagues for friendly working atmosphere. Particularly, we would like to thank Zdeněk Dvořák, Louis Esperet, Tomáš Gavenčák, Andrew Goodall, Jan van den Heuvel, Ida Kantor, Jiří Matoušek, Reza Naserasr, Melda Nešetřilová (née Hope), and Pascal Ochem for comments to parts of the book.

Paris, Prague, December 2011

Jaroslav Nešetřil
Patrice Ossona de Mendez

Contents

List of Symbols

We list here most of the symbols throughout this book, together with the page corresponding to the symbol's definition.

Variables

Γ, G, H	finite loopless undirected graphs, 21
$\vec{G}, \vec{H}$	finite directed graphs, 24
$\mathbb{A}, \mathbb{B}, \mathbb{L}$	Limits of homomorphism equivalence classes, 219
u, v, x, y	vertices, 21
e, f, g	edges, 21
$\mathcal{C}, \mathcal{F}, \mathcal{D}$	classes of graphs, 89
$\mathfrak{C}$	a sequence of infinite graph classes, 93
Σ	a surface, 31
ϱ, ς	graph parameters, 95
$\lambda_1, \lambda_2, \ldots, \lambda_n$	eigenvalues, 37
a, b, c	depth of a shallow (topological) minor, 62
$\mathcal{H}$	a hypergraph, 48
ϕ	formula or sentence, 49
σ	signature, 47
$P(X), Q(X, Y)$	polynomials, 55

Asymptotic Notations

$f = O(g)$	Landau symbol O: asymptotic domination of f by g, 55
$f = \Omega(g)$	asymptotic domination of g by f, 55
$f = \Theta(g)$	asymptotic equivalence of f and g, 55
$f = o(g)$	Landau symbol o: $f/g \to 0$, 55
$f \sim g$	Asymptotic equality, 55
$f \asymp g$	polynomial functional dependence, 55

Special Structures

C_n	cycle of order n (and length n), 21
K_n	complete graph of order n, 21
$K_{n,m}$	complete bipartite graph with parts of size n and m, 21
P_n	path of order n (and length $n-1$), 21
$\vec{P}_n$	directed path of order n (and length $n-1$), 42
$\vec{T}_n$	transitive tournament of order n, 42
$G(n, p(n))$	random graph of order n and edge probability $p(n)$, 314

Graph Parameters

$\|G\|$	order of the graph G, 21
$\|\|G\|\|$	size of the graph G, 21
$\alpha(G)$	independence number of G, 58
$\beta^*(G)$	size of a maximum induced matching of G, 344
$\beta(G)$	matching number of G, 14
$\Delta(G)$	maximum degree of G, 21
$\delta(G)$	minimum degree of G, 21
$\chi(G)$	chromatic number of G, 24
$\chi_g(G)$	game chromatic number, 352
$\chi_{rk}(G)$	vertex ranking number of G, 125
$\chi_s(G)$	star chromatic number of G, 147
$\omega(G)$	clique number of G, 39
$b_\epsilon(G)$	ϵ-boundedness of G, 39
$bw(G)$	band-width of G, 37
$col(G)$	coloring number of G, 86
$col_k(G)$	k-coloring number of G, 86
$cr(G)$	crossing number of G, 319
$cr(\vec{G})$	cycle rank of the digraph $\vec{G}$, 127
$\overline{d}(G)$	average degree of G, 21
$g(G)$	genus of the graph G, 55
$g_\alpha(G)$	α-vertex expansion of G, 37
$girth(G)$	minimum length of a cycle of G, 27
$h(G)$	Hadwiger number of G, 33
$Iso(G)$	edge expansion of G, 37
$mad(G)$	maximum average degree of G, 24
$pw(G)$	path-width of G, 34
$qn(G)$	queue number of G, 321
$r(G)$	Ramsey number of G, 53
$s(G)$	separation number of G, 37
$sn(G)$	stack number of G, 321
$tw(G)$	tree-width of G, 34

$\mathrm{wcol}_k(G)$	weak k-coloring number of G, 86
$\langle G \rangle$	profile, Lovász vector, 46

Other Voices, Other Rooms

$\alpha_r(G)$	r-independence number of G, 175
$\chi_p(G)$	p-chromatic number of G, 150
$\Phi_{\mathcal{C}}$	Scattering function of the class $\mathcal{C}$, 176
$\overline{\Phi}_{\mathcal{C}}$	Uniforam scattering function of the class $\mathcal{C}$, 177
$\mathrm{free}(F, G)$	degree of freedom of F in $\mathcal{C}$, 280
$h_i(G)$	maximal order of a clique immersion in G, 33
$h_t(G)$	maximal order of a topological clique minor of G, 33
$\ell\mathrm{dens}(G)$	logarithmic density of G, 97
$s_G(i)$	maximum minimal size of a $\frac{1}{2}$-vertex separator of a subgraph of G of order i, 37
$\mathrm{td}(G)$	tree-depth of G, 115
$\nabla_r(G)$	grad of rank r of G, 66
$\nabla(G)$	maximum edge-density of a minor of G, 66
$\widetilde{\nabla}_r(G)$	top-grad of rank r of G, 67
$\widetilde{\nabla}(G)$	maximum edge-density of a topological minor of G, 67
$\overset{\propto}{\nabla}_{p,q}(G)$	imm-grad of rank (p, q) of G, 84

Functions

$\mathcal{F}(c,t)$	maximum order of a c-colored graph of tree-depth t without non trivial involutive automorphisms, 132
$\mathcal{F}(t)$	maximum order of a graph of tree-depth t without non-trivial involutive automorphisms, 132
$R(n_1, \ldots, n_k)$	Ramsey number, 52
$E[X]$	expected value of X, 171
$\mathrm{Inj}(A, B)$	set of all injective injective mappings from A to B, 92
$\widehat{f}$	smallest upper continuous concave function greater or equal to f, 370
$f(\mathcal{C})$	$\sup_{G \in \mathcal{C}} f(G)$, 91
$\limsup_{G \in \mathcal{C}} f(G)$	limit superior of f on the class $\mathcal{C}$, 92
$\overline{f}(\mathcal{C})$	limit superior of f on the class $\mathcal{C}$, 92
$\overline{f}(\mathfrak{C})$	limit superior of f on the class sequence $\mathfrak{C}$, 93
$\delta(S)$	cut-set (or cobord) of S, 37
$d_G(v)$	degree of vertex v in the graph G, 21
$d^-(v)$	indegree of v, 24
$\mathrm{dist}_G(x, y)$	shortest path distance of x and y in G, 61
$d^+(v)$	outdegree of v, 24
$\mathrm{height}(x, F)$	height of vertex x in the rooted forest F, 115

$(\#F \subseteq G)$	number of induced copies of F in G, 278
$\sigma_\zeta(F, Y)$	number of ζ-consistent mappings from F to Y, 283
dist_L	left distance in $[\mathrm{Rel}(\sigma)]$ (and $\overline{[\mathrm{Rel}(\sigma)]}$), 208
dist_R	right distance in $[\mathrm{Rel}(\sigma)]$ (and $\overline{[\mathrm{Rel}(\sigma)]}$), 208
dist	full distance in $[\mathrm{Rel}(\sigma)]$ (and $\overline{[\mathrm{Rel}(\sigma)]}$), 208
$\mathrm{dist}_\equiv$	first-order pseudo-metric on $\mathrm{Rel}(\sigma)$, 239
dist_{FO}	first-order distance in $\mathfrak{T}$, 234
dist_{td}	tree-depth distance in $[\mathrm{Rel}(\sigma)]$, 236
d_{FO}	quotient metric of $(\mathfrak{T}_C, \mathrm{dist}_{FO})/\sim_P$, 238
$\mathrm{Hom}(G, H)$	set of all homomorphisms from G to H, 41
$\vartheta(\mathbf{A})$	bijective mapping from $\mathrm{Rel}(\sigma)$ to P, 229
$\mathbf{M}(\phi)$	bijective mapping from P to $\mathrm{Rel}(\sigma)$, 229

Operations

$A(G)$	adjacency matrix of G, 37
$D(G)$	degree matrix of G, 355
$L(G)$	Laplacian of G, 355
$N_d^G(u)$	d-neighborhood of u in G, 61
$Q_k(G_L, y)$	set of vertices that are weakly k-accessible from y, 86
$R_k(G_L, y)$	set of vertices that are k-accessible from y, 86
$G - v$	vertex deletion, 23
G/e	edge contraction, 30
$G \setminus e$	edge deletion, 30
$G[A]$	subgraph of G induced by A, 22
$G_{<k}$	subgraph of G induced by the vertices of degree strictly smaller than k, 22
$G_{\leq k}$	subgraph of G induced by the vertices of degree at most k, 22
$G/\mathcal{P}$	minor of G obtained as the adjacency graph of the parts in $\mathcal{P}$, 62
$\mathrm{clos}(F)$	closure of the rooted forest F, 115
$G \,\Box\, H$	Cartesian product of G and H, 279
$G \times H$	categorical product of G and H, 40
$G \bullet H$	lexicographic product of G and H, 80
$G + H$	disjoint union (categorical sum) of G and H, 40
$\mathrm{Gaifman}(\mathbf{A})$	Gaifman graph of $\mathbf{A}$, 49
$\mathrm{Inc}(\mathbf{A})$	incidence graph of a relational structure $\mathbf{A}$, 49
$\mathrm{Inc}(\mathcal{H})$	incidence graph of a hypergraph $\mathcal{H}$, 49
$\mathbf{A} \times \mathbf{B}$	categorical product of $\mathbf{A}$ and $\mathbf{B}$, 47
$\mathbf{A} + \mathbf{B}$	disjoint union (categorical sum) of $\mathbf{A}$ and $\mathbf{B}$, 47
$\mathbf{U}(\mathbf{A})$	Feder and Vardi function $\mathbf{U}$, 199

$[\mathcal{I}, \mathcal{F}]$	Limit in $\overline{[\mathrm{Rel}(\sigma)]}$ defined by the ideal $\mathcal{I}$ and the filter $\mathcal{F}$, 220
left $\lim_{i\to\infty}[\mathbf{G}_i]$	left limit of the $[\mathbf{G}_i]$'s, 212
right $\lim_{i\to\infty}[\mathbf{G}_i]$	right limit of the $[\mathbf{G}_i]$'s, 218
$\lim_{i\to\infty}[\mathbf{G}_i]$	full limit of the $[\mathbf{G}_i]$'s, 219
$G \triangledown 0$	class of all the subgraphs of G, 62
$G \triangledown r$	class of shallow minors of depth r of G, 62
$G \mathbin{\widetilde{\triangledown}} r$	class of shallow topological minors of depth r of G, 65
$G \mathbin{\overset{\propto}{\triangledown}} (p, q)$	class of all shallow immersions of G with complexity p and stretch q, 84
$\mathcal{C} \triangledown 0$	monotone closure of $\mathcal{C}$, 94
$H(\mathcal{C})$	hereditary closure of the class $\mathcal{C}$, 61
$\mathcal{C} \triangledown \infty$	minor closure of $\mathcal{C}$, 94
$\mathcal{C} \triangledown r$	class of shallow minors of depth r of graphs in $\mathcal{C}$, 93
$\mathcal{C} \mathbin{\widetilde{\triangledown}} \infty$	topological closure of $\mathcal{C}$, 94
$\mathcal{C} \mathbin{\widetilde{\triangledown}} r$	class of shallow topological minors of depth r of graphs in $\mathcal{C}$, 93
$\mathcal{C} \bullet F$	class which contains the lexicographic products of graphs in $\mathcal{C}$ and F, 93

Relations

$H \subseteq G$	H is a subgraph of G, 22
$G \subseteq_i H$	induced subgraph relation, 22
$(G, \gamma) \subseteq_i (H, \eta)$	labeled induced subgraph relation, 136
$G \leq_m H$	minor order, 30
$G \leq_t H$	topological minor order, 31
$G \leq_i H$	immersion order, 31
$G \leq_h H$	homomorphism quasi-order, 42
$G \to H$	existence of a homomorphism of G to H, 39
$G \nrightarrow H$	non-existence of a homomorphism of G to H, 39
$G \underset{\not\leftarrow}{\rightarrow} H$	$G \to H$ and $H \nrightarrow G$, 45
$G \cong H$	isomorphism, 39
$G \subseteq_i^\star H$	is a retract of relation, 139
$[G] \leq_h [H]$	homomorphism order, 43

Classes and Sets

$\binom{G}{H}$	set of all the induced subgraphs of G which are isomorphic to H, 22
$[G]$	hom-equivalence class of G, 43
$\mathrm{Forb}_m(\mathcal{F})$	Class of graphs with no minor in $\mathcal{F}$, 90

$\mathrm{Forb_h}(\mathcal{F})$	Class of graphs with no homomorphic image of a graph in $\mathcal{F}$, 90
$(\mathbf{A} \rightarrow)$	structures with a homomorphism from $\mathbf{A}$, 205
$(\rightarrow \mathbf{A})$	structures with a homomorphism to $\mathbf{A}$, 205
$\mathrm{Inc}(\mathcal{C})$	Class of the incidence graphs of relational structures in $\mathcal{C}$, 111
$\mathcal{T}_t$	class of all graphs with tree-depth at most t, 134
$\mathcal{G}raph$	class of all (isomorphism types of) finite graphs, 40
$\mathrm{Rel}(\sigma)$	Category of all finite σ-structures, 47
$\mathfrak{Rel}(\sigma)$	class of all (finite or infinite) σ-structures, 233
$\mathrm{Tree}(\sigma)$	Class of all finite σ-trees, 49
$[\mathcal{G}raph]$	poset of all hom-equivalence classes of graphs, 43
$\overline{[\mathrm{Rel}(\sigma)]}_L$	left completion of $[\mathrm{Rel}(\sigma)]$, 212
$\overline{[\mathrm{Rel}(\sigma)]}_R$	right completion of $[\mathrm{Rel}(\sigma)]$, 218
$\overline{[\mathrm{Rel}(\sigma)]}$	full completion of $[\mathrm{Rel}(\sigma)]$, 219
$\mathcal{C}^{\nabla}$	resolution of $\mathcal{C}$, 94
$\mathcal{C}^{\widetilde{\nabla}}$	topological resolution of $\mathcal{C}$, 94
$\mathcal{C}^{\overset{\propto}{\nabla}}$	immersion resolution of $\mathcal{C}$, 94
$\mathfrak{P}$	class of all closed PP-theories, 231
$\mathfrak{T}$	class of all theories, 233
$\mathfrak{T}_C$	class of all complete theories, 233
$\mathfrak{T}_F$	class of all complete theories with a finite model, 233
$\mathfrak{T}_{FMP}$	class of all complete theories with finite model property, 233

Posets

$x \vee y$	join of x and y, 203
$x \wedge y$	meet of x and y, 203
$x \leq_F y$	partial order induced by a rooted forest F, 115
A^{ℓ}	set of lower bounds of A, 204
$\mathcal{F}^{\star}$	Ideal dual to the filter $\mathcal{F}$, 205
$\mathcal{I}^{\star}$	Filter dual to the ideal $\mathcal{I}$, 205
A^{u}	set of upper bounds of A, 204
${}_{\downarrow}[\mathbf{A}]$	lower set of $[\mathrm{Rel}(\sigma)]$ defined as the elements $\leq_h \mathbf{A}$, 203
$[\mathbf{A}]^{\uparrow}$	upper set of $[\mathrm{Rel}(\sigma)]$ defined as the elements $\geq_h \mathbf{A}$, 203

Logic

$\phi \vdash \psi$	entailment relation, 231
$G \equiv H$	elementary equivalence, 50
$G \equiv^n H$	n-back-and-forth equivalence, 50
$\mathbf{A} \models \phi$	$\mathbf{A}$ satisfies ϕ, 49

qcount(ϕ)	quantifier count of ϕ, 49
qrank(ϕ)	quantifier rank of ϕ, 49
Mod(T)	class of the models of the theory T, 233
Th(**A**)	theory of **A**, 233
FO	Class of all σ-sentences, 229
FO^n	Class of all σ-sentences with quantifier rank at most n, 229
P	Class of all primitive positive sentences, 229
P^n	Class of all primitive positive sentences with quantifier rank at most n, 229

Complexity Classes

AC^i	unbounded fanin $O(\log^i(n))$-depth circuits, 57
AW[*]	alternating W[*], 399
FO	first-order logic, 57
FPT	fixed-parameter tractable, 399
L	deterministic logarithmic space, 56
NC^i	Nick's Classes: $O(\log^i(n))$ time on a a polynomial number of processors, 57
NL	non-deterministic logarithmic space, 56
NP	non-deterministic polynomial time, 56
P	deterministic polynomial time, 56
PSPACE	polynomial space, 56
W[1]	weighted analogue of NP, 399
W[t]	non-deterministic fixed-parameter hierarchy, 399
W[*]	union of the W[t]'s, 399

Presentation

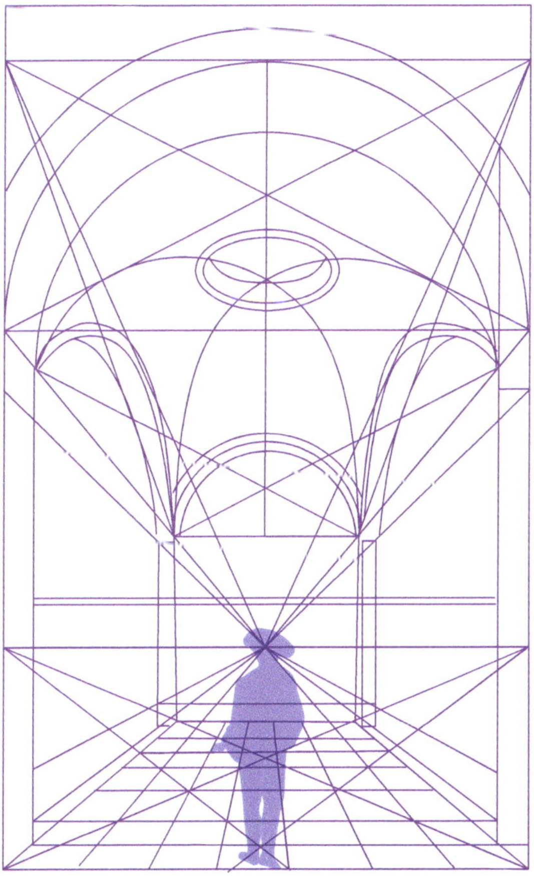

Chapter 1
Introduction

Where the reader will learn why we wrote this book.

Combinatorics is a long story. But we believe that we live in the unique point of scientific history when combinatorics is becoming an essential part of mathematics and when the rich techniques developed in isolated and specific contexts are put to service in solving problems which are of general mathematical and scientific interest.

Combinatorics is already widely regarded as a set theory for computer science. By now, it is easy to find other specific examples. For instance, who would have thought just a few years ago that questions related to Ramsey theory (and combinatorial number theory) could be proposed under the same umbrella as both asymptotic and convergence properties and structural properties of homomorphisms, next to the analysis of partition functions of statistical physics and even next to constraint satisfaction problems (CSPs)? Who would have thought that abstract properties of finite structures would be tested using limit objects and advanced probabilistic techniques? These are not mere analogies. They are rooted in deep results of mathematicians from many fields. They are results of many mathematicians with a growing combinatorial expertise. The references here cannot be exhaustive, for a sample see [78, 293, 453].

Various qualitative questions in mathematics (and in science in general) are dichotomies: probabilistic versus deterministic, polynomial versus exponential, local versus global, easy versus hard. Such dichotomies are often not well defined. Rather, they reflect the experience gained by researchers in a particular area. One such dichotomy is "sparse versus dense", which is the main topic of this book.

J. Nešetřil and P.O. de Mendez, *Sparsity*, Algorithms and Combinatorics 28,
DOI 10.1007/978-3-642-27875-4_1, © Springer-Verlag Berlin Heidelberg 2012

The notion of sparsity of course depends on the particular field of study (sparse matrix, sparse graph, sparse set). However, a bit surprisingly, many of these concepts and problems fit into a grand picture which we introduce. We aim for generality yet we shall illustrate most of the topics on a very simple model: undirected graphs. It is a specific feature of combinatorics that it allows this: Generality, like in a good fairy tale, is illustrated by means of a key case.

What is a sparse graph? This is a fuzzy notion which has to be understood in the context. A sparse graph is not defined by itself, it is sparse relative to other graphs. Sparsity is a robust notion which does not change by a small change, like the notion of a heap of hay which does not change by a adding or removing a few straws. Hence this notion should not apply to a single graph, but rather to a *class of graphs*.

As opposed to dense situations, the geometry (or rather topology) plays a key role when one studies sparse structures, as these structures may have intricate global properties (shape) which we do not want to destroy. Thus our work evolved from the study of several areas, such as constrained colorings and homomorphisms of geometrically defined graphs, Ramsey theory, and constrained orientations of topological graphs.

On the one hand, we consider graph transformations which are related to topology and geometry, like *minors* (i.e. contractions), *topological minors* (i.e. subdivisions), and even *immersions*, and we restrict their use with further local constraints (bounded radius of contracted parts, bounded number of subdivision vertices) to ensure the stability of the global shape. On the other hand, we mention in this book many generalizations of chromatic numbers, the homomorphism paradigm, limits and universal structures, structural Ramsey theory, various tree- and branching- structures as a measure of approximation, and various notions related to topological graph theory. This mixture of topology and geometry on one the one hand and of combinatorics and algebra on the other hand is central to our analysis.

The sparse versus dense dichotomy takes in this book the form of a classification of arbitrary (infinite) classes of finite graphs (or finite structures) into *somewhere dense* and *nowhere dense* classes (see Fig. 1.1). We study this dichotomy in great detail.

Several of our general results include celebrated results on computational complexity, separators of meshes, Ramsey numbers, and homomorphism preservation as special cases. In spite of its generality our approach is very effective and in many instances yields very fast algorithms and structural results. The basic techniques (for example: fraternal augmentation and transitive fraternal augmentation) lead to very fast algorithms (which became almost linear in the sparse case) and they can be viewed as master algorithms for many problems considered in the literature for special classes (such

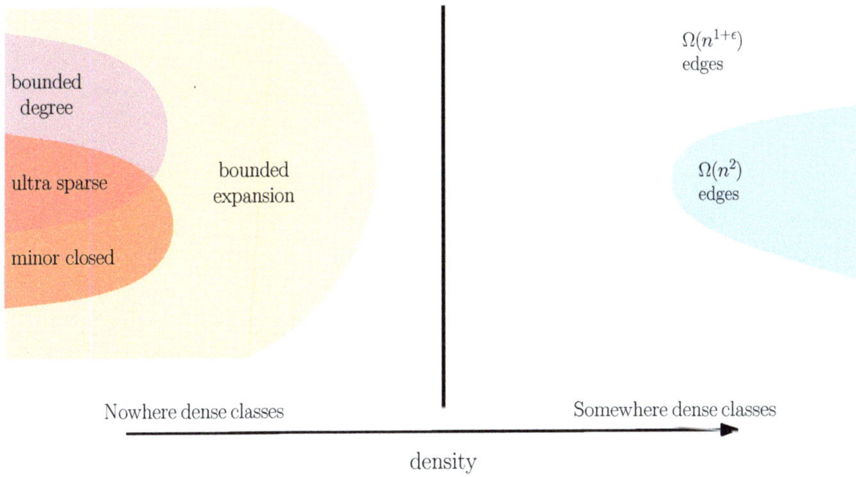

Fig. 1.1 Infinite classes of graphs are either nowhere dense or somewhere dense

as various algorithms for planar graphs and proper minor closed classes of graphs). We devote the whole third part of this book to such applications.

The topic of this book developed gradually and we believe that the whole area is calling for a synthesis and for a proper organization. We also feel that the time is right because of the relationship (and interest) of this subject with other parts of combinatorics, and with mathematics and computer science in general: this topic lies at the crossroads of several disciplines and this, we believe, leads to new perspectives.

In Chap. 2, we illustrate potential applications by a selection of a few particular problems. An informed reader may skip this motivating chapter and proceed to the beginning of Part *The Theory*, i.e. to Chap. 3.

Chapter 2
A Few Problems

A solution without a problem is an ill-stated solution.

2.1 Breaking a Mesh

Meshes are a standard support for finding approximate solutions to partial differential equations (PDE) as well as of integral equations. Computational techniques working with a Divide-and-Conquer scheme need an initial mesh to be recursively broken into pieces of comparable size by cutting along small set of points (called a *vertex separator*).

This leads to the following well known problem, formulated here in graph theoretical terms.

Vertex Separator Problem

Given a graph G with n vertices, what is the smallest size of a *vertex separator* of G, that is a subset of vertices whose deletion separates G into two parts, each including at least one third of the vertices?

This problem (illustrated by Fig. 2.1) has been extensively investigated. By a famous result of Lipton and Tarjan [304, 305] it is known that (yes, only!) about $\sqrt{n}$ vertices suffice for any planar graph with n vertices (i.e. for graphs on n vertices drawn on the plane without crossings) and this has been extended to more general situations [24, 25, 218]. A similar problem concerns

J. Nešetřil and P.O. de Mendez, *Sparsity*, Algorithms and Combinatorics 28, DOI 10.1007/978-3-642-27875-4_2, © Springer-Verlag Berlin Heidelberg 2012

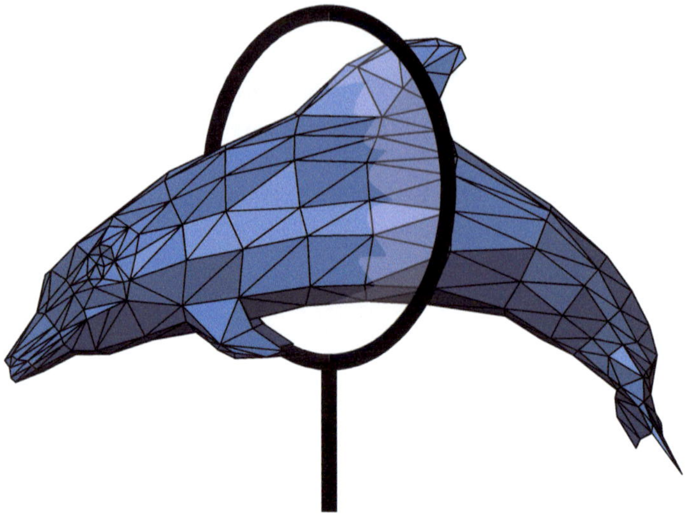

Fig. 2.1 How to cut a mesh into two parts, both of approximately equal size, by removing a few vertices?

geometric meshes: The *aspect ratio* of a d-dimensional body is the ratio of its diameter by the d-th root of its volume—for instance, the aspect ratios of a d-simplex, d-cube, and d-ball are respectively $\alpha_s = 2^{1/2}(d!)^{1/d}(d+1)^{-1/(2d)} \sim \sqrt{2}d/e$, $\alpha_c = \sqrt{d}$, and $\alpha_b = 2\pi^{-1/2}(d/2)!^{1/d} \sim \sqrt{2d/(e\pi)}$. A *d-dimensional mesh of aspect ratio* α is a 1-skeleton of a complex in which the aspect ratio of every d-simplex is at most α. It is of particular importance in finite element methods to find small vertex separators for d-dimensional meshes with bounded aspect ratio. Miller and Thurston first proved that such separators may be found in dimension 2 or 3 [333]. Then, Plotkin, Rao and Smith, using a notion of "limited depth minors", proved that for arbitrary dimension d the number of vertices needed to break a mesh into two parts of similar size is bounded by approximately $(n \log n)^{1-1/2d}$ [386], and this result was eventually improved by Miller and Thurston who proved that vertex separators meeting the optimal bound $O(n^{1-1/d})$ can be found in linear time [332].

Can these results be further generalized? There are some limits to do this: for instance, random regular graphs with n vertices almost surely need a positive fraction ϵn of the whole set of vertices to be removed in order to be split into two parts, each containing at least one third of the vertices. These graphs, called *expanders*, present a natural barrier to effective recursion [22, 259, 316, 317, 324, 383].

Where lies the border line for the existence of a small (say sub-linear) separator? It is related to expansion properties of the graph and we shall add to this area yet another line by relating this to our study of "class expansion". We treat this in detail in Chap. 16, and particularly in Sect. 16.3.

2.2 Forging Alliances

The traditional peaceful (and definitely politically correct) interpretation of the Four Color Theorem reads as follows: one can color the regions of any political map of states by using no more than four colors in such a way that no two adjacent regions receive the same color (assuming that each state is connected, i.e. has no colonies). In a more aggressive interpretation we may consider contiguous countries in a fictious world where countries sharing a common frontier naturally tend to be in conflict (Fig. 2.2). The Four Color Theorem would state that the countries may be partitioned into four conflict-free alliances.

Fig. 2.2 The fictive conflict graph of a fictitious world

Actually, in this fictitious world, things can be even more complicated: each country is inclined to consider those of its enemies' enemies which are not direct enemies as objective allies and, similarly, those enemies of its objective allies which are neither direct enemies nor objective allies as objective enemies. From a mathematical point of view, this example highlights a specific notion of transitivity adapted to the so-called *conflict graphs*: direct enemies are at distance 1, objective allies are at distance 2 and objective enemies are at distance 3 in the conflict graph (see Fig. 2.3). Considering similar arguments, one could consider that countries at distance four are loose allies and those at distance five are loose enemies and one can go on with the larger distances (of course we should stop as countries would be so distant that they would not really care one way or the other).

Fig. 2.3 The fictive conflict graph of the fictitious world extended by distance three fictive conflicts

Let us go back to our alliance problem. Is it possible to partition the world into a hopefully small number of alliances such that no direct enemies or objective enemies or even loose enemies would belong to the same alliance? Although this seems hopeless in view of Fig. 2.3 the answer is actually yes: for any number of countries in our fictitious world, one can always find a partition with a bounded number of alliances (presently at most $2^{10^{10}}$), as we will see in Sect. 11.9.3.

We could also consider yet another notion adapted to "shy" countries. In our situation with objective enemies and allies, we would not like to position objective allies against each other. In other words, if two of my objective allies are in conflict, then they are both my enemies, in the sense that I don't want any alliance with either one of them. From the point of view of a graph theorist, this means that the existence of an induced path of length 3 (i.e. a path with 4 vertices) should prevent two end vertices from belonging to a common alliance. In this case it is still possible to distribute the countries into a fixed (albeit large) number of alliances. However, a bit surprisingly, this in no longer the case when we try to go further by considering induced paths of length 5 (cf Fig. 2.4).

Fig. 2.4 Coloring planar graphs according to odd induced paths. If vertices linked by an induced path of length 5 are required to get different colors (induced conflicts shown in the *lower left*), an unbounded number of colors may be necessary. However, if vertices linked by an induced path of length 3 are required to get different colors (induced conflicts shown in the *lower right*) a bounded number of colors is sufficient (in the particular example shown here, three colors suffice)

2.3 Are Symmetries Frequent?

Groups and graphs are closely related: the automorphisms (symmetries) of a graph form a group and to each group one can associate a Cayley graph. For instance, a Cayley graph of the group of the rotations of a cube is shown Fig. 2.5.

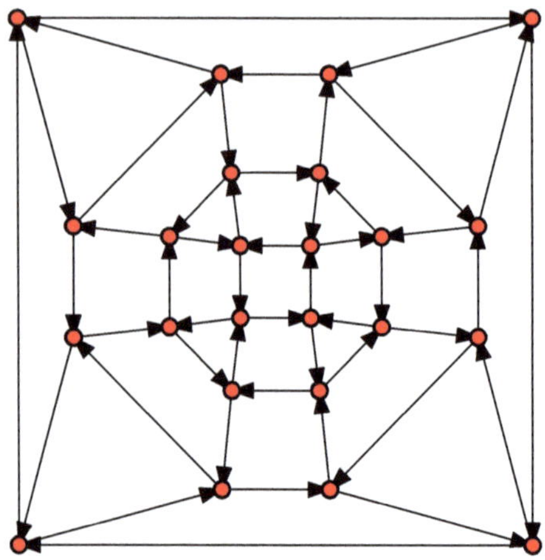

Fig. 2.5 A Cayley graph of the group S_4

Of course, not every highly symmetric graph is a Cayley graph. This does not hold, even if we demand that any two vertices are equivalent under some element of its automorphism group (i.e. if the graph is a *vertex transitive graph*), as shown Fig. 2.6.

Does every large graph have many symmetries? Obviously not: there are arbitrarily large graphs with just one automorphism, namely the identity (a graph with such a property is called *asymmetric*, a *unity graph* or sometimes a *rigid graph*). Actually almost all graphs have no other symmetries than the trivial one, see e.g. [253]. This result may look counterintuitive as most small examples of graphs have a non-trivial automorphism. Could this counterintuitive feeling be made precise? A result which goes in this direction is proved in Chap. 8 (Corollary 8.1): every large graph either has a non-trivial symmetry, or it contains a long induced path, or it contains a shallow subdivision of a large complete graph (Fig. 2.7). (By a shallow subdivision we mean a subdivision where each edge is subdivided just a few times.)

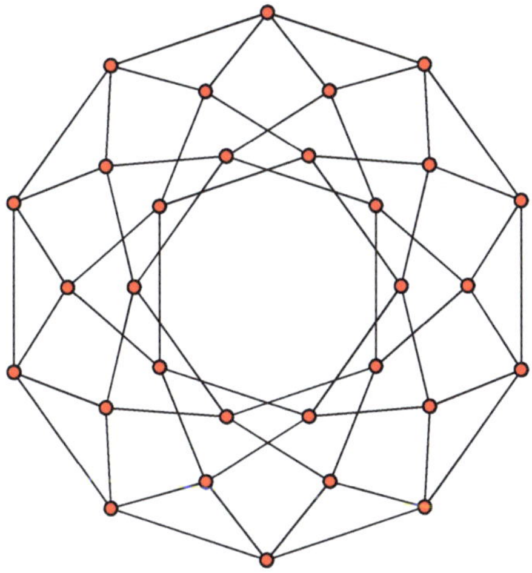

Fig. 2.6 A vertex transitive non Cayley graph, the icosidodecahedral graph

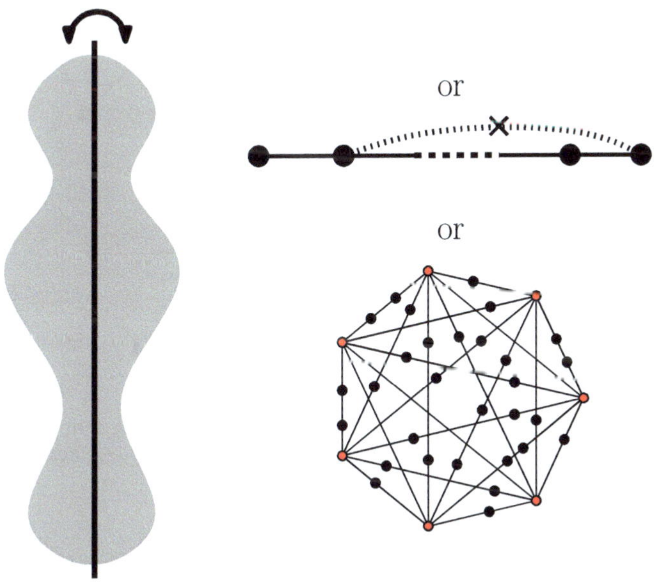

Fig. 2.7 Every large graph either has a non-trivial automorphism, or includes a long induced path, or includes a shallow subdivision of a large complete graph

An informed reader may say that this result has a Ramsey-like flavor. This is correct and Ramsey theory is one of our motivations: to find patterns appearing in large sparse graphs is one of the topics of this book.

2.4 Large Matchings on a Torus

A matching is a subset of the edges of a graph such that no two of them have a common endpoint. The *matching number* $\beta(G)$ of a graph is the maximum size of a matching of the graph. For instance, the matching number of a star graph is 1 (as any two edges are adjacent). Although many graphs have a matching number which is linear with respect to the order of the graph, there are also arbitrarily large graphs for which this is not the case, like stars, or even asymmetric ones (as shown on Fig. 2.8).

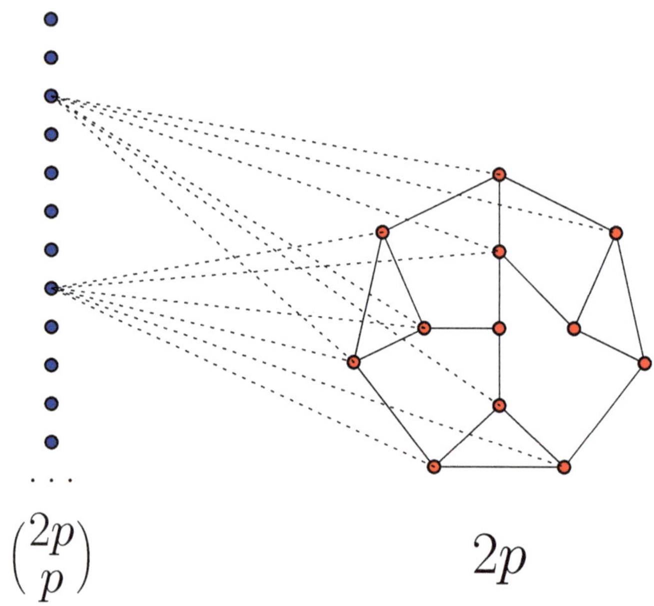

Fig. 2.8 Asymmetric graphs exist with order n and matching number $\approx \log n$: To an asymmetric graph of order $2p$ and maximum degree $\leq p-1$ add $\binom{2p}{p}$ vertices adjacent to p vertices each in all the possible ways

However, consider a graph drawn on a surface, for instance on the torus. To avoid pathological cases like the stars and double stars ($n-2$ vertices adjacent to the same two vertices), we require that our graph has no vertex of degree smaller than 3. Then, as we will prove in Sect. 15.1, there exists a constant $\alpha > 0$ such that the matching number of such a graph of order n is at least equal to αn (Fig. 2.9).

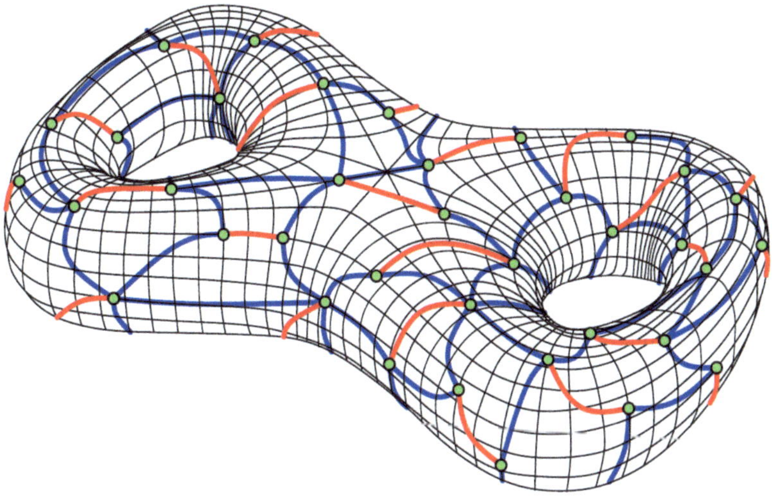

Fig. 2.9 Every graph on the torus with order n and minimum degree at least 3 has a matching of size at least αn

This result can also be proved as a consequence of known results of matching theory (cf [312]). However, we show that it is also a consequence of much more general results dealing with the existence of a linear number of disjoint short paths in sufficiently sparse large graphs with certain forbidden automorphisms. Particularly every large asymmetric sparse graph has such a generalized matching of linear size, see Sect. 15.1.

2.5 Homomorphism Dualities

The Grötzsch's celebrated theorem (see e.g. [459]) says that every triangle-free planar graph is 3-colorable. In the language of homomorphisms this says that for every triangle-free planar graph G there is a homomorphism of G into K_3, that is. Recall that a *homomorphism* from a graph G to a graph H is a mapping from the vertices of G to the ones of H such that two vertices that are adjacent in G have images that are distinct and adjacent in H. Using the partial order terminology, Grötzsch's theorem says that K_3 is an upper bound (in the homomorphism order) for the class $\mathcal{P}_3$ of all planar triangle-free graphs. That $K_3 \notin \mathcal{P}_3$ suggests a natural question (first formulated in [341]): Is there yet a smaller bound? The answer, which may be viewed as a strengthening of Grötzsch's theorem, is positive: there exists a triangle free 3-colorable graph H such that $G \longrightarrow H$ for every graph $G \in \mathcal{P}_3$. This has been proved in [349, 352] in a stronger version for minor-closed classes (see Sect. 11.5). The underlying theory of homomorphism dualities is developed in Chap. 9.

The case of triangle-free planar graphs is interesting in its own. It seems to find a proper setting in the context of $\mathcal{T}\mathcal{T}$-continuous mappings, [367] and it has been related to a conjecture by Seymour and to Guenin's theorem [237] by Naserasr, who proved that every triangle-free planar graph has a homomorphism to the *Clebsch graph* (see Fig. 2.10). (This bound has then been proved to hold for the class of all triangle free graphs without K_5 minor [338].) This graph is triangle free and (only) 4-colorable.

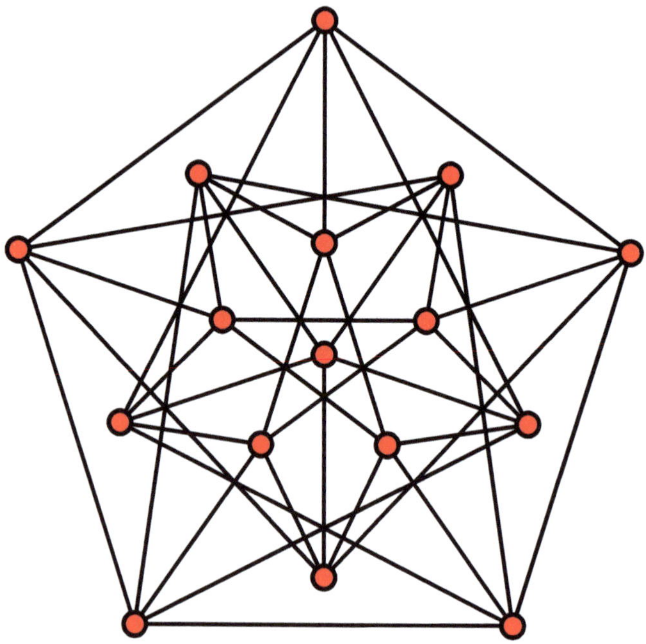

Fig. 2.10 The Clebsch graph is a bound of triangle-free planar graphs

As every triangle-free planar graph has a homomorphism to both K_3 and the Clebsch graph, every triangle-free planar graph has also a homomorphism to the categorical product of K_3 and the Clebsch graph (see Sect. 3.7 for basic properties of homomorphisms). This means that every triangle-free planar graph has a homomorphism to a triangle-free graph that is 3-colorable. More: every properly 3-colored triangle-free planar graphs has a color-preserving homomorphism to the triangle-free properly 3-colored graph depicted Fig. 2.11.

It is noticeable that no triangle free bound of triangle-free planar graphs is planar and that, more generally [344], for any fixed graph F, no planar graph exists with no homomorphism of F which is a bound of the planar graphs that have no homomorphisms from F (except if the bound is trivial, i.e. K_1 or K_4).

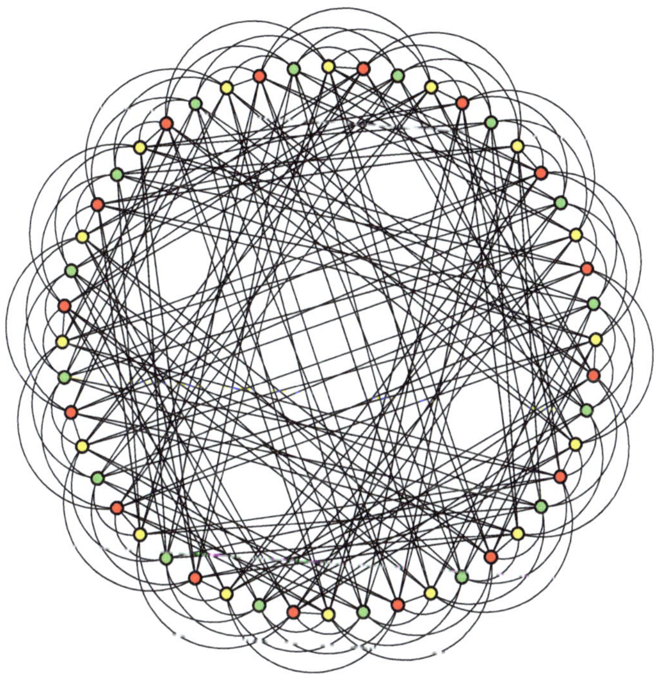

Fig. 2.11 A 3-colorable bound of triangle-free planar graphs. This graph is the product of K_3 and the Clebsch graph

The above examples are just samples. But perhaps they illustrate the versatility of the questions considered in this book. A more detailed commented contents follows in the next brief Chapter.

The Theory

Chapter 3
Prolegomena

Prolegomena to Any Future Metaphysics
That Will Be Able to Present Itself as a Science
(Immanuel Kant)

In this chapter we introduce the relevant concepts and techniques, and prove some basic results which will be used later on.

3.1 Graphs

We mostly deal with graphs. By a graph G we mean a finite undirected simple (loopless) graph, i.e. a pair (V, E) where $V = \{v_1, \dots, v_n\}$ is a finite set of *vertices* and $E = \{e_1, \dots, e_m\}$ is the set of *edges*, which is a subset of the set $\binom{V}{2}$ of all 2-element subsets of V. We denote by P_n (resp. C_n) the path (resp. the cycle) of order n and by K_n (resp. $K_{n,m}$) the complete graph of order n (resp. the complete bipartite graph with parts of size n and m).

Our terminology is standard (we refer to [60] and [328]) and we denote by $d_G(v)$ the degree of a vertex v in G. We use the concise notation $|G|$ for the *order* of G (i.e. the number of vertices of G), and $\|G\|$ for the *size* of G (i.e. the number of edges of G). As G is a simple graph we have $\sum_{v \in V} d_G(v) = 2|E|$ (sometimes called the *handshaking lemma*). This lemma not only establishes that the number of vertices with odd degree is even (hence the name) and that the *average degree* $\overline{d}(G)$ of a vertex of G is $2|E|/|V|$. We denote as usual by $\delta(G)$ and $\Delta(G)$ the *minimum degree* and the *maximum degree* of the vertices of G. Of course we have

J. Nešetřil and P.O. de Mendez, *Sparsity*, Algorithms and Combinatorics 28,
DOI 10.1007/978-3-642-27875-4_3, © Springer-Verlag Berlin Heidelberg 2012

$$\delta(G) \le \overline{d}(G) \le \Delta(G), \tag{3.1}$$

and

$$\delta(G) \le \frac{2\|G\|}{|G|}. \tag{3.2}$$

We shall find it convenient to consider also the ratio $\|G\|/|G|$ which we call the *edge density* of G.

We can obtain more information about the degrees of G when we consider subgraphs. Recall that a graph $G' = (V', E')$ is a *subgraph* of a graph $G = (V, E)$ (this being denoted by $G' \subseteq G$) if $V' \subseteq V$ and $E' \subseteq E$. If E' is the set of all the edges of G on the set V', i.e. $E' = E \cap \binom{V'}{2}$, then we say that G' is an *induced subgraph*, or more precisely the subgraph of G induced by V'. We usually denote this by $G[V']$ or simply $G' \subseteq_i G$. The following nice notation originated in Ramsey theory (cf [340]): For graphs G, H let $\binom{G}{H}$ denote the set of all the induced subgraphs of G which are isomorphic to H.

In terms of subgraphs we give in the next section quantitative refinements to the basic inequalities (3.1), (3.2).

3.2 Average Degree and Minimum Degree

We denote by $G_{<k}$ (resp. $G_{\le k}$, etc.) the subgraph of G induced by the vertices of degree strictly smaller than k (resp. at most k, etc.). An easy result, which is nothing other than Markov's inequality, then states that for any positive integer k and any graph G of average degree $\overline{d}(G)$ at most $(\overline{d}(G)/k)|G|$ vertices of G have degree at least k, that is:

$$|G_{<k}| \ge \left(1 - \frac{\overline{d}(G)}{k}\right)|G|. \tag{3.3}$$

Indeed, we have

$$k\ |\{v, d(v) \ge k\}| \le \sum_{v} d(v) = \overline{d}(G)\ |G|.$$

Hence

$$|G| - |G_{<k}| \le \frac{\overline{d}(G)}{k}|G|.$$

Another standard method involves the iterative removal of small vertices until some condition is fulfilled.

For instance, assume a graph G contains a vertex v with $d(v) < \frac{\|G\|}{|G|}$ and let $H = G - v$ be the induced subgraph of G obtained by deleting v. Clearly H is not empty. Moreover,

$$\begin{aligned}\frac{\|H\|}{|H|} - \frac{\|G\|}{|G|} &= \frac{\|G\| - d(v)}{|G| - 1} - \frac{\|G\|}{|G|} \\ &= \frac{\|G\| - |G|d(v)}{|G|(|G| - 1)} \\ &= \left(1 - \frac{d(v)}{\|G\|/|G|}\right)\frac{\|G\|}{|G|(|G| - 1)} > 0.\end{aligned}$$

Thus by iteratively repeating this operation we eventually obtain a non-empty induced subgraph H of G such that

$$\delta(H) \geq \frac{\|H\|}{|H|} \geq \frac{\|G\|}{|G|}, \tag{3.4}$$

Instead of requiring a bound on $d(v)$ which changes after each iteration, we can consider a fixed constant bound which depends on the original graph. Consider a graph G and a positive real $\epsilon > 1/|G|$. Iteratively delete the vertices v such that $d(v) < (1 - \epsilon)\frac{\|G\|}{|G|}$ until no deletion will be possible any more, and let H be the obtained induced subgraph of G. Then the number of deleted edges is at most $|G|\,(1 - \epsilon)\frac{\|G\|}{|G|} \leq (1 - \epsilon)\|G\|$, hence H is not empty. Moreover, H is such that

$$\delta(H) \geq (1 - \epsilon)\frac{\|G\|}{|G|} \qquad \text{and} \qquad \|H\| \geq \epsilon\|G\|. \tag{3.5}$$

Together with the trivial bound (3.2) this shows that the minimum degree is well approximated by the edge density of some subgraph.

3.3 Graph Degeneracy and Orientations

Recall that a graph G is k-*degenerate* if each nonempty subgraph of G contains a vertex of degree at most k. For instance, trees are 1-degenerate, planar graphs are 5-degenerate. The *maximum average degree* of G, denoted mad(G), is the maximum average degrees of the subgraphs of G:

$$\mathrm{mad}(G) = \max_{H \subseteq G} \overline{d}(H) = \max_{H \subseteq G} \frac{2\|H\|}{|H|}$$

These two concepts are related by the following folklore result:

Proposition 3.1. *Let* G *be a graph and let* k *be an integer. Then we have the following implications:*

$$k \geq \lfloor \mathrm{mad}(G) \rfloor \quad \Rightarrow \quad G \textit{ is } k\textit{-degenerate} \quad \Rightarrow \quad \mathrm{mad}(G) < 2k.$$

Thus the maximum average degree relates to graph degeneracy, and it can be used to give a simple bound for the *chromatic number* $\chi(G)$ of a graph G (which is the minimum number of colors needed to color the vertices of G in such a way that no two adjacent vertices get the same color): for every graph G it holds

$$\chi(G) \leq \lfloor \mathrm{mad}(G) \rfloor + 1.$$

From an algorithmic point of view, the computation of the maximum average degree (and more generally the problems of finding a densest subgraph) is known to be solvable in polynomial time [210].

It is important to note that there is yet another approach to degeneracy of graphs by means of graph orientations. Recall that every undirected graph G can be transformed into a directed graph $\vec{G}$ by choosing for each undirected edge $\{u, v\}$ one of the two possible arcs: either (u, v) or (v, u). The resulting directed graph is called an *orientation* of G (There are $2^{\|G\|}$ orientations of $G = (V, E)$; any of them will be denoted by $\vec{G} = (V, \vec{E})$). An orientation is *acyclic* if it does not contain an oriented cycle (also called *circuit*); alternatively this means that vertices of G can be ordered $v_1, \ldots, v_n$ in such a way that the arcs (v_i, v_j) of $\vec{G}$ satisfy $i < j$. For oriented graphs we use only basic terminology. Recall that the *indegree* $d^-(v)$ of a vertex v is the number of arcs which terminate in v: $d^-(v) = |\{(u, v) : (u, v) \in \vec{E}\}|$, while the *outdegree* $d^+(v)$ is the number of arcs which start in v: $d^+(v) = |\{(v, w) : (v, w) \in \vec{E}\}|$.

One of the properties of k-degenerate graphs is to have an orientation with small maximum indegree. A basic result of this type is the following:

Proposition 3.2. *Let G be a graph and let k be an integer. Then the following conditions are equivalent:*

- *G is k-degenerate,*
- *G has an acyclic orientation such that every vertex has indegree at most k.*

Proof. Assume G is k-degenerate. Order the vertices of G as $x_1, \dots, x_n$ according to the following rule: x_1 has minimum degree in G, x_2 has minimum degree in $G - x_1$, etc. and orient each edge $\{x_i, x_j\}$ with $i < j$ from x_j to x_i. This orientation is clearly acyclic and each vertex x_i will have indegree at most k as $G - x_1 - x_2 - \dots - x_{i-1}$ has minimum degree at most k (as G is k-degenerate).

Conversely, assume G has an acyclic orientation such that the maximum indegree is at most k and let $<$ be a linear order on the vertex set of G compatible with the acyclic orientation (that is: (x, y) is an arc implies $x < y$). Then, for each subgraph H of G, the maximum vertex of H has at most k neighbors in H (for otherwise its indegree would be greater than k), hence H contains a vertex of degree at most k. It follows that G is k-degenerate. □

We note that k-degenerate graphs have many useful properties. For example they have linearly many cliques (a *clique* is a complete subgraph):

Lemma 3.1. *Let G be a k-degenerate graph. Then G includes at most $\binom{k}{t-1}|G|$ cliques of order t, thus at most $2^k|G|$ cliques.*

Proof. Consider an acyclic orientation of G with indegree at most k. Then the vertices of any clique of size t are naturally ordered as $x_1, x_2, \dots, x_t$ (with all arcs oriented from x_i to x_j whenever $i < j$). We have at most $|G|$ choices for x_t. The vertex x_t being given, we have at most $\binom{d^-(x_t)}{t-1}$ choices for $\{x_1, \dots, x_{t-1}\}$. It follows that the number of cliques of order t is bounded by $\binom{\Delta^-(G)}{t-1}|G| = \binom{k}{t-1}|G|$ and by summing we get that G includes at most $2^k|G|$ cliques. □

This property has been used from an algorithm point of view to list cliques in a planar graph [96], and from a more theoretical point of view, to prove that every proper minor-closed class of graphs includes at most $n!c^n$ graphs with vertex-set $\{1, 2, \dots, n\}$, that is are *small* classes [372].

One may use a similar argument to give an upper bound on the number of copies of an arbitrary graph H in a graph G (see Exercise 3.2).

Proposition 3.2 has a variant when acyclicity is not required. The following condition gives necessary and sufficient conditions for an orientation to exist in which each vertex has its indegree bounded by a given function.

Proposition 3.3. *Let* G *be a graph and* $\lambda : V(G) \to \mathbb{N}$ *be a mapping. Then, there exists an orientation of* G *such that each vertex* v *has indegree* $d^-(v)$ *smaller than or equal to* $\lambda(v)$ *if and only if, for every induced subgraph* H *of* G,

$$\|H\| \leq \sum_{v \in V(H)} \lambda(v) \tag{3.6}$$

Moreover, if $\|G\| = \sum_{v \in V(G)} \lambda(v)$ *there exists an orientation of* G *such that each vertex* v *has indegree exactly* $\lambda(v)$.

Proof. The proof uses an argument which is standard in the context of network flows and combinatorial optimization (see [188, 376]). Actually, (3.6) is the Hoffman condition of the trivial max-flow problem associated to the orientation computation. For completeness, let us give a short proof.

Assume that there exists an orientation of G such that each vertex v has indegree $d^-(v)$ smaller or equal to $\lambda(v)$. Then, for every induced subgraph H of G we have

$$\|H\| \leq \sum_{v \in V(H)} d^-(v) \leq \sum_{v \in V(H)} \lambda(v).$$

Conversely, assume that for every induced subgraph H of G we have

$$\|H\| \leq \sum_{v \in V(H)} \lambda(v).$$

To any orientation $\mathcal{O}$ of G we associate the value

$$S(\mathcal{O}) = \sum_{v \in V(G), d^-(v) > \lambda(v)} (d^-(v) - \lambda(v)),$$

where the indegrees are computed according to the orientation $\mathcal{O}$. Let $\mathcal{O}$ be an orientation of G such that $S(\mathcal{O})$ is minimum. If $S(\mathcal{O}) = 0$, we are done. Let us prove by contradiction that this is indeed the case. So assume that $S(\mathcal{O}) > 0$. Then there exists a vertex $a \in V(G)$ such that $d^-(a) > \lambda(a)$. Let H_a be the subgraph of G induced by the set of the vertices v such that there exists in G a (maybe empty) directed path from v to a (with respect to the orientation $\mathcal{O}$). By construction, the indegree in H_a of a vertex in H_a is the same as its indegree in G. Hence we have

$$\|H_a\| = \sum_{v \in V(H_a)} d^-(v).$$

As $\|H_a\| \leq \sum_{v \in V(H)} \lambda(v)$ and as $d^-(a) > \lambda(a)$ there exists in H_b at least a vertex b such that $d^-(b) < \lambda(b)$. By the definition of H_a there exists a directed path $\vec{P}$ in G from b to a. Consider the orientation $\mathcal{O}'$ of G obtained from $\mathcal{O}$ by reorienting the edges of $\vec{P}$. The indegrees of all but two vertices (a and b) are the same in $\mathcal{O}$ and $\mathcal{O}'$; the indegree of a in $\mathcal{O}'$ is one less than its indegree in $\mathcal{O}$ while the indegree of b in $\mathcal{O}'$ is one more than its indegree in $\mathcal{O}$. It follows that the indegree of b in $\mathcal{O}'$ is at most $\lambda(b)$. Altogether we deduce that $S(\mathcal{O}') = S(\mathcal{O}) - 1 < S(\mathcal{O})$, a contradiction.

If G is oriented in such a way that each vertex v has indegree at most $\lambda(v)$, it is clear that we have equality for every vertex if and only if the global sum of the indegrees (that is: $\|G\|$) equals the global sum of the $\lambda(v)$. □

We remark that two orientations have the same indegrees if and only if they differ by the reorientation along a sequence of directed cycles. In the planar case, the set of these orientations has a distributive lattice structure [190, 374–376] (see Exercise 3.3).

From an algorithmic point of view, orientations with bounded indegree (or equivalently, with bounded out-degree) are interesting as they allow checking for the adjacency of two vertices in constant time. This simple remark seems to have first been used by Chrobak and Eppstein [96]. Orientations with bounded indegrees play a particular role in topological graph theory, especially in relation to planarity [187, 188, 192, 193, 195] and contact representations of graphs [186, 189, 191, 196–201].

3.4 Girth

The *girth* of a graph G, denoted girth(G), is the minimum length of a cycle of G (or ∞ if G is acyclic). In several contexts, it will be useful to find a dense subgraph with high girth in a dense graph. This is not an easy task, but it might be always possible:

Conjecture of Thomassen [457]:

For all integers c, g there exists an integer $f(c, g)$ such that every graph G of average degree at least $f(c, g)$ contains a subgraph of average degree at least c and girth at least g.

The case $g = 4$ of this conjecture is a direct consequence of the simple fact that every graph can be made bipartite by deleting at most half of its edges (see Exercise 3.1). The case $g = 6$ has been proved in [292]. The conjecture has also been proved for graphs whose average degree is not too small compared to their maximum degree. Improving an earlier similar result [390] (with single log), the following has been proved in [114].

Theorem 3.1. *For every $c \geq 1, g \geq 3$ there exist $\alpha, \beta > 0$ and $d_0 > c$ for which the following holds.*

Suppose that G is a graph with average degree

$$\overline{d}(G) \geq \max\{\alpha(\log\log\Delta(G))^{\beta}, d_0\}.$$

Then G contains a subgraph H with $\overline{d}(H) \geq c$ and $\text{girth}(H) \geq g$.

As already mentioned, average degree and chromatic number are related notions. Actually, Thomassen's conjecture has a similar flavor to the following conjecture formulated earlier:

Conjecture of Erdős and Hajnal [164]:

For all integers c, g there exists an integer $f(c, g)$ such that every graph G of chromatic number at least $f(c, g)$ contains a subgraph of chromatic number at least c and girth at least g.

The case $g = 4$ of this conjecture was proved by Rödl [420], and this is presently the only non-trivial case known.

It should be noticed that the mere existence of a graph with high chromatic number and high girth is a well known result of Erdős [163], which was at the origin of the use of probabilistic methods in graph theory:

Theorem 3.2. *For all integers* c, g *there exists a graph with girth at least* g *and chromatic number at least* c.

Erdős showed that a random graph on n vertices and edge-probability $n^{(1-g)/g}$ has, with high probability, at most $n/2$ cycles of length at most g, but no independent set of size $n/2c$. Removing one vertex in each short cycle leaves a graph with girth at least g and chromatic number at least c. Constructive proofs of Theorem 3.2 are known, the simplest beeing perhaps [364].

Forcing a high girth is actually a way to bound the density of a graph. In other terms, there exists no small graph with both high girth and high average degree. For d-regular graphs, a bound is easily derived from the fact that the ball of radius $\lfloor \frac{g-1}{2} \rfloor$ around a vertex or an edge (depending on the parity of g) is a tree. This bound, denoted by $n_0(d, g)$, is called *Moore bound* and its value (see for instance [62]) is given by:

$$n_0(d, 2r+1) = 1 + d\sum_{i=0}^{r-1}(d-1)^i$$

$$n_0(d, 2r) = 2\sum_{i=0}^{r-1}(d-1)^i.$$

The fact that this bound should hold in general has been conjectured by Bollobás, and proved by Alon et al. [19]:

Theorem 3.3. *The order* n *of a graph of girth* g *and average degree at least* $d \geq 2$ *is greater or equal to the Moore bound* $n_0(d, g)$:

$$n \geq n_0(d, g).$$

On the other hand, for every positive integer n and an "expected degree" k (where $k < n/3$), there exists a graph G of order n, size $\lfloor nk/2 \rfloor$, vertex degrees in $\{k-1, k, k+1\}$ and whose girth g is such that $g > \log_k(n) + O(1)$ (see for instance [91]).

3.5 Minors

Geometry and topology are essential to this book; even when we do not speak about drawings or topology these concepts are present behind the scene. The notion of a minor is a prime example: A graph H is a *minor* of a graph G (denoted $H \leq_m G$) if we can obtain H from G by repeating the following three operations: $G - v$ (*vertex deletion*), $G \setminus e$ (*edge deletion*) and G/e (*edge contraction*). If multiple edges are created when contracting an edge, we simplify the graph—all our graphs are simple.

This recursive definition can be equivalently expressed as follows: A graph H with vertex set $\{v_1, \ldots, v_n\}$ is a minor of a graph G if there are pairwise vertex disjoint connected subgraphs $G_1, \ldots, G_n$ of G such that for every edge $\{v_i, v_j\}$ of H there is at least one edge of G joining G_i and G_j in G (by this we mean that there exists an edge $\{x_i, x_j\}$ of G such that $x_i \in V(G_i)$ and $x_j \in V(G_j)$). In other words H arises from a subgraph of G by contracting connected subgraphs. The requirement that the subgraph should be connected, highlighting the difference with homomorphisms, is chiefly responsible for the "geometric flavor" of this concept.

As connected unions of connected graphs are connected we see that the minor relation $\leq_m$ is a quasi-order on $\mathcal{G}raph$, called the *minor order*. This quasi-order was intensively studied and maybe the most important properties of this quasi-order stands in the following important and difficult theorem of Robertson and Seymour [398]:

Theorem 3.4. *Any infinite sequence $G_1, G_2, \ldots$ of finite graphs contains two members G_i, G_j with $i < j$ and $G_i \leq_m G_j$.*

A quasi-ordering $\leq$ of a set X which satisfies a similar statement is called a *well-quasi-ordering* (or *wqo*). Well-quasi-ordering may be equivalently defined by the following two conditions (this is yet another consequence of Ramsey's theorem):

1. $(X, \leq)$ does not contain an infinite descending chain (i.e. an infinite sequence $x_1 > x_2 > \cdots > x_i > \ldots$)
2. $(X, \leq)$ does not contain an infinite antichain (i.e. an infinite subset of X containing no two distinct elements x and y such that $x \leq y$)

As for the minor relation (for finite graphs) the condition (3.1) obviously holds, the contents of the theorem lies in (3.2), i.e. the non-existence of an infinite antichain. This old problem (known as Wagner's conjecture) was then solved by Robertson and Seymour in their landmark series of papers [397, 399–417].

This result is important in topological graph theory as it implies that for each surface Σ, a graph G may be embedded on Σ if and only if it avoids finitely many minors, the forbidden minors of the surface Σ. (The class of the graphs which may be embedded is closed under minors. Thus the forbidden minors for the surface are the minor minimal graphs which cannot be embedded on the surface and there are finitely many ones according to the theorem.) For the plane and the projective plane, an explicit list of forbidden minors is known. For instance, for the plane these are *Kuratowski's graphs* $K_{3,3}$ and K_5 [296, 467].

Actually, Kuratowski's theorem is stated in terms of forbidden subdivisions. This leads to the following notion.

A graph H with vertex set $\{v_1, \ldots, v_n\}$ is said to be a *topological minor* of a graph G if there are distinct vertices $x_1, \ldots, x_n$ of G and, for each edge $e = \{v_i, v_j\}$ of H there exists a path P_e in G with endpoints x_i and x_j where the P_e's are pairwise *internally vertex disjoints* (i.e. may only share a common end-vertex). Each vertex $x_1, \ldots, x_n$ is a *principal vertex* and each path P_e is a *branch* of the subdivision. In other words, H is a topological minor of G if a subdivision of H is isomorphic to a subgraph of G. This is denoted by $\leq_t$. The quasi-order $\leq_t$ is called the *topological minor order*. The topological order $\leq_t$ is contained in the minor order $\leq_m$. It is important to realize that these orders are very different and that $\leq_t$ is generally much more restrictive than $\leq_m$, although these quasi-orders may coincide when restricted to some graph classes (for instance, to graphs with maximum degree 3). For example the relation $G \leq_t H$ implies $\Delta(G) \leq \Delta(H)$ which is obviously not true for the minor order. This difference is exemplified by the fact that the topological minor order contains an infinite antichain and thus fails to be a well-quasi-ordering (see Fig. 3.1). However, the property that a graph G may be embedded on a fixed surface can always be characterized by finitely many forbidden topological minors.

We introduce yet another quasi-order.

Let G be a graph and let $\{u, v\}, \{v, w\} \in E(G)$. The operation of deleting the edges $\{u, v\}$ and $\{v, w\}$ and then adding a new edge between u and w is called a *split*. We say that a graph G *immerses* a graph H if a graph isomorphic to H may be obtained from G by repeatedly making splits and deleting vertices and edges; we denote this by $H \leq_i G$. It has been conjectured by Nash-William [339] and recently proved by Robertson and Seymour [418] that this quasi-order is actually a well-quasi-ordering.

Another difference between minors and topological minors is exemplified by two classical conjectures. Both are very easy to state:

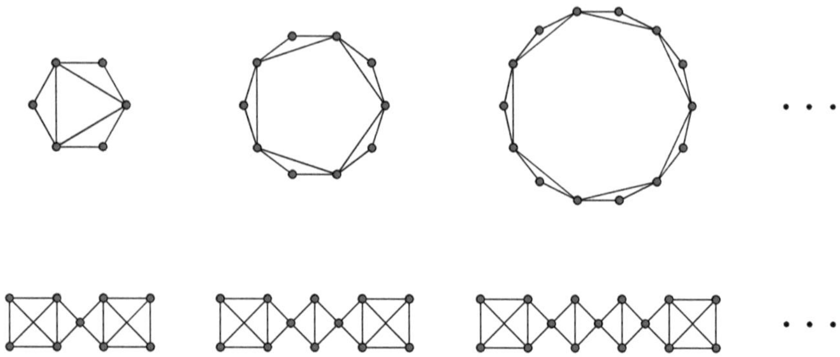

Fig. 3.1 Two examples of infinite antichains for the topological minor order

Hadwiger's conjecture:
$K_k \leq_m G$ for every graph G with $k \leq \chi(G)$.

Hajós' conjecture:
$K_k \leq_t G$ for every graph G with $k \leq \chi(G)$.

The present status of these conjectures is very different: Hadwiger's conjecture is true for $k \leq 6$ [419] (for both $k = 5$ and $k = 6$ this is related to the Four Color Theorem) and open for $k \geq 7$. Hajós' conjecture is true for $k \leq 4$ and false for all $k \geq 7$. Actually, for almost all graphs Hadwiger's conjecture is true and Hajós' conjecture is false (as observed by Erdős and Fajtlowicz [166]). Thomassen [460] proves that many examples known for a long time in different areas of graph theory can serve as counterexamples to Hajós' conjecture. For immersions we have the following:

Conjecture of Abu-Khzam and Langston [1]:
$K_k \leq_i G$ for every graph G with $k \leq \chi(G)$.

This conjecture has been recently proved for $k \le 7$ [120]. Also on the positive side let us mention that Hajós' conjecture has been validated for graphs with large girth ($\geq$186 [291], recently improved to $\geq$27). In this book we shall see that, a bit surprisingly, our main results are not sensitive to whether we consider minors, topological minors, or immersions. In this we shall make use of the following quantitative results.

The *Hadwiger number* $h(G)$ of a graph G is the maximum integer k such that $K_k \le_m G$. Similarly, by $h_t(G)$ (resp. $h_i(G)$) we denote the maximum integer k such that $K_k \le_t G$ (resp. $K_k \le_i G$). Then the following extremal results are known: The first of these results, which concerns minors, was obtained independently by Kostochka [284] and Thomason [455] (extending earlier work of Mader [322]; see [456] for a tight value of constant γ).

Theorem 3.5. *There exists a constant γ such that every graph G with minimum degree at least $\gamma k\sqrt{\log(k)}$ satisfies $h(G) \ge k$.*

For topological minors (i.e. subdivisions) an analog result was proved independently by Komlós and Szemerédi [280, 281] and by Bollobás and Thomason [77].

Theorem 3.6. *There exists a constant c such that every graph G with minimum degree at least ck^2 satisfies $h_t(G) \ge k$.*

For immersions, the analog has been recently provided by DeVos et al. [119].

Theorem 3.7. *Every simple graph of minimum degree at least $200\,k$ satisfies $h_i(G) \ge k$.*

3.6 Width, Separators and Expanders

Tree-width [241, 397, 467] is a fundamental graph invariant with many applications in graph structure theory and graph algorithms. For instance, it is known that *graph isomorphism* can be checked in polynomial time when restricted to a class with bounded tree-width [68]. Also, if the tree-width of the primal graph of the instance is at most k and if n is the size of the input then constraint satisfaction problems (CSP) can be solved in time $n^{O(k)}$; no algorithm can be significantly better than this [326]. The concept is central to Robertson and Seymour's analysis of graphs with forbidden minors and we recall here basics about tree-width. For general properties of tree-width we refer the reader to [71]. We recall the definition here.

A tree decomposition represents the vertices of the given graph as subtrees of a tree, in such a way that vertices are adjacent only when the corresponding subtrees intersect.

Formally, a *tree decomposition* of a graph G is a pair (X, T), where $X = X_1, \dots, X_n$ is a family of subsets of $V(G)$, and T is a tree whose nodes are the subsets X_i such that (see Fig. 3.2):

- $V(G) = \bigcup_{i=1}^{n} X_i$;
- $E(G) \subseteq \bigcup_{i=1}^{n} \binom{X_i}{2}$;
- $\forall v \in V(G), \quad T[\{X_i : v \in X_i\}]$ is connected.

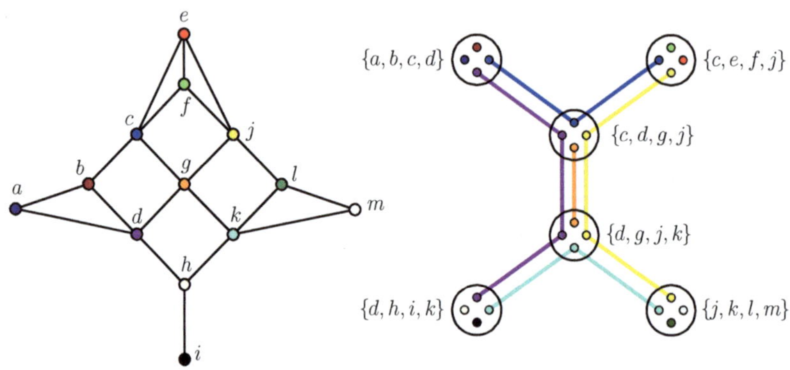

Fig. 3.2 Tree decomposition of width 3 of a graph

The *tree-width* tw(G) is the minimum over all tree decompositions (X, T) of G of $\max_i |X_i| - 1$. For instance, the tree-width of a $n \times n$ grid is equal to n. Tree-width leads to a variety of tree-like parameters. For instance, if T is required to be a path, the decomposition is called a *path decomposition*. The *path-width* pw(G) of a graph G is the minimum over all path decompositions (X, T) of G of $\max_i |X_i| - 1$.

An alternative definition of tree-width may be given in terms of partial k-trees: A k-*tree* is a graph which may be obtained from a clique of order k by a sequence of operations consisting of adding a vertex to the graph and making it adjacent to the vertices of a clique of size at most k already present in the graph. A *partial* k-*tree* is a subgraph of a k-tree. According to this definition, the minimum integer k for which a graph G is a partial k-tree is exactly the tree-width of G [432, 469].

One can characterize minor closed classes $\mathcal{C}$ with bounded tree-width [401]:

Theorem 3.8. *For every minor closed class $\mathcal{C}$, the following properties are equivalent:*

- *$\mathcal{C}$ has bounded tree-width (i.e.* $\sup_{G\in\mathcal{C}} \mathrm{tw}(\mathcal{C}) < \infty$*),*
- *$\mathcal{C}$ does not include all planar graphs,*
- *$\mathcal{C}$ excludes some grid.*

A dual approach to tree-width may be achieved through the notion of a bramble. A *bramble* in a graph G is a family of connected subgraphs of G such that any two of these subgraphs have a nonempty intersection or are joined by an edge. The *order* of a bramble is the least number of vertices required to cover (or hit) every subgraph in the bramble (see Fig. 3.3). Seymour and Thomas proved that the maximum order of a bramble in a graph G equals $\mathrm{tw}(G)+1$ [55, 434] (see also [394] and [230] for a discussion on brambles' size). An extension and common generalization of these "dual characterizations" for various width parameters may be found in [329].

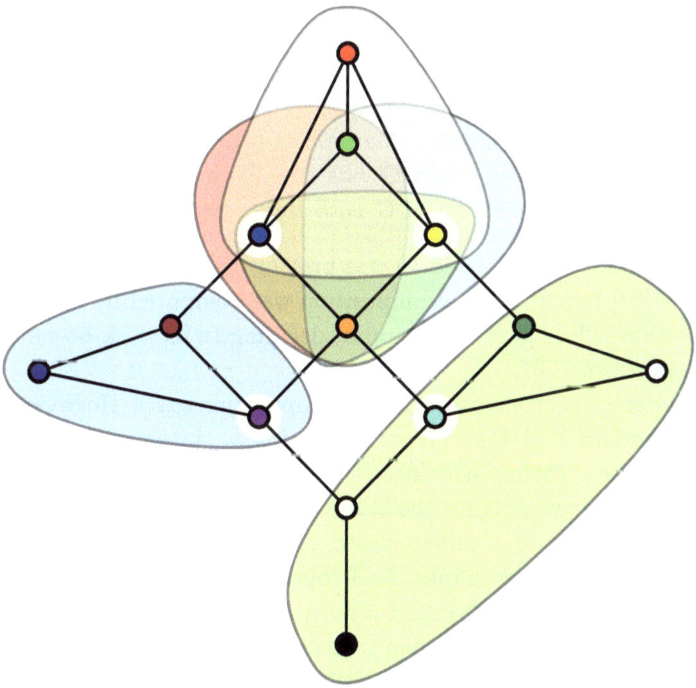

Fig. 3.3 A bramble of order 4

It has been proved by Robertson and Seymour that belonging to a minor closed class with bounded tree-width may be checked in linear time [398]. It is well known that in classes of graphs of bounded tree-width, every monadic second-order property is decidable in linear time [102, 103] (see also [32, 67]). Since many important graph properties are easily expressible in this logic,

Courcelle's theorem [102, 103] yields a unified framework for showing that numerous problems on graphs of bounded tree width are solvable in linear time.

Theorem 3.9. *Let $\mathcal{K}$ be class of finite graphs $G = \langle V, E, R \rangle$ represented as a structure with two sorts of elements (vertices V and edges E) and an incidence relation R. Let ϕ be a monadic second order sentence. If $\mathcal{K}$ has bounded tree width and $G \in \mathcal{K}$, then checking whether $G \models \phi$ can be done in linear time.*

Courcelle's theorem has been extended by Arnborg et al. [32], who considered a counting version, and by Flum et al. [178], who considered enumeration problems.

However, these algorithms need a tree decomposition of the input graph. Such a decomposition can be computed in linear time, thanks to the following result of Bodlaender [70]:

Theorem 3.10. *For all $k \in \mathbb{N}$, there exists a linear time algorithm, that tests whether a given a graph G has tree-width at most k, and if so, outputs a tree decomposition of G with tree-width at most k.*

Note that $tw(G) \leq k$ (for fixed k) was previously known to be decidable in linear time [31] but no tree decomposition was computed by the algorithm. Also note that if k is part of the input, deciding $tw(G) \leq k$ becomes a NP-complete problem [30].

Bodlaender's algorithm uses a linear computation space. However, a recent result of Elberfeld et al. [146] shows that a similar statement holds for deterministic log-space Turing machines (*log-space DTM*), making the power of MSO-definability available for the study of logarithmic space.

As noticed above, if a minor closed class $\mathcal{C}$ has unbounded tree-width then it contains all planar graphs. As 3-colorability is NP-complete for planar graphs, we get that if $P \neq NP$ and if for a class $\mathcal{C}$ every existential monadic second-order property is in P, then $\mathcal{C}$ has bounded tree-width.

The tree-width of a graph is closely related to the size of its vertex separators.

Let G be a graph of order n and let $0 < \alpha < 1$. An *α-vertex separator* of G is a subset S of vertices such that every connected component of $G - S$ contains at most αn vertices. For instance, it is easy to observe that every graph G has a $1/2$-vertex separator of size $tw(G) + 1$ [73, 219, 306, 400].

Vertex separators are a central tool in divide and conquer algorithms, where searches for balanced vertex separators are frequently repeated recursively on smaller and smaller subgraphs (see also Chap. 16). This justifies the introduction of a hereditary measure of balanced vertex separators in graphs:

Let G be a graph of order n. We define $s_G : \{1, \ldots, n\} \to \mathbb{N}$ by

$$s_G(i) = \max_{\substack{|A| \leq i, \\ A \subseteq V(G)}} \min\{|S| : S \text{ is a } \tfrac{1}{2}\text{-vertex separator of } G[A]\}$$

We shall deal with this measure extensively in Chap. 16. Also, the *separation number* $s(G)$ of a graph G is the smallest s such that all subgraphs of G have an $(s, 2/3)$-separator.

The *band-width* of a graph G of order n, denoted by $bw(G)$, is the minimum positive integer b, such that there exists a numbering $f : V(G) \to [n]$ of the vertices of G so that the labels of every pair of adjacent vertices differ by at most b.

Having a vertex of high degree is sufficient to have a large band-width: $bw(G) \geq \lceil \Delta(G)/2 \rceil$. However, this condition is not necessary. Consider a random bipartite graph G with bounded maximum degree. With high probability, G does not have small band-width since in any linear ordering of its vertices there will be an edge between the first $n/3$ and the last $n/3$ vertices [81].

Definition 3.1. Let $\epsilon > 0$ be a real number and let G be a graph. We say that G is an *ϵ-expander* if all subsets U of vertices of G with $|U| \leq |G|/2$ satisfy $|N_G(U) \setminus U| \geq \epsilon|U|$.

This definition may be compared to two standard definitions of expansions: For $0 < \alpha < 1$, the *α-vertex expansion* $g_\alpha(G)$ of a graph G is defined by

$$g_\alpha(G) = \min_{1 \leq |S| \leq \alpha|G|} \frac{|N_G(S)|}{|S|}$$

Also, the *edge expansion* (or *isoperimetric number*, or *Cheeger constant*) $\mathrm{Iso}(G)$ of G is defined by

$$\mathrm{Iso}(G) = \min_{1 \leq |S| \leq |G|/2} \frac{|\delta(S)|}{|S|},$$

where $\delta(S)$ denotes the set of all the edges of G linking S to $V(G) \setminus S$, i.e. the *cut-set* (or *cobord*) of S (see Fig. 3.4).

It is easily checked that $\mathrm{Iso}(G) \geq g_{1/2}(G) - 1$ and that G is an ϵ-expander if and only if $0 < \epsilon \leq g_{1/2}(G)$.

It is important that the edge expansion $\mathrm{Iso}(G)$ of G is related to its spectral properties, and particularly to the second largest eigenvalue: if G is d-regular and if the adjacency matrix $A(G)$ of G has eigenvalues $d = \lambda_1 \geq \lambda_2 \geq \cdots \geq \lambda_n$, then [11, 22, 125]:

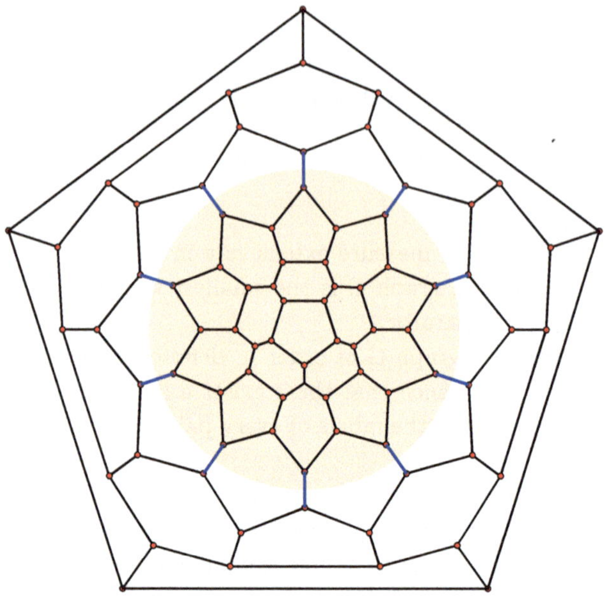

Fig. 3.4 The Ramanujan graph C_{80} has edge expansion one fourth as witnessed by the shown cut

$$\frac{1}{2}(d - \lambda_2) \le \mathrm{Iso}(G) \le \sqrt{2d(d-\lambda_2)}.$$

This relationship goes back to Cheeger [92] and Fiedler [177]. This inequality leads to the following important results which found many applications in various branches of mathematics and computer sciences, see [13, 63] and the excellent survey article [259]. The first of these results is known as the *Expander Mixing Lemma*: Let G be a d-regular graph of order n and let λ be the maximum of the absolute values of the second largest eigenvalue and of the minimum eigenvalue of G (i.e. $\lambda = \max(|\lambda_2|, |\lambda_n|)$). Then for all $S, T \subseteq V(G)$ it holds:

$$\left| |\omega(S,T)| - \frac{d\,|S|\,|T|}{n} \right| \le \lambda\sqrt{|S|\,|T|},$$

where $\omega(S,T)$ denotes the set of edges of G with one endpoint in S and the other in T. The second result shows that this relation between λ and the expansion properties of G is best possible in the following sense: if

$$\left| |\omega(S,T)| - \frac{d\,|S|\,|T|}{n} \right| \le \rho\sqrt{|S|\,|T|}$$

holds for all every two disjoints $S, T \subseteq V(G)$ and for some positive ρ. Then $\lambda \leq O(\rho\,(1+\log(d/\rho)))$ (and the bound is tight).

Definition 3.2. Let $\epsilon > 0$ be a real number, $b \in \mathbb{N}$ and let G be a graph. We say that G is (b, ϵ)-*bounded* if no subgraph $G' \subseteq G$ with $|G'| \geq b$ is an ϵ-expander. The ϵ-*boundedness* $b_\epsilon(G)$ is the minimum b for which G is $(b+1, \epsilon)$-bounded.

We shall make use of the following result [81]:

Theorem 3.11. *Let* $\epsilon > 0$ *be constant and let* G *be a graph of order* n. *Then:*

$$\begin{aligned} bw(G) &\leq \frac{6n}{\log_{\max(\Delta(G),2)}(n/s(G))}, \\ tw(G) &\leq 2b_\epsilon(G) + 2\epsilon n, \\ b_\epsilon(G) &\leq 2bw(G)/\epsilon. \end{aligned}$$

Thus these parameters are not independent. In particular we have [230]:

Theorem 3.12. *Let* $\epsilon > 0$ *be constant and let* G *be a graph has order* n. *If* G *satisfies* $b_\epsilon(G) < \epsilon n$, *then* $tw(G) \leq 2\epsilon n$.

Thus the expansion properties of graphs are linked to the tree width. We shall make use of these connections in Chap. 16.

3.7 Homomorphisms

For graphs $G = (V, E)$ and $G' = (V', E')$, a *homomorphism* from G to G' is a mapping $f : V \to V'$ satisfying

$$\{u, v\} \in E \quad \Longrightarrow \quad \{f(u), f(v)\} \in E'.$$

This fact is denoted by $f : G \to G'$ and the existence (resp. the non-existence) of a homomorphism is denoted by $G \longrightarrow G'$ (resp. $G \not\longrightarrow G'$). Note that no other condition is imposed (non edges can be mapped to edges, vertices can be identified). An *isomorphism* is then a bijective homomorphism whose inverse is also a homomorphism. The existence of an isomorphism $G \to G'$ is denoted by $G \cong G'$ and we say that G and G' are isomorphic; this relation is obviously an equivalence relation. A homomorphism $f : G \to G'$ may be one-to-one in which case G is isomorphic to a subgraph of G'. Thus $K_k \to G$ when G contains a clique of size k. The *clique number* $\omega(G)$ is the maximum order of a clique of G, that is the maximum order of a complete subgraph of G. Hence $\omega(G) = \max\{k : K_k \longrightarrow G\}$.

On the other hand every homomorphism $f : G \to K_k$ corresponds to a coloring of the vertices of G by k colors such that no two vertices colored the same are adjacent. Thus $\chi(G) = \min\{k : G \to K_k\}$ is the chromatic number of G.

The class of all finite graphs (up to isomorphism) will be denoted by $\mathcal{G}raph$. This class together with homomorphisms forms a *category* (cf e.g. [253]). This amounts to the following:

- For every graph G the identity mapping $V(G) \to V(G)$ is a homomorphism;
- Whenever $f : G_1 \to G_2$ and $g : G_2 \to G_3$ are homomorphisms then their composition $g \circ f : G_1 \to G_3$ is a homomorphism.

We can consider standard categorical constructions. Particularly we shall make use of the *categorical sum* $G + H$ (also called *sum* or *disjoint union* of G and H) and of the *categorical product* $G \times G'$ (also called *direct product*) which is defined as follows:

$$V(G \times G') = V(G) \times V(G')$$
$$E(G \times G') = \{\{(u, u'), (v, v')\} : \quad \{u, v\} \in E(G) \text{ and } \{u', v'\} \in E(G')\}$$

An example is given on Fig. 3.5.

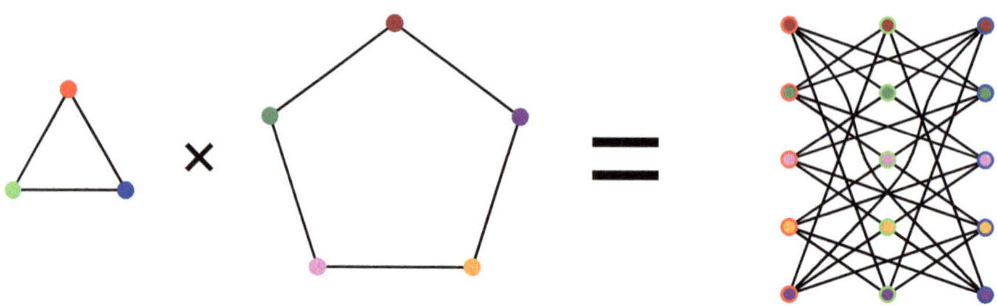

Fig. 3.5 The categorical product $K_3 \times C_5$

The *injection* $i_G : V(G) \to V(G + H)$ defined by $i_G(u) = u$ is obviously a homomorphism of G to $G + H$ to G, and similarly from the injection $i_H : G \to G + H$.

The *projection* $\pi_G : V(G \times H) \to V(G)$ defined by $\pi_G(u, v) = u$ is a homomorphism of $G \times H$ to G, and similarly from the projection $\pi_H : G \times H \to H$.

These injections and projections determine the sum and the product uniquely and thus the above definitions are a particular case of sum and product and in any category. This can be expressed by the following diagrams:

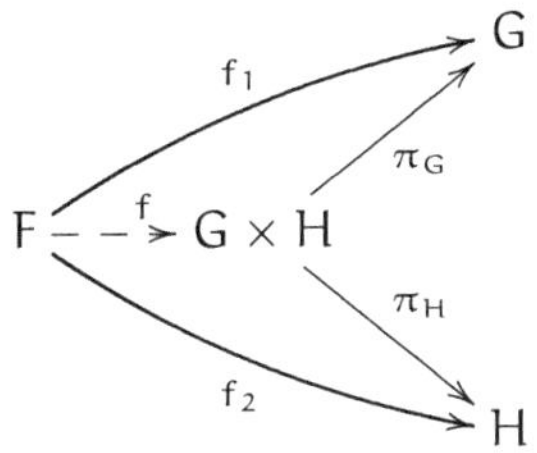

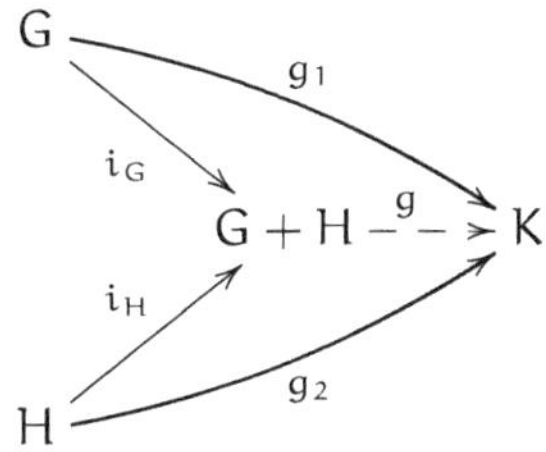

In words: For every graph F and homomorphisms $f_1 : F \to G$ and $f_2 : F \to H$ there exists uniquely defined homomorphism $f : F \to G \times H$ such that $f_1 = \pi_G \circ f$ and $f_2 = \pi_H \circ f$; For every graph K and homomorphisms $g_1 : G \to K$ and $g_2 : H \to K$ there exists uniquely defined homomorphism $g : G+H \to K$ such that $g_1 = g \circ i_G$ and $g_2 = g \circ i_H$. This implicit definition of $G \times H$ and $G+H$ may be very useful. For example one can prove easily the following proposition which is the basis of *product dimension* (see [311]). See also Exercise 3.7.

Proposition 3.4. *Every graph* G *is an induced subgraph of a power of a complete graph, i.e. of a graph of the form*

$$K_n^p = \underbrace{K_n \times K_n \times \cdots \times K_n}_{p}.$$

Proof. Let us prove the first statement: let G be a graph of order n and denote by $\mathrm{Hom}(G, K_n)$ the set of all homomorphisms $G \longrightarrow K_n$. Put explicitly $\mathrm{Hom}(G, K_n) = \{f_1, \ldots, f_t\}$. We consider the graph K_n^t and denote by $\pi_1, \ldots, \pi_t$ its projections. By the above property there exists an unique homomorphism $f : G \to K_n^t$ such that $\pi_i \circ f = f_i$ for every $i = 1, \ldots, t$. As at least one f_i is one-to-one so is f. Moreover, if $\{u, v\}$ is not an edge of G then $\{f(u), f(v)\}$ is not an edge of K_n^t, as there exists $f_i \in \mathrm{Hom}(G, K_n)$ which only identifies u and v. Thus f is actually an isomorphism of G with the subgraph of K_n^t induced by the set $\{f(v) : v \in V(G)\}$. □

The direct product is a powerful construction and we shall make use of it below.

Homomorphisms lead us also to homomorphism dualities, which are statements of the following type:

$$\mathcal{F} \not\longrightarrow G \iff G \longrightarrow \mathcal{D} \tag{3.7}$$

where $\mathcal{F}$ and $\mathcal{D}$ are finite sets of graphs. Here the undefined symbols are natural extensions of our earlier notations:

$$\begin{aligned} \mathcal{F} \not\longrightarrow G \quad &\text{means} \quad F \not\longrightarrow G \text{ for every } F \in \mathcal{F} \\ G \longrightarrow \mathcal{D} \quad &\text{means} \quad G \longrightarrow D \text{ for some } D \in \mathcal{D} \end{aligned}$$

If (3.7) holds for sets $\mathcal{F}$ and $\mathcal{D}$ then we call the pair $(\mathcal{F}, \mathcal{D})$ a *finite duality*. This notion captures the fact that the existence of a homomorphism into a given set $\mathcal{D}$ of graphs (called *duals* or *templates*) can be alternatively expressed dually by the non-existence of homomorphisms from a given set $\mathcal{F}$ of (forbidden) graphs. This notion has not much of a meaning for undirected graphs but for richer structures (including directed graphs) we have an abundance of examples. Here is one for directed graphs: Denote by $\vec{P}_n$ the directed path of order n (and length $n-1$) and by $\vec{T}_n$ the transitive tournament of order n. If we extend the notion of homomorphism to oriented graphs then we have (for every directed graph $\vec{G}$):

$$\vec{P}_{k+1} \not\longrightarrow \vec{G} \iff \vec{G} \longrightarrow \vec{T}_k$$

This (easy) theorem was discovered several times and it is known as Gallai-Hasse-Roy-Vitaver theorem [209, 244, 426, 466]. It is usually stated as a result about orientations:

Theorem 3.13. *A graph* G *has chromatic number at least* k *if and only if every orientation* $\vec{G}$ *of* G *contains a directed path of length* $k+1$.

In Chap. 9 we generalize these results to relational structures and we also consider dualities for graphs restricted to a fixed class, for special classes having applications to logic.

Existence of homomorphisms induce a quasi-order: for two graphs G, H we also write $G \leq_h H$ if $G \longrightarrow H$. The relation $\leq_h$ is obviously a quasi-order on the class $\mathcal{G}raph$ of all finite graphs, which we call the *homomorphism order*. The relation $\leq_h$ is clearly not a partial order as we may have non isomorphic graphs G, H such that $G \leq_h H \leq_h G$. Such graphs are called *hom-equivalent*. For instance, every non-discrete bipartite graph G is hom-equivalent to K_2 and thus any two graphs with chromatic number 2 are hom-equivalent. The equivalence class of a graph G for hom-equivalence will be denoted by [G]. The homomorphism order defines a partial order $\leq_h$ on the set $[\mathcal{G}\mathrm{raph}]$ of all these equivalence classes.

As we are discussing finite graphs only, the homomorphism equivalence takes a particularly simple form.

Proposition 3.5. *For any graph* G *there is (up to isomorphism) a unique graph* G′ *which is hom-equivalent to* G *and which has the minimal number of vertices. Such a graph* G′ *is called the* core *of* G*, and it is isomorphic to an induced subgraph of* G *(see Fig. 3.6).*

Proof. Let G′ be a graph hom-equivalent to G and which has the minimal number of vertices. Let $f : G \to G'$ and $g : G' \to G$ be the corresponding homomorphisms. Because of the minimality of G′, the homomorphism g is injective, $f \circ g$ is an automorphism of G′ hence G′ is actually isomorphic to the subgraph $g(G')$ of G. From this also follows that the graph G′ is uniquely determined (up to isomorphism). □

This is a categorical argument which can be used for any finite structure. It is also easy to prove that the core of G can be alternatively defined as the minimal *retract* of G, i.e. as the smallest subgraph G′ of G such that there exists a homomorphism $f : G \to G'$ which is the identity on G′ (i.e. $f(v) = v$ for every $v \in V(G')$).

Actually, any two graphs are hom-equivalent if and only if they have isomorphic cores. A graph which is a core (of some graph or, equivalently, of itself) is called a *core graph.* Hence each equivalence class for hom-equivalence contains exactly one core graph, which we can consider as a standard representative for the class.

The homomorphism order $\leq_h$ is very complex. Merely testing the relation $G \leq_h G'$ is an NP-complete problem as exemplified by the case $G \leq_h K_3$ which amounts to testing whether a graph G is 3-colorable.

The complexity of the homomorphism order is demonstrated by its (countable) universality and density:

Theorem 3.14 (Universality). *The homomorphism order is a* countably universal order*, that is: for every countable poset* $P = (X, \leq_P)$ *there exists and injective mapping* $x \mapsto G_x$ *of* X *into* $\mathcal{G}\mathrm{raph}$ *satisfying*

$$x \leq_P y \quad \Longleftrightarrow \quad G_x \leq_h G_y.$$

(Such a mapping is called an embedding *of* P *into the homomorphism order.)*

Universality of the homomorphism order has been proved by several different methods in [250, 264, 389]. Resolving an old conjecture it has been proved

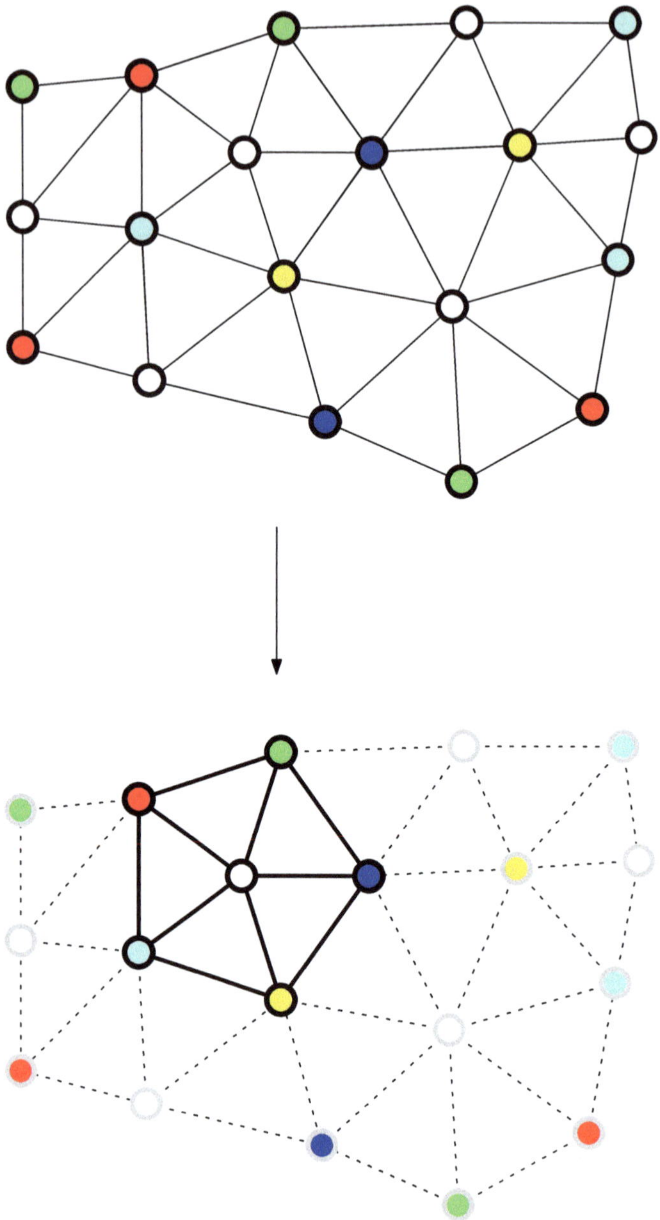

Fig. 3.6 The core of a graph

in [263] that even the class of all finite oriented paths is a countably universal partial order. It follows that also the homomorphism order of all finite planar graph (and even series-parallel graphs) is universal. The essence (and difficulty) of these results is that the elements of a countable partial order are represented by finite objects.

Theorem 3.15 (Density). *For every pair of graphs* G_1, G_2 *such that* $G_1 \to G_2, G_2 \nrightarrow G_1$ *and* $G_2 \nrightarrow K_2$ *there exists a graph* G *such that*

$$G_1 \underset{\not\leftarrow}{\to} G \underset{\not\leftarrow}{\to} G_2. \tag{3.8}$$

In other words, the homomorphism order $(\mathcal{G}\mathrm{raph}, \leq_h)$ *is a* dense partial order *with the exception of the* gap (K_1, K_2).

The density theorem was proved (in the context of computer science) by Welzl [468]. A short proof by Nešetřil and Perles [342] was the start of new investigations and the one-to-one correspondence between exceptional cases for the density (the *gaps*) and homomorphism dualities was established [370].

The density theorem has an interesting consequence. First we state a technical lemma.

Lemma 3.2. *Let* G_1 *and* G_2 *be homomorphically incomparable graphs, where* G_1 *is non bipartite (i.e.* $G_1 \nrightarrow K_2$*).*

Then there exist graphs G_L *and* G_R *such that the following diagram holds:*

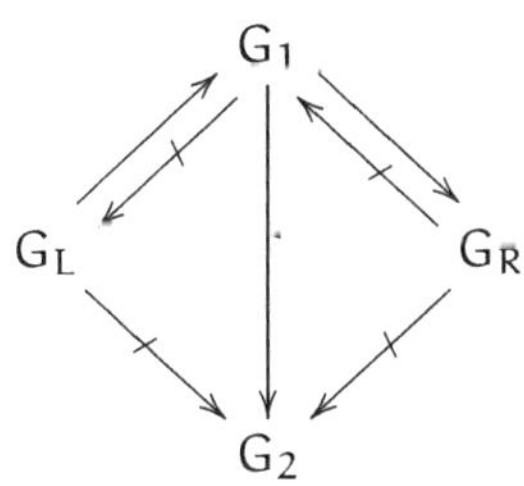

Proof. As $G_1 \nrightarrow G_2$ we have $G_1 \times G_2 \underset{\not\leftarrow}{\to} G_1 \underset{\not\leftarrow}{\to} G_1 + G_2$. Hence by density there exist G_L and G_R such that

$$G_1 \times G_2 \underset{\not\leftarrow}{\to} G_L \underset{\not\leftarrow}{\to} G_1 \underset{\not\leftarrow}{\to} G_R \underset{\not\leftarrow}{\to} G_1 + G_2.$$

Obviously $G_L \nrightarrow G_2$ (for otherwise $G_L \to G_1 \times G_2$) and $G_2 \nrightarrow G_R$ (for otherwise $G_1 + G_2 \to G_R$). □

Here is the promised corollary:

Corollary 3.1. *The homomorphism order defines every graph both from left and right:*

Up to a homomorphism equivalence, every graph G *can be reconstructed from the set* $\{\mathrm{H} : \mathrm{H} < \mathrm{G}\}$ *and also from the set* $\{\mathrm{H} : \mathrm{H} > \mathrm{G}\}$.

Note that for the injective mappings (instead of homomorphisms) a similar theorem does not hold for oriented graphs (even for tournaments). However for undirected graphs this is the old and well known *Ulam's conjecture.*

For graphs F, G denote by $\hom(\mathrm{F}, \mathrm{G})$ the number of homomorphisms from F to G. Let $\mathrm{F}_1, \dots, \mathrm{F}_n, \dots$ be a fixed enumeration of all isomorphism types of finite graphs. To every graph H we can associate the (infinite) sequence $\langle \mathrm{G} \rangle = (a_1, a_2, \dots)$ by putting $a_i = \hom(\mathrm{F}_i, \mathrm{G})$. The sequence $\langle \mathrm{G} \rangle$ is called the *profile* or *Lovász vector* of G because of the following fundamental result proved by (then very young) Lovász [309]:

Theorem 3.16. *For any two graphs* G, G'*, we have* $\mathrm{G} \cong \mathrm{G}'$ *if and only if* $\langle \mathrm{G} \rangle = \langle \mathrm{G}' \rangle$.

This theorem has many consequences and it motivated much of the recent developments. So, if we can "count" homomorphisms, we can have (admittedly a rather inefficient) graph parameter describing isomorphism.

Counting of homomorphisms led to a very interesting development, particularly for dense graphs. This is related to the Szemerédi's regularity lemma [151, 314, 315], property testing [79], partition functions in statistical physics [79, 310, 448], limit objects in probability [56, 314], non-standard analysis, ergodic theory [151], quasi-random structures [313]. We refer the reader to [78] for a nice survey of this development and return to this topic in Chap. 12.

In this book we are motivated by optimization problems (such as the existence of homomorphisms and constrained satisfaction problems) and by the analysis (and classification) of sparse classes of graphs. These questions are more existential in nature and lead to more deterministic settings. However the techniques are often similar, problems and their context often resemble each other. Interestingly, nowhere dense classes may be defined by counting homomorphisms in a proper scaling [356, 357], see Chap. 12.

3.8 Relational Structures and First-Order Logic

The topic of this book draws from experience with many areas of mathematics. The main body of material deals with graphs, graph theory and combinatorics. But this predominance may be misleading: although most of the material belongs to algebraic graph theory and structural combinatorics, most of the covered material is not special to graphs. The whole setting generalizes to *hypergraphs* and *relational structures*. What do we mean by these structures and how can these generalizations be achieved? This will be briefly explained in this section.

Our notation and terminology is standard, see e.g. [141].

3.8.1 Relational Structures

Given a set σ of relation symbols, each with a specified *arity*, which we call *signature* (or *vocabulary*), we define a *σ-structure* $\mathbf{A}$ as a set A (the *universe*, *domain*, or *carrier* of $\mathbf{A}$) together with the interpretation of each symbol R of σ with arity r as a relation $R^A \subseteq A^r$. In this book, we only consider finite signatures. A *relational structure* is a σ-structure for some signature σ. A relational structure is finite if its universe is finite. When we shall consider infinite structures we shall explicitly precise it; otherwise, the considered relational structures will implicitly be assumed to be finite. We shall denote relational structures by boldface letters such as $\mathbf{A}$, $\mathbf{B}$.

A σ-structure $\mathbf{B}$ is a *substructure* of $\mathbf{A}$ if $B \subseteq A$ and $R^B \subseteq R^A$ for every $R \in \sigma$. It is an *induced substructure* if $R^B = R^A \cap B^r$ for every $R \in \sigma$ of arity r. A substructure $\mathbf{B}$ of $\mathbf{A}$ is *proper* if $\mathbf{A} \neq \mathbf{B}$. If $\mathbf{B}$ is an induced substructure of $\mathbf{A}$, we say that $\mathbf{A}$ is an *extension* of $\mathbf{B}$. If $\mathbf{B}$ is a proper induced substructure, then $\mathbf{A}$ is a *proper extension*. If $\mathbf{A}$ is the disjoint union of $\mathbf{B}$ with another σ-structure, we say that $\mathbf{A}$ is a *disjoint extension* of $\mathbf{B}$. If $S \subseteq A$ is a subset of the universe of $\mathbf{A}$, then $\mathbf{A} \cap S$ denotes the *induced substructure generated by* S; in other words, the universe of $\mathbf{A} \cap S$ is S, and the interpretation in $\mathbf{A} \cap S$ of the r-ary relation symbol R is $R^A \cap S^r$.

A *homomorphism* $\mathbf{A} \to \mathbf{B}$ between two σ-structure is defined as a mapping $f : A \to B$ which satisfies for every relational symbol $R \in \sigma$ the following:

$$(x_1, \ldots, x_k) \in R^A \quad \Longrightarrow \quad (f(x_1), \ldots, f(x_k)) \in R^B.$$

The class of all finite σ-structures is denoted by $\mathrm{Rel}(\sigma)$. The same symbol will be used for the category of all σ-structures and all homomorphisms between them, and also for the corresponding quasi-order defined by the existence of a homomorphism. The particular meaning will be always clear from the context. As for graphs we can consider standard categorical constructions: the *categorical sum* $\mathbf{A} + \mathbf{B}$ (also called *sum* or *disjoint union* of $\mathbf{A}$ and $\mathbf{B}$) and the *categorical product* $\mathbf{A} \times \mathbf{B}$ of two σ-structures (also called *direct product*) which is defined as follows: Let $\mathbf{A}$ and $\mathbf{B}$ be two σ-structures with respective universes A and B. Then:

- The universe of $\mathbf{A} \times \mathbf{B}$ is $A \times B$,
- For every k-ary relational symbol $R \in \sigma$, $((x_1, y_1), \ldots, (x_k, y_k)) \in R^{A \times B}$ if both $(x_1, \ldots, x_k) \in R^A$ and $(y_1, \ldots, y_k) \in R^B$.

The *chromatic number* of a σ-structure is the minimum number of colors needed to color its universe in such a way that no tuple in any relation is monochromatic. (Clearly this requires that arities are at least 2.)

Note that a somewhat different formalism emerged for σ-structures from the theory of structures and the theory of categories (see e.g. [253, 370, 389]). A finite set of positive integers $\Delta = (\delta_i : i \in I)$ is called a *type*. A *relational structure* $\mathbf{A}$ of type Δ is a pair $(X, (R_i : i \in I))$ where X is a finite set and $R_i \subseteq X^{\delta_i}$. Both approaches are of course equivalent and are just slightly different formalizations of the same thing. We use alternatively both approaches depending on the particular context.

A *hypergraph* (or *set system*) $\mathcal{H}$ is a pair $(X, \mathcal{M})$ where X is a finite set and $\mathcal{M}$ is a family of non-empty subsets of X whose union is X [59]. The elements of X and $\mathcal{M}$ are respectively the *vertices* (or *points*) and the *edges* (or *hyperedges*). Given two hypergraphs $\mathcal{H} = (X, \mathcal{M})$ and $\mathcal{H}' = (X', \mathcal{M}')$ a mapping $f : X \to X'$ is said to be a *homomorphism* $\mathcal{H} \to \mathcal{H}'$ if for every $M \in \mathcal{M}$ $f(M) = \{f(x), x \in M\}$ belongs to $\mathcal{M}'$.

Many of the results for graphs have analogous form and proofs for general systems. For instance, the density Theorem 3.15 can be proved for general σ-structures and in fact this correspondence between gaps and dualities is the key to the proof, see [370] and Sect. 9.2.

3.8.2 First-Order Logic

Let σ be a relational vocabulary. The *atomic formulas* of σ are those of the form $R(x_1, \ldots, x_r)$, where $R \in \sigma$ is a relation symbol of arity r, and $x_1, \ldots, x_r$ are first-order variables that are not necessarily distinct. Formulas of the form $x = y$ are also atomic.

The collection of *first-order formulas* is obtained by closing the atomic formulas under negation, conjunction, disjunction, universal and existential first-order quantification. The collection of *existential first-order formulas* is obtained by closing the atomic formulas and the negated atomic formulas under conjunction, disjunction, and existential quantification. The semantics of first-order logic is standard.

The *quantifier count* of a formula ϕ is the total number of quantifiers in ϕ. The *quantifier rank* of a first-order formula is the maximum nesting of quantifiers of its sub-formulas. These quantities are denoted by $\mathrm{qcount}(\phi)$ and $\mathrm{qrank}(\phi)$, respectively. Quantifier-rank is obviously at most quantifier-count and often strictly less.

Let $\mathbf{A}$ be a σ-structure, and let $a_1, \ldots, a_n$ be points in $\mathbf{A}$. If $\phi(x_1, \ldots, x_n)$ is a formula with free variables $x_1, \ldots, x_n$, we denote by $\mathbf{A} \models \phi(a_1, \ldots, a_n)$ the fact that ϕ is true in $\mathbf{A}$ when x_i is interpreted by a_i. If m is an integer, the first-order m-*type* of $a_1, \ldots, a_n$ in $\mathbf{A}$ is the collection of all first-order formulas $\phi(x_1, \ldots, x_n)$ of quantifier rank at most m, up to logical equivalence, for which $\mathbf{A} \models \phi(a_1, \ldots, a_n)$.

3.8.3 Derived Graphs

The *Gaifman graph* of a σ-structure $\mathbf{A}$, denoted by $\mathrm{Gaifman}(\mathbf{A})$, is the (undirected) graph whose set of nodes is the universe of $\mathbf{A}$, which is denoted by A, and whose set of edges consists of all pairs (a, a') of distinct elements of A such that a and a' appear together in some tuple of a relation in $\mathbf{A}$. This notion coincides with the combinatorial notion of 2-*section* in the sense of Berge, see e.g. [59]. The *degree* of a structure is the maximum degree of its Gaifman graph, that is, the maximum number of neighbors of nodes of the Gaifman graph. Other notions (such as shallow minor or distance) are defined analogously via Gaifman graphs.

Let us remark that alternatively we may convert any hypergraph $\mathcal{H} = (X, \mathcal{M})$ to a graph by means of *incidence graph* $\mathrm{Inc}(\mathcal{H})$ which may be defined as the following graph (V, E) where $V = X \cup \mathcal{M}$, $E = \{(x, e) : x \in e \in \mathcal{M}\}$.

For relational structures we can proceed analogously: Given a relational structure $\mathbf{A} = (A, (R_i : i \in I))$ the *incidence graph* $\mathrm{Inc}(\mathbf{A})$ is the bipartite graph (A, B, E) where B is the set of *blocks*, i.e. pairs $(i, (x_1, \ldots, x_{\delta_i}))$ where $i \in I$ and $(x_1, \ldots, x_{\delta_i}) \in R_i$. The edges are then all the incidences "$x \in (x_1, \ldots, x_{\delta_i})$". (this is interpreted as a multigraph, thus x forms two edges with $(i, (x, x, y))$).

This construction allows us to import some basic concepts from graph theory (see e.g. [297]): the *distance* $d_{\mathbf{A}}(a, b)$ between two elements a and b of $\mathbf{A}$ is defined as half their distance in $\mathrm{Inc}(\mathbf{A})$, the *diameter* of $\mathbf{A}$ is defined as half the diameter of $\mathrm{Inc}(\mathbf{A})$, and the *girth* of $\mathbf{A}$ is defined as half the shortest length of a cycle in $\mathrm{Inc}(\mathbf{A})$. In particular, $\mathbf{A}$ has girth 1 if and only if $\mathrm{Inc}(\mathbf{A})$ has parallel edges, and infinite girth if and only if $\mathrm{Inc}(\mathbf{A})$ is acyclic. Notice in particular that tuples with repeated entries (such as (a, a, b)) create parallel edges and hence cycles; this property is not captured in the Gaifman graph.

A σ-structure **T** is called a *σ-tree* (or *relational tree*, or *tree* for short) if Inc(**T**) is a (graph) tree, i.e. it is acyclic and connected. We denote by Tree(σ) the set of all σ-trees.

Incidence graphs and Gaifman graphs do not exhaust all the possibilities to transform general structures to graphs: other possibilities are the use of *star selectors* or the replacement of k-tuples by oriented paths. When dealing with unbounded arities we have to use incidence graphs (or star selectors) as Gaifman graphs would then have unbounded clique number. On the other hand, Gaifman graphs seem to be preferred by logicians (see e.g. [42, 321]) and in a way this concept presents the first approximation. As we shall see the star selectors are often more sensitive approach to problems treated in this book.

3.8.4 Ehrenfeucht-Fraïssé Games

The *Ehrenfeucht-Fraïssé game* is a well-known technique for determining whether two structures are "equivalent". Suppose that we are given two graphs G and H and a fixed natural number n. We can then define the Ehrenfeucht-Fraïssé game $\partial_n(G, H)$ to be a game between two players, Spoiler and Duplicator, played as follows: Start with $A_0 = B_0 = \emptyset$ and let π_0 be the (empty) mapping from A_0 to B_0. Notice that π_0 is an isomorphism from $G[A_0]$ to $G[B_0]$. For each $1 \leq i \leq n$, Spoiler picks either a vertex a in G or a vertex b in H. In the first case, the Duplicator choose a vertex b in H; in the second case he chooses a vertex a in G. Let $A_i = A_{i-1} \cup \{a\}$ and $B_i = B_{i-1} \cup \{b\}$. If no isomorphism $\pi_i : G[A_i] \to G[B_i]$ extending π_{i-1} exists such that $\pi(a) = b$ then Spoiler wins the game. Otherwise, the game continues until $i = n$ (notice that π_i is uniquely determined by π_{i-1} and vertices a and b). If i reaches n and π_n is an isomorphism from $G[A_n]$ to $H[B_n]$ then Duplicator wins the game.

If Duplicator has a winning strategy for n then we note $G \equiv^n H$ and we say that G and H are *n-back-and-forth equivalent* (see Fig. 3.7). If $G \equiv^n H$ for every n then G and H are *elementarily equivalent*, what is denoted by $G \equiv H$. Notice that although two finite graphs are clearly elementarily equivalent if and only if they are isomorphic, elementary equivalence is coarser than isomorphism for general (i.e. non necessarily finite) graphs.

The meaning of the n-back-and-forth equivalence is clarified by the classical results of Fraïssé [184, 185] and Ehrenfeucht [145].

Fig. 3.7 Back-and-Forth Game: $G \equiv^2 H$ but $G \not\equiv^3 H$

Theorem 3.17. *Two graphs (and more generally two structures) are* n*-back and forth equivalent if and only if they satisfy the same first order sentences of quantifier rank* n*.*

The quantifier rank plays in this book a prominent role. Not only because of our use of homomorphism preservation theorems (see Chap. 10) but one of our principal decomposition techniques involves the notion of tree-depth which is related to the notion of quantifier rank (see [425] and Sect. 6.9).

3.8.5 Interpretation

Let $\mathcal{L}$ and $\mathcal{L}'$ be two languages and let T be a theory in $\mathcal{L}$. Recall (see, for instance [299], p. 178–180) that an *interpretation* I of $\mathcal{L}'$ in $\mathcal{L}$ is defined by:

- An integer n,
- An $\mathcal{L}$-formula $U[v_1, \ldots, v_n]$ with n free variables,
- An $\mathcal{L}$-formula $E[\overline{w}_1, \overline{w}_2]$ with $2n$ free variables ($\overline{w}_1, \overline{w}_2$ represent each a sequence of n variables),
- And an $\mathcal{L}$-formula $F_R[\overline{w}_1, \ldots, \overline{w}_k]$ with kn free variables for each relational symbol R with arity k,

which satisfy the following conditions:

1. The theory T entails that E is an equivalence relation;
2. The theory T entails that U is a union of E-classes;
3. For every integer k and every symbol R of arity k in $\mathcal{L}'$, T entails that F_f is interpreted by a set which is closed for the relation E.

Then, if $\mathbf{A}$ is a model of T, we can *interpret* in $\mathbf{A}$ the $\mathcal{L}'$-structure $\mathbf{A}'$ defined as follows:

- The universe A' of $\mathbf{A}'$ is $U[\mathbf{A}]/E[\mathbf{A}]$;
- Let R be a symbol of arity k of $\mathcal{L}'$ and $(a_1, \ldots, a_k) \in A'^k$; then $(a_1, \ldots, a_k) \in R^{A'}$ if and only if there exists $\overline{b}_1 \in a_1, \ldots, \overline{b}_k \in a_k$ such that $\mathbf{A} \models F_R[\overline{b}_1, \ldots, \overline{b}_k]$.

In such a case, $\mathbf{A}'$ is an *interpretation* of $\mathbf{A}$ by I, what we denote by $\mathbf{A}' = I(\mathbf{A})$. A main interest of such an interpretation lies in the following property (See, for instance [299], p. 180):

Lemma 3.3. *For every formula* $F[v_1, \ldots, v_k]$ *of* $\mathcal{L}'$ *there exists a formula* $I(f)[\overline{w}_1, \ldots, \overline{w}_k]$ *of* $\mathcal{L}$ *with* kn *free variables (each* $\overline{w}_i$ *represents a succession of* n *free variables) such that for every model* $\mathbf{A}$ *of* T*, if* $\mathbf{A}' = I(\mathbf{A})$ *and if* $(a_1, \ldots, a_k) \in A'^k$ *then the three following conditions are equivalent:*

1. $\mathbf{A}' \models F[a_1, \ldots, a_k]$;
2. *There exist* $\overline{b}_1 \in a_1, \ldots, \overline{b}_k \in a_k$ *such that* $\mathbf{A} \models I(F)[\overline{b}_1, \ldots, \overline{b}_k]$;
3. *For all* $\overline{b}_1 \in a_1, \ldots, \overline{b}_k \in a_k$ *it holds* $\mathbf{A} \models I(F)[\overline{b}_1, \ldots, \overline{b}_k]$.

3.9 Ramsey Theory

The reader can rightly raise his/her eyebrows: what does a typical area dealing with "dense structures" such as Ramsey theory have to do with sparsity in the sense of the main theme of this book? We shall apply Ramsey theory (particularly in Chap. 8) as a tool to obtain shapes of *unavoidable configurations* for certain special classes of graphs. This is in line with the original motivation of Ramsey [393] who discovered and applied his theorem in the context of a problem (*Entscheidungsproblem*, i.e. decision problem) of logic and used it to produce canonical (i.e. unavoidable, in todays terms) models. Thus let us recall the basics of Ramsey theory for graphs which will be needed in the sequel. The Ramsey theorem [393] we formulate as follows (see an example on Fig. 3.8):

Theorem 3.18. *For every choice of positive integers* $n_1, \ldots,$ n_k *(*$k \geq 1$*) there exists a number* $R = R(n_1, \ldots, n_k)$ *(*Ramsey number*), which is minimum with the following property:*

For every set X *of cardinality at least* R *and every coloring of the set* $\binom{X}{2}$ *by* k *colors there exists* i, $1 \leq i \leq k$, *and a subset* $Y \subseteq X$ *such that* $|Y| \geq n_i$ *and* $\binom{Y}{2}$ *is monochromatic of color* i.

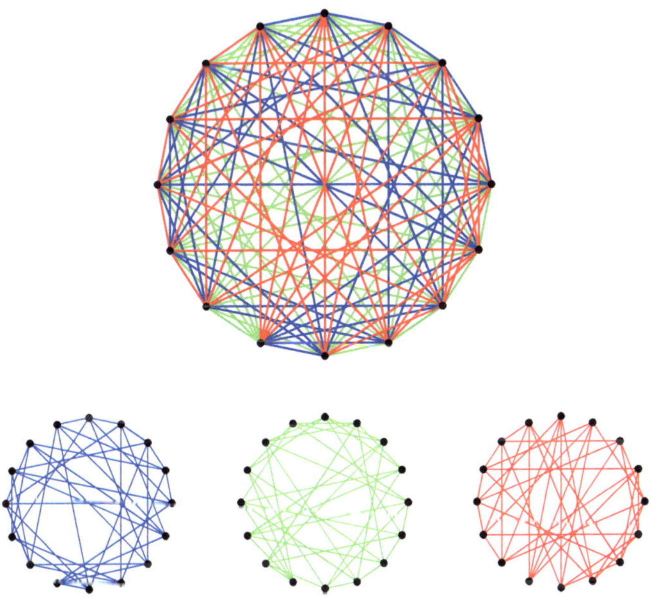

Fig. 3.8 The Ramsey number $R(3,3,3)$ is 17; the edges of K_{16} may be colored by three colors without creating monochromatic triangles

In the important case $k = 2$ the theorem can be restated by saying that every sufficiently large graph contains either a large independent set or a large complete graph. In other words complete graphs or independent sets of a given size are unavoidable for large graphs.

One can restate the above property in a slightly different way:

Proposition 3.6. *Let* G *be a graph and let* c, n *be integers. Then either* G *contains a subgraph isomorphic to the complete graph* K_c *or every subset* A *of at least* $R(c, n)$ *vertices of* G *includes an independent set of size* n. □

Instead of considering independent sets in graphs, one can consider in bipartite graphs sets of vertices having no common neighbors.

Other related Ramsey-type problems relate to *graph Ramsey numbers* (sometimes called *generalized Ramsey numbers*): for a graph G we denote by $r(G)$ the minimum integer N such that for every blue–red coloring of the edges of the complete graph K_N there exists a (of course not induced) subgraph of K_N which is isomorphic to G and which has all its edges of the same color. In many cases the numbers $r(G)$ are known (which is in sharp contrast with $R(n, n) = r(K_n)$ which is only known for $n \leq 4$). A class of graphs $\mathcal{C}$ is a *Ramsey linear class* if there exists a constant $c = c(\mathcal{C})$ such that $r(G) \leq cn$ for every $G \in \mathcal{C}$ of order n.

Conjecture of Burr and Erdős [87]:
For every integer d there exists a constant c_d such that for every d-degenerate graph G holds:

$$r(G) \leq c_d|G|.$$

In other words: degenerated graphs form a linear Ramsey class. Many partial results toward this conjecture were obtained. In Sect. 15.2 we present a unified view using our theory.

3.10 Graph Parameters

In mathematics objects of study are often treated indirectly: We do not study the objects in their entire complexity (leaving this to philosophy) but rather relate them to particular aspects. These aspects are (in a mild mathematical form) expressed as parameters. A *graph parameter* (or *graph invariant*) ϱ is a function defined on finite graphs which is invariant under isomorphisms (actually we only consider here graph parameters defined on simple graphs, sometimes called *simple graph parameters* [78]). The value $\varrho(G)$ may be a number, a collection of numbers, a graph, a set of graphs, a set of numbers indexed by graphs, etc. One should realize that the actual complexity of $\varrho(G)$ (in the intuitive sense) may sometimes be much larger than the one of G. In the first chapter we introduced several graph parameters, for instance: the order $|G|$ of a graph G (that is: the number of vertices of G), the size $\|G\|$ (that is: the number of edges of G), the maximum degree $\Delta(G)$, the minimum degree $\delta(G)$, and the maximum average degree $\mathrm{mad}(G)$. Another, less intuitive example, is the *isomorphism type* of G. Usual perception of this graph parameter is quite complicated. The isomorphism type of G is the class of all graphs which are isomorphic to G, that is $\{G',\ G' \cong G\}$. Thus isomorphism types are infinite sets, indeed proper classes. But this complexity is only illusionary as we may select from each isomorphism type a representative. For example this can be defined as follows: given G, let G' be the isomorphic image of G with vertex set $\{1, 2, \dots, n\}$ (where $n = |G|$) by an isomorphism ϕ such that the value $\sum_{\{u,v\}\in E(G)} 2^{\phi(i)+\phi(j)}$ attains the minimal possible value (this corresponds to the lexicographically minimal set of edges $E(G') = \{\{\phi(u), \phi(v)\},\ \{u, v\} \in E(G)\}$) So, alternatively, an isomorphism type may be viewed as a class represented by a single graph.

By definition, every graph parameter p only depends on the isomorphism type of its argument. Thus the parameter study naturally relates to the following:

Isomorphism Problem

Given two graphs G, G', can one decide polynomially whether $G \cong G'$?

The isomorphism problem was solved positively for special classes of graphs (such as planar and bounded degree graphs) but in general, despite many efforts, it is still an open problem. Part of our research is related to this problem (see Chap. 18).

We study graphs by means of graph parameters. We compare parameters, investigate their mutual dependence and relative growth. The choice of graph parameters is of course principal and leads to very different questions and answers. For example we may investigate the dependence of $\|G\|$ on $|G|$. This is easy for all graphs but for special classes (such as classes not containing a given graph) this leads to the *extremal graph theory* pioneered by Turán who solved the case of graphs not containing a complete graph K_k. However if we investigate the dependence of $\|G\|$ and the *genus* $g(G)$ of G then we get a very different theory which goes back to Euler formula. Also, the graph parameter $\|G\|$ in combination with $|G|$ and the independence number $\alpha(G)$ leads to density questions in Ramsey theory. Many such dependencies are exact and hold for all values of the parameters (such as in Turán theorem), others are asymptotic where we can determine the growth. In such cases we use classical Landau notation $O(f), o(f)$. Recall that we write $g = O(f)$ if there exists a constant K such that $g(x) \leq Kf(x)$ for all x in the common domain of g and f; as usual in computer science, we also write $g = \Omega(f)$ if there exists a constant K such that $g(x) \geq Kf(x)$ for all x in the common domain of g and f and we write $g = \Theta(f)$ if both $g = O(f)$ and $g = \Omega(f)$ hold. For $f : \mathbb{R} \to \mathbb{R}$ and $g : \mathbb{R} \to \mathbb{R} \setminus \{0\}$ we write $g = o(f)$ if $\frac{f(x)}{g(x)} \to 0$ as $x \to \infty$, and $f \sim g$ if $\frac{f(x)}{g(x)} \to 1$ as $x \to \infty$.

Here we add yet another (weaker) notion: For real-valued graph parameters ϱ, ς we write $\varrho \asymp \varsigma$ if there exists polynomials $P(X)$ and $Q(X)$ such that for every graph G hold $\varrho(G) \leq P(\varsigma(x))$ and $\varsigma(x) \leq Q(\varrho(x))$. Such a dependence we call *polynomial functional equivalence*. Notice that $\varrho \asymp \varsigma$ if and only if $\log \varrho = \Theta(\log \varsigma)$.

A weaker notion is the one of *functional equivalence* which is defined analogously. As an example of this, consider the definition of a topological parameter: A real-valued graph parameter ϱ is said to be *topological* [133] if there exists a function f such that for every graph G, denoting G' a 1-subdivision of G the following holds: $\varrho(G') \leq f(\varrho(G))$ and $\varrho(G) \leq f(\varrho(G'))$.

Let us take another example: are clique number $\omega(G)$ and chromatic number $\chi(G)$ functionally dependent? Clearly $\omega(G) \leq \chi(G)$. But the converse does not hold, because there is no functional dependence here as for every n there exists a graph G with $\chi(G) \geq n$ and $\omega(G) = 2$ (i.e. a triangle-free high chromatic graph). But $\omega(G)$ and $\chi(G)$ are functionally related for perfect graphs (by the definition) and also for intersection graphs of chords in a cycle. In some parts of this book, polynomial functional dependence will play the main role. In the complex interplay of our parameters (see Table 13.1 in Chap. 13), functional dependence will sometimes be the best we can ask for.

3.11 Computational complexity

The complexity of computational problems can be measured by the amount of time or space a computational model such as the deterministic Turing machine requires to perform the computation. A *complexity class* is a set of problems of related complexity, defined by

- The type of computational problem: decision problems, function problems, counting problems, optimization problems, etc.
- The model of computation: deterministic Turing machine, non-deterministic Turing machine, Boolean circuits, etc.
- The measured resource: computation time, memory space, circuit-depth, etc.

For an introduction to computational complexity we refer the reader to [34, 220, 379].

The classes of main interest for us in this book are the following classes of decision problems:

- P (resp. NP), corresponding to problems solved in polynomial time by a deterministic (resp. in a non-deterministic) Turing machine;
- PSPACE, corresponding to problems solved in polynomial space (either by deterministic or a non-deterministic Turing machine, the two possibilities defining the same class according to Savitch's theorem [429]);
- L (resp. NL), corresponding to problems solved in logarithmic space by a deterministic (resp. a non-deterministic) Turing machine;

The following relations are known between these classes:

$$\mathrm{L} \subseteq \mathrm{NL} \subseteq \mathrm{P} \subseteq \mathrm{NP} \subseteq \mathrm{PSPACE}.$$

It is also known that $\mathrm{NL} \neq \mathrm{PSPACE}$. However, it is not known whether $\mathrm{L} = \mathrm{NL}, \mathrm{NL} = \mathrm{P}, \mathrm{P} = \mathrm{NP}$ or $\mathrm{NP} = \mathrm{PSPACE}$.

Apart from time and space complexity, a natural complexity measure of complexity stands in the circuit complexity of Boolean functions, according to the size or depth of Boolean circuits that compute them. A *Boolean circuit* computes a function of its n inputs, by means of AND, OR, and NOT gates. The *fanin* of an AND or OR is the number of its inputs. The *depth* of a circuit is maximal length of a path from an input gate to the output gate. The *circuit-depth* complexity of a Boolean function f is the minimal depth of any circuit computing f. Complexity classes defined in terms of Boolean circuits include

- The classes NC^i of binary functions computed by Boolean circuits with depth $O(\log^i n)$ and a polynomial number of constant-fanin AND and OR gates;
- The classes AC^i of binary functions computed by Boolean circuits with depth $O(\log^i n)$ and a polynomial number of unlimited-fanin AND and OR gates;
- The classes TC^i of binary functions computed by Boolean circuits with depth $O(\log^i n)$ and a polynomial number of unlimited-fanin AND, OR gates, and Majority gates.

It is easily checked that $\mathrm{NC}^i \subseteq \mathrm{AC}^i \subseteq \mathrm{TC}^i \subseteq \mathrm{NC}^{i+1}$, and it is well known that $\mathrm{NC}^0 \subsetneq \mathrm{AC}^0 \subsetneq \mathrm{TC}^0$.

A connection with time a space complexity is also given by the inclusions

$$\mathrm{NC}^1 \subseteq \mathrm{L} \subseteq \mathrm{NL} \subseteq \mathrm{AC}^1.$$

It will be also useful for us to consider complexity classes from the angle of finite model theory, by the type of logic needed to express the problems. In particular:

- First-order logic defines the class FO, which corresponds to AC^0;
- First-order logic with a commutative transitive closure operator gives L;
- First-order logic with a transitive closure operator gives NL;
- First-order logic with a least fixed point operator gives P (in the presence of a linear order);
- Existential second-order logic gives NP [171];
- Second-order logic with a transitive closure gives PSPACE.

Exercises

3.1. Prove that every graph of minimum degree d contains a bipartite subgraph of minimum degree at least $\lceil d/2 \rceil$.

3.2. Let G, H be graphs. If G is k-degenerate then the number of copies of H in G is at most

$$\sum_{t=1}^{\alpha(H)} \mathrm{Acyc}_t(H) k^{|H|-t} |G|^t,$$

where $\mathrm{Acyc}_t(H)$ is the number of acyclic orientations of H with t sinks, and $\alpha(H)$ is the *independence number* (or *stability number*) of H (that is the maximum size of a subset of vertices of H without adjacent vertices).

3.3. Let G be a plane graph and let $f : V(G) \to \mathbb{N}$. The aim of this exercise is to prove that the sets of the orientations of G such that $d^-(v) = f(v)$ holds for every $v \in V(G)$ has a structure of distributive lattice [376].

- Let $\vec{G}$ be a connected acyclically oriented graph. A *cut* is the set of arcs linking a subset A of vertices of $V(\vec{G})$ to its complement $V(\vec{G}) \setminus A$. It is *positive* if all the arcs are oriented from A to $V(\vec{G}) \setminus A$. A *positive cocircuit* is a disjoint union of positive cuts. Let x_0 be a vertex of $\vec{G}$. Prove that the positive cocircuits of $\vec{G}$ are in bijection with the mappings $F : V(G) \to \mathbb{Z}$ such that $F(x_0) = 0$ and such that for every $(x, y) \in E(\vec{G})$ it holds $F(x) \leq F(y) \leq F(x) + 1$.
- Prove that the partial order on these mappings defined by $F \leq G$ if $F(v) \leq G(v)$ holds for every $v \in V(G)$ is a distributive lattice, where $F \wedge G$ (resp. $F \vee G$) is the mapping $x \mapsto \min\{F(x), G(x)\}$ (resp. the mapping $x \mapsto \max\{F(x), G(x)\}$).
- Prove that if $\vec{G}$ is a plane digraph (that is: a planar digraph embedded in the plane) the set of the positive circuits of $\vec{G}$ has a structure of distributive lattice.
- Deduce that if G is a planar graph and $f : V(G) \to \mathbb{N}$ is a mapping, the set of the orientations of G such that $d^-(v) = f(v)$ holds for every $v \in V(G)$ may be given a structure of distributive lattice.

3.4. Prove that the edges of any graph G such that $\chi(G) < R(\overbrace{3, 3, \ldots, 3}^{p})$ may be covered of at most p triangle-free subgraphs. In particular, every 5-colorable graph is the edge union of two triangle-free graphs.

3.5. Let H be a graph of order n. The *Mycielskian* of H [336] is the graph M(H) obtained from H as follows: for each vertex x of H we add a vertex x' linked to all the neighbours of x, and eventually we add a vertex z linked to all the added vertices x'.

Prove that the kth Mycielskian of K_2 is triangle-free, has order $3.2^k - 1$ and chromatic number $k + 2$.
One can also start the construction with any fixed graph instead of K_2. The fact that the chromatic number grows has the same proof. However, there is an interesting variant of this (called *cone of a graph*), which is less trivial (see [240]).

3.6. Prove that every proper minor closed class of graphs is characterized by finitely many forbidden **topological** minors.

3.7. Prove that every graph G with $\chi(G) \leq 3$ is an induced subgraph of a product of sufficiently many copies of the graph $P_4 \oplus K_1 =$ 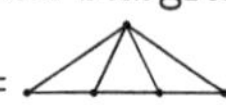.

3.8. The aim of this exercise is to follow the lines Rödl's proof [420] of the case $g = 4$ of the following conjecture of Erdős and Hajnal [164]: For all integers c, g there exists an integer $f(c, g)$ such that every graph G of chromatic number at least $f(c, g)$ contains a subgraph of chromatic number at least c and girth at least g.

Define

$$h(\omega, c) = c^{c^{\cdot^{\cdot^{\cdot^{c}}}}} \Big\}\omega - 2.$$

The proof of the case $g = 4$ of the conjecture will follow from the property that each graph G such that $\chi(G) > h(\omega(G), c)$ has a triangle free subgraph of chromatic number at least c. This last property will be proved by induction.

- Deduce from Exercise 3.5 that every graph G with $\omega(G) > 2^c$ contains a triangle free subgraph with chromatic number at least c.
- Consider a linear order on the vertices of a graph G and the corresponding natural orientation of G. Prove induction step by considering the two following cases:

  1. There exists a vertex v such that $\chi(G[N^-(v)]) > h(\omega(G) - 1, c)$;
  2. For every vertex v, $\chi(G[N^-(v)]) \leq h(\omega(G) - 1, c)$.

3.9. Prove that for every class of graphs $\mathcal{C}$ there exists a interpretation

$$I : \mathcal{C} \to \mathrm{Sub}_2(\mathcal{C})$$

from the class $\mathcal{C}$ to the class of the 2-subdivisions of the graphs in $\mathcal{C}$, which is such that

$$G \to H \quad \Longrightarrow \quad I(G) \to I(H)$$

(such an interpretation is a *functorial interpretation*). In fact, a similar proof hold for general $\mathrm{Sub}_{2p}(\mathcal{C})$ (see Lemma 10.6).

Chapter 4
Measuring Sparsity

What does "sparse" mean?
Let me show the ways.

In the introduction we described the big picture of our theory. Here we begin with a more formal treatment. We define shallow minors, topological minors, and immersions as the basic local changes in graph classes. We show that edge densities in the iteration of these local changes are related and that they are also related to other parameters such as chromatic number and generalized coloring numbers.

4.1 Basic Definitions

We work with unlabeled finite simple graphs, except when explicitly stated otherwise. Recall that we denote by $\mathcal{G}\mathrm{raph}$ the class of all unlabeled finite simple graphs.

The *distance* in a graph G between two vertices x and y is the minimum length of a path linking x and y (or ∞ if x and y do not belong to the same connected component of G) and is denoted by $\mathrm{dist}_G(x, y)$. Let $G = (V, E)$ be a graph and let d be an integer. The d-*neighborhood* $N_d^G(u)$ of a vertex $u \in V$ is the subset of vertices of G at distance at most d from u in G: $N_d^G(u) = \{v \in V : \mathrm{dist}_G(u, v) \leq d\}$.

A class $\mathcal{C}$ of graphs is *hereditary* if every induced subgraph of a graph in $\mathcal{C}$ belongs to $\mathcal{C}$, and it is *monotone* if every subgraph of a graph in $\mathcal{C}$ belongs to $\mathcal{C}$. For a class of graphs $\mathcal{C}$, we denote by $H(\mathcal{C})$ the class containing all the

J. Nešetřil and P.O. de Mendez, *Sparsity*, Algorithms and Combinatorics 28, DOI 10.1007/978-3-642-27875-4_4, © Springer-Verlag Berlin Heidelberg 2012

induced subgraphs of graphs in $\mathcal{C}$, that is the inclusion-minimal hereditary class of graphs containing $\mathcal{C}$.

4.2 Shallow Minors

For any graphs H and G and any integer d, the graph H is said to be a *shallow minor* of G at *depth* d if there exists a collection $\mathcal{P}$ of disjoint subsets $V_1 \dots, V_p$ of $V(G)$ such that:

- Each graph $G[V_i]$ has *radius* at most d: there exists in each set V_i a vertex x_i (a *center*) such that every vertex in V_i is at distance at most d from x_i in $G[V_i]$,
- H is a subgraph of the graph $G/\mathcal{P}$: each vertex v of H corresponds (in an injective way) to a set $V_{i(v)} \in \mathcal{P}$ and two adjacent vertices u and v of H correspond to two sets $V_{i(u)}$ and $V_{i(v)}$ linked by at least one edge.

[386] attributes this notion, then called *low depth minor*, to Leiserson and Toledo. Notice that the requirement that $G[V_i]$ has radius at most d and center x_i is not equivalent to $V_i \subseteq N_d^G(x_i)$: the distance between x_i and every vertex in V_i should be at most d in $G[V_i]$, and not only in G.

The set of all shallow minors of G at depth d is denoted by $G \triangledown d$. In particular, $G \triangledown 0$ is the set of all subgraphs of G. The choice of the notation $G \triangledown r$ is motivated by cases where the depth is a more involved expression (such as $G \triangledown \left(\frac{(2p+1)^q-1}{2}\right)$).

In this book we shall study shallow minors in a great depth and thus we introduce a more specific terminology.

A *ramification* of H is any minimal graph $\widehat{H}$ (with respect to inclusion) such that H is a minor of $\widehat{H}$. It is easily shown that ramifications have a special structure: Denoting $h_1, \dots, h_p$ the vertices of H, the graph $\widehat{H}$ may be (vertex) covered by a collection of vertex disjoint rooted trees $Y_1, \dots, Y_p$ in such a way that the remaining set F of edges of $\widehat{H}$ is such that:

- No edge in F is adjacent to two vertices in a same Y_i,
- At most one edge is incident to a vertex in Y_i and a vertex in Y_j (for every $i \neq j$),
- Every leaf of every tree Y_i has at least one edge incident to it and to a vertex not in Y_i,
- An edge is incident to a vertex in Y_i and a vertex in Y_j if and only if $\{h_i, h_j\}$ is an edge of H.

Such a decomposition of $\widehat{H}$ is called a H-*decomposition* of $\widehat{H}$. In this decomposition Y_i is the *bush* of h_i, the root of Y_i being the *center* of the bush, and F is the set of the *external* edges of the H-decomposition. The *radius* of the H-decomposition is the maximum distance from a vertex v to the center of the bush to which it belongs. It is straightforward that $\widehat{H}$ admits an H-decomposition of radius d if and only if $H \in \widehat{H} \triangledown d$.

It may happen that a ramification $\widehat{H}$ of a graph H of radius d has the additional property that there exists no external edge $\{x, y\}$ such that the distance of x to the center of its bush and the distance of y to the center of its bush are both equal to d. In such a circumstance, the H-decomposition of $\widehat{H}$ is said to be *asymmetric*.

This allows us to extend our definition of $G \triangledown d$ to half-integer values: For a graph G and an integer $d \geq 1$, $G \triangledown (d - \frac{1}{2})$ is the set of the graphs H having an asymmetric ramification $\widehat{H}$ of radius d isomorphic to a subgraph of G (see Fig. 4.1).

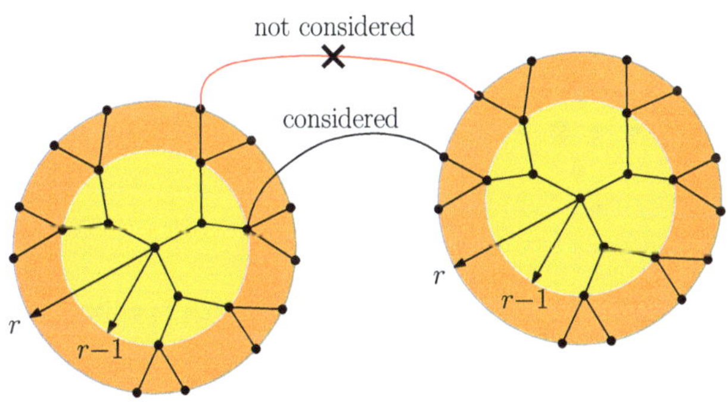

Fig. 4.1 A shallow minor of depth $r - 1/2$

For a graph G, a half-integer d and a graph $H \in G \triangledown d$, an d-*witness* of H in G is a subgraph G' of G which is a ramification of H of radius $\lceil d \rceil$ and which is asymmetric if d is not an integer. Equivalently, a d-witness of H in G is a subgraph G' of G which is a ramification of H so that $H \in G' \triangledown d$ (see Fig. 4.2).

Notice that for every two vertices x, y of $H \in G \triangledown d$, the distance in a witness W of H in G between the centers of the bushes corresponding to x and y is at most $(2d+1)\text{dist}_H(x, y)$ (see Fig. 4.3).

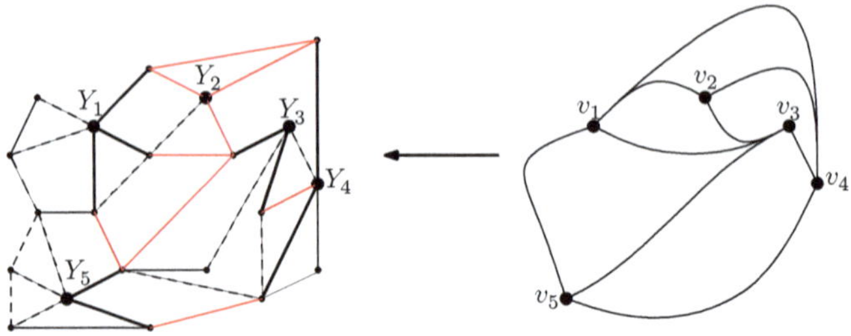

Fig. 4.2 A 1-witness of a shallow minor

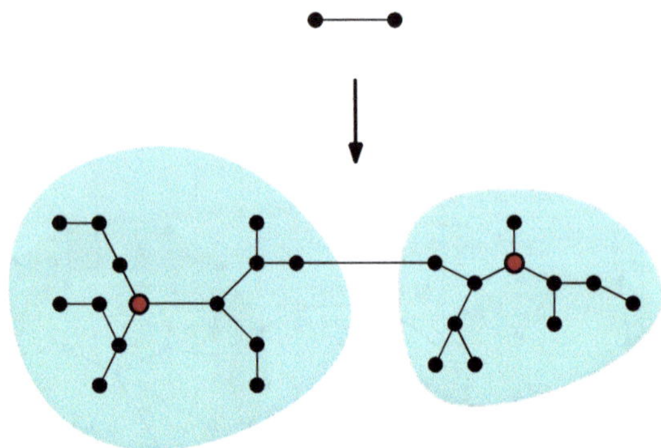

Fig. 4.3 In an d-witness, centers of bushes corresponding to adjacent vertices are at distance at most $2d + 1$ (here $d = 5/2$)

Thus for every graph G we have the following non decreasing sequence of sets:

$$G \in G \triangledown 0 \subseteq G \triangledown \frac{1}{2} \subseteq G \triangledown 1 \subseteq \cdots \subseteq G \triangledown d \subseteq \dots G \triangledown \infty$$

Here is a little " $\triangledown$ -arithmetic".

Proposition 4.1. *Let* a, b *be half-integers and let* c *be the half-integer defined by*

$$(2c+1) = (2a+1)(2b+1). \tag{4.1}$$

Then for every graph G*:*

$$G \triangledown ((\lceil a \rceil + 1)b) \subseteq (G \triangledown a) \triangledown b \subseteq G \triangledown c \tag{4.2}$$

As a corollary, we deduce that for any integers a, b and any graph G we have:

$$((\dots(G \underbrace{\triangledown a) \triangledown a) \dots) \triangledown a}_{b \text{ times}} \subseteq G \triangledown \left(\frac{(2a+1)^b - 1}{2}\right). \tag{4.3}$$

In particular:

$$G \triangledown a \subseteq \underbrace{((\ldots(G \triangledown 1) \triangledown 1)\ldots) \triangledown 1}_{a \text{ times}} \subseteq G \triangledown \left(\frac{3^a - 1}{2}\right). \tag{4.4}$$

Thus we could have defined shallow minors by means of iterations of the operation $G \mapsto G \triangledown 1$ (i.e. by iterated star forest contractions). But the single step approach by $G \triangledown a$ seems to suit better to our parametrization.

4.3 Shallow Topological Minors

A *shallow topological minor* of a graph G of depth a (where a is a half-integer) is a graph H obtained from G by taking a subgraph and then replace an internally vertex disjoint family of paths of length at most $2a+1$ by single edges (see Fig. 4.4). In other words, H is a shallow topological minor of a G of depth a if a $\leq 2a$-subdivision of H is a subgraph of G.

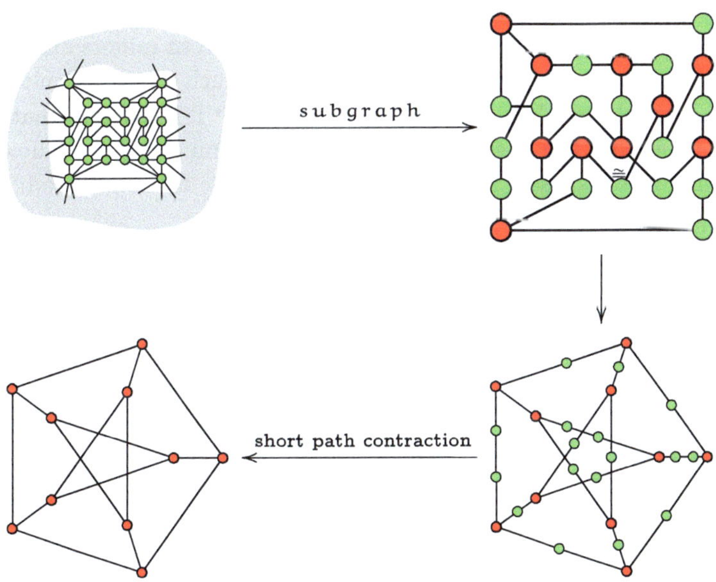

Fig. 4.4 A Petersen topological minor of depth 1 in a graph

For a graph G and a half-integer a we define $G \mathbin{\widetilde{\triangledown}} a$ as the class of the graphs which are topological minors of G at depth a. As a special case, $G \mathbin{\widetilde{\triangledown}} 0$ is the class of the subgraphs of G (no contraction allowed). As $G \mathbin{\widetilde{\triangledown}} a$ is obviously included in $G \triangledown a$ we have:

$$\begin{array}{ccccccccc} & G \triangledown 0 & \subseteq & G \triangledown 1/2 & \subseteq \dots & G \triangledown a & \subseteq \dots & \subseteq G \triangledown \infty \\ & \parallel & & \cup\!\! \mid & & \cup\!\!\mid & & \cup\!\!\mid \\ G \in & G \widetilde{\triangledown} 0 & \subseteq & G \widetilde{\triangledown} 1/2 & \subseteq \dots & G \widetilde{\triangledown} a & \subseteq \dots & \subseteq G \widetilde{\triangledown} \infty \end{array}$$

Not surprisingly, " $\widetilde{\triangledown}$ -arithmetic" is similar to " $\triangledown$ -arithmetic".

Proposition 4.2. *Let* a, b *be half-integers and let* c *be the half-integer defined by*

$$(2c+1) = (2a+1)(2b+1). \tag{4.5}$$

Then for every graph G*:*

$$(G \widetilde{\triangledown} a) \widetilde{\triangledown} b = G \widetilde{\triangledown} c \tag{4.6}$$

As a corollary, for any integers a, b and any graph G:

$$((\dots(G \underbrace{\widetilde{\triangledown} a) \widetilde{\triangledown} a) \dots) \widetilde{\triangledown} a}_{b \text{ times}} = G \widetilde{\triangledown} \left(\frac{(2a+1)^b - 1}{2} \right). \tag{4.7}$$

In Graph Theory, there is a great difference between minors and topological minors. In general, topological minors are much more difficult to deal with (this we also explained in the introduction). It is important that on our level of generality the basic properties of shallow minors and shallow topological minors are closely related. This in fact is one of the main advantages of our theory.

4.4 Grads and Top-Grads

The *greatest reduced average density* (shortly *grad*) with rank r of a graph G [354] is defined by formula

$$\nabla_r(G) = \max \left\{ \frac{\|H\|}{|H|} : H \in G \triangledown r \right\} \tag{4.8}$$

Also we denote by $\nabla(G) = \nabla_\infty(G) = \max_{r \geq 0} \nabla_r(G)$ the maximum edge-density of a minor of G. Notice that we have:

$$\frac{\text{mad}(G)}{2} = \nabla_0(G) \leq \nabla_1(G) \leq \dots \leq \nabla_{|G|}(G) = \nabla(G). \tag{4.9}$$

and that $\nabla(G)$ is polynomially equivalent to the order of the largest complete graph which is a minor of G, i.e. to the *Hadwiger number* $h(G)$ of G. Symbolically (using notation introduced in Sect. 3.10):

$$h \asymp \nabla \tag{4.10}$$

Precisely, we have:

Lemma 4.1. *For every graph* G *it holds*

$$\frac{h(G)-1}{2} \leq \nabla(G) = O(h(G)\sqrt{\log h(G)}), \tag{4.11}$$

Proof. Let $h = h(G)$. As K_h is a $(h-1)$-regular minor of G , $\frac{h-1}{2} \leq \nabla(G)$. Moreover, there exists a constant C such that if $\nabla(G) > C(h+1)\sqrt{\log(h+1)}$ then G has a minor with minimum degree at least $\gamma(h+1)\sqrt{\log(h+1)}$ hence a minor K_{h+1} by Theorem 3.5. □

Similarly we define the *topological greatest reduced average density* (*top-grad*) with rank r of a graph G as:

$$\widetilde{\nabla}_r(G) = \max\left\{\frac{\|H\|}{|H|} : H \in G \mathbin{\widetilde{\triangledown}} r\right\} \tag{4.12}$$

Also, we denote by $\widetilde{\nabla}(G)$ the limit value $\widetilde{\nabla}_\infty(G)$.

A simple but useful fact is that if a graph has a shallow topological minor at depth r which is not too sparse (say of average degree greater than 4) then it has a subgraph which average degree is bounded away from 2. Precisely:

Lemma 4.2. *For every graph* G *and every integer* r, *if* $\widetilde{\nabla}_r(G) > 2$ *then*

$$\widetilde{\nabla}_0(G) > 1 + \frac{1}{4r+1} \ . \tag{4.13}$$

Proof. For some $H \in G \mathbin{\widetilde{\triangledown}} r$, we have $\widetilde{\nabla}_r(G) = \widetilde{\nabla}_0(H)$. Let G' be a $\leq 2r$-subdivision of H that is a subgraph of G. Let $2\overline{r}$ be the average number of subdivision vertices of G' per branch. Then $|G'| = |H| + 2\overline{r}\|H\|$ and $\|G'\| = \|H\| + 2\overline{r}\|H\|$. Hence

$$\widetilde{\nabla}_0(G) \geq \frac{\|G'\|}{|G'|} = \frac{\|H\| + 2\overline{r}\|H\|}{|H| + 2\overline{r}\|H\|} = \frac{1+2\overline{r}}{1/\widetilde{\nabla}_r(H) + 2\overline{r}} > 1 + \frac{1}{4r+1} \ .$$

□

This property will be one of the tools used in Chap. 14 to give a characterization of classes with bounded expansion.

4.5 Polynomial Equivalence of Grads and Top-Grads

It is an unexpected result that grads and top-graphs are not only functionally equivalent, but even polynomially equivalent. Why is this unexpected? Well, there are numerous differences of behavior between minors and topological minors. We mentioned some of them earlier and some others will be highlighted later in this book. However, many of these strong divergences seem to disappear when one considers shallow minors and shallow topological minors. For example, in his thesis [134], Dvořák proved that every graph with a sufficiently large grad actually includes a shallow subdivision of a graph with large minimum degree:

Theorem 4.1. *Let* $r, d \geq 1$ *be arbitrary integers and let* $p = 4(4d)^{(r+1)^2}$. *If* $\nabla_r(G) \geq p$, *then* G *contains a subgraph* F' *that is a* $\leq 2r$*-subdivision of a graph* F *with minimum degree* d.

As we noticed in Chap. 3, minimum degree and average degree are sometimes close notions, and we easily deduce from Theorem 4.1 the following functional equivalence of grads and top-grads:

Corollary 4.1. *For every graph* G *and every integer* $r \geq 1$ *holds*

$$\widetilde{\nabla}_r(G) \leq \nabla_r(G) \leq 4(4\widetilde{\nabla}_r(G))^{(r+1)^2}$$

Notice that the ranks are preserved and that, for any fixed r, the dependency is polynomial. In other words and signs:

$$\nabla_r \asymp \widetilde{\nabla}_r.$$

As $\widetilde{\nabla}_0(G \,\triangledown\, r) = \nabla_r(G)$ and $\widetilde{\nabla}_0(G \,\widetilde{\triangledown}\, r) = \widetilde{\nabla}_r(G)$, this polynomial dependency could be alternatively written as

$$\widetilde{\nabla}_0(G \,\triangledown\, r) \asymp \widetilde{\nabla}_0(G \,\widetilde{\triangledown}\, r).$$

Instead of proving Theorem 4.1, we shall prove the slightly more general Theorem 4.2, which explicits a polynomial dependency

$$\widetilde{\nabla}_s(G \,\triangledown\, r) \asymp \widetilde{\nabla}_s(G \,\widetilde{\triangledown}\, r)$$

for every integer r and every half-integer s.

First we take time for a technical lemma which could be called a "tree-to-spider ramification reduction". A *spider* is just a subdivision of a star rooted at its center, and its *height* is the maximum length of a path from its center.

Lemma 4.3. *Let* G *be a graph with a partition* $V(G) = \bigcup_{1\le i\le n_1} A_i \cup \bigcup_{1\le j\le n_2} B_j$ *and special vertices* $a_i \in A_i$ *and* $b_j \in B_j$*, and let* r_1, r_2 *be integers. Assume that:*

- *Each* $G[A_i]$*, rooted at* a_i*, is a tree of height at most* r_1*;*
- *Each* $G[B_j]$*, rooted at* b_j*, is a tree of height at most* r_2*;*
- $\forall 1 \le i < i' \le n_1,\ e(A_i, A_{i'}) = 0$*;*
- $\forall 1 \le j < j' \le n_2,\ e(B_j, B_{j'}) = 0$*;*
- $\forall 1 \le i \le n_1\ \forall 1 \le j \le n_2,\ e(A_i, B_j) \le 1$.

(Recall that for $X, Y \subseteq V(G)$*,* $e(X, Y)$ *is the number of edges with one incidence in* X *and one incidence in* Y*.) Let* H *be the bipartite graph with vertex set* $\{a_i\}_{1\le i\le n_1} \cup \{b_j\}_{1\le j\le n_2}$ *and edges* $\{a_i, b_j\}$ *for* i, j *such that* $e(A_i, B_j) > 0$*. Let*

$$b = \lceil \delta(H)^{1/(r+1)} \rceil.$$

Then there exists a subgraph G' *of* G*, partitions* $(X_{i,k})$ *of a subset of* A_i *and some special vertices* $x_{i,k}$ *in each* $X'_{i,k}$ *such that*

- *Each* $G'[X_{i,k}]$*, when rooted at* $x_{i,k}$*, is a spider of height at most* r_1 *and the degree of* $x_{i,k}$ *in* G' *is at least* b*;*
- $\forall (i,k) \ne (i',k'), e(X_{i,k}, X_{i',k'}) = 0$*;*
- $\forall (i,k)$*, every leaf of the spider* $G[X_{i,k}]$ *is adjacent to exactly one vertex in* $\bigcup B_j$ *and no other vertex of* $X_{i,k}$ *except* $x_{i,k}$ *is adjacent to a vertex in* $\bigcup B_j$*;*
- *The bipartite graph with vertex set* $\{x_{i,k}\} \cup \{b_j\}$ *and edges* $\{x_{i,k}, b_j\}$ *for* i, k, j *such that* $e(X_{i,k}, B_j) > 0$ *has density*

$$\frac{\|H'\|}{|H'|} \ge \frac{b}{3}.$$

Proof. Let $d = \delta(H)$. First we inductively delete from G the vertices of degree 1 (this does not change the assumed properties of the graph). For each value $1 \le i \le n_1$, inductively split the rooted tree $G[A_i]$ as follows: Let Y be the rooted tree $G[A_i]$ and let $\mathcal{A}_i = \emptyset$. For a vertex x in Y, denote by Y_x the subtree of Y rooted at x and denote by $d^+(x)$ the sum of the number of descendant of x in Y and of the number of edges incident to both x and a vertex in $\bigcup B_j$. While there exists a vertex x in Y such that $d^{(x)} \ge b$ but no descendant y of x in Y is such that $d^+(y) \ge b$ then add the tree Y_x (rooted at x) to $\mathcal{A}_i$ and delete its vertices from Y. Otherwise, Y is either a tree rooted at a_i (with $d^+(a_i) \le b$) or it is empty. As $b^{r+1} \ge d$ we have $|\mathcal{A}_i| \ge 1$. For $Y_x \in \mathcal{A}_i$ define $f(x)$ as the sum of the number of descendant of x in Y_x and of the number of edges incident to both x and a vertex in $\bigcup B_j$. Obviously we have

$$\sum_{Y_x \in \mathcal{A}_i} f(x)b^r + b^{r+1} \geq \sum_j e(A_i, B_j).$$

It follows that

$$\sum_i \left(\sum_{Y_x \in \mathcal{A}_i} f(x)b^r + b^{r+1} \right) \geq \|H\|$$

hence, as $|\mathcal{A}_i| \geq 1$:

$$b^r \sum_i \sum_{Y_x \in \mathcal{A}_i} (f(x) + b)b^r \geq \|H\|.$$

Let $n_1' = \sum_i |\mathcal{A}_i|$, let α the average value of $f(x)$ for $Y_x \in \bigcup_i \mathcal{A}_i$, and let $\beta = \|H\|/n_2$ (that is: the average degree of the b_j's in H). Then we get:

$$b^r n_1' (\alpha + b) \geq n_2 \beta$$

from which follows

$$n_2 \leq \frac{b^r(\alpha + b)}{\beta} n_1.$$

Let $(Y_{i,k}, x_{i,k})$ be defined as the pairs (Y_x, x) for Y_x ranging in $\mathcal{A}_i$. For each pair (i, k), select arbitrarily a (maximum number of) disjoint paths in $G[Y_{i,k}]$ from $x_{i,k}$ to leaves of $G[Y_{i,k}]$. The selected vertices form $X_{i,k}$. Now, G' is a subgraph of G obtained from $G[\bigcup_{i,k} X_{i,k} \cup \bigcup_j B_j]$ by selecting at each leaf of each $G[X_{i,k}]$ exactly one edge incident to a vertex in $\bigcup_j B_j$. Then, to the partition $V(G') = \bigcup_{i,k} X_{i,k} \cup \bigcup_j B_j$ correspond a bipartite graph H' such that $\|H'\| = \sum f(x)$ and $|H'| = n_1' + n_2$. Hence

$$\begin{aligned} \frac{\|H'\|}{|H'|} &= \frac{\sum f(x)}{n_1' + n_2} \\ &\geq \frac{\alpha n_1'}{b^r(\alpha + b) + \beta} \\ &= \frac{\beta}{b^r} \frac{\alpha}{\alpha + b + \beta/b^r}. \end{aligned}$$

As this expression increases with α and as $\alpha \geq b$ we get:

$$\frac{\|H'\|}{|H'|} \geq \frac{\beta}{b^r} \frac{\alpha}{2b + \beta/b^r}.$$

As this expression increases with β and as $\beta \geq b^{r+1}$ we get:

$$\frac{\|H'\|}{|H'|} \geq \frac{b}{3}.$$

□

Here is the promised refinement of Theorem 4.1.

Theorem 4.2. *Let* G *be a graph, let* r *be an integer and let* s *be a half-integer. Then*

$$\widetilde{\nabla}_s(G \,\widetilde{\triangledown}\, r) \leq \widetilde{\nabla}_s(G \triangledown r) \leq 2^{r+2}3^{(r+1)(r+2)}\widetilde{\nabla}_s(G \,\widetilde{\triangledown}\, r)^{(r+1)^2}. \tag{4.14}$$

In particular, $\widetilde{\nabla}_s(G \triangledown r) \asymp \widetilde{\nabla}_s(G \,\widetilde{\triangledown}\, r)$. *Moreover, notice that the polynomial dependence is independent of* s.

Proof. The inequality $\widetilde{\nabla}_s(G \,\widetilde{\triangledown}\, r) \leq \widetilde{\nabla}_s(G \triangledown r)$ directly follows from $G \,\widetilde{\triangledown}\, r \subseteq G \,\widetilde{\triangledown}\, r$. We consider now the second inequality.

Let $H \in G \triangledown r$ be a 2s-subdivision of a graph $\widehat{H}$ such that such that $\|\widehat{H}\|/|\widehat{H}| = d = \widetilde{\nabla}_s(G \triangledown r)$ and let R be a ramification of H in G. Obviously, the ramification R has the following special structure: denoting $\widehat{h}_1, \dots, \widehat{h}_n$ the vertices of $\widehat{H}$, the ramification R may be decomposed into

- A collection of vertex disjoint rooted trees $Y_1, \dots, Y_n$ (with roots $c_1, \dots, c_n$) of height at most $r + 1$,
- For $\widehat{h}_i$ and $\widehat{h}_j$ adjacent in $\widehat{H}$, some path $P_{i,j}$ (of length at most $2s(2r+1)+1$) linking a vertex $x_{i,j} \in Y_i$ to a vertex $x_{j,i} \in Y_j$. (all these paths and trees are vertex disjoints).

It is folklore that there exists a bipartition $\widehat{X} \cup \widehat{Y}$ of $V(\widehat{H})$ such that at least half of the edges of $\widehat{H}$ have one incidence in $\widehat{A}$ and one incidence in $\widehat{B}$. By deleting small degree vertices, we obtain a bipartite subgraph $\widehat{H}_1$ of $\widehat{H}$ (with bipartition $\widehat{A} \cup \widehat{B}$) such that

$$\delta(\widehat{H}_1) \geq d/2.$$

Let H_1 be the corresponding subgraph of H and let R_1 be the corresponding ramification of H_1 in G.

Let G'_1 be the graph obtained from R_1 by contracting each $P_{i,j}$ into a single edge $e_{i,j}$, and let F be the set of all edges $e_{i,j}$ obtained by such path contractions. Let $A_1, \dots, A_{n_1}$ (resp. $B_1, \dots, B_{n_2}$) be the vertex sets of the trees corresponding to vertices in $\widehat{A}$ (resp. $\widehat{B}$). By applying Lemma 4.3 we obtain, as a subgraph of G'_1 a new ramification of a graph $\widehat{H}_2$ such that the bushes in the first part are spiders and such that

$$\frac{\|\widehat{H}_2\|}{|\widehat{H}_2|} \geq \frac{1}{3}(d/2)^{1/(r+1)}.$$

By deleting small degree vertices, we obtain a subgraph $\widehat{H}_3$ of $\widehat{H}_2$ with minimum degree at least $\frac{1}{3}(d/2)^{1/(r+1)}$ and we consider the corresponding subgraph R_3 of R_2. Applying again Lemma 4.3 (with exchanged parts), we eventually obtain a subgraph G_3' of R_3 which is a $\leq (2r)$-subdivision of a graph $\widehat{H}_4$ with

$$\frac{\|\widehat{H}_4\|}{|\widehat{H}_4|} \geq \frac{1}{3}\left(\frac{1}{6}(d/2)^{1/(r+1)}\right)^{1/(r+1)}.$$

Notice that each branch of G_3' contains exactly one edge in F hence G_3' corresponds to a $\leq (2r+1)(2s+1)$-subdivision of $\widehat{H}_4$ in G. It follows that

$$\widetilde{\nabla}_{s(2r+1)+r}(G) \geq \frac{\|\widehat{H}_4\|}{|\widehat{H}_4|} \geq \frac{1}{3}\left(\frac{1}{6}(\widetilde{\nabla}_s(G \triangledown r)/2)^{1/(r+1)}\right)^{1/(r+1)},$$

that is:

$$\widetilde{\nabla}_s(G \triangledown r) \leq 2^{r+2}3^{(r+1)(r+2)}\widetilde{\nabla}_{s(2r+1)+r}(G)^{(r+1)^2}.$$

According to Proposition 4.2, $(G \,\widetilde{\triangledown}\, r) \,\widetilde{\triangledown}\, s = G \,\widetilde{\triangledown}\, (2rs + r + s)$ hence

$$\widetilde{\nabla}_{s(2r+1)+r}(G) = \widetilde{\nabla}_s(G \,\widetilde{\triangledown}\, r),$$

what completes the proof. □

It is possible to extend this result for half-integer r by a slight modification of the proof. In such a case, for integer r and $s = 0$, we obtain

$$\nabla_{r+1/2}(G) \leq c_r \widetilde{\nabla}_{r+1/2}(G)^{\lfloor (r+3/2)^2 \rfloor},$$

and for integer r and for half-integer $s > 0$:

$$\widetilde{\nabla}_s(G \triangledown (r+1/2)) \leq d_r \widetilde{\nabla}_s(G \,\widetilde{\triangledown}\, (r+1/2))^{(r+2)^2}$$

(where c_r and d_r are suitable constants depending on r). As a result, the polynomial dependency

$$\widetilde{\nabla}_s(G \triangledown r) \asymp \widetilde{\nabla}_s(G \,\widetilde{\triangledown}\, r)$$

actually holds for every half-integers r and s (and the polynomial dependency only depends on r).

We shall need in Sect. 7.3 the special cases where $r = 1/2$ and $s > 0$ is a half integer. We give here a better bound for these cases:

Theorem 4.3. *Let* G *be a graph and let* $s \geq 1/2$ *be a half-integer. Then*

$$\widetilde{\nabla}_s(G \mathbin{\widetilde{\triangledown}} (1/2)) \leq \widetilde{\nabla}_s(G \triangledown (1/2)) \leq 8\widetilde{\nabla}_s(G \mathbin{\widetilde{\triangledown}} (1/2))^2, \tag{4.15}$$

that is:

$$\widetilde{\nabla}_{2s+\frac{1}{2}}(G) \leq \widetilde{\nabla}_s(G \triangledown (1/2)) \leq 8\widetilde{\nabla}_{2s+\frac{1}{2}}(G)^2 \tag{4.16}$$

Proof. We have $\widetilde{\nabla}_s(G \triangledown (1/2)) \geq \widetilde{\nabla}_s(G \mathbin{\widetilde{\triangledown}} (1/2))$ and, according to Proposition 4.2, we have $\widetilde{\nabla}_s(G \mathbin{\widetilde{\triangledown}} (1/2)) = \widetilde{\nabla}_{2s+\frac{1}{2}}(G)$ hence

$$\widetilde{\nabla}_{2s+\frac{1}{2}}(G) \leq \widetilde{\nabla}_s(G \triangledown (1/2)).$$

Now consider a graph $H \in (G \triangledown 1/2) \mathbin{\widetilde{\triangledown}} s$. Let $u_1, \ldots$ be the vertices of H. There exists rooted trees $Y_1, \ldots$ of G and paths $P_{i,j}$ (for $(i,j) \in A$) linking the root of Y_i to a vertex of Y_j in such a way that the following conditions hold:

- For every i, $(i,i) \notin A$,
- For every $i \neq j$, $(i,j) \in A \Rightarrow (j,i) \notin A$,
- For every $i \neq j$, $((i,j) \in A$ or $(j,i) \in A) \iff \{u_i, u_j\} \in E(H)$,
- For every i there exists j such that $((i,j) \in A$ or $(j,i) \in A)$.

The directed graph $\vec{H}$ is obtained by orienting H in such a way that $(u_i, u_j) \in \vec{E}(\vec{G})$ if $(i,j) \in A$. Let $\vec{G}'$ be the graph induced by the Y_i's and the $P_{i,j}$'s, each arc of Y_i being directed to the root of Y_i and each path $P_{i,j}$ being directed from Y_i's end to Y_j's end (that is: consistently with H). By construction, the outdegree of the root of Y_i in $\vec{G}'$ equals the outdegree of u_i in $\vec{H}$.

Delete from $\vec{H}$ all the vertices with indegree less than $\nabla_0(H)/2$. At most half of the edges may have been deleted so we get a non empty subgraph $\vec{H}'$ of $\vec{H}$ such that every vertex has indegree at least $\nabla_0(H)/2$. We consider accordingly the subset of the Y_i's and the $P_{i,j}$'s. Let n be the order of $\vec{H}'$.

If a vertex of Y_i is the target of strictly less than $\sqrt{d^-_{\vec{H}'}(u_i)}$ directed paths, remove arbitrarily all the incoming paths but one and delete accordingly the corresponding arcs of $\vec{H}$. As for each vertex of Y_i the number of incoming paths $P_{j,i}$ has been (at most) divided by $\sqrt{d^-_{\vec{H}'}(u_i)}$, we infer that the indegree of u_i will be at least $\sqrt{d^-_{\vec{H}'}(u_i)}$ after all the path deletions. At the end of this process, we augment each path $P_{i,j}$ linking the root of Y_i to a vertex of degree 2 of Y_j to a path linking the root of Y_i to the root of Y_j. Consider the graph $\vec{G}''$ obtained by contracting all the paths into edges. The vertices of $\vec{G}''$ may be indexed as $r_1, \ldots, r_n, v_{1,1}, \ldots, v_{1,k_1}, v_{2,1}, \ldots, v_{2,k_2}, \ldots v_{n,1}, \ldots, v_{n,k_n}$ as follows: r_i is the root of Y_i and $v_{i,1}, \ldots, v_{i,k_i}$ are the other (remaining) vertices of Y_i. Notice that if $k_i = 0$ then the root of Y_i is the target of at least $\sqrt{d^-(u_i)}$ incoming paths hence $d^-_{\vec{G}''}(r_i) + \sum_{j=1}^{k_i} d^-_{\vec{G}''}(v_{i,j}) \geq (k_i+1)/2\sqrt{\nabla_0(H)/2}$. Thus:

$$\begin{aligned}\frac{\|\vec{G}''\|}{|\vec{G}''|} &= \frac{\sum_{i=1}^{n}\left(d^-_{\vec{G}''}(r_i) + \sum_{j=1}^{k_i} d^-_{\vec{G}''}(v_{i,j})\right)}{\sum_{i=1}^{n}(k_i+1)}\\ &\geq \frac{\sum_{i=1}^{n}(k_i+1)\sqrt{\nabla_0(H)/8}}{\sum_{i=1}^{n}(k_i+1)}\\ &= \sqrt{\nabla_0(H)/8}\end{aligned}$$

By considering $H \in (G \triangledown (1/2)) \,\widetilde{\triangledown}\, s$ such that $\nabla_0(H) = \widetilde{\nabla}_r(G \triangledown 1/2)$ we conclude that

$$\widetilde{\nabla}_s(G \triangledown (1/2)) \leq 8\widetilde{\nabla}_{2s+\frac{1}{2}}(G)^2.$$

□

As mentioned in Sect. 3.2, average degrees and minimum degrees are closely related invariants. We shall now prove that minimum degrees are, in some sense, preserved when considering shallow topological minors. We need to prove a technical lemma first.

Lemma 4.4. *Let $G = (X, Y, E)$ be a bipartite graph and let $1 \leq r \leq s \leq |X|$. Assume that each vertex in Y has degree at least r.*

Then there exists a subset $X' \subseteq X$ and a subset $Y' \subseteq Y$ such that $|X'| = s$ and $|Y'| = |Y|/2$ and every vertex in Y' has at least $r\frac{|X'|}{|X|}$ neighbors in X'. (see Fig. 4.5)

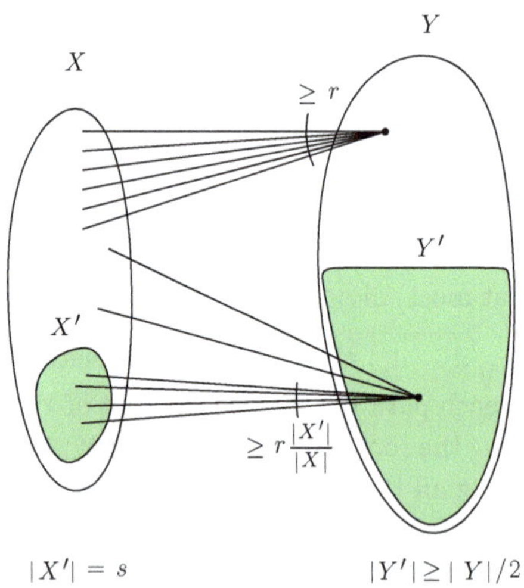

Fig. 4.5 A general regularity property

Proof. First notice that we may assume, without loss of generality, that every vertex in Y has exactly r neighbors in X, for otherwise we may safely delete surplus edges.

Consider a vertex $y \in Y$. The number of subsets of X of size s including exactly i neighbors of y is $\binom{|X|-r}{s-i}\binom{r}{i}$ (choose $s-i$ non-neighbors and i neighbors). Define

$$F(i) = \frac{1}{|Y|} \cdot \frac{\sum_{A \in \binom{X}{s}} |\{y \in Y : |N(y) \cap A| = i\}|}{\binom{|X|}{s}}.$$

Then:

$$F(i) = \frac{1}{|Y|} \cdot \frac{\sum_{A \in \binom{X}{s}} \sum_{y \in Y} \mathfrak{I}_{|N(y) \cap A| = i}(y)}{\binom{|X|}{s}}$$

(where $\mathfrak{I}_{|N(y) \cap A| = i}(y)$ is 1 if $|N(y) \cap A| = i$ and 0 otherwise), thus:

$$\begin{aligned} F(i) &= \frac{1}{|Y|} \cdot \frac{\sum_{y \in Y} \sum_{A \in \binom{X}{s}} \mathfrak{I}_{|N(y) \cap A| = i}(y)}{\binom{|X|}{s}} \\ &= \frac{1}{|Y|} \cdot \frac{\sum_{y \in Y} \binom{n-r}{s-i}\binom{r}{i}}{\binom{|X|}{s}} \\ &= \frac{\binom{|X|-r}{s-i}\binom{r}{i}}{\binom{|X|}{s}} \end{aligned}$$

It follows that $F(i)$ is exactly the probability mass function of a hypergeometric distribution with mean $\frac{rs}{|X|}$ and variance $\sigma^2 = \frac{|X|}{|X|-1}\left(1 - \frac{r}{|X|}\right)\left(1 - \frac{s}{|X|}\right)\frac{rs}{|X|} \leq \frac{rs}{|X|}$. □

Proposition 4.3. *Let G be a graph and let d be such that*

$$\frac{\delta(G)}{2} \geq d > 1.$$

Then there exists a graph H such that an exact 1-subdivision of H is a subgraph of G (hence $H \in G \,\widetilde{\nabla}\, \frac{1}{2}$) and

- *Either* $H \cong K_{\delta(G)/2d}$,
- *Or* H *has minimum degree at least* d *and order*

$$\sqrt{\frac{d-1}{2d}} \cdot \sqrt{|G|} \leq |H| \leq \frac{1}{2d} \cdot |G|.$$

Proof. Let $n = |G|, \delta = \delta(G)$ and $\epsilon = 1/d$. Consider the bipartite graph B whose parts, denoted by X and Y, are two copies of $V(G)$ and such that $x \in X$ is adjacent to $y \in Y$ if the corresponding vertices x and y of G are adjacent. Then every vertex in Y has at least δ neighbors in X. According to Lemma 4.4 there exists a subset X' of X of size $\epsilon n/2$ and a subset Y' of Y of size $n/2$ such that every vertex in Y' has at least $\epsilon\delta$ neighbors in X'.

Consider the subset S of vertices of G corresponding to X' and the subset T of $V \setminus S$ corresponding to vertices in Y'. Hence $|S| = \epsilon n/2$ and $|T| \geq (1-\epsilon)n/2$.

Consider the graph Γ with vertex set S constructed as follows: start from the edgeless graph on S and iteratively consider the vertices in T. For each considered vertex $t \in T$, if all the neighbors of t already induce a clique in Γ stop; otherwise, add an edge e_t in Γ between two arbitrary not yet adjacent neighbors x and y of t, associate with this edge the path x, t, y of length 2 and go to the next vertex of T (cf. Fig. 4.6).

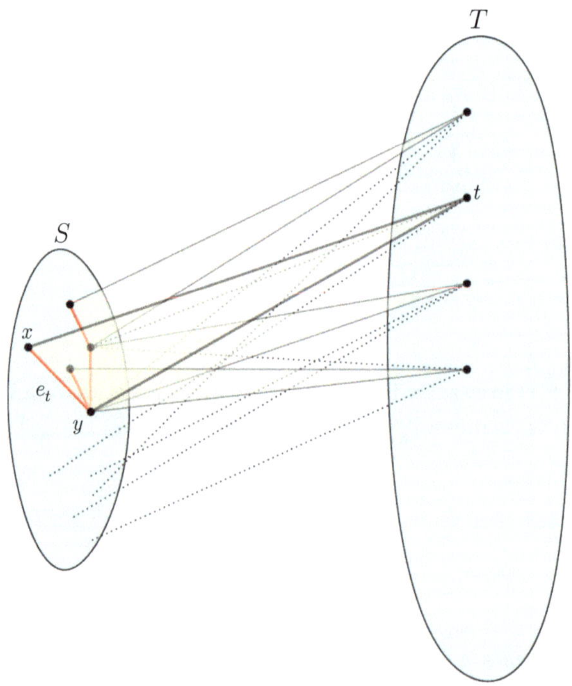

Fig. 4.6 Building Γ

If a clique of size at least $\delta\epsilon/2$ was created in Γ then the process stopped and we found a 1-subdivision of $K_{\delta\epsilon/2}$ in G. Otherwise, all the vertices of T have been used and we discovered the 1-subdivision of a graph Γ in G with $|\Gamma| = |S| = \epsilon n/2$ and $\|\Gamma\| = |T| \geq (1-\epsilon/2)n$. Let $\mu = \frac{1-\epsilon}{2-\epsilon}$ (hence $\frac{1}{2} < \mu < 1$). Then Γ includes a subgraph H of minimum degree at least $(1-\mu)\frac{\|\Gamma\|}{|\Gamma|} = \frac{1}{2-\epsilon}\frac{2-\epsilon}{\epsilon} = \frac{1}{\epsilon}$ and size $\|H\| \geq \mu\|\Gamma\| \geq \frac{1-\epsilon}{2-\epsilon}\frac{2-\epsilon}{2}n = \frac{1-\epsilon}{2}n$ (see Sect. 3.2). As $H \subseteq \Gamma$ we have $|H| \leq |\Gamma| = \epsilon n/2$. As $\|H\| \geq \frac{1-\epsilon}{2}n$ we have $|H| \geq \sqrt{2\|H\|} \geq \sqrt{(1-\epsilon)n/2}$. □

In particular, we infer that every graph G with minimum degree at least 4 includes as a subgraph the exact 1-subdivision of a graph H such that

$$\delta(H) \geq \frac{\sqrt{2\delta(G)+1}-1}{2} \tag{4.17}$$

4.6 Relation with Chromatic Number

As well as every graph G being $(\lfloor 2\nabla_0(G)\rfloor + 1)$-colorable, every graph in $G \mathbin{\widetilde{\triangledown}} a$ is $(\lfloor 2\widetilde{\nabla}_a(G)\rfloor + 1)$-colorable (as $\widetilde{\nabla}_a(G) = \nabla_0(G \mathbin{\widetilde{\triangledown}} a)$).

In the opposite direction, we take time out for the following lemma [134]:

Lemma 4.5. *Let $c \geq 4$ be an integer and let G be a graph with average degree $d > 56(c-1)^2\frac{\log(c-1)}{\log c - \log(c-1)}$. Then the graph G contains a subgraph G' that is the 1-subdivision of a graph with chromatic number c.*

Proof. Every graph contains a bipartite subgraph with at least half of the edges of the original graph, i.e., G contains a bipartite subgraph G_1 with average degree more than $d/2$. The graph G_1 cannot be $d/4$-degenerate, since otherwise the average degree of G_1 would be at most $d/2$. Let G_2 be a subgraph of G_1 with minimum degree at least $d_2 = d/4$. The graph G_2 is bipartite, let $V(G_2) = A \cup B$ be a partition of its vertices to two independent sets such that $|A| \leq |B|$. Let $a = |A|$ and $b = |B|$. Since the minimum degree of G_2 is at least d_2, it follows that $d_2 \leq a \leq b$.

Let $q = 7\frac{\log(c-1)}{\log c - \log(c-1)}$. Note that $d_2/q \geq 10$. We construct a subgraph G_3 in the following way: if $b \geq qa$, then let $G_3 = G_2, A' = A$ and $B' = B$. Otherwise, we choose sets $A' \subseteq A$ and $B' \subseteq B$ as described in the next paragraph, and let G_3 be the subgraph of G_2 induced by A' and B': Let A' be a subset of A obtained by taking each element of A randomly independently with probability $p = b/qa$. The expected size of A' is $ap = b/q$, and by Chernoff Inequality, the size of A' is more than $2b/q$ with probability less than $e^{-\frac{3b}{8q}} \leq e^{-\frac{3d_2}{8q}} \leq e^{-15/4} < 0.5$. Consider a vertex v of B with degree $s \geq d_2$ in G_2, and let s' be the number of neighbors of v in A' and $r(v) = s'/s$. The expected number of neighbors of v in A' is ps. By Chernoff Inequality, the probability that $s' < \frac{p}{2}s$ is less than $e^{-\frac{3ps}{28}} \leq e^{-\frac{3}{28}\cdot\frac{b}{a}\cdot\frac{d_2}{q}} \leq e^{-\frac{15}{14}} < 0.35$.

Let B' be the set of vertices v of B such that $r(v) \geq p/2$. The expected value of $|B \setminus B'|$ is less than $0.35b$, and by Markov Inequality, $\mathrm{Prob}[|B \setminus B'| \geq 0.7b] \leq 0.5$. Therefore, the probability that the set A' has size at most $2b/q$ while the set B' has size at least $0.3b$ is greater than zero. We let A' and B' be a pair of sets that satisfies these properties.

Let $a' = |A'|$ and $b' = |B'|$. Observe that the degree of every vertex of B' in G_3 is at least $\frac{b}{2qa} d_2 \geq \frac{1}{2q} d_2 = (c-1)^2 = d_3$, and that $b' \geq 0.3b \geq 0.15qa'$. Let $D_1, \ldots, D_{b'} \geq d_3$ be the degrees of vertices of B'.

We show that the graph G_3 contains as a subgraph the 1-subdivision of a graph with chromatic number c. Suppose for contradiction that each graph whose 1-subdivision is a subgraph of G_3 has chromatic number at most $c-1$. Let $\mathcal{S}$ be the set of all partial graphs of G_3 such that every vertex in B' has degree 2 (these are the 1-subdivisions having their principal vertices in A' and using all the vertices in B' as subdivision vertices). Let $N_H = |\mathcal{S}|$. Then $N_H = \prod_{i=1}^{b'} \binom{D_i}{2}$ such subgraphs. Let N_C be the number of colorings of A' by $c-1$ colors. Then $N_C = (c-1)^{a'}$.

Let φ be a coloring of A' by $c-1$ colors. We determine the number of subgraphs $H \in \mathcal{S}$ such that φ is a proper coloring of the graph obtained from H by contracting the vertices in B'. Let us consider a vertex v in B' of degree D. Since φ is proper, the two edges incident with v in H lead to vertices with different colors. Let M be the neighborhood of v, $|M| = D$. Let m_i be the number of vertices of M colored by φ with the color i. The number s of the pairs of neighbors of v that have different colors satisfies

$$\begin{aligned} s &= \sum_{1 \leq i < j \leq c-1} m_i m_j \\ &= \frac{1}{2} \sum_{1 \leq i,j \leq c-1,\ i \neq j} m_i m_j \\ &= \frac{1}{2} \sum_{i=1}^{c-1} m_i (D - m_i) \\ &= \frac{1}{2} \left(D^2 - \sum_{i=1}^{c-1} m_i^2 \right) \leq \frac{1}{2} \left(D^2 - \frac{D^2}{c-1} \right). \end{aligned}$$

Therefore, the number of the subgraphs in $\mathcal{S}$ for which φ is proper is at most $N_P = (1 - \frac{1}{c-1})^{b'} \prod_{i=1}^{b'} \frac{D_i^2}{2}$. For each $H \in \mathcal{S}$ there exists at least one proper coloring, hence $N_C N_P \geq N_H$, and we obtain

$$(c-1)^{a'} \left(1 - \frac{1}{c-1} \right)^{b'} \prod_{i=1}^{b'} \frac{D_i^2}{2} \geq \prod_{i=1}^{b'} \binom{D_i}{2}$$

Hence

$$(c-1)^{a'}\left(1-\frac{1}{c-1}\right)^{b'} \geq \prod_{i=1}^{b'}\left(1-\frac{1}{D_i}\right) \geq \left(1-\frac{1}{(c-1)^2}\right)^{b'}$$

As $(1-\frac{1}{(c-1)^2})(1-\frac{1}{c-1})^{-1} = \frac{c}{c-1}$, it follows that

$$(c-1)^{a'} \geq \left(\frac{c}{c-1}\right)^{b'}$$

This is a contradiction since from $b' \geq 0.15qa'$ it follows that

$$(c-1) \geq \left(\frac{c}{c-1}\right)^{0.15q} > \left(\frac{c}{c-1}\right)^{\frac{\log(c-1)}{\log c - \log(c-1)}} = c-1.$$

□

We deduce the following connection between the maximum chromatic number of shallow topological minors and top-grads. By $\chi(G \mathbin{\widetilde{\triangledown}} a)$ we mean the maximum of $\chi(G')$ for $G' \in G \mathbin{\widetilde{\triangledown}} a$.

Proposition 4.4. *For every graph* G *and every half-integer* a *it holds:*

$$\frac{\chi(G \mathbin{\widetilde{\triangledown}} a)-1}{2} \leq \widetilde{\nabla}_a(G) = O(\chi(G \mathbin{\widetilde{\triangledown}} (2a+1/2))^4).$$

Thus, consequently:

Proposition 4.5. *There exists a constant* C *such that for every graph* G *and every half-integer* a *holds:*

$$\frac{\chi(G \triangledown a)-1}{2} \leq \nabla_a(G) \leq C^{(a+1)^2}\chi(G \triangledown (2a+1/2))^{4(a+1)^2}.$$

Proof. According to Theorem 4.1, $\nabla_a(G) = O((4\widetilde{\nabla}_a(G))^{(a+1)^2})$. Hence, according to Proposition 4.4 there exits a constant C such that

$$\nabla_a(G) \leq (C \cdot \chi(G \mathbin{\widetilde{\triangledown}} (2a+1/2))^4)^{(a+1)^2}.$$

As $G \mathbin{\widetilde{\triangledown}} r \subseteq G \triangledown r$ hence $\chi(G \mathbin{\widetilde{\triangledown}} r) \leq \chi(G \triangledown r)$, which concludes the proof. □

4.7 Stability of Grads by Lexicographic Product

Let G and H be graphs. The *lexicographic product* $G \bullet H$ is defined by

$$V(G \bullet H) = V(G) \times V(H)$$
$$E(G \bullet H) = \{\{(x, y), (x', y')\} : \{x, x'\} \in E(G) \text{ or } (x = x' \text{ and } \{y, y'\} \in E(H))\}.$$

Figure 4.7 perhaps expresses this better.

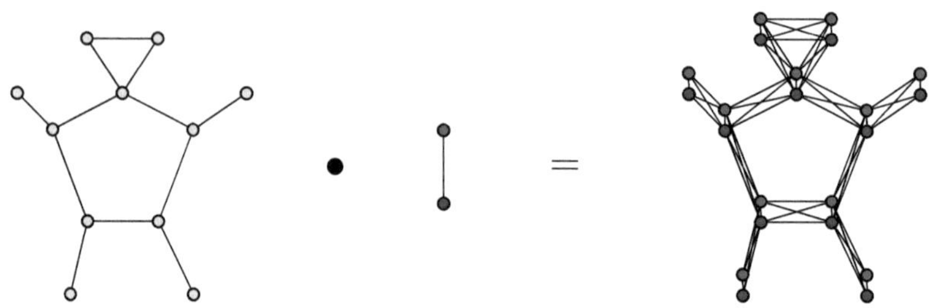

Fig. 4.7 Lexicographic product of a graph and K_2

The lexicographic product $G \bullet K_k$ is also called *replication graph*: each vertex is replaced by k mutually adjacent clones of it. Another common name for this operation is *multiplication of vertices*.

The fact that $\widetilde{\nabla}_r(G \bullet \overline{K}_p)$ is bounded almost linearly by $\widetilde{\nabla}_r(G)$ is a cornerstone of the decomposition results we present in Chap. 7.

Proposition 4.6. *Let G be a graph, let $p \geq 2$ be a positive integer and let r be a half-integer. Then*

$$\widetilde{\nabla}_r(G \bullet K_p) \leq \max(2r(p-1)+1, p^2)\widetilde{\nabla}_r(G) + p - 1$$

Proof. Let $\{a_1, a_2, \dots, a_p\}$ be the vertices of the K_p. The vertices of $G \bullet K_p$ are then the pairs (v, a_i) for v vertex of G and $1 \leq i \leq p$. For every vertex v of G, we say that (v, a_i) and (v, a_j) are twins in $G \bullet K_p$ and that v is the projection on G of these twin vertices.

Let $H \in (G \bullet K_p) \,\widetilde{\nabla}\, r$ be such that $\frac{\|H\|}{|H|} = \widetilde{\nabla}_r(G \bullet K_p)$ and let $S(H) \subseteq G \bullet K_p$ be the corresponding $(\leq 2r)$-subdivision of H in $G \bullet K_p$. Notice that we may assume that no branch of S(H) contains two twin vertices, except if the branch is a single edge path linking two twin vertices (otherwise we shorten the branch without changing $\|H\|$ and $|H|$, see Fig. 4.8).

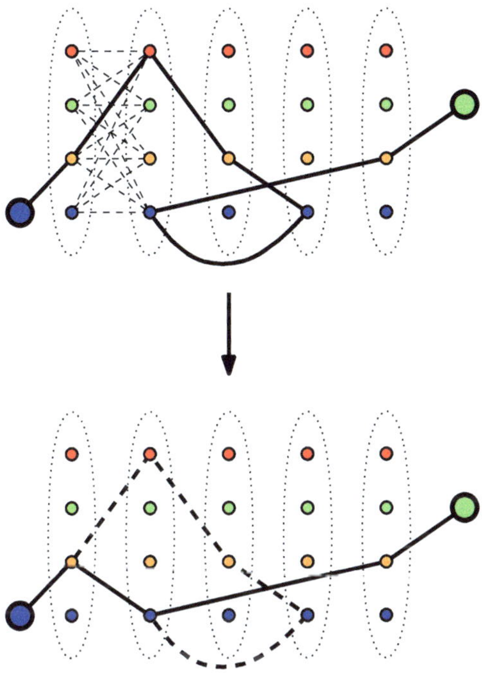

Fig. 4.8 If a branch contains twin vertices, we shorten it

We define the graph H_1 and its $(\leq 2r)$-subdivision $S(H_1)$ by the following procedure: Start with $H_1 = H$ and $S(H_1) = S(H)$. Then, for each subdivision vertex $v \in S(H_1)$ having a twin which is a principal vertex of $S(H)$, delete the branch of $S(H_1)$ containing v and the corresponding edge of H_1. In this way, we delete at most $(p-1)|H|$ edges and thus $\frac{\|H_1\|}{|H_1|} \geq \frac{\|H\|}{|H|} - (p-1)$. $S(H_1)$ is such that no subdivision vertex is a twin of a principal vertex (see Fig. 4.9).

Given H_1 we construct the conflict graph C of H_1 as follows: the vertex set of C is the edge set of H_1 and the edges of C are the pairs of edges $\{e_1, e_2\}$ such that:

- Either e_1 and e_2 are not subdivided in $S(H_1)$ and their endpoints are equal or twins,
- Or e_1 and e_2 are subdivided in $S(H_1)$ and one of the subdivision vertices of the branch corresponding to e_1 is a twin of one of the subdivision vertex of the branch corresponding to e_2.

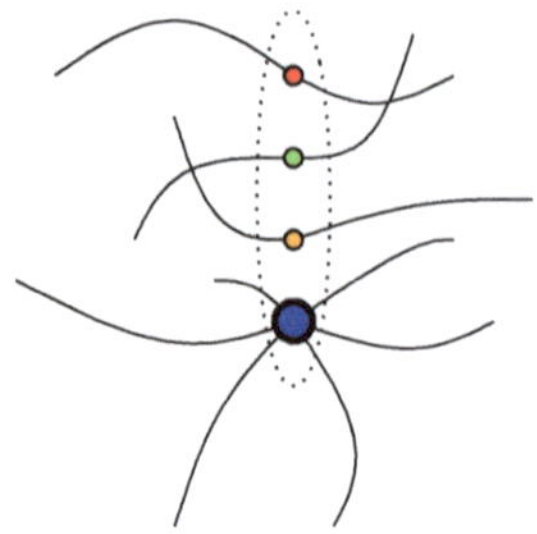

Fig. 4.9 Each principal vertex of $S(H)$ may be the twin of at most $(p-1)$ subdivision vertices of $S(H)$

Note that graph C has maximum degree at most $\max(p^2-1, 2r(p-1))$ (see Fig. 4.10) hence it is $\max(p^2, 2(p-1)r+1)$-colorable. Fix such a coloring. Consider a monochromatic set of vertices of C (i.e. of edges of H_1) of size at least $\frac{\|H_1\|}{\max(2r(p-1)+1, p^2)}$. Let H_2 be the partial graph of H_1 defined by these edges and let $S(H_2)$ be the corresponding subgraph of $S(H_1)$.

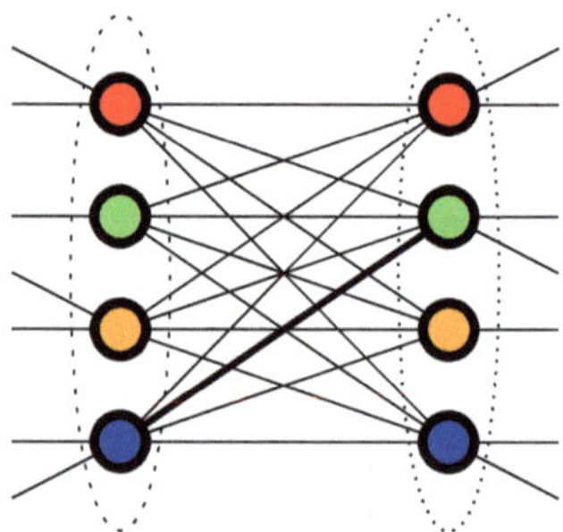

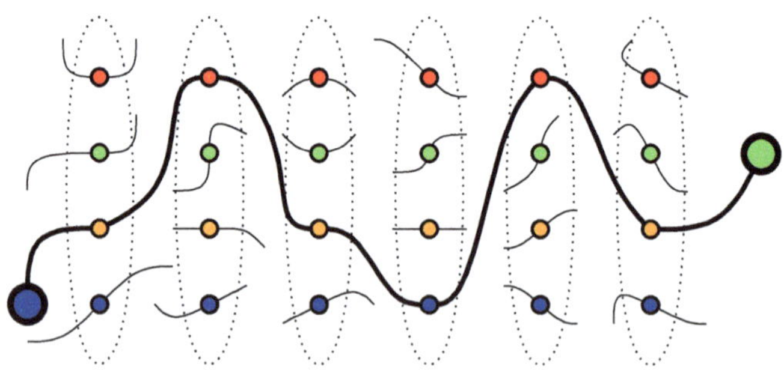

Fig. 4.10 Each edge of H_1 may be in conflict with most $\max(p^2-1, 2r(p-1))$ other edges

Let v be a principal vertex of $S(H_2)$. Then two edges incident to v cannot have their other endpoints equal or twins (because of the coloration).

Let H_3 be the projection of H_2 on G. Because of the above coloration, no two edges of H_2 are projected on a same edge of H_3 and only the edges linking twin vertices may have been removed (simultaneously to the removal of all but one of the twins). As the surplus twins then have degree at most $p-1 \leq \max(2r(p-1)+1, p^2)\widetilde{\nabla}_r(G)+p-1$ they can be removed safely. As we have $\widetilde{\nabla}_r(G) \geq \frac{\|H_3\|}{|H_3|} \geq \frac{\|H_2\|}{|H_2|}$ the result follows. □

We just proved that $\widetilde{\nabla}_r(G \bullet K_p)$ and $\widetilde{\nabla}_r(G)$ are polynomially equivalent. As a corollary, of Proposition 4.6 and Corollary 4.1 we have that also $\nabla_r(G \bullet K_p)$ and $\nabla_r(G)$ are polynomially equivalent. This we proved earlier directly by a more complicated argument [354].

4.8 Shallow Immersions

We now introduce of a shallow version of immersion relation which nicely fits to our framework.

Recall that an *immersion* of a graph H in a graph G (see [339]) is a function ι with domain $V(H) \cup E(H)$, such that:

- $\iota(v) \in V(G)$ for all $v \in V(H)$, and $\iota(u) \neq \iota(v)$ for all distinct $u, v \in V(H)$;
- For each edge $e = \{u, v\}$ of H, $\iota(e)$ is a path of G with ends $\iota(u), \iota(v)$;
- For all distinct $e, f \in E(H)$, $E(\iota(e)) \cap E(\iota(f)) = \emptyset$.

Alternatively, a graph H is an immersion of the graph G if H can be obtained from G by a sequence of vertex deletions, edge deletions and *edge lift* (An edge lift consists of replacing a pair of adjacent edges $\{u, v\}$ and $\{v, w\}$ by a single edge $\{u, w\}$), see Fig. 4.11.

The *stretch* of an immersion ι is the maximum over the edges $e \in E(H)$ of $(\|\iota(e)\| - 1)/2$; the *complexity* of ι is the maximum number of times a vertex of G appears as a vertex of the paths $\iota(e)$ (possibly as an end vertex). A *shallow immersion* of *depth* (p, q) is an immersion of stretch at most q and complexity at most p.

The main motivations for introducing these two parameters are the following facts:

- Immersions at depth $(1, p)$ are exactly topological minors at depth p,
- Every graph may be immersed into a very sparse graph with a stretch of $3/2$ if one does not bound the complexity of the immersion (as shown in Fig. 4.11).

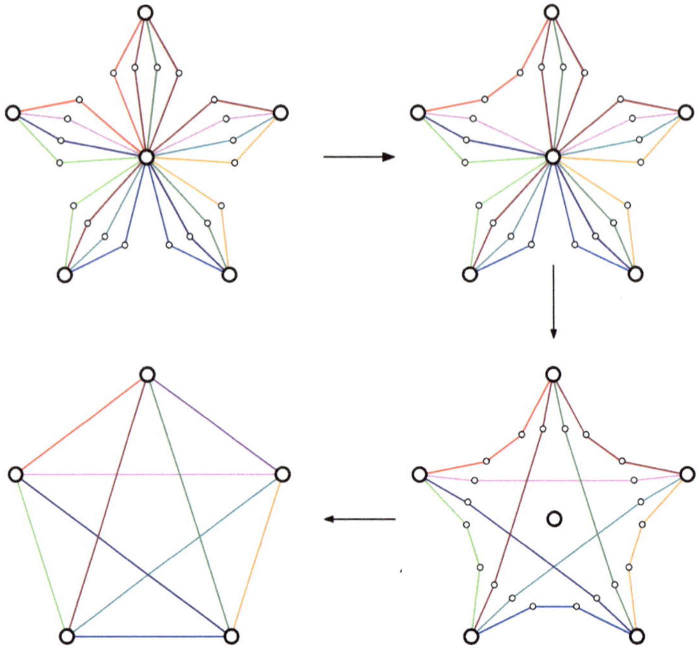

Fig. 4.11 A complete graph as an immersion of a very sparse graph

Obviously, if ι has stretch q and complexity p then

$$H \in (G \bullet \overline{K}_p) \,\widetilde{\triangledown}\, q.$$

We define the set $G \overset{\propto}{\triangledown} (p, q)$ as the class of all shallow immersions of G with complexity p and stretch q and we get

$$G \,\widetilde{\triangledown}\, q \subseteq G \overset{\propto}{\triangledown} (p, q) \subseteq (G \bullet \overline{K}_p) \,\widetilde{\triangledown}\, q. \tag{4.18}$$

Similarly, one can introduce the *imm-grad* $\overset{\propto}{\nabla}_{p,q}(G)$ by

$$\overset{\propto}{\nabla}_{p,q}(G) = \max_{H \in G \overset{\propto}{\triangledown} (p,q)} \frac{\|H\|}{|H|}.$$

Then, according to Proposition 4.6 we have:

Corollary 4.2.

$$\widetilde{\nabla}_q(G) \le \overset{\propto}{\nabla}_{p,q}(G) \le \max(2q(p-1)+1, p^2)\widetilde{\nabla}_q(G) + p - 1. \tag{4.19}$$

Thus all of ∇_r, $\widetilde{\nabla}_r$ *and* $\overset{\propto}{\nabla}_{P(r),r}$ *(for a fixed polynomial* P*) are polynomially equivalent:*

$$\nabla_r \asymp \widetilde{\nabla}_r \asymp \overset{\propto}{\nabla}_{P(r),r}.$$

Actually we can prove a better bound for $\overset{\propto}{\nabla}_{p,q}(G)$ and further connections between shallow immersions and shallow topological minors:

Proposition 4.7. *Let* p *and* q *be respectively an integer and a half-integer and let* G *be a graph. Then, for every* $H \in G \overset{\propto}{\triangledown} (p,q)$ *there exists* $H_1, \dots, H_{2q(p-1)+1} \in G \,\widetilde{\triangledown}\, q$ *with the same vertex set such that*

$$H = H_1 \cup \dots \cup H_{2q(p-1)+1}.$$

Hence we have:

$$\widetilde{\nabla}_q(G) \le \overset{\propto}{\nabla}_{p,q}(G) \le (2q(p-1)+1)\widetilde{\nabla}_q(G) \tag{4.20}$$

$$\chi(G \,\widetilde{\triangledown}\, q) \le \chi(G \overset{\propto}{\triangledown} (p,q)) \le \chi(G \,\widetilde{\triangledown}\, q)^{2q(p-1)+1} \tag{4.21}$$

$$\omega(G \,\widetilde{\triangledown}\, q) \le \omega(G \overset{\propto}{\triangledown} (p,q)) < R(\overbrace{\omega(G \,\widetilde{\triangledown}\, q)+1, \dots, \omega(G \,\widetilde{\triangledown}\, q)+1}^{2q(p-1)+1}) \tag{4.22}$$

(Here $R(n_1, \dots, n_k)$ *denotes the Ramsey number introduced in Sect. 3.9.)*

Proof. A branch of the immersion of H in G can cross at most $2q(p-1)$ other branches. Hence one can color the branches into $2q(p-1)+1$ colors in such a way that no two branches with the same colors cross. Each color hence define a shallow topological minor of G at depth q.

The inequalities follow as for every k graphs $H_1, \dots, H_k$ on the same vertex set,

- The average degree of $H_1 \cup \dots \cup H_k$ is at most the sum of the average degrees of $H_1, \dots, H_k$;
- The chromatic number of $H_1 \cup \dots \cup H_k$ is at most the Product of the chromatic numbers of $H_1, \dots, H_k$;
- If $\omega(H_1 \cup \dots \cup H_k) = N$ and K is a clique of order N in $H_1 \cup \dots \cup H_k$ and $N \ge R(\omega_1, \dots, \omega_k)$ then there exists an integer i such that $\omega(H_i) \ge \omega_i$.

Thus, according to Corollary 4.2, we have one of our highlights: densities arising from iterated local operations of minor, topological minor and immersion are all functionally equivalent.

4.9 Generalized Coloring Numbers

The *coloring number* $\mathrm{col}(G)$ of a graph G (introduced at least as early as [167]) is the minimum integer k such that there is a linear ordering L of the vertices of G for which each vertex v has indegree at most $k-1$, i.e. v has at most $k-1$ neighbors u with $u <_L v$. According to Proposition 3.2, a graph G has coloring number at most k if and only if it is $(k-1)$-degenerate. Hence

$$\lfloor \nabla_0(G) + 1 \rfloor \leq \mathrm{col}(G) = \lceil 2\nabla_0(G) + 1 \rceil. \tag{4.23}$$

Generalizations of the coloring number include the *arrangeability* [93] (in the context of Ramsey numbers of graphs), the *admissibility* [273] and the *rank* [272] (in the context of game chromatic numbers of graphs). However, for our purpose, it seems that a natural generalization is the k-*coloring number* of a graph introduced recently by Kierstead and Yang [274]. We proceed as follows:

For a graph G, let $\Pi(G)$ be the set of all linear orderings of the vertices of G. For $L \in \Pi(G)$, denote by G_L the graph G with vertices linearly order by L.

For $L \in \Pi(G)$ and x and y vertices of G, we say that x is k-*weakly accessible* from y if $x <_L y$ and there is an $x-y$-path P of length at most k (i.e., with at most k edges) so that for any $z \in P, x <_L z$. If every internal vertex z of P satisfies the condition $y <_L z$ then we say x is k-*accessible* from y. Let $Q_k(G_L, y)$ be the set of vertices that are weakly k-accessible from y and let $R_k(G_L, y)$ be the set of vertices that are k-accessible from y.

Definition 4.1. The k-*coloring number* $\mathrm{col}_k(G)$ and the *weak* k-*coloring number* $\mathrm{wcol}_k(G)$ (illustrated on Fig. 4.12) of G are defined by

$$\mathrm{col}_k(G) = 1 + \min_{L\in\Pi(G)} \max_{v\in V(G)} |R_k(G_L, v)| \tag{4.24}$$

and

$$\mathrm{wcol}_k(G) = 1 + \min_{L\in\Pi(G)} \max_{v\in V(G)} |Q_k(G_L, v)|. \tag{4.25}$$

Observe that

$$\mathrm{wcol}_1(G) = \mathrm{col}_1(G) = \mathrm{col}(G).$$

The invariants col_i and wcol_i form, for each graph G, two non decreasing sequences:

$$\mathrm{col}_1(G) \leq \cdots \leq \mathrm{col}_{|G|}(G) = \mathrm{col}_\infty(G)$$

and

$$\mathrm{wcol}_1(G) \leq \cdots \leq \mathrm{wcol}_{|G|}(G) = \mathrm{wcol}_\infty(G) = \\ 1 + \min_{L \in \Pi(G)} \max_{v \in V(G)} |\{u \neq v : \exists u\text{–}v \text{ path } P \subseteq G, \min P = u\}.$$

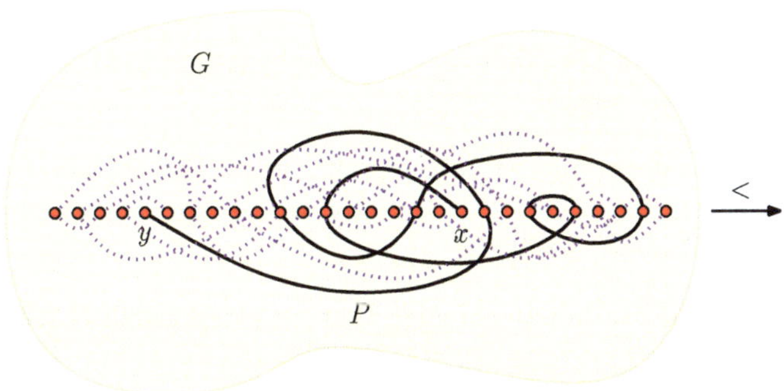

Fig. 4.12 The weak k-coloring number is one more than the minimum over all the linear orderings of the vertices of the maximum over x of the number of vertices y which can be reached from x by a path of length at most k with minimum vertex y

Kierstead and Yang introduced the k-coloring number of a graph for the purpose of studying coloring games and marking games on graphs.

As noticed in [274], these two invariants are easily shown to be polynomially equivalent via the monotone path segmentation:

Proposition 4.8.

$$\mathrm{col}_k(G) \leq \mathrm{wcol}_k(G) \leq \mathrm{col}_k(G)^k$$

Proof. Observe that both $Q_k(G_L, v)$ and $R_k(G_L, v)$ are defined by paths and that any path of length k under consideration in $Q_k(G_L, v)$ can be broken into at most k paths under consideration in $R_k(G_L, v)$. This easy observation implies that

$$\max_{v \in V(G)} |Q_k(G_L, v)| \leq \max_{v \in V(G)} |R_k(G_L, v)|^k.$$

□

In [476], Zhu proved that these invariants are also polynomially equivalent to the grads. Precisely, for each integer k there exists a polynomial F_k such that

$$\nabla_{\frac{k-1}{2}}(G) + 1 \leq \mathrm{wcol}_k(G) \leq \mathrm{col}_k(G)^k \leq F_k(\nabla_{\frac{k-1}{2}}(G)). \tag{4.26}$$

The proof of [476] of these functional dependencies is included in Sect. 7.5 in the context of decomposition theorems (Theorems 7.10 and 7.11).

Exercises

4.1. Compute an upper bound for $\nabla_r(G)$ and $\widetilde{\nabla}_r(G)$ when

- G is planar,
- G has maximum degree D,
- G may be drawn in the plane in such a way that every edge is crossed by at most one other edge.

4.2. Let d be an integer and let G be a graph. Prove that if $\text{girth}(G) \geq 8d+3$ and $\delta(G) \geq 3$ then $\nabla_{2d}(G) > 2^d$. (Actually, with some efforts one can prove a slightly better bound, see [121]).

4.3. Does there exist a function f such that for every graph G it holds

$$\chi(G \,\nabla\, \frac{1}{2}) \leq f(\chi(G \,\widetilde{\nabla}\, \frac{1}{2}))?$$

4.4. Edges densities $\nabla_r, \widetilde{\nabla}_r$, and $\overset{\propto}{\nabla}_{P(r),r}$ are related by polynomial functions. We noted that the Hadwiger number $h(G)$ is related to $\nabla(G)$ (see Lemma 4.1).

Deduce a similar relation between $h_t(G)$ and $\widetilde{\nabla}(G)$ from Theorem 3.6;
Deduce a similar relation between $h_i(G)$ and $\overset{\propto}{\nabla}(G)$ from Theorem 3.7.

4.5. Let G be a graph and let L be a linear ordering of $V(G)$. The k-*backconnectivity* $b_k(v)$ of a vertex v is the maximum number of paths from v of length at most k that intersect only at v, such that all the endvertices of these paths distinct from v are smaller than v; the k-*admissibility* $\text{adm}_k(G)$ is the minimum over the linear orders L of the maximum over $v \in V(G)$ of $b_k(v)$:

$$\text{adm}_k(G) = \min_{L} \max_{v \in V(G)} b_k(v).$$

Prove the following properties that relate adm_k to the generalized coloring numbers col_k and wcol_k [136]:

Prove that in the definition of $b_k(v)$ we can assume that all internal vertices of the paths are greater than v hence $\text{adm}_k(G) < \text{col}_k(G)$;
Let $R_k(v)$ be the set of all the vertices that are k-accessible from v. Show that there exists a tree Y rooted at v, with height at most k, with set of leaves $R_k(v)$, and all internal vertices of which are greater than v;
Prove that every non-leaf vertex of Y has degree at most $\text{adm}_k(G)$;
Deduce that

$$\text{col}_k(G) \leq \text{adm}_k(G)\,(\text{adm}_k(G) - 1)^{k-1} + 1.$$

Chapter 5
Classes and Their Classification

Do classes matter?
The class struggle within graph theory.

This chapter starts on an abstract level, by dealing with classes and their properties. As such it belongs to model theory (and general theory of categories). However our approach is very concrete and we deal with classes of graphs although many of the concepts and results carry over a more general setting. This will be made explicit in Sect. 5.8.

We denoted by $\mathcal{G}\mathrm{raph}$ the class of all finite graphs. In most of our book we do not distinguish between isomorphic graphs and thus $\mathcal{G}\mathrm{raph}$ will be considered as the (countable) class of isomorphism types of all finite graphs (often and less precisely, this is formulated as the class of all finite non-isomorphic graphs). In other words, we consider graphs up to an isomorphism.

A *class of graphs* is a finite or infinite set (isomorphism types) of graphs. Classes will be denoted by letters like $\mathcal{C}, \mathcal{F}, \mathcal{D}$. Although isomorphism closed classes are standard, the following restrictions (although frequent) are not always assumed:

- A class $\mathcal{C}$ is *hereditary* if for every graph $G \in \mathcal{C}$,

$$G' \subseteq_i G \quad \Longrightarrow \quad G' \in \mathcal{C};$$

- A class $\mathcal{C}$ is *monotone* if for every graph $G \in \mathcal{C}$,

$$G' \subseteq G \quad \Longrightarrow \quad G' \in \mathcal{C}.$$

The following restrictions are less frequent:

- A class $\mathcal{C}$ is *minor closed* if for every graph $G \in \mathcal{C}$,

J. Nešetřil and P.O. de Mendez, *Sparsity*, Algorithms and Combinatorics 28, DOI 10.1007/978-3-642-27875-4_5, © Springer-Verlag Berlin Heidelberg 2012

$$G' \leq_m G \quad \Longrightarrow \quad G' \in \mathcal{C}.$$

- A class $\mathcal{C}$ is *topologically minor closed* if for every graph $G \in \mathcal{C}$,

$$G' \leq_t G \quad \Longrightarrow \quad G' \in \mathcal{C}.$$

- A class $\mathcal{C}$ is *homomorphism closed* if for every graph $G \in \mathcal{C}$,

$$G' \leq_h G \quad \Longrightarrow \quad G' \in \mathcal{C}.$$

Examples of classes which are minor closed, topologically minor closed, or homomorphism closed are abundant. For example the class $\mathcal{G}raph$ itself is both homomorphism, minor, and topologically minor closed. If a class $\mathcal{C}$ is (minor, topologically minor, or homomorphism) closed and $\mathcal{C} \neq \mathcal{G}raph$ we say that $\mathcal{C}$ is a *proper* (minor, topologically minor, or homomorphism) closed class.

The class $\mathrm{Forb_h}(\mathcal{F})$ of the graphs with no homomorphic image of a graph in $\mathcal{F}$ deserves a specific notation, according to the importance of such classes in the context of homomorphism dualities studied in Chap. 9. Formally:

$$\begin{aligned}\mathrm{Forb_h}(\mathcal{F}) &= \{G : \ \forall F \in \mathcal{F}, \ F \nrightarrow G\} \\ &= \{G : \ \forall F \in \mathcal{F}, \ F \not\leq_h G\}.\end{aligned}$$

Similarly, we define the class $\mathrm{Forb_m}(\mathcal{F})$ of all graphs which do not have any graph $F \in \mathcal{F}$ as a minor:

$$\mathrm{Forb_m}(\mathcal{F}) = \{G : \ \forall F \in \mathcal{F}, \ F \not\leq_m G\}.$$

In such a situation, the set $\mathcal{F}$ is said to be a set of *forbidden minors* of $\mathcal{C}$. Clearly a class $\mathcal{C}$ is minor closed if and only if there exists a (possibly infinite) set $\mathcal{F}$ such that $\mathcal{C} = \mathrm{Forb_m}(\mathcal{F})$ (for example we can put $\mathcal{F} = \mathcal{G}raph \setminus \mathcal{C}$). Also, a class $\mathcal{C}$ is homomorphism closed if and only if there exists a (possibly infinite) set $\mathcal{F}$ such that $\mathcal{C} = \mathrm{Forb_h}(\mathcal{F})$. It follows from the well-quasi-ordering of graphs by the minor order that the following much stronger statement holds for minor closed classes [397, 399–417]:

Theorem 5.1. *For every minor closed class $\mathcal{C}$ there exists a finite set $\mathcal{F}$ of graphs such that $\mathcal{C} = \mathrm{Forb_m}(\mathcal{F})$.*

For homomorphism closed classes such a result does not hold. In fact, the opposite is true: most natural homomorphism closed classes of graphs are **not** of the form $\mathrm{Forb_h}(\mathcal{F})$ for some finite set $\mathcal{F}$. For example there are uncountably many homomorphism closed classes. However the question when a class $\mathcal{C}$ is determined by a finite set $\mathcal{F}$ is very interesting and leads to the notion of *homomorphism duality* which is one of the notions studied thoroughly in this book, see Chap. 9.

One could argue that there is no need to introduce the "class of graphs" denomination. Indeed classes of graphs are in one-to-one correspondence with graph properties: the property P corresponds to the class $\mathcal{C}$ of all graphs with property P. Thus perhaps the class terminology is superfluous if not misleading: instead of saying "consider the class of all triangle-free graphs" we can say simply "consider all graphs not containing a triangle". However, in this book we deal with statements which are far more complex and we use class-terminology abundantly. This not only reflects our preferences and taste but we find it necessary. We shall consider classes constructed from other classes, derived classes and derived classes from derived classes, etc. Such discussions could soon become cumbersome and would obscure the dynamics of what we are doing. The resolution of a class, which will be introduced in Sect. 5.1, is a typical example.

The importance of classes also lies in considering individual properties of graphs in the context to state results which apply to graphs belonging to some class or even to graphs belonging to some type of classes. This kind of study is sometimes called *relativization.*

Let us take two examples which motivate a large part of our research. Two examples of relativization will be considered in greater detail: The homomorphism preservation theorems in Chap. 10, and restricted dualities in Chap. 9.

5.1 Operations on Classes and Resolutions

5.1.1 Class Suprema and Class Limits

If $\mathsf{f} : \mathcal{G}\mathit{raph} \to \mathbb{R}$ is a graph invariant and $\mathcal{C}$ is a class, we define the *supremum* of f on $\mathcal{C}$ by

$$\mathsf{f}(\mathcal{C}) = \sup_{\mathsf{G}\in\mathcal{C}} \mathsf{f}(\mathsf{G}).$$

This definition is consistent with standard uses like the "chromatic number of planar graphs".

As we are mainly interested in infinite classes of graphs, we will consider that two classes of graphs $\mathcal{C}$ and $\mathcal{C}'$ are *asymptotically equivalent* if they differ by a finite set. For instance, any finite class of graphs is asymptotically equivalent to the empty class.

Let $\mathcal{C}$ be an infinite class of graphs and let $f : \mathcal{C} \to \mathbb{R}$ be a graph parameter. Let $\mathrm{Inj}(\mathbb{N}, \mathcal{C})$ be the set of all injective mappings from $\mathbb{N}$ to $\mathcal{C}$. Then we define the *class limit* of a graph parameter f on the class $\mathcal{C}$ as

$$\limsup_{G \in \mathcal{C}} f(G) = \sup_{\phi \in \mathrm{Inj}(\mathbb{N}, \mathcal{C})} \limsup_{i \to \infty} f(\phi(i))$$

and we also introduce the concise notation

$$\overline{f}(\mathcal{C}) = \limsup_{G \in \mathcal{C}} f(G).$$

An advantage of this definition (as compared to $f(\mathcal{C}) = \sup_{G \in \mathcal{C}} f(G)$) is that if two infinite classes $\mathcal{C}$ and $\mathcal{C}'$ are asymptotically equivalent then $\overline{f}(\mathcal{C}) = \overline{f}(\mathcal{C}')$.

Notice that $\limsup_{G \in \mathcal{C}} f(G)$ always exist and is either a real number or $\pm\infty$. Of course, $\mathrm{Inj}(\mathbb{N}, \mathcal{C})$ corresponds to the set of all sequences of distinct members of $\mathcal{C}$ (and recall that distinct members are non-isomorphic). We could also define the class limit $\limsup_{G \in \mathcal{C}} f(G)$ as

$$\limsup_{G \in \mathcal{C}} f(G) = \lim_{i \to \infty} \sup\{f(G) : \ G \in \mathcal{C}, |G| \geq i\}.$$

For $\alpha \in \overline{\mathbb{R}} = \mathbb{R} \cup \{-\infty, \infty\}$, the property $\limsup_{G \in \mathcal{C}} f(G) = \alpha$ holds if and only if the following two properties hold:

- For every $\phi \in \mathrm{Inj}(\mathbb{N}, \mathcal{C})$, $\limsup_{i \to \infty} f(\phi(i)) \leq \alpha$;
- There exists $\phi_M \in \mathrm{Inj}(\mathbb{N}, \mathcal{C})$, $\limsup_{i \to \infty} f(\phi(i)) = \alpha$.

Assuming $\limsup_{G \in \mathcal{C}} f(G) = \alpha$ the existence of ϕ_M is easy to prove: consider a sequence $\phi_1, \dots, \phi_i, \dots$ such that

$$\lim_{i \to \infty} \limsup_{j \to \infty} f(\phi_i(j)) = \alpha.$$

For each i, let $s_i(1) < \dots < s_i(j) < \dots$ be such that

$$\limsup_{j \to \infty} f(\phi_i(j)) = \lim_{j \to \infty} f(\phi_i(s_i(j))).$$

Then iteratively define $\phi_M \in \mathrm{Inj}(\mathbb{N}, \mathcal{C})$ by $\phi_M(1) = \phi_1(s_1(1))$ and $\phi_M(i) = \phi_i(s_i(j))$, where j is the minimal integer greater or equal to i such that $\phi_i(s_i(j))$ will be different from $\phi(1), \dots, \phi(i-1)$. Then $\limsup_{j \to \infty} f(\phi_M(j)) = \alpha$.

In other words, $\limsup_{G \in \mathcal{C}} f(G) = \alpha$ means that for every $\epsilon > 0$, infinitely many $G \in \mathcal{C}$ are such that $f(G) \geq \alpha - \epsilon$ but only finitely many $G \in \mathcal{C}$ are such that $f(G) \geq \alpha + \epsilon$.

Let now $\mathfrak{C} = (\mathcal{C}_1, \dots, \mathcal{C}_i, \dots)$ be a nested sequence of infinite graph classes: $\mathcal{C}_1 \subseteq \mathcal{C}_2 \subseteq \cdots \subseteq \mathcal{C}_i \subseteq \dots$. Let $f : \mathcal{G}raph \to \mathbb{R}$ be a graph invariant. We define

$$\overline{f}(\mathfrak{C}) = \lim_{i\to\infty} \overline{f}(\mathcal{C}_i) = \sup_{i\to\infty} \limsup_{G\in\mathcal{C}_i} f(G).$$

The limit is well defined as the values $\limsup_{G\in\mathcal{C}_i} f(\mathcal{C}_i)$ form a non decreasing sequence.

Again let us note that $\overline{f}(\mathfrak{C}) = \alpha$ means that for every $\epsilon > 0$:

- On the one hand, one can find some i_0 such that infinitely many $G \in \mathcal{C}_{i_0}$ satisfy $f(G) \geq \alpha - \epsilon$;
- On the other hand, for every i, there are only finitely many $G \in \mathcal{C}_i$ so that $f(G) \geq \alpha + \epsilon$.

5.1.2 Class Operations

We shall make a recurrent use of the *lexicographic product* of a class $\mathcal{C}$ by a graph F, which is elementwise defined by

$$\mathcal{C} \bullet F = \{G \bullet F : \ G \in \mathcal{C}\}.$$

The following notions are principal for this book. They extend the definition of the shallow minor (Sect. 4.2) to graph classes.

Definition 5.1. Let $\mathcal{C}$ be a class of graphs, let a be a half-integer and let b be an integer. We define

$$\mathcal{C} \,\triangledown\, a = \bigcup_{G\in\mathcal{C}} G \,\triangledown\, a,$$

$$\mathcal{C} \,\widetilde{\triangledown}\, a = \bigcup_{G\in\mathcal{C}} G \,\widetilde{\triangledown}\, a,$$

and

$$\mathcal{C} \,\overset{\propto}{\triangledown}\, (b, a) = \bigcup_{G\in\mathcal{C}} G \,\overset{\propto}{\triangledown}\, (b, a).$$

Remark 5.1. It follows from these definitions that for every class $\mathcal{C}$ we have the following inclusions:

$$
\begin{array}{cccccccccc}
 & \mathcal{C} \triangledown 0 & \subseteq & \mathcal{C} \triangledown \frac{1}{2} & \subseteq \dots \subseteq & \mathcal{C} \triangledown a & \subseteq \dots & \subseteq \mathcal{C} \triangledown \infty \\
 & \| & & \cup\!\!| & & \cup\!\!| & & \cup\!\!| \\
\mathcal{C} \subseteq & \mathcal{C} \widetilde{\triangledown} 0 & \subseteq & \mathcal{C} \widetilde{\triangledown} \frac{1}{2} & \subseteq \dots \subseteq & \mathcal{C} \widetilde{\triangledown} a & \subseteq \dots & \subseteq \mathcal{C} \widetilde{\triangledown} \infty \\
 & \| & & \cap\!\!| & & \cap\!\!| & & \cap\!\!| \\
 & \mathcal{C} \overset{\propto}{\triangledown} (1,0) & \subseteq & \mathcal{C} \overset{\propto}{\triangledown} (1, \frac{1}{2}) & \subseteq \dots \subseteq & \mathcal{C} \overset{\propto}{\triangledown} (2a+1, a) & \subseteq \dots & \subseteq \mathcal{C} \overset{\propto}{\triangledown} (\infty, \infty)
\end{array}
$$

Some of these classes have a particular meaning:

- $\mathcal{C} \triangledown 0$ is the *monotone closure* of $\mathcal{C}$ (that is: the smallest monotone class including $\mathcal{C}$),
- $\mathcal{C} \triangledown \infty$ is the *minor closure* of $\mathcal{C}$ (that is: the smallest minor closed class including $\mathcal{C}$),
- $\mathcal{C} \widetilde{\triangledown} \infty$ is the *topological closure* of $\mathcal{C}$ (that is: the smallest topologically closed class including $\mathcal{C}$).

For the sake of completeness, recall that the *hereditary closure* of $\mathcal{C}$ (the smallest hereditary class including $\mathcal{C}$) is denoted by $\mathrm{H}(\mathcal{C})$.

5.1.3 Class Resolutions

As the reader surely observed, the time evolving parametrization is the key to our analysis. This propagates to classes and the following definitions: The *resolution* of a class $\mathcal{C}$ is the nested sequence of graph classes

$$\mathcal{C}^{\triangledown} = (\mathcal{C} \triangledown 0, \mathcal{C} \triangledown \frac{1}{2}, \mathcal{C} \triangledown 1, \dots),$$

its *topological resolution* is the nested sequence of graph classes

$$\mathcal{C}^{\widetilde{\triangledown}} = (\mathcal{C} \widetilde{\triangledown} 0, \mathcal{C} \widetilde{\triangledown} \frac{1}{2}, \mathcal{C} \widetilde{\triangledown} 1, \dots),$$

and its *immersion resolution* is the nested sequence of graph classes

$$\mathcal{C}^{\overset{\propto}{\triangledown}} = (\mathcal{C} \overset{\propto}{\triangledown} (1,0), \mathcal{C} \overset{\propto}{\triangledown} (2, \frac{1}{2}), \mathcal{C} \overset{\propto}{\triangledown} (3,1), \dots),$$

It is convenient to interpret these resolutions as sequences developing in time (in time scaled by half-integer units). These infinite sequences will be at the heart of a surprisingly stable classification of general infinite classes of graphs. The main ingredient of this classification will be the concentration properties of logarithmic densities of graphs, which we shall now present. We find it convenient to start with topological minors, the classes $\mathcal{C}\,\widetilde{\nabla}\,i$, and with the corresponding resolution $\mathcal{C}^{\widetilde{\nabla}}$.

5.1.4 Topological Parameters

Recall that a *graph parameter* is a function ϱ for which $\varrho(G)$ is a non-negative real number for every graph G. Examples include minimum degree, average degree, maximum degree, connectivity, chromatic number, tree-width, etc.

This is a very large concept, which needs to be refined. Particularly, we say that a graph parameter ϱ is

- *Subdivision bounded* if for some function f, for every graph G and every subdivision H of G it holds

$$\varrho(H) \leq f(\varrho(G)).$$

- *Weakly topological* if for some function f, for every graph G and every ≤ 1-subdivision H of G,

$$\varrho(G) \leq f(\varrho(H)) \text{ and } \varrho(H) \leq f(\varrho(G)).$$

Many graph parameters are subdivision bounded but not weakly topological. This is the case, for instance, of the clique number $\omega(G)$, the average degree $\overline{d}(G)$, or the chromatic number $\chi(G)$ (if H is a subdivision of G then $\chi(H) \leq \chi(G) + 1$).

Recall that the graph parameter ϱ is *monotone* (respectively, *hereditary*) if $\varrho(H) \leq \varrho(G)$ for every subgraph (respectively, every induced subgraph) H of G.

The following statement shows that for an infinite class $\mathcal{C}$ a subdivision bounded graph parameter is bounded at each time of a topological resolution if and only if some related monotone weakly topological graph parameter is bounded on the class $\mathcal{C}$.

Proposition 5.1. *Let ϱ be a subdivision bounded graph parameter and let $\mathcal{C}$ be an infinite class. Then the two following conditions are equivalent:*

- *For every integer r it holds $\varrho(\mathcal{C}\,\widetilde{\nabla}\,r) < \infty$,*
- *There exists a weakly topological monotone graph parameter $\widetilde{\varrho}$ functionally bounding ϱ such that $\widetilde{\varrho}(\mathcal{C}) < \infty$.*

Proof. As ϱ is a subdivision bounded parameter, there exists a function g such that for every graph G and every subdivision H of G it holds $\varrho(H) \leq g(\varrho(G))$.

Assume there exists no positive integer r such that $\varrho(\mathcal{C}\,\widetilde{\nabla}\,r) = \infty$. Let $\mathcal{C}^+$ be the class of all (possibly trivial) subdivisions of graphs in $\mathcal{C}$. Then for every integer r, $\varrho(\mathcal{C}^+\,\widetilde{\nabla}\,r) \leq g(\varrho(\mathcal{C})) < \infty$. Let $f(r) = \varrho(\mathcal{C}^+\,\widetilde{\nabla}\,r)$. Notice that f is a non-decreasing function.

- If the function f is bounded, then define $\widetilde{\varrho}(G) = \varrho(G\,\widetilde{\nabla}\,\infty)$. This parameter is clearly monotone and obviously functionally bounds ϱ. If H is a ≤ 1-subdivision of G then each graph in $H\,\widetilde{\nabla}\,\infty$ is a ≤ 1-subdivision of a graph in $G\,\widetilde{\nabla}\,\infty$ hence

$$\widetilde{\varrho}(H) = \varrho(H\,\widetilde{\nabla}\,\infty) \leq g(\varrho(G\,\widetilde{\nabla}\,\infty)) = g(\widetilde{\varrho}(G))$$

  Also, as $G\,\widetilde{\nabla}\,\infty \subseteq H\,\widetilde{\nabla}\,\infty$ it holds

$$\widetilde{\varrho}(H) \geq \widetilde{\varrho}(G).$$

  It follows that $\widetilde{\varrho}$ is weakly topological.
- Otherwise, as f is unbounded, there exists for each graph G an integer $w(G)$ such that $g(\varrho(G\,\widetilde{\nabla}\,\infty)) \leq f(w(G))$. Hence there exists for each graph G a minimum non-negative real number $\widetilde{\varrho}(G) \leq w(G)$ such that for (possibly trivial) subdivision G' of G and every integer r it holds

$$\varrho(G'\,\widetilde{\nabla}\,r) \leq f(\widetilde{\varrho}(G)(r+1)).$$

  The graph parameter $\widetilde{\varrho}$ functionally bounds ϱ as

$$\varrho(G) \leq \varrho(G\,\widetilde{\nabla}\,0) \leq f(\widetilde{\varrho}(G)).$$

  It is also clear that $\widetilde{\varrho}$ is monotone: if H is a subgraph of G then $\varrho(H\,\widetilde{\nabla}\,r) \leq \varrho(G\,\widetilde{\nabla}\,r)$, hence $\widetilde{\varrho}(H) \leq \widetilde{\varrho}(G)$.
  Let G be a graph and let H be a ≤ 1-subdivision of G. Then every subdivision of H is also a subdivision of G hence $\widetilde{\varrho}(H) \leq \widetilde{\varrho}(G)$. Conversely, if G' is a subdivision of G, there exists a ≤ 1-subdivision H' of G' which is a subdivision of H. Hence

$$\begin{aligned}\varrho(G'\,\widetilde{\nabla}\,r) \leq \varrho(H'\,\widetilde{\nabla}\,(2r+1)) &\leq f(\widetilde{\varrho}(H)(2r+2)) \\ &= f(2\widetilde{\varrho}(H)(r+1)).\end{aligned}$$

  Hence $\widetilde{\varrho}(G) \leq 2\widetilde{\varrho}(H)$ and thus $\widetilde{\varrho}$ is a strong topological graph parameter.
  Finally, it is directly checked from the definition that $\widetilde{\varrho}(G) \leq 1$ for every graph $G \in \mathcal{C}$.

Now assume that $\widetilde{\varrho}$ is a weakly topological and monotone parameter bounding ϱ. By definition, there exists some function f such that for every graph G hold $\varrho(G) \leq f(\varrho(G'))$ and $\varrho(G') \leq f(\varrho(G))$ where G' is the

1-subdivision of G. Moreover, the function f may obviously be chosen non decreasing. Let $\mathcal{C} = \{G : \widetilde{\varrho}(G) \leq c\}$ for some constant c. Let r be an integer. Let $G \in \mathcal{C}$. For some $H \in G \mathbin{\widetilde{\triangledown}} r$, we have $\varrho(H) = \varrho(G \mathbin{\widetilde{\triangledown}} r)$. Let S be a $\leq r$-subdivision of H isomorphic to a subgraph of G. Let $p = \lceil \log_2(2r) \rceil$. There is a sequence $H = H_0, H_1, \ldots, H_p = S$ such that H_{i+1} is a ≤ 1-subdivision of H_i, for each $i \in \{0, \ldots, p-1\}$. By induction, $\widetilde{\varrho}(H) \leq f^p(\widetilde{\varrho}(S))$ where f^p is f iterated p times. Since f is non-decreasing, $\widetilde{\varrho}(H) \leq f^p(c)$. Since $\widetilde{\varrho}$ bounds ϱ there exists a function g such that $\varrho(G) \leq g(\widetilde{\varrho}(G))$ holds for every graph G. Hence

$$\varrho(G \mathbin{\widetilde{\triangledown}} r) = \varrho(H) \leq g(\widetilde{\varrho}(H)) \leq g(f^p(c)) = g(f^{\lceil \log_2(2r) \rceil}(c)).$$

Thus

$$\varrho(\mathcal{C} \mathbin{\widetilde{\triangledown}} r) \leq g(f^{\lceil \log_2(2r) \rceil}(c)) < \infty.$$

□

5.2 Logarithmic Density and Concentration

A simple and classical distinction between "dense" and "sparse" graphs is that the former have a quadratic number of edges and the latter have "much fewer edges". This intuitive classification suggests a new invariant, the *logarithmic density* $\ell\mathrm{dens}(G)$ of a graph G, which we define as

$$\ell\mathrm{dens}(G) = \begin{cases} -\infty, & \text{if } \|G\| = 0 \\ \frac{\log \|G\|}{\log |G|}, & \text{otherwise.} \end{cases} \tag{5.1}$$

Notice that, according to this definition, the logarithmic density of a single vertex graph or an empty graph is $-\infty$. Also, it is immediate that $\ell\mathrm{dens}(G)$ is either $-\infty$ (if G is edgeless) or a real value in the interval $[0;2]$. This is clear if we note that $\ell\mathrm{dens}(G) = \alpha \geq 0$ if and only if $\|G\| = |G|^\alpha$. So the logarithmic density expresses the size of the graph as a power of its order.

As we have seen in Sect. 3.2, one of the important properties of the minimum degree and the average degree is that every graph G has a large subgraph whose minimum degree is not significantly smaller than half of the average degree of G: For every graph G and every $\epsilon > 1/|G|$ there exists a subgraph H of G of minimum degree at least $(1-\epsilon)\frac{\|G\|}{|G|}$ and size $\|H\|$ at least $\epsilon\|G\|$. It follows that for such a graph H it holds

$$\frac{\log \delta(H)}{\log |H|} \geq \frac{\log \|G\| - \log |G| + \log(1-\epsilon)}{\log |G|} \tag{5.2}$$

$$= \ell\text{dens}(G) - 1 + \frac{\log(1-\epsilon)}{\log |G|}. \tag{5.3}$$

The logarithmic density of such a subgraph H cannot be significantly smaller than the one of G. Indeed, as $\delta(H) \leq 2\|H\|/|H|$ it holds

$$\ell\text{dens}(H) \geq \frac{\log \frac{\delta(H)|H|}{2}}{\log |H|} = \frac{\log \delta(H)}{|H|} + 1 - 2\frac{\log 2}{\log |H|}$$
$$\geq \ell\text{dens}(G) + \frac{\log(1-\epsilon)}{\log |G|} - \frac{2\log 2}{\log |H|}$$

Hence

$$\ell\text{dens}(H) \geq \ell\text{dens}(G) + o(\log |H|) \tag{5.4}$$

The ratio $\frac{\log \delta(G)}{\log |G|}$ appears to be particularly suitable to prove concentration results in conjunction with Proposition 4.3. In particular, by a suitable choice of the constants, we obtain the following result:

Lemma 5.1. *Let $\rho > 1$. There is a positive $N(\rho)$ such that for every graph G of order $|G| \geq N(\rho)$ which satisfies*

$$\frac{\log \delta(G)}{\log |G|} \geq \frac{1}{\rho}$$

there exists a graph H such that the 1-subdivision of H is a subgraph of G and

- *Either* $H \cong K_{|G|^{1/3\rho^2}}$,
- *or* $\dfrac{\log \delta(H)}{\log |H|} > \dfrac{1}{\rho - 1/2}$ *and* $|H| \geq \sqrt{|G|/3}$.

Proof. Put $\mu = 1/\rho$. We apply Proposition 4.3 with $\delta = n^\mu$ and $\epsilon = 2n^{\mu^2/3-\mu}$ (with the notations of Proposition 4.3). We get that there exists a graph H such that an exact 1-subdivision of H is a subgraph of G and either $H \cong K_{n^{\mu^2/3}}$ or H has minimum degree at least $\frac{1}{2}n^{\mu-\mu^2/3}$ and order

$$\sqrt{(1-2n^{-\mu+\mu^2/3})n/2} \leq |H| \leq n^{1-\mu+\mu^2/3}.$$

For sufficiently large n, $n^{-\mu+\mu^2/3} \leq 1/6$ hence $|H| \geq \sqrt{n/3}$. As $|H| \leq n^{1-\mu\mu^2/3}$ we deduce

$$\delta(H) \geq \frac{1}{2}|H|^{\frac{\mu-\mu^2/3}{1-\mu+\mu^2/3}} = \frac{1}{2}|H|^{\mu+\frac{\mu^2}{3}\cdot\frac{2-\mu}{1-\mu+\mu^2/3}} \geq |H|^{\mu+\frac{\mu^2}{2}}$$

(for sufficiently large n). As

$$\left(\frac{1}{\rho}+\frac{1}{2\rho^2}\right)^{-1}=\rho-\frac{1}{2}+\frac{1/2}{2\rho+1}>\rho-\frac{1}{2}$$

we conclude the proof. □

This result shows that if the logarithmic density of a large graph is bounded away from 1 (i.e. at least $1+\epsilon$ for some positive real ϵ) then it contains a shallow subdivision of a large clique (of depth $k(\epsilon)$). Precisely, we have:

Lemma 5.2. *Let $\rho > 1$ and let G be a graph of order $n \geq N''(\rho)$ and minimum degree at least $n^{1/\rho}$. Then the complete graph of order $\frac{1}{3}(3n)^{2^{-4\rho}}$ belongs to $G \,\widetilde{\triangledown}\, (9^\rho - 1)/2$:*

$$\omega\left(G \,\widetilde{\triangledown}\, \frac{9^\rho-1}{2}\right) \geq \frac{1}{3}(3|G|)^{2^{-4\rho}}$$

Proof. We construct a sequence of graphs $G_0, G_1, \ldots, G_k$ such that for each $0 \leq i \leq k$ the graph G_i has order n_i and minimum degree at least $n_i^{1/(\rho-i/2)}$ as follows: Put $G_0 = G$. Iteratively, for each $i \geq 0$, if G_i is not a complete graph we apply Lemma 5.1 to G_i. Then we get a graph H_i whose 1-subdivision is a subgraph of G_i. If H_i is a 1-subdivision of a complete graph we stop. Otherwise we let $G_{i+1} = G_i$. Notice that $n_{i+1} \geq \sqrt{n_i/3}$ hence $\log(3n_{i+1}) \geq \log(3n_i)/2$. As obviously the process stops after at most 2ρ iterations (because of the increase of $\log \delta(G_i)/\log|G_i|$), we will obtain a complete graph of order at least $\frac{1}{3}(3n)^{2^{-4\rho}}$ at depth $\frac{9^\rho-1}{2}$. □

Hence for every $\rho > 1$ and every graph G of order at least $N''(\rho)$ holds

$$\ell\text{dens}(G) > 1 + \frac{1}{\rho} \quad\Longrightarrow\quad \omega\left(G \,\widetilde{\triangledown}\, \frac{9^\rho-1}{2}\right) \geq \frac{1}{3}(3|G|)^{2^{-4\rho}}. \tag{5.5}$$

(Actually we have the stronger conclusion that the $(9^\rho-1)/2$-subdivision of a clique of size at least $(3|G|^{2^{-4\rho}})/3$ is a subgraph of G). This result will be the basis of our trichotomy Theorem 5.4.

The following result is due to Kostochka and Pyber [283]:

Theorem 5.2. *Let ϵ be a positive real such that $0 < \epsilon < 1$. Let n, t be positive integers, and let G be a graph on n vertices and $m \geq 4^{t^2} n^{1+\epsilon}$ edges. Then*

$$K_t \in G \,\widetilde{\triangledown}\, \frac{2(1+2\log_2 t)}{\epsilon}.$$

This result has been improved by Jiang [268]:

Theorem 5.3. *Let* t *be a positive integer and let* $0 < \epsilon < 1/2$ *be a positive real. There exists* $n_0 = n_0(t, \epsilon)$ *such that for all integer* $n \geq n_0$*, if* G *is a graph on* n *vertices and* $m \geq n^{1+\epsilon}$ *edges, then* G *contains a* $\lfloor 10/\epsilon \rfloor$*-subdivision of* K_t*, that is:*

$$\omega\left(G \mathbin{\widetilde{\nabla}} \frac{5}{\epsilon}\right) \geq t.$$

This result solves an open problem of Kostochka–Pyber paper. The proof is an elaboration of earlier results of [172, 283] and is quite technical. We do not include it here. From the point of view of this book this result gives improvements on dependencies between logarithmic density and clique number of shallow topological minors. However it does not change the classification of classes we are going to introduce next.

5.3 Classification of Classes by Clique Minors

The resolution of the infinite classes of graphs allows a direct classification of these classes in two types. This is one of our principal definitions.

Definition:

- Classes $\mathcal{C}$ such that there exist a (finite) half-integer a such that $\mathcal{C} \nabla a = \mathcal{G}raph$ will be called *somewhere dense*.
- Classes $\mathcal{C}$ such that $\mathcal{C} \nabla a \neq \mathcal{G}raph$ for every half-integer a will be called *nowhere dense*.

As $\mathcal{C} \nabla a$ is a monotone class for each a, we notice that the statement "$\mathcal{C} \nabla a = \mathcal{G}raph$" is equivalent to the statement "$\omega(\mathcal{C} \nabla a) = \infty$". Recall that we put $\omega(\mathcal{C}) = \sup\{\omega(G) : G \in \mathcal{C}\}$.

Let us remark that we can extend this definition easily to (relational) structures and hypergraphs by means of their incidence graphs and we shall state explicitly these results in Sect. 13.6.

We may also consider a classification based on topological minors instead of minors, i.e. a dichotomy between the classes $\mathcal{C}$ such that there exists an integer t_0 such that $\omega(\mathcal{C} \mathbin{\widetilde{\nabla}} t_0) = \infty$ and those classes $\mathcal{C}$ such that $\omega(\mathcal{C} \mathbin{\widetilde{\nabla}} t) < \infty$ for every t. However, it appears that these two approaches (minor resolution and topological minor resolution) lead to the same classification. This is a consequence of the following inequalities:

Proposition 5.2. *Let* G *be a graph and let* a *be a half-integer. Then*

$$\omega(G \mathbin{\widetilde{\nabla}} a) \leq \omega(G \nabla a) \leq 2\omega(G \mathbin{\widetilde{\nabla}} (3a+1))^{\lfloor a \rfloor + 1}$$

Proof. As $G \mathbin{\widetilde{\nabla}} a \subseteq G \nabla a$, the inequality $\omega(G \mathbin{\widetilde{\nabla}} a) \leq \omega(G \nabla a)$ is straightforward.

Let a be a half-integer and let $p = \omega(G \nabla a)$. Let Ω_p be an a-witness of K_p in G and let $Y_1, \ldots, Y_p$ be the bushes of the corresponding K_p-decomposition of Ω_p.

Orient the external edges of the K_p-decomposition of Ω_p arbitrarily if a is an integer and otherwise (if a is not an integer) in such a way that for every external edge $\{x, y\}$, if x is at distance $\lceil a \rceil$ from the center of its bush then $\{x, y\}$ is oriented from x to y (this is consistent as the decomposition is asymmetric). Note that the orientation of the external edges of the decomposition naturally induces an orientation of K_p.

Let $q = (p/2)^{1/(r+1)}$. For $i = 1, \ldots, q$ we will consider a subset A_i of vertices of K_p of size at least $2(q-1)^{\lfloor a \rfloor + 1} + 1$. Initially, we set A_1 to be the vertex set of K_p. At each step, q vertices will be removed from A_{i-1} to form A_i.

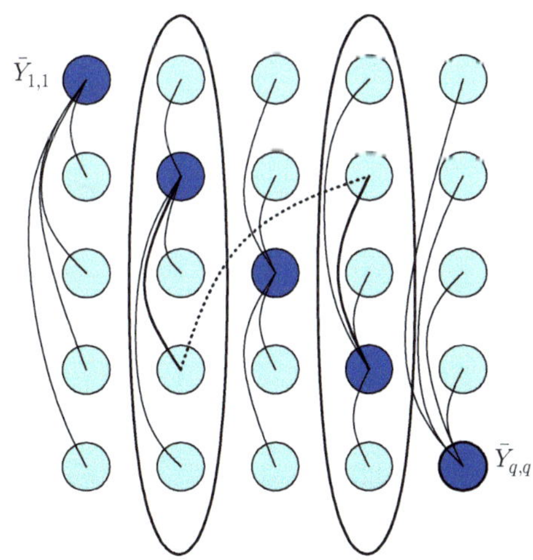

Fig. 5.1 Finding a subdivision of K_q

Consider A_i for some $1 \leq i \leq q$. At least one vertex in A_i has at least $(|A_i| - 1)/2$ in-neighbors in A_i. An easy counting shows this value is the average number of in-neighbors in A_i for the vertices of A_i. Let Y be the bush associated with such a vertex and let Y' be the rooted subtree of Y (with the same root as Y) which is the union of all the branches of Y from its root to a leaf having at least one incoming external edge incident to another bush corresponding to a vertex in A_i. Orient the edges of Y' toward its root. First notice that, by construction, the distance of the leaves of Y' to the root is at most $\lfloor a \rfloor$. It follows, by a simple counting argument, that at least one vertex c of Y' has indegree at least $((|A_i|-1)/2)^{1/(r+1)} \geq q-1$. Let $\bar{Y}_{i,i}$ be the the subtree of Y' rooted at c. By construction, the tree $\bar{Y}_{i,i}$ has at least $q-1$ leaves, each being the terminal endpoint of an edge incident to some bush corresponding to a vertex in A. Let $\bar{Y}_{i,j}$ $(j \in \{1, \dots, i-1, i+1, q\})$ be these $q-1$ bushes. The corresponding vertices of A_i are removed from A_i to form the set A_{i+1}. For $i < q$, this set has size at least $p-(q-1)q \geq 2(q-1)^{\lfloor a \rfloor+1}+1$

At the end of the process, we have q^2 bushes $\bar{Y}_{i,j}$ $(1 \leq i, j \leq q)$ which allow us to exhibit a $(6a+2)$-subdivision of K_q (Fig. 5.1). □

Thus the class $\mathcal{C}$ is somewhere dense if and only if there exists time t_0 such that the class of all topological minors at depth t_0 (i.e. $\mathcal{C} \mathbin{\widetilde{\nabla}} t_0$) is the class of all graphs. In the latter case we can say that $\mathcal{C}$ is *topologically somewhere dense*. Thus $\mathcal{C}$ is somewhere dense if and only if $\mathcal{C}$ is topologically somewhere dense. We shall see in the next section that $\mathcal{C}$ is somewhere dense if and only if it is "immersion somewhere dense".

5.4 Classification by Density—Trichotomy of Classes

Recall that according to Sect. 5.1.1

$$\overline{\ell\mathrm{dens}}(\mathcal{C}^{\widetilde{\nabla}}) = \sup_{i \to \infty} \limsup_{G \in \mathcal{C} \widetilde{\nabla} i} \frac{\log \|G\|}{\log |G|}.$$

A consequence of the results of Sect. 5.2 is the following concentration result:

Proposition 5.3. *Let $\mathcal{C}$ be an infinite class of graphs. Then the following conditions are equivalent:*

(i) $\overline{\ell\mathrm{dens}}(\mathcal{C}^{\widetilde{\nabla}}) > 1$,
(ii) $\overline{\ell\mathrm{dens}}(\mathcal{C}^{\widetilde{\nabla}}) = 2$
(iii) There exists a half-integer i_0 such that $\mathcal{C} \mathbin{\widetilde{\nabla}} i_0 = \mathcal{G}\mathrm{raph}$.

It follows that infinite classes of graphs follow the following trichotomy (we do not consider the case of edgeless graphs as distinct from the one of bounded size graphs):

Theorem 5.4 (Class trichotomy). *Let $\mathcal{C}$ be an infinite class of graphs. Then the limit*

$$\overline{\ell\mathrm{dens}}(\mathcal{C}^{\widetilde{\nabla}}) = \lim_{r\to\infty} \limsup_{G\in\mathcal{C}\,\widetilde{\nabla}\,r} \frac{\log\|G\|}{\log|G|}$$

can only take four values, namely $-\infty, 0, 1$ or 2 (see Fig. 5.2).

Thus every infinite class $\mathcal{C}$ may be categorized using the following three types:

- *A* bounded size *class (or* asymptotically edgeless *class(i.e. $\|G\|$ is globally bounded on $\mathcal{C}$) if and only if $\overline{\ell\mathrm{dens}}(\mathcal{C}^{\widetilde{\nabla}}) = 0$ or $-\infty$,*
- *A* nowhere dense *class if and only if $\overline{\ell\mathrm{dens}}(\mathcal{C}^{\widetilde{\nabla}}) \in \{-\infty, 0, 1\}$,*
- *A* somewhere dense *class if and only if $\overline{\ell\mathrm{dens}}(\mathcal{C}^{\widetilde{\nabla}}) = 2$.*

Proof. According to Proposition 5.3, either $\mathcal{C}$ is somewhere dense and $\overline{\ell\mathrm{dens}}(\mathcal{C}^{\widetilde{\nabla}}) = 2$ or $\mathcal{C}$ is nowhere dense and $\overline{\ell\mathrm{dens}}(\mathcal{C}^{\widetilde{\nabla}}) \leq 1$.

If $\mathcal{C}$ has unbounded size then $\mathcal{C}\,\widetilde{\nabla}\,0$ contains arbitrarily large graphs G without isolated vertices (hence such that $\ell\mathrm{dens}(G) \geq 1 - \frac{\log 2}{\log|G|}$) thus $\overline{\ell\mathrm{dens}}(\mathcal{C}^{\widetilde{\nabla}}) \geq 1$. Otherwise, each $\mathcal{C}\,\widetilde{\nabla}\,r$ has bounded size (as topological minors cannot have more edges than the original graphs) but is infinite (as $\mathcal{C}$ is infinite) hence contains graphs of arbitrarily large orders. It follows that for each r, $\overline{\ell\mathrm{dens}}(\mathcal{C}^{\widetilde{\nabla}}) = 0$ if one can find arbitrarily large graphs with at least one edge, and $\overline{\ell\mathrm{dens}}(\mathcal{C}^{\widetilde{\nabla}}) = -\infty$, otherwise. □

At this level of class hierarchies again, topological shallow minors, shallow minors an immersions play equivalent roles:

Proposition 5.4. *For every infinite class of graphs $\mathcal{C}$ holds*

$$\overline{\ell\mathrm{dens}}(\mathcal{C}^{\nabla}) = \overline{\ell\mathrm{dens}}(\mathcal{C}^{\overset{\propto}{\nabla}}) = \overline{\ell\mathrm{dens}}(\mathcal{C}^{\widetilde{\nabla}}). \tag{5.6}$$

Proof. As noticed in Sect. 4.8, we have for every $p, q \geq 1$

$$G\,\widetilde{\nabla}\,q \subseteq G\,\overset{\propto}{\nabla}\,(p,q) \subseteq (G \bullet \overline{K}_p)\,\widetilde{\nabla}\,q.$$

Moreover for every $p \geq 1$

$$\begin{aligned}\overline{\ell\mathrm{dens}}((\mathcal{C} \bullet K_p)^{\widetilde{\nabla}}) &= \limsup_{G \in \mathcal{C} \bullet K_p} \frac{\log \|G\|}{\log |G|} = \limsup_{G \in \mathcal{C}} \frac{\log \|G \bullet K_p\|}{\log |G \bullet K_p|} \\ &= \limsup_{G \in \mathcal{C}} \frac{\log \|G\| + 2 \log p}{\log |G| + \log p} \\ &= \limsup_{G \in \mathcal{C}} \frac{\log \|G\|}{\log |G|} = \overline{\ell\mathrm{dens}}(\mathcal{C}^{\widetilde{\nabla}}).\end{aligned}$$

Thus $\overline{\ell\mathrm{dens}}(\mathcal{C}^{\overset{\propto}{\nabla}}) = \overline{\ell\mathrm{dens}}(\mathcal{C}^{\widetilde{\nabla}})$.

Let us now prove that $\overline{\ell\mathrm{dens}}(\mathcal{C}^{\nabla}) = \overline{\ell\mathrm{dens}}(\mathcal{C}^{\widetilde{\nabla}})$.

As a consequence of the inclusions displayed in Remark 5.1 we have $\overline{\ell\mathrm{dens}}(\mathcal{C}^{\widetilde{\nabla}}) \leq \overline{\ell\mathrm{dens}}(\mathcal{C}^{\nabla}) \leq 2$.

Hence if $\overline{\ell\mathrm{dens}}(\mathcal{C}^{\widetilde{\nabla}}) = 2$ then $\overline{\ell\mathrm{dens}}(\mathcal{C}^{\nabla}) = 2$. Also, it is immediate that if $\overline{\ell\mathrm{dens}}(\mathcal{C}^{\widetilde{\nabla}})$ is either $-\infty$ or 0 then $\overline{\ell\mathrm{dens}}(\mathcal{C}^{\nabla}) = \overline{\ell\mathrm{dens}}(\mathcal{C}^{\widetilde{\nabla}})$.

Assume for contradiction that $\overline{\ell\mathrm{dens}}(\mathcal{C}^{\widetilde{\nabla}}) = 1$ and that there exists $\epsilon > 0$ such that $\overline{\ell\mathrm{dens}}(\mathcal{C}^{\nabla}) \geq 1 + \epsilon$. Then there exists an integer t such that

$$\limsup_{G \in \mathcal{C} \,\nabla\, t} \frac{\log \|G\|}{\log |G|} \geq 1 + \epsilon/2.$$

According to Proposition 5.3, there exists an integer i_0 such that $(\mathcal{C} \,\nabla\, t) \,\nabla\, i_0 \supseteq (\mathcal{C} \,\nabla\, t) \,\widetilde{\nabla}\, i_0 = \mathcal{G}\mathit{raph}$. Put $c = ((2t+1)(2i_0+1)-1)/2$. According to Proposition 4.1, we have $\mathcal{C} \,\nabla\, c = \mathcal{G}\mathit{raph}$ thus $\mathcal{C}$ is somewhere dense, what contradicts our assumption that $\overline{\ell\mathrm{dens}}(\mathcal{C}^{\widetilde{\nabla}}) = 1$. It follows that that $\overline{\ell\mathrm{dens}}(\mathcal{C}^{\nabla}) = 1$. □

It follows that our classification of classes (bounded size versus nowhere dense versus somewhere dense) is the same if we consider topological minors or immersions instead of minors in the computation of asymptotic logarithmic densities of resolutions. This stability of our classification is not singular. We shall see other examples of this. An interested reader can jump forward to Chap. 13 to see the stability in the full light. This jump may also provide him a motivation to return and read the details of the next section.

5.5 Classes with Bounded Expansion

Thus the nowhere dense classes are those classes $\mathcal{C}$ such that for every t the edge density of every graph G in $\mathcal{C} \,\nabla\, t$ is bounded by $|G|^{o(1)}$. This particularly includes the case when the edge density of the graphs in $\mathcal{C} \,\nabla\, t$ is bounded by a constant $c(t)$. Such classes deserve a name on their own and they were our original motivation.

A class $\mathcal{C}$ of graphs has *bounded expansion* if for every t there exists $c(t)$ such that $\frac{\|G\|}{|G|} \leq c(t)$ for every graph $G \in \mathcal{C} \,\nabla\, t$. In other words, for every t, we have $\nabla_t(\mathcal{C}) \leq c(t)$. The function $t \mapsto \nabla_t(\mathcal{C})$ is called the *expansion function* of the class $\mathcal{C}$. Classes with bounded expansion (also called *bounded expansion classes*) may be alternatively defined as those classes $\mathcal{C}$ where for each t the class $\mathcal{C} \mathbin{\triangledown} t$ is a class of degenerate graphs.

Bounded expansion classes are of course nowhere dense. But this special case allows a finer classification.

The expansion function of a bounded expansion class $\mathcal{C}$ is (uniformly) bounded if and only if $\mathcal{C}$ is contained in a proper minor closed class. The expansion function of the class of all k-regular graphs is exponential. Actually, any integral non decreasing function greater than 2 is the expansion function of some bounded expansion class (see Exercise 5.1).

A leitmotiv of this book are various equivalent formulations of our classification. We shall see that this carry over to bounded expansion classes. We start with the following:

Proposition 5.5. *Let $\mathcal{C}$ be a class of graphs. Then the following conditions are equivalent:*

1. *$\mathcal{C}$ has bounded expansion,*
2. *For every integer* t, $\sup_{G \in \mathcal{C} \triangledown t} \chi(G) < \infty$,
3. *For every integer* t, $\sup_{G \in \mathcal{C} \widetilde{\triangledown} t} \chi(G) < \infty$,

Proof. This result follows immediately from Propositions 4.4 and 4.5. □

This gives a possibility to compare the notions of "nowhere dense" and the one of "bounded expansion":

Nowhere dense classes are characterized by the property $\omega(\mathcal{C} \mathbin{\triangledown} t) < \infty$ for every t;
Bounded expansion classes are characterized by the stronger assumption that $\chi(\mathcal{C} \mathbin{\triangledown} t) < \infty$ for every t.

Which nowhere dense classes fail to have bounded expansion? These are exactly those classes $\mathcal{C}$ where for some t_0 and every $t > t_0$ the class $\mathcal{C} \mathbin{\triangledown} t$ has unbounded chromatic number but bounded clique number. Such classes form a classical part of modern graph theory and we call them *Erdős classes* (see [163] for a seminal result in this direction). This notion is also justified by the following example:

Example 5.1. Let $\mathcal{E}$ be the class of graphs G such that $\Delta(G) \leq \mathrm{girth}(G)$.

It is easy to see that the class $\mathcal{E}$ is nowhere dense: assume that $K_n \in \mathcal{E} \,\widetilde{\nabla}\, t$. Then there exists a graph G in $\mathcal{E}$ with maximum degree at least $n-1$ and girth at most $3(2t+1)$, hence $\omega(\mathcal{E} \,\widetilde{\nabla}\, t) \leq 6t+4$.

The class $\mathcal{E}$ does not have bounded expansion as graphs in $\mathcal{E}$ have unbounded average degree (in particular $\mathcal{E} \,\nabla\, 0$ is not a class of degenerate graphs).

We now give yet another characterization of classes with bounded expansion, this time by controlling dense parts which may be useful for classes that are neither addable (that is, closed under disjoint unions) nor hereditary. This will be used in Sect. 14.1 (Chap. 14) and thus it may be skipped on first reading.

Proposition 5.6. *Let class $\mathcal{C}$ be a class of graphs. Then $\mathcal{C}$ has bounded expansion if and only if there exists functions $F_{\mathrm{ord}}, F_{\mathrm{deg}}, F_{\nabla}, F_{\mathrm{prop}} : \mathbb{R}^+ \to \mathbb{R}$ such that the following two conditions hold:*

$$\forall \epsilon > 0,\ \forall G \in \mathcal{C}, \quad |G| > F_{\mathrm{ord}}(\epsilon) \Longrightarrow \frac{|\{v \in G : d(v) \geq F_{\mathrm{deg}}(\epsilon)\}|}{|G|} \leq \epsilon$$

$$\forall r \in \mathbb{N},\ \forall H \subseteq G \in \mathcal{C}, \quad \widetilde{\nabla}_r(H) > F_{\nabla}(r) \Longrightarrow |H| > F_{\mathrm{prop}}(r)|G|$$

If the above conditions hold then for every r, $\widetilde{\nabla}_r(\mathcal{C})$ is finite and bounded by

$$2\max\left(F_{\mathrm{ord}}\left(\frac{F_{\mathrm{prop}}(r)}{(r+1)F_{\nabla}(r)}\right), F_{\mathrm{deg}}\left(\frac{F_{\mathrm{prop}}(r)}{(r+1)F_{\nabla}(r)}\right), (r+1)F_{\nabla}(r)\right).$$

Proof. Assume $\mathcal{C}$ has bounded expansion. Then the average degree of graphs in $\mathcal{C}$ is bounded by $2\nabla_0(\mathcal{C})$. Hence, for every $G \in \mathcal{C}$ and every integer $k \geq 1$,

$$\begin{aligned} 2\nabla_0(\mathcal{C}) \geq \frac{\sum_{i\geq 1} i\ |\{v \in G : d(v) = i\}|}{|G|} &= \frac{\sum_{i\geq 1} |\{v \in G : d(v) \geq i\}|}{|G|} \\ &\geq k\frac{|\{v \in G : d(v) \geq k\}|}{|G|}. \end{aligned}$$

Hence $\frac{|\{v\in G: d(v)\geq k\}|}{|G|} \leq \frac{2\nabla_0(\mathcal{C})}{k}$. Thus $F_{\mathrm{ord}}(\epsilon) = 0$ and $F_{\mathrm{deg}}(\epsilon) = \left\lceil \frac{2\nabla_0(\mathcal{C})}{\epsilon} \right\rceil$ suffice. The second property is straightforward: put $F_{\nabla}(r) = \widetilde{\nabla}_r(\mathcal{C})$ and $F_{\mathrm{prop}}(r) = 1$.

Now assume that the two conditions of Proposition 5.6 hold. Fix r. Let $G \in \mathcal{C}$ and let S be a subset of vertices of G of cardinality $t \leq \frac{F_{\mathrm{prop}}(r)}{(r+1)F_{\nabla}(r)} n$. Let $F_r(S)$ denote a vertex subset formed by adding paths of length at most $r+1$ with interior vertices in $V \setminus S$ and endpoints in S (not yet linked by a path), one by one until no path of length at most $r+1$ has interior vertices in $V \setminus S$ and endpoints in S. Then $|F(S)| \leq (r+1)F_{\nabla}(r)t$. Suppose not, and consider the set T of the first $(r+1)F_\nabla(r)t \leq F_{prop}(r)n$ vertices of $F(S)$. By definition the subgraph of G induced by T contains a $\leq r$-subdivision of a graph H of order t and size at least $\frac{|T\setminus S|}{r} = F_\nabla(r)t$. It follows that $\widetilde{\nabla}_r(G[T]) \geq F_\nabla(r)$ hence $|T| > F_{prop}(r)n$, a contradiction.

Let $D_0 = F_{deg}(\frac{F_{prop}(r)}{(r+1)F_\nabla(r)})$. Then for sufficiently big graphs G (of order greater than $N = F_{ord}(\frac{F_{prop}(r)}{(r+1)F_\nabla(r)}))$, $\frac{|\{v\in G: d(v)\geq D_0\}|}{|G|} < \frac{F_{prop}(r)}{(r+1)F_\nabla(r)}$. Let $D = \max(D_0, (r+1)F_\nabla(r))$. Now assume that there exists in G a $\leq r$-subdivision G' of a graph H with minimum degree at least D. As $|H|$ is the number of vertices of G' having degree at least D, we infer that $|H| \leq \frac{F_{prop}(r)}{(r+1)F_\nabla(r)}n$. It follows that $|G'| \leq (r+1)F_\nabla(r)|H|$ hence $D \leq \|H\|/|H| < (r+1)F_\nabla(r) \leq D$, a contradiction. It follows that $\widetilde{\nabla}_r(G) < 2D$. Hence, for every graph $G \in \mathcal{C}$ (including those of order at most N) we have $\widetilde{\nabla}_r(G) < 2\max(F_{ord}(\frac{F_{prop}(r)}{(r+1)F_\nabla(r)}),$ $F_{deg}(\frac{F_{prop}(r)}{(r+1)F_\nabla(r)}), (r+1)F_\nabla(r))$. □

5.6 Classes with Locally Bounded Expansion

For particular applications, it may be possible that some structural properties are needed only locally. For instance, several approximation algorithms have been developed [47] to solve NP-complete problems for planar graphs, based on the property that the tree-width of a planar graph is bounded by a function of its diameter. Such an idea popularized through the notion of *local tree-width* and classes with *bounded local tree-width.* Minor closed classes with bounded local tree-width have then been characterized by Eppstein [159]: a minor closed class $\mathcal{C}$ of graphs has bounded local tree-width if, and only if, it does not contain all apex graphs. In their study of first-order definable decision problems, Dawar et al. [111] generalized the notion of class with bounded local tree-width to the notion of class *locally excluding a minor*, allowing to make use of the rich and deep theory developed by Robertson and Seymour on graph minors. Similarly, Dvořák et al. [138] introduced the notion of class with locally bounded expansion to allow a local use of low tree-depth decompositions in the resolution of first-order definable decision problems (see Chap. 18). Formally, a class $\mathcal{C}$ of graphs is said to have *locally bounded expansion* if there exists function $f : \mathbb{N} \times \mathbb{N} \to \mathbb{R}^+$ such that $\nabla_r(G) \leq f(d,r)$ for every graph G that is the d-neighborhood of a vertex in a graph from $\mathcal{C}$.

The following follows immediately from the definition:

Corollary 5.1. *Assume that there exist a function* $f\mathbb{N} \times \mathbb{N} \to \mathbb{R}^+$ *such that for every graph* $G \in \mathcal{C}$ *and every integer* d, r *it holds*

$$\max_{v \in V(G)} \omega(B_d(G, v) \triangledown r) \leq f(d, r),$$

where $B_d(G, v)$ *is the subgraph of* G *induced by the vertices at distance at most* d *from* v *(equivalently, we assume that* $\mathcal{C}$ *is locally nowhere dense).*

Then $\mathcal{C}$ *is nowhere dense.*

The containment relations between the different kind of classes we discussed is displayed on Fig. 5.2.

5.7 A Historical Note on Connection to Model Theory

Implicitly, the notion of nowhere dense graphs appears early in a model theoretic context. This will be briefly explained in this section.

Shelah's classification theory programme for infinite model theory [435] makes in particular use of two key dividing lines, corresponding to the notions of stability and independence. A theory (that is: a set of formulas) is *stable* if it does not contain a first-order formula that codes an infinite linear order on a set of tuples. A theory has the *independence property* if it contains a formula that codes every subset of some infinite set, that is if it contains a formula with infinite *Vapnik-Chervonenkis dimension* (or *VC-dimension*), which is a key notion in computational learning theory [464]. The notions of stability and independence (or the opposite notion of *dependence*) can be translated to infinite classes $\mathcal{C}$ of finite graphs by considering the countable graphs formed as the disjoint union of the graphs in $\mathcal{C}$.

As noticed by Adler and Adler [2], the notion of nowhere dense class is then essentially the stability theoretic notion of superflateness introduced by Podewski and Ziegler [387] because of its connection to stability. A class $\mathcal{C}$ of (non-necessarily finite) graphs is *superflat* if for every r there is an m such that the r-subdivision of K_m is not a subgraph of a graph in $\mathcal{C}$. By an easy application of Ramsey theorem (Theorem 3.18) it is easily seen that a class $\mathcal{C}$ of finite graphs is superflat if and only if it is nowhere dense.

Adler and Adler [2] formalized this by the following:

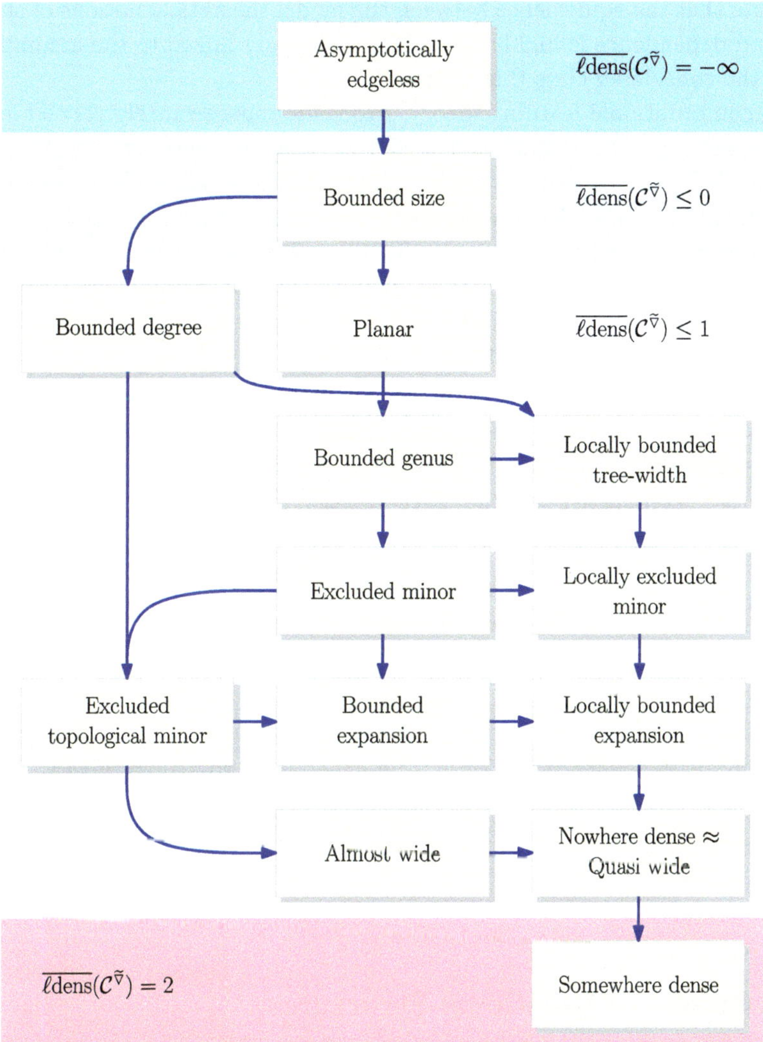

Fig. 5.2 The trichotomy of infinite classes

Theorem 5.5. *Let $\mathcal{C}$ be a monotone class of finitely colored digraphs and let $\underline{\mathcal{C}}$ be the class of the underlying undirected graphs. Then the following eight model theoretical conditions are equivalent:*

1. $\underline{\mathcal{C}}$ is nowhere dense,
2. $\underline{\mathcal{C}}$ is superflat,
3. $\underline{\mathcal{C}}$ is stable,
4. $\underline{\mathcal{C}}$ is dependent,
5. $\mathcal{C}$ is nowhere dense,
6. $\mathcal{C}$ is superflat,
7. $\mathcal{C}$ is stable,
8. $\mathcal{C}$ is dependent.

Note that the equivalence between the model theoretical notions of stability and dependence is highly unusual and strongly linked to the assumption that the considered class $\mathcal{C}$ is monotone.

In computational learning theory, *Probably approximately correct* learning model (or *PAC* model) is a model for machine learning introduced by Valiant [463]. As a corollary of Theorem 5.5, the authors of [2] obtain the following connection to computational learning theory, by making use of the standard connection between VC-dimension and sample complexity in the PAC model (see [232], where an analogous result is obtained, which connects finite VC-dimension of monadic second order formulas on a monotone class to bounded tree-width):

Corollary 5.2. *Let $\mathcal{C}$ be a monotone class of graphs. Every concept class definable in first-order logic on $\mathcal{C}$ has bounded sample complexity in the PAC model if and only if $\mathcal{C}$ is nowhere dense.*

However, Theorem 5.5 does not immediately generalize to relational structures. Actually, the following open problem is posed in [2]:

Problem 5.1. Can Theorem 5.5 be generalized to arbitrary relational structures with finite signatures?

5.8 Classes of Relational Structures

How to do with relational structures and hypergraphs?

Let $\mathcal{C}$ be an infinite class of relational structures, and let Gaifman($\mathcal{C}$) denote the class of the Gaifman graphs of the structures in $\mathcal{C}$. We say that the class $\mathcal{C}$ is *G-nowhere dense* (resp. *G-somewhere dense*) if Gaifman($\mathcal{C}$) is nowhere dense (resp. somewhere dense). Also, we can follow the same to define the *G-bounded expansion* classes of structures as the classes of structures $\mathcal{C}$ such that Gaifman($\mathcal{C}$) has bounded expansion.

This definition leads to a trichotomy for classes of relational structures, as in Sect. 5.4. Also other characterizations of nowhere dense classes of graphs can be directly translated to classes of nowhere dense structures.

However this direct transposition of results about nowhere dense classes to σ-structures are just results about classes of underlying Gaifman graphs. Nevertheless such results still have several applications in logic and model theory (cf. [42]). But of course they do not capture the real complexity of higher arities (a triple system with only $O(n^2)$ triples may lead to a complete graph of order n). Also the direct computation of edge densities for systems of k-tuples leads to no trichotomy but rather to $k+1$ possibilities. This will follow from the results in Chap. 12.

A second approach stands in considering the incidence graphs of relational structures. It is immediate that two relational structures have the same Gaifman graph if they have the same incidence graph, but that the converse does not hold in general. For a class of relational structures $\mathcal{C}$, denote by $\mathrm{Inc}(\mathcal{C})$ the class of all the incidence graphs $\mathrm{Inc}(\mathbf{A})$ of the relational structures $A \in \mathcal{C}$. (see Sect. 3.8.3). We say that a class $\mathcal{C}$ is *I-nowhere dense* (resp. *I-somewhere dense*) if $\mathrm{Inc}(\mathcal{C})$ is nowhere dense (resp. somewhere dense), and say that $\mathcal{C}$ has *I-bounded expansion* if $\mathrm{Inc}(\mathcal{C})$ has bounded expansion. We have the following result, which shows the equivalence of these encodings in the most frequent cases:

Proposition 5.7. *Assume that the arities of the relational symbols in σ are bounded, and let $\mathcal{C}$ be an infinite class of σ-structures. Then*

- *$\mathcal{C}$ is G-somewhere dense if and only if $\mathcal{C}$ is I-somewhere dense,*
- *$\mathcal{C}$ is G-nowhere dense if and only if $\mathcal{C}$ is I-nowhere dense,*
- *$\mathcal{C}$ has G-bounded expansion if and only if $\mathcal{C}$ has I-bounded expansion.*

Proof. First notice that the unary relations have obviously no repercussion on the classifications hence we can assume without loss of generality that all the relations will have arity at least 2. Let k be the maximum arity of a relational symbol in σ. Let $\mathbf{A} \in \mathcal{C}$.Let $F(\mathbf{A}) = \mathrm{Inc}(A) \bullet \overline{K}_{\binom{k}{2}}$. Because of Proposition 4.6, and using the characterizations of nowhere dense classes and bounded expansion classes bases on top-grads (Theorem 5.4 and item (6) of Theorem 13.2) we get that the class $\mathcal{F} = \{F(\mathbf{A})\}$ is nowhere dense (resp. has bounded expansion) if and only if the class $\mathrm{Inc}(\mathcal{C})$ is nowhere dense (resp. has bounded expansion). Also, $\mathrm{Gaifman}(\mathbf{A}) \in \mathcal{F}(\mathbf{A}) \,\widetilde{\triangledown}\, \frac{1}{2}$ hence if $\mathcal{F}$ is nowhere dense (resp. has bounded expansion) then $\mathrm{Gaifman}(\mathcal{C})$ is also nowhere dense (resp. has also bounded expansion). Last, it is easily checked that for every integer r we have $\nabla_r(\mathrm{Gaifman}(\mathbf{A})) \geq \nabla_r(\mathrm{Inc}(\mathbf{A}))$ hence $\mathrm{Gaifman}(\mathcal{C})$ is nowhere dense (resp. has bounded expansion) then $\mathrm{Inc}(\mathcal{C})$ is also nowhere dense (resp. has also bounded expansion). Altogether, we obtain the requested equivalences (the equivalence of G-somewhere dense and I-somewhere dense following from the equivalence of G-nowhere dense and I-nowhere dense). □

If one consider hypergraphs instead of relational structures, the assumption that the size of the edges is bounded may become irrelevant, in which case the classification by Gaifman graphs and incidence graphs do not coincide anymore.

Exercises

5.1. Prove that for every non-decreasing function $f : \mathbb{N} \to \mathbb{N}$ such that $f(0) \geq 2$, there exists a bounded expansion class $\mathcal{C}_f$ such that

$$\forall r \in \mathbb{N}, \qquad \nabla_r(\mathcal{C}_f) = f(r).$$

5.2. Let $f : \mathbb{N} \to \mathbb{R}^+$ be such that $\lim_{n\to\infty} f(n) = 0$ (i.e. $f(n) = o(1)$).

Prove that there exists a nowhere dense class $\mathcal{C}_f$ such that for every integer n there exists a graph $G_n \in \mathcal{C}_f$ such that

$$|G_n| \geq n \qquad \text{and} \qquad \|G_n\| \geq |G_n|^{1+f(|G_n|)}.$$

5.3. The *limiting density* of a minor closed class $\mathcal{C}$ is $\nabla(\mathcal{C})$. The set of possible values of $\nabla(\mathcal{C})$ (while considering all possible minor closed classes of graphs) has been investigated by Eppstein in [161].

- Prove that for every real $a \geq 0$ the class $\mathcal{C}_a = \{G, \nabla(G) \leq a\}$ is minor closed;
- Using Robertson-Seymour Theorem 5.1, prove that there are only countably many minor closed classes, hence countably many possible limiting densities;
- Robertson and Seymour [418] proved that the immersion quasi-order is also a well-quasi-ordering. Deduce that for every real $a \geq 0$ there exists $\epsilon > 0$ such that not limiting density belongs to the open interval $[a; a+\epsilon]$;
- Deduce that the set of the possible limiting values is well-ordered.

5.4. Prove that the results of Exercise 5.3 extend, when considering immersions and $\overset{\propto}{\nabla}$ instead of minors and ∇.

5.5. Galluccio et al. [211] proved that for any fixed graph H, H-minor free graphs with high enough girth has a circular-chromatic number arbitrarily close to 2. The aim of this Exercise is to prove a similar results for graphs with sub-exponential expansion.

A graph G is p-*path degenerate* if there is a sequence $G = G_0, G_1, \ldots, G_t$ of subgraphs G where G_t is bipartite and for each $1 \leq i \leq t$, G_{i-1} is obtained from G_i by attaching a path of length at least p (which is then a *handle* of length p in G_{i-1}).

Let $\mathcal{C}$ be a class of graphs such that

$$\lim_{r\to\infty} \frac{\log \nabla_r(\mathcal{C})}{r} = 0.$$

- Prove that if P is a handle of length p in a graph G and G′ is the graph obtained from G by deleting the internal vertices of G then for every odd-integer $l \leq p+1$ it holds

$$G \to C_l \qquad \Longleftrightarrow \qquad G' \to C_l.$$

- Deduce that if G is p-path degenerate then $G \to C_l$ for every odd integer $l \leq p+1$ hence
$$\chi_c(G) \leq 2 + \frac{1}{\lfloor p/2 \rfloor}.$$
- Let $G \in \mathcal{C}$ be a graph of girth at least $k =$. Assume that G does not contain a handle of length $p-1$. Then there exists $H \in G \triangledown \frac{p-2}{2}$ with minimum degree 3 and girth at least $k/(p-1)$. Deduce a contradiction using Exercise 4.2 and Proposition 4.1.

Chapter 6
Bounded Height Trees and Tree-Depth

Where the reader will discover that short branches can make large trees beautiful.

After treating graph classes and class resolutions we return to the basics: the structure of finite trees as the true measure of our things.

6.1 Definitions and Basic Properties

Tree-depth is of particular importance to us. The original non-recursive definition [352] reads as follows:

Definition 6.1. The *tree-depth* $\mathrm{td}(G)$ of a graph G is the minimum height of a rooted forest F such that $G \subseteq \mathrm{clos}(F)$ (see Fig. 6.1).

Here, a *rooted forest* is a disjoint union of rooted trees. The *height* of a vertex x in a rooted forest F is the number of vertices of a path from the root (of the component of F to which x belongs) to x and is noted $\mathrm{height}(x, F)$. Thus $\mathrm{height}(x, F)$ is one more than the length of the path from x to the root of F. The *height* of F is the maximum height of the vertices of F. Let x, y be vertices of F. The vertex x is an *ancestor* of y in F if x belongs to the path of F linking y to the corresponding root. The *closure* $\mathrm{clos}(F)$ of a rooted forest F is the graph with vertex set $V(F)$ and edge set $\{\{x, y\} : x \text{ is an ancestor of } y \text{ in } F, x \neq y\}$ (actually, in model theory a tree usually means a relation which is the closure of a tree). A rooted forest F defines a partial order on its set of vertices: $x \leq_F y$ if x is an ancestor of y in F. The comparability graph of this partial order is obviously $\mathrm{clos}(F)$.

J. Nešetřil and P.O. de Mendez, *Sparsity*, Algorithms and Combinatorics 28,
DOI 10.1007/978-3-642-27875-4_6,

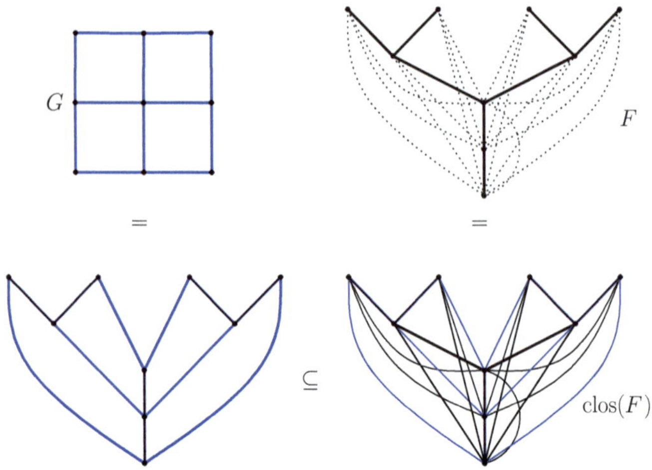

Fig. 6.1 The tree-depth of the 3×3 grid is 4

The *tree-depth* of a graph is an intuitively easy and very useful invariant. It has been defined in [352], but equivalent or similar notions include the *rank function* (used for the analysis of countable graphs, see e.g. [368]), the *vertex ranking number*, the minimum height of an *elimination tree* [72, 117, 431], etc. (see Sect. 6.3) Recall that an *elimination tree* for a connected graph G is a rooted tree Y with vertex set $V(\mathrm{G})$ defined recursively as follows: If $V(\mathrm{G}) = \{x\}$ then Y is just $\{x\}$. Otherwise a vertex $r \in V(\mathrm{G})$ is chosen as the root of Y and the branches of Y at r are the elimination trees of the connected components of $\mathrm{G} - r$ (whose roots are the sons of r in Y). For instance, every depth-first search tree of a connected graph G is an elimination tree for G. Thus the minimum height of an elimination tree for a connected graph G is at most equal to the minimum height of a depth-first search tree of G.

Tree-depth can also be seen as an analog for undirected graphs of the cycle rank defined by Eggan [144], which will be introduced in Sect. 6.6. The cycle-rank of a digraph is a parameter relating digraph complexity to other areas such as regular language complexity and asymmetric matrix factorization.

The equivalence between tree-depth and minimum height of an elimination tree is not difficult to establish:

Lemma 6.1. *Let* G *be a connected graph. A rooted tree* Y *is an elimination tree for* G *if and only if* $\mathrm{G} \subseteq \mathrm{clos}(\mathrm{Y})$. *Hence* $\mathrm{td}(\mathrm{G})$ *is the minimum height of an elimination tree for* G.

Proof. We prove the lemma by induction on the order of G. The statement is true if $V(\mathrm{G}) = \{x\}$. Otherwise let r be the root of Y and let $\mathrm{G}_1, \ldots, \mathrm{G}_p$

be the connected components of $G - r$. Then Y is an elimination tree for G if and only if the connected components of $Y - r$ may be labeled $Y_1, \dots, Y_p$ in such a way that, for any $1 \le i \le p$, Y_i is an elimination tree for G_i. By induction, this is equivalent to the existence of a labeling $Y_1, \dots, Y_p$ for each of the connected components of $G - r$. It follows that $G_i \subseteq \mathrm{clos}(Y_i)$, for any $1 \le i \le p$ which is equivalent to $G \subseteq \mathrm{clos}(Y)$. □

It follows that the tree-depth could be alternatively defined by the following recursive formula:

$$\mathrm{td}(G) = \begin{cases} 1, & \text{if } |G| = 1; \\ 1 + \min_{v \in V(G)} \mathrm{td}(G - v), & \text{if } G \text{ is connected and } |G| > 1; \\ \max_{i=1}^{p} \mathrm{td}(G_i), & \text{otherwise;} \end{cases} \qquad (6.1)$$

(where $G_1, \dots, G_p$ are the connected components of G). This recursive definition can be used to design an easy dynamic algorithm deciding $\mathrm{td}(G) \le k$ (for fixed k) in time $O(|G|^k)$.

6.2 Tree-Depth, Minors and Paths

One of the fundamental properties of tree-depth is its monotonicity according to the operation of taking a minor:

Lemma 6.2. *If* H *is a minor of* G *then* $\mathrm{td}(H) \le \mathrm{td}(G)$.

Proof. Let F be a rooted forest of height $\mathrm{td}(G)$ such that $G \subseteq \mathrm{clos}(F)$ and let e be an edge of G. Observe that both $G - e$ and G/e are subgraphs of $\mathrm{clos}(F)$. □

It follows that the class of graphs with tree-depth at most k is minor closed hence there exists a finite set of minor obstructions for $\mathrm{td}(G) \le k$. A study of these obstructions may be found in [134, 216]. In particular, it is known that exactly $\frac{1}{2}2^{2^{k-1}-k}(1 + 2^{2^{k-1}-k})$ of these obstructions are trees. However we shall show more in Sect. 6.8: the class of graphs with tree-depth at most k can be described by finitely many forbidden subgraphs (not just minors).

The tree-depth of a path P_n of order n is logarithmic in n (see Fig. 6.2). More precisely: The tree-depth of a path of order n is

$$\mathrm{td}(P_n) = \lceil \log_2(n+1) \rceil \qquad (6.2)$$

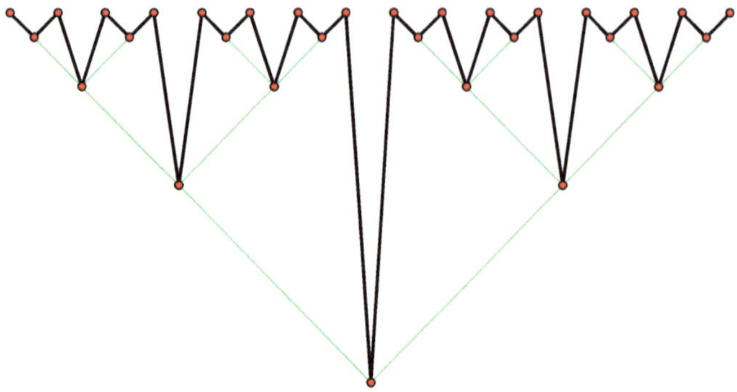

Fig. 6.2 The tree-depth of a path is logarithmic in the order of the path

Indeed, $\mathrm{td}(P_1) = 1$ and

$$\begin{aligned} \mathrm{td}(P_n) &= \min_{1 \le i \le n-2} (1 + \max(\mathrm{td}(P_i), \mathrm{td}(P_{n-1-i}))) && \text{by (6.1)} \\ &= 1 + \mathrm{td}(P_{\lceil n/2 \rceil}) && \text{(as td is monotone)} \end{aligned}$$

hence the result follows from easy induction. The relationship with path inclusion gives us another characterization of bounded tree-depth:

Proposition 6.1. *Assume* G *includes no path of order greater than* n. *Then the tree-depth of* G *is at most* n.

Proof. According to (6.1), we may restrict our attention to the component of G including a longest path. Hence we may assume without loss of generality that G is connected. We construct an elimination tree by performing a depth-first search on G. As G has no path of length greater than n, the height of a depth-first search tree of G is at most n hence the tree-depth of G is at most n. □

Notice that the complete graph shows that this bound is tight. Also, we see that the relationship between tree-depth and the order of a longest path of a graph is analogous (but easier) to the relation of the tree-width and the size of a largest grid in a graph [399].

Also, we relate the tree-depth of a biconnected graph to the length of its longest cycles.

Proposition 6.2. *Let* G *be a biconnected graph, and let* L *be the length of a longest cycle of* G. *Then*

$$1 + \lceil \log_2 L \rceil \le \mathrm{td}(G) \le 1 + (L-2)^2. \tag{6.3}$$

Proof. As $\mathrm{td}(C_L) = 1 + \mathrm{td}(P_{L-1}) = 1 + \lceil \log_2 L \rceil$, the first inequality follows. Now consider a DFS traversal of the graph G and the corresponding rooted DFS-tree Y. Let $x_1, \ldots, x_h$ be a longest tree branch starting from the root x_1 of Y. As $G \subseteq \mathrm{clos}(Y)$ we have $\mathrm{td}(G) \leq \mathrm{height}(Y) = h$. We construct two decreasing sequences $h = a_1 > a_2 > \cdots > a_k = 1$ and $b_1 > b_2 > \cdots > b_k = 1$ as follows: Let $a_1 = h$. As G is biconnected, x_h has degree at least two thus there exists $j < h-1$ such that x_i is adjacent to x_h. Let b_1 be the minimum such j. Assume $b_1 \neq 1$. Then, as G is biconnected, there exists i such that $b_1 < i < a_1$ such that x_i is adjacent to some x_j with $j < b_1$. Choose such a pair (i, j) with j minimum and then (according to this choice of j) i minimal. Let $a_2 = i$ and $b_2 = j$. Again, if $b_2 \neq 1$, there exists i such that $b_2 < i < a_2$ such that x_i is adjacent to some x_j with $j < b_2$. We choose such a pair (i, j) with j minimum and then i minimal and we let $a_3 = i$ and $b_3 = j$. This way we define iteratively the sequences $a_1, \ldots, a_k$ and $b_1, \ldots, b_k$ and we stop when $b_k = 1$. Let γ be the cycle (we assume k odd but the construction is similar for k even) formed by the edge $\{a_1, b_1\}$, the (possibly empty) tree chain between b_1 and a_3, the edge $\{a_3, b_3\}$, the (possibly empty) tree chain between b_3 and a_5, ..., the edge $\{a_k, b_k\}$, the (non empty) tree chain between b_k and b_{k-1}, the edge $\{b_{k-1}, a_{k-1}\}$, the (possibly empty) tree chain between a_{k-1} and $b_{k-3}, \ldots$, the edge $\{b_2, a_2\}$ and the (non empty) tree chain between a_2 and a_1. The length of γ is at least $k+2$. Hence $k+2 \leq L$. Moreover, as $a_i - b_i \leq L - 1$ we have (by construction) $h \leq 1 + k(L-2)$ thus $h \leq 1 + (L-2)^2$. □

We view the graphs with bounded tree depth as the building blocks of our decomposition theorems and thus we investigate the notion of tree-depth in a greater detail. Particularly we give several characterizations. The first one is in terms of a separation property.

Proposition 6.3. *Let $\mathcal{C}$ be a hereditary class of graphs. Then the two following properties are equivalent:*

(i) There is an integer s and a mapping $N : \mathbb{N} \to \mathbb{N}$ such that for every integer p and every graph $G \in \mathcal{C}$ of order at least $N(p)$ there exists a subset S of (at most) s vertices of G such that $G - S$ has at least p connected components.
(ii) $\mathcal{C}$ has bounded tree-depth

Proof. Assume $\mathcal{C}$ has tree-depth at most t and let $N(p) = p^{t+1}$. Let p be an integer and let $G \in \mathcal{C}$ be a graph of order at least $N(p)$. As $\mathrm{td}(G) \leq t$ there exists a rooted forest F such that $G \subseteq \mathrm{clos}(F)$. If F has at least p connected components, we are done. Otherwise, some component Y of F has order $|Y| > N(p)/p$ and thus includes a vertex x of degree at least $p+1$. Let

S be the set including x and all its ancestors in Y. Then $G - s$ has at least p connected components thus proving (i).

Conversely, assume that (i) holds and assume for contradiction that $\mathcal{C}$ does not have tree-depth at most $N(s+2)$. Let $G \in \mathcal{C}$ be a graph of tree-depth at least $N(s+2)+1$. According to Lemma 6.1 G includes a path of order at least $N(s+2)+1$. Let A be the vertex set of this path. Consider the subgraph $G[A]$ of G induced by A. Then $|G[A]| > N(s+2)$ hence by (i) there exists $S \subseteq A$ such that $|S| \le s$ and $G[A] - S$ has at least $s+2$ connected components. Then this later property has to be true for the Hamiltonian path of $G[A]$, although a path cannot be cut into $s+2$ connected components by removing at most s vertices, a contradiction. We infer that $\mathrm{td}(\mathcal{C}) \le N(s+2)$. □

We shall now give a characterization of classes with bounded tree-depth in terms of induced paths. Particularly, we generalize Proposition 6.1, which applies to classes of degenerate graphs. We need the following two lemmas which are interesting in themselves and maybe part of graph theory folklore.

Lemma 6.3. *Let G be a graph, let $d \in \mathbb{N}$ and let $P = (x_1, \dots, x_L)$ be a path of G of order L. Assume that each x_i is adjacent to at most d vertices x_j with $j < i-1$. Then the graph $G[P]$ includes an induced path $(x_{i_1}, \dots, x_{i_l})$ of order l where $1 \le i_1 < i_2 < \dots < i_l = L$ and*

$$l \ge \frac{\log(dL+1)}{\log(d+1)}$$

Proof. We prove the lemma by induction on $\lceil \frac{\log(dL+1)}{\log(d+1)} \rceil$ (see Fig. 6.3). If $L = 1$ then $\lceil \frac{\log(dL+1)}{\log(d+1)} \rceil = 1$ and P is an induced path of G; if $1 < L \le d+2$, then $\lceil \frac{\log(dL+1)}{\log(d+1)} \rceil = 2$ and (x_{L-1}, x_L) is an induced path of order 2 of G.

Assume the lemma has been proved for all d, L with $\lceil \frac{\log(dL+1)}{\log(d+1)} \rceil \le c$ for some integer c and assume $\lceil \frac{\log(dL+1)}{\log(d+1)} \rceil = c+1$. By assumption, x_L is adjacent to $p \le d$ vertices $x_{j_1}, \dots, x_{j_p}$ with $1 \le j_1 < j_2 < \dots < j_p < L-1$. Define $j_0 = 0$ and $j_{p+1} = L-1$. Let $0 \le a \le p$ be such that $j_{a+1} - j_a$ is maximal. Then $j_{a+1} - j_a \ge (L-1)/(p+1) \ge (L-1)/(d+1)$. Let P' be the subpath of P with vertices $(x_{j_a+1}, x_{j_a+2}, \dots, x_{j_{a+1}})$. By induction, P' includes an induced subpath P'' $(x_{j_{i_1}}, \dots, x_{j_{i_q}})$ with

$$q \geq \left\lceil \frac{\log(d(L-1)/(d+1)+1)}{\log(d+1)} \right\rceil$$
$$= \left\lceil \frac{\log((dL+1)/(d+1))}{\log(d+1)} \right\rceil$$
$$= \left\lceil \frac{\log(dL+1)}{\log(d+1)} \right\rceil - 1$$
$$= c.$$

and $j_{i_1} < j_{i_2} < \cdots < j_{i_q} = j_{a+1}$. As no vertex of P'' but $x_{j_{a+1}}$ is adjacent to x_L we deduce that $(x_{j_{i_1}}, \ldots, x_{j_{i_q}}, x_L)$ is an induced path of G of length $c+1$. □

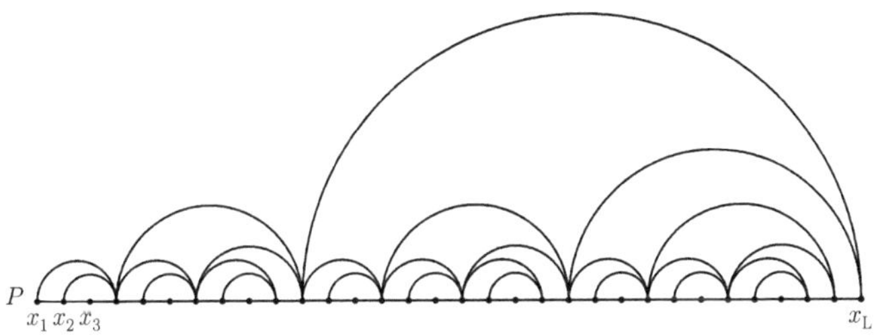

Fig. 6.3 Illustration for Lemma 6.3

The next lemma gives a more concise information about induced paths in degenerate graphs containing long (not necessarily induced) paths.

Lemma 6.4. *Let* G *be a graph and let* $P = (x_{i_1}, \ldots, x_{i_l})$ *be a path of* G *of order* L. *Then* G[P] *includes an induced path of* G *of order* l *where:*

$$l \geq \frac{\log \log L}{\log(\lceil \nabla_0(G) \rceil + 1)}$$

Proof. We first prove the following: let $P = (x_1, \ldots, x_L)$ and let $d = \lceil \nabla_0(G) \rceil$. Then G has an orientation with maximum indegree at most d. Consider the subgraph G_1 of G obtained from the subgraph of G induced by the vertex set of P by deleting all the arcs oriented from x_j to x_i with $1 \leq i \leq j-2 \leq L-2$. According to Lemma 6.3, there exists $1 \leq i_1 < i_2 < \cdots < i_p \leq L$ such that $(x_{i_1}, \ldots, x_{i_p})$ is an induced path P' of G_1 and $p = \lceil \frac{\log(dL+1)}{\log(d+1)} \rceil$. In G_1, all the edges linking non-consecutive vertices of P' are oriented from the lower index vertex to the higher index vertex. Consider the (reversed) path $\overline{P'} = (v_1, \ldots, v_p)$ where $v_j = x_{i_{p+1-j}}$. Then we can apply Lemma 6.3 to $\overline{P'}$ to obtain an induced path P'' of G of length $\left\lceil \frac{\log(d\lceil\frac{\log(dL+1)}{\log(d+1)}\rceil+1)}{\log(d+1)} \right\rceil \geq \frac{\log\log L}{\log(d+1)}$.

□

From the last lemma we immediately deduce:

Proposition 6.4. *Let $\mathcal{C}$ be a class of graphs. Then $\mathcal{C}$ has bounded tree-depth if and only if $\mathcal{C}$ is degenerate and graphs in $\mathcal{C}$ exclude some path P_n as an induced subgraph.*

Despite its elementary setting the following problem seems be interesting and seems to be not yet investigated.

Problem 6.1. What is the largest function $f : \mathbb{N} \times \mathbb{N} \to \mathbb{N}$ such that every graph G which contains a path of length L also contains an induced path of length $f(L, \nabla_0(G))$?

The tree-depth of a graph is a natural invariant which appears in many situations. We consider a few more equivalences which document this. These equivalences are not just interesting per se. In fact they present miniatures of our general results for nowhere dense classes.

6.3 Compact Elimination Trees and Weak-Coloring

As we noticed, the tree-depth of a graph G equals the minimum height of an elimination tree for G. This property can be strengthened by showing that one may demand that the elimination tree has some additional property. Say that an elimination tree F of a connected graph G is *compact* if for every non-leaf vertex v and every son w of v, the subgraph $T_{v \to w}$ of G induced by v and the vertex set of the branch of F rooted at w is connected. The equality of the tree-depth of a connected graph G and the minimum height of a compact elimination tree of G follows directly from (6.1). However, an immediate property of a compact elimination tree F of a connected graph G is that for every vertex x and every ascendant y of x (i.e. $y <_F x$) there exists a path P in G with endpoints x and y whose internal vertices are descendants of y (consider the son z of y such that $x \in T_{y \to z}$ and use the connectivity of the subgraph induced by $V(T_{y \to z})$). This is the property which we will use to make the connection between tree-depth and weak colorings (see Sect. 4.9 for the definition of weak coloring numbers).

Lemma 6.5. *For every non-empty connected graph* G *we have the equality*

$$\mathrm{td}(G) = \mathrm{wcol}_\infty(G). \tag{6.4}$$

Proof. Consider a linear extension $<$ of $<_F$ for a compact elimination tree F. Then, for every path P with endpoints x and y such that $y = \min P$ (with respect to the order $<$) we have that y is also smaller than all the vertices of P with respect to $<_F$. According to the properties of compact elimination trees, we deduce that $\mathrm{td}(G) \geq 1 + \min_< \max_x \{y \neq x : \exists x\text{–}y \text{ path } P \subseteq G, \min P = y\}$. In order to prove the reverse inequality, consider a linear order on $V(G)$ for which the value $\max_x \{y \neq x : \exists x\text{–}y \text{ path } P \subseteq G, \min P = y\}$ is minimum. Then build an elimination tree using this linear order as an elimination order. As G belongs to the closure of this elimination tree we deduce $\mathrm{td}(G) \leq 1 + \max_x \{y \neq x : \exists x\text{–}y \text{ path } P \subseteq G, \min P = y\}$. Hence $\mathrm{td}(G) = \mathrm{wcol}_\infty(G)$. □

6.4 Tree-Depth, Tree-Width and Vertex Separators

The following inequalities between tree-width, pathwidth and tree-depth are easy to prove [73]: for any graph G holds

$$\mathrm{tw}(G) \leq \mathrm{pw}(G) \leq \mathrm{td}(G) - 1.$$

Let G be a graph of order n. Recall that, for $0 < \alpha < 1$, an *α-vertex separator* of G is a subset S of vertices such that every connected component of $G - S$ contains at most αn vertices.

Lemma 6.6. *Let* G *be a graph of order* n *and let* $s_G : \{1, \ldots, n\} \to \mathbb{N}$ *be defined by*

$$s_G(i) = \max_{\substack{|A| \leq i, \\ A \subseteq V(G)}} \min\{|S| : S \textit{ is a } \tfrac{1}{2}\textit{-vertex separator of } G[A]\}$$

Then:

$$s_G(n) \leq \mathrm{td}(G) \leq \sum_{i=0}^{\log_2 n} s_G\left(\frac{n}{2^i}\right)$$

Proof. The first inequality is easy: if G is a graph of order n which is a subgraph of the closure of rooted tree Y of height $td(G)$, it is possible to separate G into parts of size at most $n/2$ by deleting the vertices of a path of Y: consider a traversal of Y and let $v_1, \dots, v_n$ be the vertices of G ordered by this traversal (v_1 is the root of Y). Let k be the smallest index of a leaf of Y which is a descendant of $x_{\lceil n/2 \rceil}$. Delete the chain from the root of Y to x_k (thus at most $td(G)$ vertices). In the obtained graph, no vertex of index smaller than or equal to $x/2$ is adjacent to a vertex of index greater than $x/2$. Hence $s_G(n) \le td(G)$.

We prove the other inequality of the lemma by induction on n. The lemma is straightforward if $n = 1$. Assume the lemma has been proved for graphs of order at most $n - 1$.

By definition of s_G, G has a $\frac{1}{2}$-vertex separator S of size at most $s_G(n)$. Let $G_1, \dots, G_p$ be the connected components of $G - S$. Then, according to (6.1) and the fact that the function s_{G_i} corresponding to G_i is obviously bounded by s_G:

$$td(G) \le |S| + \max_i td(G_i) \le s_G(n) + \sum_{i=1}^{\log_2(n/2)} s_G\left(\frac{n/2}{2^i}\right) \le \sum_{i=0}^{\log_2 n} s_G\left(\frac{n}{2^i}\right)$$

□

Corollary 6.1. *For any connected graph* G *of order* n, $td(G) \le (tw(G)+1)\log_2 n$.

Proof. It is proved in [400] that any graph of tree-width at most k has a $\frac{1}{2}$-vertex separator of size at most $k+1$. Hence $s_G(i) \le tw(G)+1$ for all i. The result follows. □

Notice that this result is optimal for tree-depth, as shown by the example of paths of length n: $tw(P_n) = 1$, but $td(P_n) = \lceil \log_2(n+1) \rceil$.

Proposition 6.5. *Every graph* G *of order* n *with no minor isomorphic to* K_h *has tree-depth at most* $(2+\sqrt{2})\sqrt{h^3 n}$.

Proof. It is proved in [25] that a graph of order i with no K_h minor has a $\frac{1}{2}$-vertex separator of size at most $\sqrt{h^3 i}$. Hence $s_G(i) \le \sqrt{h^3 i}$ and:

$$td(G) \le \sum_{i=0}^{\log_2 n - 1} s_G\left(\frac{n}{2^i}\right) \le \sqrt{h^3 n} \sum_{i=0}^{\log_2 n - 1} \left(\frac{1}{\sqrt{2}}\right)^i \le (2+\sqrt{2})\sqrt{h^3 n}$$

□

Generally, we have (as a direct consequence of Lemma 6.6):

Corollary 6.2. *Let* $0 < \alpha < 1$ *and let* $\mathcal{C}$ *be an hereditary class of graphs such that each graph* $G \in \mathcal{C}$ *of order* n *has tree-width at most* Cn^{α}.

Then, every graph $G \in \mathcal{C}$ *of order* n *has tree-depth at most* $\frac{C}{1-2^{-\alpha}} n^{\alpha}$.

As we have seen, the notion of the tree-depth is closely connected to the tree-width. Consider the Erdős-Rényi model $G(n, p(n))$ (see also Sect. 14.1) for random graph. A random graph $G \in G(n, p(n))$ has n vertices and every pair of vertices is chosen independently to be an edge with probability $p(n)$. We say that a property P holds *asymptotically almost surely* (*a.s.s.*) for random graphs $G \in G(n, p(n))$ if $\lim_{n\to\infty} \Pr(G \text{ has } P) = 1$.

Perarnau and Serra proved the following bounds for the tree-depth of random graphs [381]:

Theorem 6.1. *Let* $G \in G(n, p(n))$.

- *If* $p(n) = \omega(n^{-1})$ *then a.a.s.* $\mathrm{td}(G) = n - o(n)$;
- *If* $p(n) = c/n$ *with* $c > 0$:
  - *if* $c < 1$, *then a.a.s.* $\mathrm{td}(G) = \Theta(\log\log n)$;
  - *if* $c = 1$, *then a.a.s.* $\mathrm{td}(G) = \Theta(\log n)$;
  - *if* $c > 1$, *then a.a.s.* $\mathrm{td}(G) = \Theta(n)$.

Note that the last part of this theorem gives an alternative proof of a conjecture of Kloks [277] on the linear behaviour of tree-width for random graphs with $c > 1$ (originally proved by Lee et al. [300]). Related results can be found in [74].

6.5 Centered Colorings

Definition 6.2. A *centered coloring* of a graph G is a vertex coloring such that, for any (induced) connected subgraph H, some color $c(H)$ appears exactly once in H (see Fig. 6.4).

Note that a centered coloring is necessarily proper. We relate the minimum number of colors in a centered coloring to yet another coloring notion, the vertex ranking number of a graph which has been investigated in [117, 431]: The *vertex ranking* (or *ordered coloring*) of a graph is a vertex coloring by a linearly ordered set of colors such that for every path in the graph with end vertices of the same color there is a vertex on this path with a higher color. A vertex-coloring $c : V(G) \to \{1, \ldots, t\}$ with this property is a *vertex*

Fig. 6.4 A centered coloring of the $(2^n - 1) \times (2^n - 1)$ grid using $2^{n+1} - 1$ colors

t-*ranking* of G. The minimum t such that G has a vertex t-ranking is the *vertex ranking number* of G and is noted $\chi_{rk}(G)$ (see [117, 431]).

Lemma 6.7. *Any vertex ranking is a centered coloring and conversely any centered coloring defines a vertex ranking with the same number of colors. Thus* $\chi_{rk}(G)$ *is the minimum number of colors in a centered coloring of* G.

Proof. Assume c is a vertex ranking of a graph G and let H be a connected subgraph of G. Let $i = \max_{v \in V(H)} c(v)$. Then H has at most one vertex colored i for otherwise the path linking them would include a vertex with color $j > i$.

Conversely, assume f is a centered coloring of G using t colors. We shall prove by induction on t that f defines a vertex t-ranking of G. As we may consider each connected component of G separately, we may assume G is connected. As f is a centered coloring there exists a color α which appears exactly once in G, at a vertex v. Hence if $t = 1$ the graph G has only one vertex v and we define a 1-ranking c of G by defining $c(v) = 1$. Assume $t > 1$ and that any centered coloring of a graph using $t - 1$ colors defines a

$(t-1)$-ranking of the graph. As the restriction of f to $G-v$ is a centered coloring using $t-1$ colors, it defines (by induction) a vertex $(t-1)$-ranking c of $G-v$. We extend c to G by defining $c(v)=t$. Now any path linking two vertices with the same c-color i is either a path of $G-v$ (so includes a vertex of c-color $j>i$) or includes v which has c-color t. □

Proposition 6.6. *For every graph* G, $\mathrm{td}(G)$ *is the minimum number of colors in a centered coloring of* G.

Proof. Notice that the minimum number of colors in a centered coloring of G is the maximum of the minima computed on the connected components of G. As $\mathrm{td}(G)$ is the maximum tree-depth of the connected components of G, we may restrict our proof to the case where G is connected.

First we prove that $\mathrm{td}(G)$ is at most equal to the number of colors in any centered coloring of G. We proceed by induction on the number k of colors in the centered coloring. If $k=1$, $G=K_1$ and thus $\mathrm{td}(G)=1$. Assume we have proved $\mathrm{td}(G)\leq k$ if $k\leq k_0$, and assume $k=k_0+1$. There exists a color c_0 which appears only once in G, at a vertex v_0. Each of the connected components $G_1,\dots,G_p$ of $G-v_0$ has a centered coloring using $k-1$ colors, and thus has depth at most $k-1$. Let $Y_1,\dots,Y_p$ be trees rooted at $r_1,\dots,r_p$, such that $G_i\subseteq \mathrm{clos}(Y)$ and $\mathrm{height}(Y_i)=\mathrm{td}(G_i)$. Then the tree Y with root v_0 and subtrees $Y_1,\dots,Y_p$ is such that $G\subseteq \mathrm{clos}(Y)$ and $\mathrm{height}(Y)\leq k+1$. Thus, $\mathrm{td}(G)\leq k+1$.

Now, we prove the opposite inequality: $\mathrm{td}(G)$ is at least equal to the number of colors in some centered coloring of G. Towards this end let Y be a rooted tree of height $\mathrm{td}(G)$, such that $G\subseteq \mathrm{clos}(Y)$. Color each vertex by its height in Y, thus using $\mathrm{td}(G)$ colors. According to the structure of $\mathrm{clos}(Y)$, any connected subgraph H of $\mathrm{clos}(Y)$ (and thus any connected subgraph of G) has a vertex which is minimum in Y. The color assigned to this vertex hence appears exactly once in H, and the constructed coloring is thus a centered coloring of G. □

According to the construction used above, if G has a centered k-coloring and $e=\{x,y\}$ is an edge of G, then the graph G/e also has a centered k-coloring. This can be deduced by modifying a centered coloring of G: the vertex corresponding to x and y has either the color of x or the color of y and all the other vertices of G/e have the same color they have in G.

In the case G is connected, we obtain:

Corollary 6.3. *Let* G *be a connected graph. Then,* $\mathrm{td}(G)$, $\chi_{\mathrm{rk}}(G)$, *the minimum height of an elimination tree for* G *and the minimum number of colors in a centered coloring of* G *are equal.*

We remark that the equality of $\chi_{rk}(G)$ and of the minimum height of an elimination tree already appears in [117].

6.6 Cycle Rank

The recursive definition (6.1) of tree-depth is an analog of the definition of the cycle rank of a digraph, as defined by Eggan [144].

Definition 6.3 (cycle rank). The *cycle rank* of a digraph $\vec{G}$, denoted $\mathrm{cr}(\vec{G})$ is inductively defined as follows.

$$\mathrm{cr}(\vec{G}) = \begin{cases} 0, & \text{if } \vec{G} \text{ is acyclic;} \\ 1 + \min_{v \in V(\vec{G})} \mathrm{cr}(\vec{G} - v), & \text{if } \vec{G} \text{ is strongly connected;} \\ \max_{i=1}^{p} \mathrm{cr}(\vec{G}_i), & \text{otherwise;} \end{cases} \tag{6.5}$$

(where $\vec{G}_1, \dots, \vec{G}_p$ are the strongly connected components of G).

As observed by Giannopoulou et al. [215], the tree depth of every graph G is related by the cycle rank of the digraph $\vec{G}_d$ obtained by replacing each edge by two opposite arcs by $\mathrm{td}(G) = \mathrm{cr}(\vec{G}_d) + 1$. Actually, if one consider the full symmetrization $\vec{G}_\circ$ of G obtained replacing each edge by two opposite arcs and attaching a loop at each vertex, then the connected subgraphs of G correspond to the strongly connected sub-digraphs of $\vec{G}$ and hence

$$\mathrm{td}(G) = \mathrm{cr}(\vec{G}_\circ).$$

As noticed by Gelade [214], the cycle rank is that it is monotone by minor. This is a consequence of its monotony under sub-digraphs [98] and of a result of McNaughton [331] implying its monotony under contractions. (Here, minors of digraphs are interpreted in the weakest sense as minors of underlying multigraphs with proper orientation). We do not include the proof here.

Theorem 6.2. *If* $\vec{H}$ *is a minor of* $\vec{G}$ *then* $\mathrm{cr}(\vec{H}) \le \mathrm{cr}(\vec{G})$.

Although computing the cycle rank of a digraph is a NP-complete problem (NP-hardness follows from the one of the computation of tree-depth), checking whether a digraph of order n has cycle rank at most t can be done in time $O(n^t)$ (by the recursive definition and the use of any linear-time algorithm to compute strongly connected components).

Cycle rank has been introduced in the context of language theory, precisely in the course of investigating the star height of regular languages. This connection to language theory deserves some short description here, hence we take time for recalling some definitions and facts about formal language and automata theory. For a thorough treatment, we refer the reader to [260].

Let Σ be a finite alphabet and let $\Sigma^\star$ be the set of all words over the alphabet Σ, including the *empty word* ϵ. The *length of a word* w is denoted by $|w|$, where $|\epsilon| = 0$. A *formal language* over the alphabet Σ is a subset of $\Sigma^\star$. The *regular expressions* over an alphabet Σ are recursively defined as follows:

- $\emptyset$, ϵ and every $a \in \Sigma$ is a regular expression;
- If r_1 and r_2 are regular expressions, so are their alternation $(r_1|r_2)$, their concatenation $c(r_1 \cdot r_2)$, and their Kleene closure $(r_1)^\star$.

The language defined by a regular expression r, denoted by $L(r)$, is defined as follows:

$$\begin{aligned}
L(\emptyset) &= \emptyset \\
L(\epsilon) &= \{\epsilon\} \\
L(a) &= \{a\} \\
L(r_1|r_2) &= L(r_1) \cup L(r_2) \\
L(r_1 \cdot r_2) &= L(r_1) \cdot L(r_2) \\
L(r_1^\star) &= L(r_1)^\star.
\end{aligned}$$

The *star height* of a regular expression r over Σ, denoted by $h(r)$, is a structural complexity measure inductively defined by:

- $h(\emptyset) = h(\epsilon) = h(a) = 0$,
- $h(r_1|r_2) = h(r_1 \cdot r_2) = \max(h(r_1), h(r_2))$,
- $h(r_1^\star) = 1 + h(r_1)$.

The *star height* (or *loop complexity*) of a regular language L, denoted by $h(L)$, is the minimum star height among all regular expressions describing L.

The following relation between cycle rank of automata and star height of regular languages became known as Eggan's Theorem [144] (see also [308])

Theorem 6.3. *The star height of a regular language* L *equals the infimum of the cycle ranks among all nondeterministic finite automata accepting* L.

In the same paper, Eggan asked whether the star height of a rational language is computable. This problem was considered for a long time as one of the most difficult problems in the theory of automata. A positive answer for pure-group languages was given by McNaughton in 1967 [331], and the problem was eventually solved (positively) by Hashiguchi in 1988 [243]. The algorithm given by Hashiguchi had a non-elementary complexity and even for very small examples the involved computations were by far impossible. Kirsten [275] drastically improved this complexity and gave an algorithm which runs, for a given nondeterministic finite automaton as input, within double-exponential space.

6.7 Games and a Min-Max Formula for Tree-Depth

It is possible to translate the recursive definition of tree depth into a game. Apart from the several variants of the cops and robber game investigated in [215], the following simple game derives directly from the form of (6.1).

Definition 6.4. Given an integer k, the k-*step selection-deletion game* on a graph G is defined as follows: two players, Alice and Bob, confront one another. Bob tries to destroy a graph that Alice tries to preserve. Alice plays first. At each round:

- Alice selects a connected component of the graph, which is kept as the new graph,

- Bob deletes a vertex in this graph. If he deletes the last vertex of the graph, Bob wins.

If Bob wins within k rounds, we say that Bob wins the k-step selection-deletion games. Otherwise, we say that Alice wins the k-step selection-deletion games.

Lemma 6.8. *Let G be a graph and let Y be a rooted forest of height at most t such that $G \subseteq \text{clos}(Y)$.*

Then Y encodes a winning strategy for Bob in the t-step selection-deletion game.

Proof. Let G be the playground graph. Alice selects a connected component H of G, corresponding to a component Y_i of Y. Then Bob deletes the vertex v corresponding to the root of Y_i. The new playground graph is $G' = H - v$ and the rooted forest obtained from Y_i by deleting v and rooting the connected components of the obtained forest at the former son vertices of v is such that $G' \subseteq \text{clos}(Y')$ and $\text{height}(Y') \leq t - 1$. By an immediate induction, Bob will eventually finish the game within at most t steps.

We shall see that a winning strategy for Alice can conversely be defined by means of the following notion [215].

Definition 6.5. A *shelter* in a graph G is a family $\mathcal{S}$ of non-empty connected subgraphs of G partially ordered by inclusion such that every subgraph $H \in \mathcal{S}$ not minimal in $\mathcal{F}$ and for every $x \in H$ there exists $H' \in \mathcal{S}$ covered by H such that $x \notin H'$. (recall that in a poset a covers b if $a > b$ but no c exist with $a > c > b$.)

The *thickness* of a shelter $\mathcal{S}$ is the minimal length of a maximal chain of $\mathcal{S}$.

Lemma 6.9. *Let G be a graph, let $\mathcal{S}$ be a shelter in G, and let t be the thickness of $\mathcal{S}$.*

Then $\mathcal{S}$ encodes a winning strategy for Alice in the $(t-1)$-step selection-deletion game.

Proof. If $t = 1$ the lemma obviously holds. So assume $t > 1$. Let G be the playground graph. Let $H \in \mathcal{S}$ be a maximal element in $\mathcal{S}$. As H is connected, it is included in some connected component G_i of G, which is the one that Alice selects. Let x be the vertex deleted by Bob. As $t > 1$, H is not a minimal element of $\mathcal{S}$. Hence there exists $H' \in \mathcal{S}$ such that $x \notin H'$ and H covers H'. It follows that $G_i - x \supseteq H'$. Let $\mathcal{S}' = \{X \in \mathcal{S}, X \subseteq G'\}$. The family $\mathcal{S}'$ is clearly a shelter in G' with thickness at least $t - 1$ (as these chains can be extended by adding H into a maximal chain of $\mathcal{S}$). Thus the lemma follows from an immediate induction.

From these lemmas we obtain immediately the following min-max characterization [215]:

Theorem 6.4. *Let* G *be a non-empty graph and* k *be a positive integer. Then the following are equivalent.*

1. G *has tree-depth at most* k,
2. *Every shelter in* G *has thickness at most* k.

Hence the minimum height of a rooted forest Y *such that* $G \subseteq \mathrm{clos}(Y)$ *equals the maximum thickness of a shelter in* G.

Notice that this min-max relation relating tree-depth and shelters in similar to the relation between tree-width and brambles mentioned in Sect. 3.6.

6.8 Reductions and Finiteness

In this section we prove two powerful reduction theorems (and finiteness results) related to tree-depth. As preparation for the next result we introduce the following notions:

A c-*colored graph* is a graph $G = (V, E)$ together with a mapping γ from V to a finite set (of colors) C of cardinality c. Notice that γ does not need to be a proper coloring. Two c-colored graphs (G, C, γ_G) and (H, C, γ_H) are isomorphic if there exist an isomorphism $f : G \to H$ such that $\gamma_H \circ f = \gamma_G$.

Definition 6.6. Let c, n be positive integers. The value $r_c(n)$ is the number of c-colored unlabeled rooted trees of order n.

Knuth [278] proved that the number $r(n)$ of unlabeled rooted trees of order n is asymptotically

$$r(n) \sim AB^n n^{-3/2}$$

for some constants A and B.

Using an encoding of c-colored unlabeled trees by uncolored unlabeled trees (increasing the order by a factor depending on c), we get the following bound on $r_c(n)$:

$$r_c(n) \lesssim A_c B_c^n n^{-3/2}$$

for some constants A_c and B_c depending on c.

Define $T(c, t)$ inductively by:

$$T(c,t) = \begin{cases} c, & \text{if } t = 1, \\ \sum_{i=1}^{T(c,t-1)+1} r_c(i), & \text{otherwise.} \end{cases}$$

and also define

$$\mathcal{F}(c,t) = T(c2^{t-1},t) \text{ and } \mathcal{F}(t) = \mathcal{F}(1,t).$$

Lemma 6.10. *Every* c*-colored rooted forest of height* t *and of order strictly greater than* $T(c,t)$ *has an involutive automorphism exchanging two branches of the same component or two rooted trees forming two isomorphic components.*

Proof. We prove the lemma by induction on t. If $t = 1$, the result is obvious. Assume the result has been proved for rooted forests of height t and let F be a c-colored rooted forest of height $t+1$, with order strictly greater than $T(c,t+1)$ and with connected components $Y_1,\dots,Y_p$. If some connected component Y_i has order greater than $T(c,t)+1$, then it follows from the induction hypothesis that the c-colored forest obtained by deleting the root of Y_i has an involutive automorphism exchanging two branches or two rooted trees, hence the same holds for Y_i and for F. Otherwise, all the connected components of F have order at most $T(c,t)+1$. As the order of F is strictly greater than $\sum_{i=1}^{T(c,t)+1} r_c(i)$ we deduce that two connected components are isomorphic (as colored rooted unlabeled trees). □

Lemma 6.10 has a number of immediate consequences. In most of our applications we shall not need T to be optimal. Rather we shall be satisfied by its existence and thus with the fact that for each integer k there is an absolute bound on the order of involution free k-colored trees with bounded heights. This simplifies the discussion. From this, we deduce:

Theorem 6.5. *For any integer* c*, any graph* G *of order* $n > \mathcal{F}(c,\mathrm{td}(G))$ *and any coloring* $g : V(G) \to \{1,\dots,c\}$*, there exists a partition* X, Y, Z *of* $V(G)$ *and a* g*-preserving automorphism* $\mu : G \to G$ *such that:*

- *The parts* Y *and* Z *are non-empty;*
- *There is no edge between vertices of* Y *and vertices of* Z*;*
- *The restriction of* μ *to* X *is the identity;*
- *The restriction of* μ *to* Y *is a bijection from* Y *to* Z*;*
- *The automorphism* μ *is involutive (that is:* $\mu \circ \mu$ *is the identity).*

Proof. Let $t = \mathrm{td}(G)$ and let Y be a rooted forest with vertex set $V(G)$ such that $G \subseteq \mathrm{Clos}(G)$. Define a coloring $\gamma : V(Y) \to [c] \times [2]^{t-1}$ as follows: for a vertex x at height $1 \leq h \leq t$ in Y let

$$\gamma(x) = (g(x), \epsilon_1(x), \epsilon_2(x), \dots, \epsilon_{h-1}(x), 0, \dots, 0),$$

where

$$\epsilon_i(x) = \begin{cases} 1, & \text{if } x \text{ is adjacent to its ancestor at height } h - i; \\ 0, & \text{otherwise.} \end{cases}$$

As $n > F(c, \mathrm{td}(G)) = T(c2^{\mathrm{td}(G)-1}, \mathrm{td}(G))$ Lemma 6.10 ensures the existence of a γ-preserving involutive automorphism μ of Y exchanging two branches of the same components or two rooted trees forming two isomorphic components. This automorphism μ exchanges sets of vertices Y and Z while fixing a set X. By the definition of γ, the mapping μ is actually a g-preserving involutive automorphism of G. □

The following two consequences indicate that tree-depth is a good "scale" for asymmetric graphs and even cores: For each given tree-depth we get only finitely many cores. (Note that this does not hold for tree-width in a very strong sense: According to [253] the class of series parallel graphs, that is the class of graphs with no K_4 minor, is even (countably) universal, see Theorem 3.14 in Sect. 3.7.)

Corollary 6.4. *Any asymmetric graph of tree-depth* t *has order at most* $F(t)$.

Corollary 6.5. *Let* c *be an integer. For any graph* G *and any* c*-coloring of the vertices of a graph* G *of tree depth* t*, there exists a subset* A *of* $V(G)$ *of cardinality at most* $F(c, t)$*, such that* G *has a color-preserving homomorphism to* $G[A]$.

In particular, the core of any graph G *has at most* $F(\mathrm{td}(G)) = F(1, \mathrm{td}(G))$ *vertices.*

For an integer $t \geq 1$, define $\mathcal{T}_t$ as the class of all graphs G with $\mathrm{td}(G) \leq t$.

Corollary 6.6. *Let* $t \geq 1$ *be an integer. Then, the class* $\mathcal{D}_t$ *includes a finite subset* $\hat{\mathcal{T}}_t$ *such that every graph* $G \in \mathcal{T}_t$ *has core in* $\hat{G} \in \hat{\mathcal{T}}_t$.

It should be noted that although the set $\hat{\mathcal{T}}_t$ is finite it is very large, its cardinality being *non-elementary* (i.e. not bounded by any tower-exponential function $2^{2^{\cdot^{\cdot^{\cdot^{2^n}}}}}$ of fixed height), see [239, 425]. We find it useful to further refine this finiteness.

Lemma 6.11. *Let* G *be a tree of size* m *having* p *leaves and tree-depth* t. *Then,* $m \leq (2^{t-1} - 1)p$.

Proof. We prove the inequality by induction on t. The inequality is obviously true for $t = 1$ and we now assume it is true for $t-1$. Also, we assume without loss of generality that G is connected. Let Y be a rooted tree of height t such that $G \subseteq \mathrm{clos}(Y)$ and let v be the root of Y. The graph $G - v$ has connected components $G_1, \ldots, G_{d(v)}$ which are trees of size $m_1, \ldots, m_{d(v)}$ having $p_1, \ldots, p_{d(v)}$ leaves, where $m = d(v) + \sum_i m_i$ and $p = \sum_i (p_i - 1)$. By induction, $m_i \leq (2^{t-2} - 1)p_i$. Hence, $m \leq d(v) + (p + d(v))(2^{t-2} - 1) = (p + d(v))2^{t-2} - p$. Moreover, $p \geq d(v)$ as each G_i includes at least one leaf of G. Thus, $m \leq 2^{t-1}p - p = (2^{t-1} - 1)p$. □

Theorem 6.6. *There exists a function* $\mu : \mathbb{N} \to \mathbb{N}$, *such that any graph* G *has a connected subgraph* $H \subseteq G$, *so that* $\mathrm{td}(H) = \mathrm{td}(G)$ *and* $|E(H)| \leq \mu(\mathrm{td}(G))$.

Proof. We define μ by induction. If $\mathrm{td}(G) = 1$ then G is isomorphic to $\overline{K_n}$, thus we can choose any vertex subgraph for H and put $\mu(1) = 0$. Assume $\mathrm{td}(G) \geq 2$ and let $t = \mathrm{td}(G)$. According to Lemma 6.2, the class $\mathcal{T}_{t-1} = \{G : \mathrm{td}(G) \leq t - 1\}$ is a proper minor closed class of graphs. Thus, (using Robertson – Seymour minor graph theorem) there exists a finite set $\mathcal{F}_{t-1}$ of forbidden minors for the class $\mathcal{T}_{t-1}$. As $G \notin \mathcal{T}_{t-1}$, there exists $F \in \mathcal{F}_{t-1}$, so that F is a minor of G. Moreover, we may assume that G is minimal in the sense that any edge deletion decreases the tree depth of G. Thus G is connected and, for any edge e, F is not a minor of $G - e$. Hence, F is obtained from G by contracting some connected trees into single vertices, and deleting at most one edge for any connected components of F but one. By minimality of G, each vertex v of F is obtained by contracting a tree G_v of G of tree-depth at most $t - 1$ having at most $d(v)$ extremal vertices. According to Lemma 6.11, G_v has size at most $(2^{t-2} - 1)d(v)$. Altogether, G has at most $2^{t-2}|E(F)| + c_0(F) - 1$ edges, where $c_0(F)$ is the number of connected components of F. Putting $\mu(t) = \max_{F \in \mathcal{F}_{t-1}} 2^{t-2}|E(F)| + c_0(F) - 1$ completes the proof. □

6.9 Ehrenfeucht-Fraïssé Games

The Ehrenfeucht-Fraïssé game definition has been introduced in Sect. 3.8.4, as well as the definition of n-back-and-forth equivalence.

In general, a graph is not n-equivalent to one of its proper subgraph. For instance a cycle is not n-equivalent to one of its subgraphs if $n \geq 3$ (as one can express that every vertex has degree at least 2 by a using at most four nested quantifiers). However, the situation for graphs with bounded tree-depth is very different:

Theorem 6.7. *For every integers* t, n, c *there exists an integer* $N(t, n, c)$ *such that every graph with tree-depth at most* t *with vertices colored using* c *colors is* n*-equivalent to one of its induced subgraphs of order at most* $N(t, n, c)$.

Proof. We prove the theorem by induction on the tree-depth t. If $t = 1$, it is clear that we only have to keep at most n vertices of each color to get an induced subgraph n-equivalent to G. Hence we can let $N(1, n, c) = cn$.

Assume that the theorem holds for graphs with tree-depth at most $t \geq 1$ and let G be a graph of tree-depth $t + 1$. First consider the case where G is connected. Let r be a vertex of G such that $\text{td}(G - r) = t$. Recolor the vertices of $G - r$ with the product of the original coloring of G by the coloring $\gamma : V(G) - r \to \{0, 1\}$ defined by $\gamma(v) = 1$ if v is adjacent to r and $\gamma(v) = 0$ otherwise. Let A be a subset of vertices of $G - r$ with minimum cardinality such that $(G - r)[A]$ is n-equivalent to G (with the new coloring). Then clearly $G[A \cup \{r\}]$ is n-equivalent to G with the old coloring. This induced subgraph has order at most $N(t, n, 2c) + 1$. Now consider the case where G is disconnected. Then each connected component is n-equivalent to an induced subgraph of order $N(t, n, 2c)+1$. As there are at most $F(N(t, n, 2c)+1)$ graphs of order at most $N(t, n, 2c) + 1$ and as we clearly do not need to keep more than n non-isomorphic ones, we get that $N(t+1, n, c) \leq n\, F(N(t, n, 2c)+1)$. □

6.10 Well Quasi-orders

Classes of graphs with bounded tree-depth have finitely many cores. But they have further finitary properties. We list some of those related to very restrictive well quasi-orders (see Sect. 3.5) defined by means of inclusion and retracts.

Let $(Q, \leq)$ be a well quasi-ordered set and let t be an integer. We denote by $\mathcal{T}_t(Q)$ the class of Q-labeled graphs of tree-depth at most t, i.e. the class of couples (G, γ) where G is a graph with tree-depth at most t and γ is a map from the vertex set of G to Q.

On $\mathcal{T}_t(Q)$ we consider the naturally defined labels respecting induced subgraph relation $\subseteq_i$. Explicitly, we put $(G, \gamma) \subseteq_i (H, \eta)$ if there exists an injective function $f : V(G) \to V(H)$ such that $\{x, y\} \in E(G) \iff \{f(x), f(y)\} \in E(H)$ and such that $\eta(f(x)) \geq \gamma(x)$ for every vertex x of G.

We start with a result of Ding [123] which we formulate by means of tree-depth and which we prove by making use of the following result of Erdős and Rado [168].

Lemma 6.12. *Let A be a quasi-ordered set. Let $S(A)$ be the set of finite subsets of A, and make it into a quasi-ordered set by the rule $X \leq Y$ if there is a one-to-one increasing map of X into Y.*

If A is well quasi-ordered then so is $S(A)$.

Notice that this result, quoted by Higman in [256], can be deduced as a corollary of Higman's further result on well quasi-ordering of finite sequence of elements of a well quasi-ordered set.

Lemma 6.13. *For every integer t, the class $\mathcal{T}_t(Q)$ is well quasi ordered by induced subgraphs (i.e. by the partial order $\subseteq_i$).*

Proof. We proceed by induction on t. The case $t = 1$ is a direct consequence of Lemma 6.12.

Assume the results holds for $t = t_0 \geq 1$ and let $t = t_0 + 1$. The set $\{0, 1\} \times Q$ is well quasi-order by $(x, y) \leq (x', y')$ if $x = x'$ and $y \leq y'$. Thus the set $\mathcal{T}_{t-1}(\{0, 1\} \times Q), \subseteq_i$ is well quasi-ordered by the induction hypothesis. So is also the set $Q \times \mathcal{T}_{t-1}(\{0, 1\} \times Q)$, ordered by $(x, G) \leq (x', G')$ if $x \leq x'$ and $G \subseteq_i G'$.

Assume we are given an infinite sequence of graphs $(G_i, \gamma_i) \in \mathcal{T}_t(Q)$. For each i we consider a vertex r_i of G_i such that $\mathrm{td}(G_i - r_i) = \mathrm{td}(G_i) - 1$. We define $(x_i, (H_i, c_i)) \in Q \times \mathcal{T}_{t-1}(\{0, 1\} \times Q)$ by $x_i = \gamma_i(r_i), H_i = G_i - r_i$ and $c_i(x) = (\mathrm{Adj}_{G_i}(r_i, x), \gamma_i(x))$, where $\mathrm{Adj}_G(x, y)$ is 1 if x and y are adjacent in G, and 0 otherwise. By the induction hypothesis, there exist $i < j$ such that $x_i \leq x_j$ and such that there exists a one-to-one increasing map f from $V(H_i)$ to $V(H_j)$ with $H_j[f(V(H_i))] \cong H_i$. According to the definition of the partial order on $\{0, 1\} \times Q$, and as $x_i \leq x_j$ means $\gamma_i(r_i) \leq \gamma_j(r_j)$, we deduce that the extention $\hat{f}$ of f to $V(G_i)$ defined by $f(r_i) = r_j$ is a one-to-one increasing map from $V(G_i)$ to $V(G_j)$ with $G_j[f(V(G_i))] \cong G_i$. It follows that $(\mathcal{T}_t(Q), \subseteq_i)$ is a well quasi-ordering. □

Notice that even if Q is actually finite, the partial order $(\mathcal{T}_t(Q), \subseteq_i)$ may contain arbitrarily large finite antichains.

Also, notice that even the class of 2-colored paths (which have, of course, unbounded tree-depth) contains an infinite antichain for $\subseteq_i$, as shown Fig. 6.5.

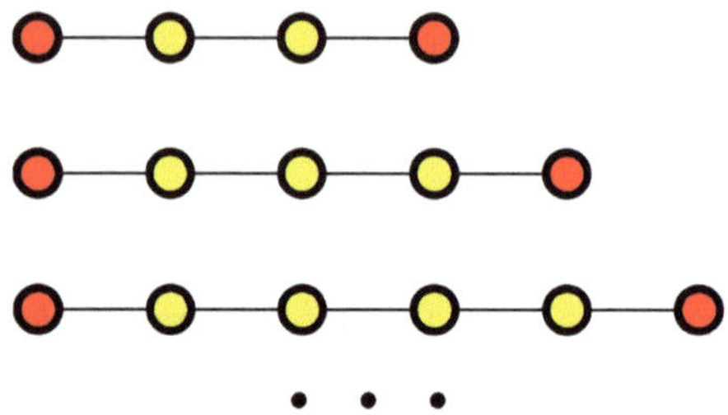

Fig. 6.5 Paths with a coloration of the vertices with two colors are not well quasi ordered by containment partial order

A direct consequence of Lemma 6.13 is that for every positive integer t there exists an existential first-order formula τ_t such that the property "The graph G has tree-depth at least t" is equivalent to the satisfaction of τ_t:

$$\mathrm{td}(G) \geq t \quad \Longleftrightarrow \quad G \models \tau_t. \tag{6.6}$$

In other words, there exists for every integer t a finite set of forbidden subgraphs characterizing the property $\mathrm{td}(G) \leq t$ (see Fig. 6.6).

From this it follows that for every positive integer t there exists a linear time algorithm which, for each input graph G, answers either that the tree-depth of G is at most t (and computes an elimination tree of height at most t) or that the tree-depth of G is strictly greater than t (and exhibits a certificate for this). The main idea of the algorithm is to start by computing a DFS-tree for G. If the DFS tree has height at least 2^t a maximal tree-path is a certificate that $\mathrm{td}(G) > t$. Otherwise, G has tree-depth at most 2^t and we may use this bound to reduce the problem of the satisfiability of τ_t from polynomial time to linear time (see Sect. 17.3).

Another proof of the existence of a linear time algorithm has been given in [72]. This proof relies on the fact that the class of graphs with tree-depth at most t is minor closed and excludes a planar graph hence belonging to this class can be determined in linear time according to a well known result

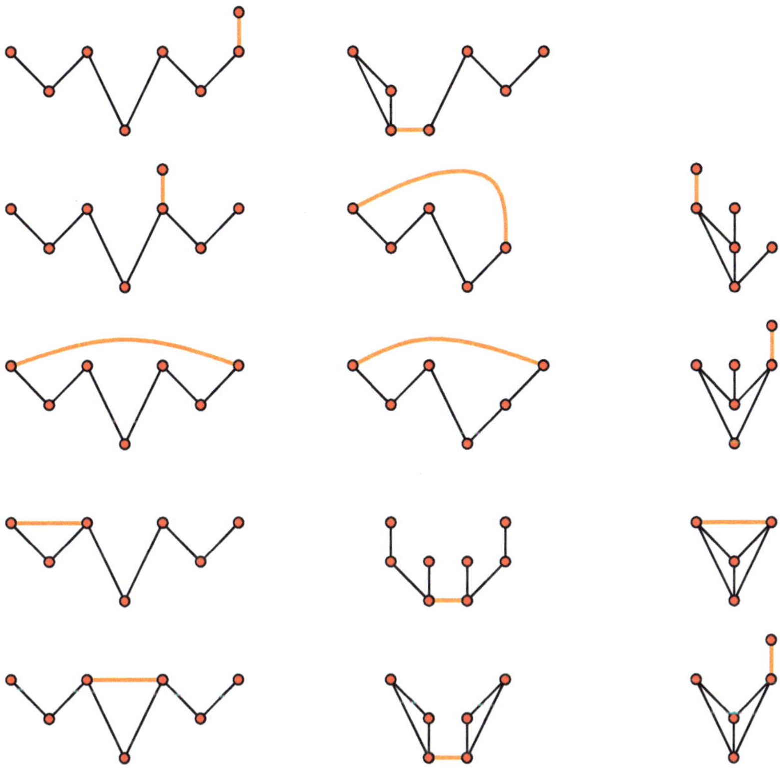

Fig. 6.6 Forbidden subgraphs for the property $\mathrm{td}(G) \leq 3$ (From [134])

of Robertson and Seymour [398]. However the direct proof we just sketched is certainly much easier.

The strongest of our results on well quasi-ordering is the following:

Theorem 6.8. *For integers* r *and* t*, let* $\mathcal{T}_t^{(r)}$ *be the class of graphs of tree-depth at most* t *whose vertices are colored using (at most)* r *colors. For* $G, H \in \mathcal{T}_t^{(r)}$ *define* $H \subseteq_i^{\star} G$ *if* H *is isomorphic to an induced subgraph of* G *as a colored graph and there exists a color preserving homomorphism of* G *to* H *(that is:* H *is isomorphic to a retract of* G*). Then* $\mathcal{T}_t^{(r)}$ *is well quasi ordered by* $\subseteq_i^{\star}$*.*

Proof. Consider a subset S of $\mathcal{T}_t^{(r)}$. Then S contains finitely many classes of graphs which are homomorphically equivalent (in a color preserving way). In each class, there are finitely many graphs which are minimal for $\subseteq_i$, according to Lemma 6.13. The result follows. □

We note that Theorem 6.8 does not extend if we relax the condition that the coloring uses a finite set of colors and consider a labeling by well-quasi ordered poset of bounded width (see Fig. 6.7).

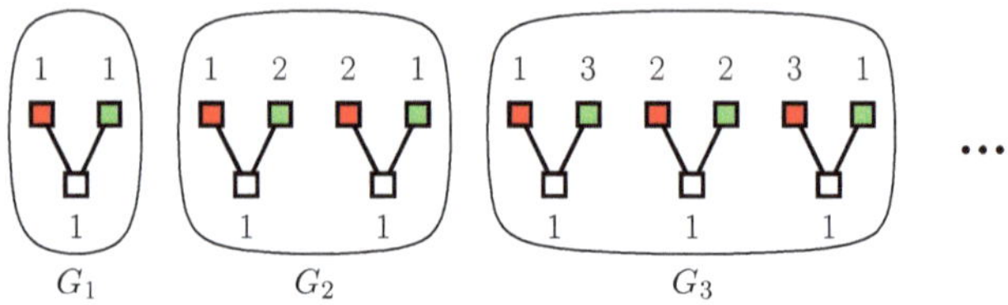

Fig. 6.7 An infinite antichain for the relation $\subseteq_i^\star$ (in a colored + labeled version)

6.11 The Homomorphism Quasi-order

Recall that the *homomorphism quasi-order* on graphs corresponds to the existence of a homomorphism. By taking the quotient of $\mathcal{G}raph$ by the *hom-equivalence* relation $G \rightleftarrows H$ (meaning the existence of both a homomorphism of G to H and a homomorphism of H to G) we obtain a partial order of the hom-equivalence classes of $\mathcal{G}raph$.

It is well known that the homomorphism order on hom-equivalence classes of graphs has a lattice structure: any two classes $[G_1]$ and $[G_2]$ have infimum $[G_1] \wedge [G_2] = [G_1 \times G_2]$ and supremum $[G_1] \vee [G_2] = [G_1 + G_2]$ (where $\times$ and $+$ respectively denote the categorical product and the disjoint union).

Denote by $\mathcal{T}_t$ (resp. Core_t) the class of graphs (resp. the class of core graphs) with tree-depth at most t.

Proposition 6.7. *The restriction of the homomorphism order to the equivalence classes of $\mathcal{T}_t$ has a lattice structure with*

$$[G_1] \vee_t [G_2] = [G_1 + G_2]$$
$$[G_1] \wedge_t [G_2] = \left[\bigvee\nolimits_t \{C \in \mathrm{Core}_t : C \to G_1 \times G_2\}\right]$$

Proof. As $[G_1] \vee [G_2] = [G_1 + G_2]$ and $\mathrm{td}(G_1 + G_2) = \max(\mathrm{td}(G_1), \mathrm{td}(G_2))$ it follows that $[G_1] \vee_t [G_2] = [G_1 + G_2]$. By construction, $\bigvee_t\{C \in \mathrm{Core}_t : C \to G_1 \times G_2\} \to G_1 \times G_2$ hence $\bigvee_t\{C \in \mathrm{Core}_t : C \to G_1 \times G_2\} \to G_1$ and $\bigvee_t\{C \in \mathrm{Core}_t : C \to G_1 \times G_2\} \to G_2$. Assume $H \in \mathcal{T}_t$ is such that $H \to G_1$ and $H \to G_2$, where $G_1, G_2 \in \mathcal{T}_t$. Then $H \to G_1 \times G_2$ thus $\mathrm{Core}(H) \to G_1 \times G_2$

and obviously $\mathrm{Core}(H) \in \mathrm{Core}_t$. As $H \to \mathrm{Core}(H) \to \bigvee_t\{C \in \mathrm{Core}_t : C \to G_1 \times G_2\}$ we deduce $[G_1] \wedge_t [G_2] = \left[\bigvee_t\{C \in \mathrm{Core}_t : C \to G_1 \times G_2\}\right]$. □

Of course, the finiteness of the Core_t shows that there exists a function $F : \mathbb{N} \to \mathbb{N}$ such that

$$\begin{aligned} \mathrm{td}(\mathrm{Core}(G_1 \times G_2)) &\leq \mathrm{td}(\mathrm{Core}(G_1))|\mathrm{Core}(G_2)| \\ &\leq \mathrm{td}(\mathrm{Core}(G_1))F(\mathrm{td}(\mathrm{Core}(G_2))). \end{aligned}$$

Hence if $G_1, G_2 \in \mathcal{T}_t$ then

$$[G_1] \wedge_t [G_2] \longrightarrow [G_1] \wedge_{t+1} [G_2] \longrightarrow \ldots \longrightarrow [G_1] \wedge_{tF(t)} [G_2] = [G_1 \times G_2].$$

Although highly probable, we have yet no example showing that $[G_1] \wedge_t [G_2]$ and $[G_1 \times G_2]$ are different in general. We are naturally led to the following question:

Problem 6.2. What is the smallest possible function $f : \mathbb{N} \to \mathbb{N}$ such that for every graphs G_1, G_2 holds

$$\mathrm{td}(\mathrm{Core}(G_1 \times G_2)) \leq f(\max(\mathrm{td}(\mathrm{Core}(G_1)), \mathrm{td}(\mathrm{Core}(G_2))))?$$

Of course, if $f(t) \neq t$, this means that hom equivalence classes of $\mathcal{T}_t$ do not form a sub-lattice of the lattice of hom-equivalence classes of $\mathcal{G}raph$.

Exercises

6.1. Let T_n be the complete binary tree of height n. Prove that $\mathrm{td}(T_n) = n$.

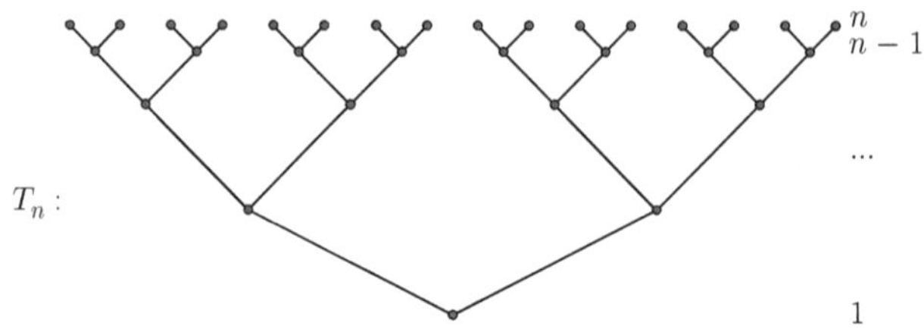

6.2. Let $m, t \in \mathbb{N}$ and let G be a graph with $\mathrm{td}(G) \leq t$. Prove that if $|G| \geq m^t$ then there exists a subset S of at most $t-1$ vertices of G, such that $G - S$ has at least m connected components.

6.3. Let G and H be two graphs with at least one edge. Then:

$$\mathrm{td}(G \times H) \geq \mathrm{td}(G) + \mathrm{td}(H) - 2.$$

6.4. Define the graph parameter cr^+ on undirected graphs by

$$\mathrm{cr}^+(G) = \max_{\vec{G} \text{ orientation of } G} \mathrm{cr}(G).$$

1. Prove that $\mathrm{cr}^+(G) \leq \mathrm{td}(G) - 1$.
2. Give an example of arbitrarily large graphs for which the equality holds.
3. Prove that cr^+ is a minor monotone parameter and deduce that for each integer k there exists a finite set of obstructions for $\mathrm{cr}^+ \leq k$.

6.5. Consider a classical cops and robbers game with c cops and r robbers [373, 391]. Prove that for every integer t there exists an integer $N = N(c, r, t)$ such that for every graph G with tree-depth at most t:

- Either the cops have no winning strategy,
- Or the maximum length of a game played with an optimal strategy is bounded by N.

6.6. The aim of this Exercise is to explicit a first-order formula ϕ_t such that $G \models \phi_t$ if and only if $\mathrm{td}(G) \leq t$.

- Prove that there exists a first-order formula $\vartheta_{t,k}(x_1, \dots, x_{k+2})$ with quantifier rank t such that for every two vertices $k+2$ distinct vertices $r_1, \dots, r_k, a, b$ of a graph G it holds $G \models \vartheta_{t,k}(r_1, \dots, r_k, a, b)$ if and only if a and b are at distance at most 2^t in $G - r_1 - \cdots - r_k$;
- Prove that there exists a formula $\delta_{t,k}(x_1, \dots, x_k)$ with quantifier rank at most $\min(2t, t+3)$ such that

$$\mathrm{td}(G - r_1 - \cdots - r_k) \le t \quad \Longrightarrow \quad G \models \delta_{t,k}(r_1, \ldots, r_k)$$

and such that if $G \models \delta_{t,k}(r_1, \ldots, r_k)$ then no connected component of $G - r_1 - \cdots - r_k$ has diameter greater than 2^t.

- Prove that there exists a formula ϕ_t of quantifier rank at most $2t$ such that $G \models \phi_t$ if and only if $\mathrm{td}(G) \le t$.

6.7. Prove that for any two (finite) graphs G and H and any integer n the following are equivalent:

- For every graph F with tree-depth at most n it holds

$$F \to G \quad \Longleftrightarrow \quad F \to H;$$

- For every existential first-order formula Φ with quantifier rank at most n it holds

$$G \models \Phi \quad \Longleftrightarrow \quad H \models \Phi.$$

The connection between tree-depth and quantifier rank in first-order logic has been studied and developed by Rossman [425].

6.8. The first-order language we consider includes two relation symbols $=$ and $\sim$, respectively representing equality and adjacency of vertices.

We also consider an extension of first-order logic **with counting quantifiers**, by allowing expressions of the type $\exists^m \Phi$ to say that there exists at least m vertices with property Φ. The *logical depth* $D_{\#}(G)$ of a graph G in this logic is the minimum depth of nested quantifiers (including counting quantifiers which also contributes 1) in a formula $\Phi(G)$ such that $H \models \Phi(G)$ if and only if H is isomorphic to G.

To ease the exposition, we define $\exists^{=m}\Phi$ as $(\exists^m \Phi) \wedge \neg(\exists^{m+1}\Phi)$, to express that there exist exactly m vertices with property Φ. Notice that the logical depth of a formula which uses such a counting quantifier is the same as the derived formula using $\exists^m$.

Also, for a graph G and a vertex $r \in V(G)$, we denote by $C_G(r)$ the vertex set of the connected component of G which contains r.

- Define quantifier free formulas $\Lambda_d[x_1, \ldots, x_{l+3}]$ (for $d \ge 1$ and $l \ge 0$) such that $G \models \Lambda_d[w_1, \ldots, w_l, u, v]$ if and only if the shortest-path distance of u and v is at most d in $G - \{w_1, \ldots, w_l\}$;
- Let G be a connected graph and let $a_1, \ldots, a_l, r$ be vertices of G. Prove that there exists a formula $\Psi[x_1, \ldots, x_{l+1}]$ such that $\mathrm{qrank}(\Psi) \le \mathrm{td}(G[C_{G-\{a_1,\ldots,a_l\}}(r)] - r) + 1$ and such that for every graph H and for every $b_1, \ldots, b_l, s$ in $V(H)$ the following conditions are equivalent:

1. The mapping $f_0 : a_i \mapsto b_i$ is an isomorphism from $G[a_1, \ldots, a_l]$ to $H[b_1, \ldots b_l]$ and $H \models \Psi[b_1, \ldots, b_l, s]$;
2. There exists an isomorphism $f : G[C_{G-\{a_1,\ldots,a_l\}}(r) \cup \{a_1, \ldots, a_l\}] \to H[C_{H-\{b_1,\ldots,b_l\}}(s) \cup \{b_1, \ldots, b_l\}]$ such that $f(r) = s$ and $f(a_i) = b_i$ (for $1 \leq i \leq l$).

- Deduce that for every graph G there exists a formula $\widehat{\Psi}$ such that $\mathrm{qrank}(\widehat{\Psi}) \leq \mathrm{td}(G) + 1$ and such that a graph H satisfies $\widehat{\Psi}$ if and only if it is isomorphic to G.

Related results on first-order definability of random graphs can be found in [74].

Chapter 7
Decomposition

Divide et impera: any few among the many will be easy to control.

7.1 Motivation, Low Tree-Width and Low Tree-Depth

The local properties of structures are frequently studied by means of decompositions: the large structure is cut into (hopefully simpler)pieces whose properties are then studied together with the interconnections between pieces. Several decomposition schemes can be considered. For example, one can stress the regularity of the interconnections of the pieces as in *modular decomposition* of graphs (a recursive partition into modules such that, for each module, the neighborhoods outside the module of the vertices within the module are all equal [206, 266, 330]). As another example, one can consider a family of overlapping simple pieces covering the structure, in the spirit of the coherency between charts in topological atlases, such as the gluing axiom of sheaves [320] or the arrow construction for the category of labeled rooted forest and the category of labeled Feynman graphs [290].

Also our decompositions are similar: We will be interested in covering a graph with pieces in such a way that:

- The number of pieces should be small,
- Each piece should be simple,
- Every small subgraph should be fully covered by at least one piece.

This approach can be described from the point of view of graph coloring. We formulate it by means of the following two ("dual") questions.

J. Nešetřil and P.O. de Mendez, *Sparsity*, Algorithms and Combinatorics 28, DOI 10.1007/978-3-642-27875-4_7, © Springer-Verlag Berlin Heidelberg 2012

How many "simple and regular" induced subgraph are needed to cover a graph G in such a way that every subset of p vertices of G belong to at least one of them?

How many colors are needed to color the vertices of a graph G such that every p color classes induce a "simple and regular" induced subgraph?

The second question is obviously related to known and intensively studied graph coloring issues. A *proper coloring* of a graph G, which is vertex coloring such that each color class induce a stable set, correspond to the extreme case where $p = 1$ (and "simpler and regular" means here edgeless). A possible next case ($p = 2$) relates to the notion of acyclic chromatic number introduced by Grünbaum introduced in [233]: the *acyclic chromatic number* of a graph is the least number of colors of a proper vertex coloring in which every 2-chromatic subgraph is acyclic. The number of required colors seemed to be small at least for sufficiently sparse graphs, and Grünbaum conjectured [233] that this variant of the chromatic number was bounded by 5 for planar graphs. This was proved by Borodin [80] and it was also proved that the acyclic chromatic number is bounded for classes with bounded genus or bounded degree [21, 23]. The "simple" structure considered here is a forest. For $p > 2$ we need a more general notion of "simple" graphs. Graphs with bounded tree-width seem to be a natural candidate. Tree-width of graphs (see Sect. 3.6) presents a powerful generalization of forests. This concept is central to Robertson and Seymour's analysis of graphs with forbidden minors, and gained much algorithmic attention, also thanks to the general complexity result of Courcelle about monadic second-order logic graph properties being decidable in linear time for graphs with bounded tree-width [102, 103]. Among the possible equivalent definitions of tree-width we recall here one based on k-trees: A k-*tree* is a graph which is either a clique of size at most k or is obtained from a smaller k-tree by adding a vertex adjacent to at most k vertices which are pairwise adjacent. The *tree-width* $\mathrm{tw}(G)$ of a graph G is the smallest integer k such that G is a subgraph of a k-tree, that is such that G is a *partial* k-*tree*.

Using tree-width, we can generalize: A p-*tree-width coloring* of a graph G is a vertex coloring of G such that any $p' \leq p$ colors induce a subgraph of tree-width at most $p - 1$. The acyclic chromatic number corresponds in our setting to the case $p = 2$.

Using the Structural Theorem of Robertson and Seymour for graphs without a particular graph as a minor [414], DeVos et al. [118] proved the following:

Theorem 7.1. *For every graph* K *and integer* $j \geq 1$, *there are integers* $i_V = i_V(K, j)$ *and* $i_E = i_E(K, j)$, *such that every graph with no* K*-minor has a vertex partition into* i_V *graphs such that any* $j' \leq j$ *parts form a graph with tree-width at most* $j' - 1$, *and an edge partition into* i_E *graphs such that any* $j' \leq j$ *parts form a graph with tree-width at most* j'.

The same paper [118] also contains a kind of dual result:

Theorem 7.2. *For every graph* K *and integer* $j \geq 1$, *there are integers* $k_V = k_V(K, j)$ *and* $k_E = k_E(K, j)$, *such that every graph with no* K*-minor has a vertex partition into* $j+1$ *graphs such that any* j *parts form a graph with tree-width at most* k_V, *and an edge partition into* $j + 1$ *graphs such that any* j *parts form a graph with tree-width at most* k_E.

A more precise and algorithmically efficient form of Theorem 7.2 has then been proved by Demaine et al. [116]:

Theorem 7.3. *For a fixed graph* H, *there is a constant* c_H *such that, for any integer* $k \geq 1$ *and for every* H*-minor free graph* G, *the vertices of* G *(or the edges of* G*) can be partitioned into* $k + 1$ *sets such that any* k *of the sets induce a graph of tree-width at most* $c_H k$. *Furthermore, such a partition can be found in a polynomial time.*

Searching for the simplest possible type of pieces, one can consider (for the case $p = 2$) a stronger variant of the acyclic chromatic number. The *star chromatic number* [233] $\chi_s(G)$ of a graph G is the least number of colors of a proper vertex coloring of G in which every 2-chromatic subgraph is a star forest (see Fig. 7.1). It is folklore that the star chromatic number is bounded on a class of finite graphs if and only if the acyclic chromatic number is bounded on the class (see for instance [176]).

For which other classes of graphs does this hold? While both the acyclic chromatic number and the star chromatic number are unbounded even for bipartite graphs, we proved in [345] that every proper minor closed class of graphs has a bounded star chromatic number (and thus a bounded acyclic chromatic number). Precisely, we proved (Theorem 2.1 of [345]) that the star-chromatic number of every graph is bounded by a function of the maximum density of its minors:

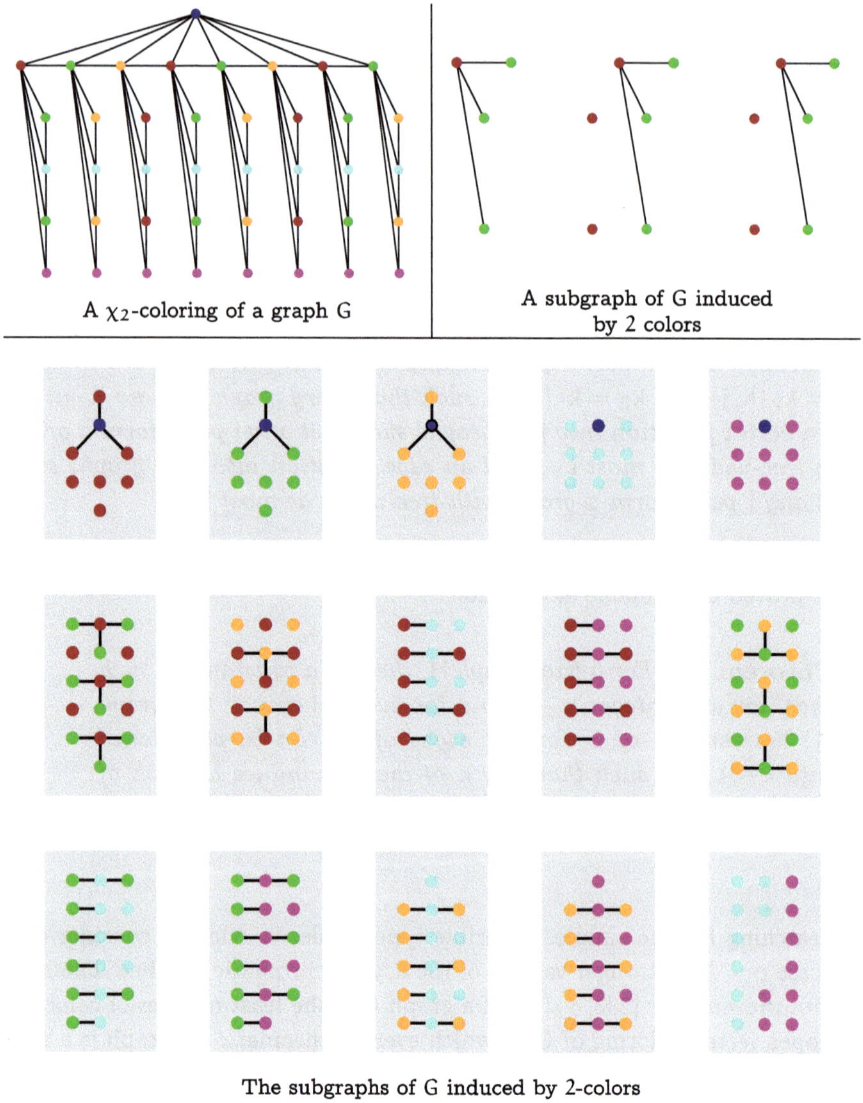

Fig. 7.1 A star-coloring of a graph; each pair of colors induces a star forest

Theorem 7.4. *For every graph* G,

$$\chi_s(G) \leq \lceil \nabla_0(G) \rceil \left(2\lceil \nabla(G) \rceil + \lceil \nabla_0(G) \rceil - 1\right) + \lfloor 2\nabla_0(G) \rfloor + 1.$$

The augmentation technique, introduced in [345], has been successfully used to improve the known upper bounds of the star chromatic number of planar graphs. For instance, every planar graph has star chromatic at most 20 [9] (which is presently the best known bound). Techniques combining our approach and discharging have been used to improve this bound for planar graphs of high girth [9, 86, 295] and sub-cubic planar graphs of high girth [94]. For instance, if G is a planar graph of girth g then the following values of $\chi_s(G)$ are known:

$$\chi_s(G) \leq \begin{cases} 18, & \text{if } g \geq 4 \quad [345] \\ 16, & \text{if } g \geq 5 \quad [9] \\ 8, & \text{if } g \geq 6 \quad [295] \\ 7, & \text{if } g \geq 7 \quad [295] \\ 6, & \text{if } g \geq 8 \quad [86, 295] \\ 5, & \text{if } g \geq 9 \quad [86] \\ 4, & \text{if } g \geq 13 \quad [86] \end{cases}$$

As we have seen in Chap. 6 a natural generalization of the notion of star forests which is a strengthening of bounded tree-width is bounded tree-depth. This leads to the following notion:

Definition 7.1. A p-*tree-depth coloring* of a graph G is a vertex coloring of G each $p' \leq p$ parts induce a subgraph with tree-depth at most p'.

As the tree-depth of a graph is at least one more than its tree-width, the low tree-depth coloring is a stronger requirement than low tree-width coloring.

Using Theorem 7.1 (hence also relying on the Structural Theorem) we proved in [352] the following strengthening:

Theorem 7.5 (Low tree-depth coloring for proper minor closed classes). *For every graph* K *and integer* $j \geq 1$*, there is an integers* $N = N(K, j)$*, such that every graph with no* K*-minor has a vertex partition into* N *graphs such that any* $j' \leq j$ *parts form a graph with tree-depth at most* j'.

A change of one word from Theorem 7.1–7.5 has profound consequences and motivated much of the material covered by this book. It also appeared that the tree-depth is the largest invariant for which an analog of Theorem 7.5 holds (see Exercise 7.5).

Motivated by Theorem 7.5 we introduce the sequence

$$\chi_1, \chi_2, \ldots, \chi_p, \ldots$$

of chromatic numbers, where χ_1 is the usual chromatic number, χ_2 is the star chromatic number and, more generally, χ_p is the minimum number of colors such that each $i \leq p$ parts induce a graph with tree-depth at most i. Obviously

$$\chi(G) = \chi_1(G) \leq \chi_2(G) \leq \cdots \leq \chi_p(G) \leq \cdots$$

The sequence stabilizes and for $p = td(G)$ one has $\chi_p(G) = td(G)$. Hence we could put $\chi_\infty(G) = td(G)$.

Theorem 7.5 may be restated concisely:

Theorem 7.6. *For every proper minor closed class of graphs $\mathcal{K}$ and for every fixed integer $p \geq 1$, $\chi_p(G)$ is bounded on $\mathcal{K}$.*

Let us add a remark concerning dual Theorem 7.2. It should be noticed that a stronger form Theorem 7.2 in which we would ask that each piece would have bounded tree-depth would be false even if we restrict ourselves to planar graphs with maximum degree 6. For an interesting example, consider the $n \times n$ hex-game graph (see Fig. 7.2). In any two coloring of the vertices, some color joins two opposite sides (this is why hex game has no draw). As the graph will include a monochromatic path of length n, one of the two monochromatic parts will have tree-depth at least $\log_2 n$. A similar statement

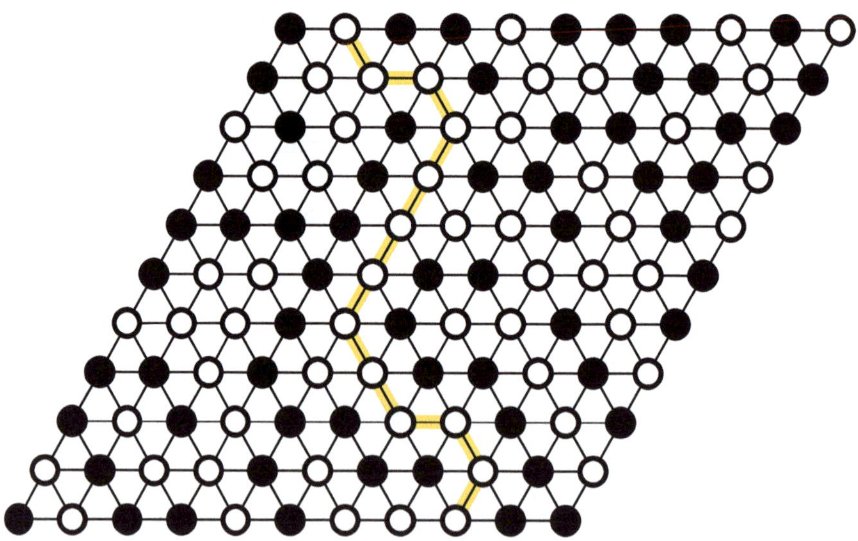

Fig. 7.2 The $n \times n$ hex game graph cannot be partitioned into two pieces of small tree-depth

would also fail to be true when restricted to graphs with tree-width 2 (i.e., to outerplanar graphs), see Exercise 7.1.

Ding, Oporowski, and Vertigan suggested to study the minimum number of parts needed to partition a graph in such a way that every monochromatic connected component has bounded size (see for instance [14, 245]). For instance, it is possible to partition every planar graphs into four parts, each of them consisting of isolated vertices (by the Four Color Theorem). However, Alon, Ding, Oporowski, and Vertigan constructed for every integer n a planar graph G_n with the property that in any coloring of the vertices with three colors, some monochromatic connected component will have order at least n. On the other hand, they conjectured that for every positive integer d there exists $f(d)$ such that one can color the vertices of every planar graph with maximum degree d with three colors in such a way that no monochromatic connected component will have order greater than $f(d)$. Interestingly, the counter-example given for planar graphs with unbounded degrees admits a coloring of the vertices with 2 colors such that no monochromatic connected component has tree-depth greater than 3. As graphs with bounded degree and bounded tree-depth also have bounded order, we are led to the following natural conjecture:

Conjecture 7.1. There exists a constant t such that one can color the vertices of every planar G by three colors in such a way that no monochromatic connected component will have tree-depth greater than t.

It is well known that χ_1 is bounded on a class of graphs if the maximum average degree of graphs in the class is bounded. In [345] we proved that χ_2 is bounded if all graphs obtained by contracting star forests have bounded maximum average degree. Also, if χ_2 is bounded then so is the maximum average degree. Here is a short proof: Assume $\chi_2(G) \leq N$. Then for every two colors $i \neq j, i, j \leq N$, orient the edges of G such that every vertex has indegree at most one in the star forest induced by colors i and j. Then the indegree of any vertex is at most $\binom{N}{2}$ and thus the graph has maximum average degree at most $2\binom{N}{2}$.

This indicates that the minor closed classes are perhaps not the most natural restriction in the context of graph partitions. One is naturally led to the study of shallow minors and their edge densities. And this is also how we arrived to this notion.

Very schematically the relationship between the χ_p's and the shallow minors naturally leads to the following two questions:

Do there exist integral functions f_1 and f_2 such that, for every integer p:

(1) If the minors of depth at most $f_1(p)$ of the graphs of a class $\mathcal{C}$ have bounded maximum average degree then the graphs in $\mathcal{C}$ have bounded χ_p?
(2) If the graphs in $\mathcal{C}$ have bounded $\chi_{f_2(p)}$ then do all the minors of depth at most p of the graphs of a class $\mathcal{C}$ have bounded maximum average degree?

Both questions have a positive answer, showing that proper minor closed classes are unnecessarily restrictive for the validity of Theorem 7.6. Perhaps more interestingly our proof does not rely on the Structural Theorem and yields an efficient algorithm (in fact a linear algorithm, see [346, 348, 350, 351, 353, 355] and Chap. 18).

Question (2) is quite easy to answer:

Proposition 7.1. *For every graph* G *and any integer* r*:*

$$\nabla_r(G) \le (2r+1)\binom{\chi_{2r+2}(G)}{2r+2} \tag{7.1}$$

Proof. Consider a vertex coloring c of G with $N = \chi_{2r+2}(G)$ colors such that any $i \le 2r+2$ colors induce a subgraph of tree-depth at most i. For every $J \in \binom{[N]}{2r+2}$, let $G_J = G[c^{-1}(J)]$ and let Y_J be a rooted forest of height $td(G_J) \le 2r+2$ such that $G_J \subseteq clos(Y_J)$.

Let $H \in G \,\nabla\, r$ be such that $\nabla_0(H) = \nabla_r(G)$, and let $\widehat{H}$ be a ramification of H included in G (as a subgraph). Let $(X_1, \dots, X_k)$ be a H-decomposition of $\widehat{H}$ and let $x_1, \dots, x_k$ be the centers of $X_1, \dots, X_k$ (corresponding to the vertices $h_1, \dots, h_p$ of H), see Sect. 4.2 for terminology. If h_i and h_j are adjacent in H then there exists a path $P_{i,j}$ of length at most $2r+1$ linking x_i and x_j. Let $I_{i,j} \in \binom{[N]}{2r+2}$ be such that $I_{i,j} \supseteq c(V(P_{i,j}))$. Then the path $P_{i,j}$ is included in some connected component of $G_{I_{i,j}}$. It follows that there exists in $P_{i,j}$ a vertex $v_{i,j}$ which is minimum with respect to the partial order defined by $Y_{I_{i,j}}$. As $\{x_i, x_j\} \subseteq V(P_{i,j}) \subseteq X_i \cup X_j$ and as $X_i \cap X_j = \emptyset$, $v_{i,j}$ either belongs to X_i or to X_j. Depending on the case, $v_{i,j}$ is a vertex of X_i which is an ancestor of x_j in $Y_{I_{i,j}} \cap X_i$ or a vertex of X_j which is an ancestor of x_i in $Y_{I_{i,j}} \cap X_j$. Thus:

$$\begin{aligned}
p\nabla_r(G) &\leq \sum_{I\in\binom{[N]}{2r+2}} \sum_{1\leq i\leq p} \sum_{\substack{1\leq j\leq p\\ j\neq i}} |\{v : v \text{ ancestor of } x_i \text{ in } Y_I \cap X_j\}| \\
&\leq \sum_{I\in\binom{[N]}{2r+2}} \sum_{i=1}^{p} |\{v : v \text{ ancestor of } x_i \text{ in } Y_I\}| \\
&\leq \binom{N}{2r+2} \times p \times (2r+1)
\end{aligned}$$

Hence

$$\nabla_r(G) \leq (2r+1)\binom{N}{2r+1}$$

□

Problem (1) is much harder and it is the subject of this chapter. It will be obtained by use of special orientations, fraternal orientations and transitive fraternal orientations studied in Sect. 7.3.

Definition 7.2. A digraph $\vec{G}$ is *fraternally oriented* if $(x, z) \in E(\vec{G})$ and $(y, z) \in E(\vec{G})$ implies $(x, y) \in E(\vec{G})$ or $(y, x) \in E(\vec{G})$.

This concept was introduced by Skrien [438] and a characterization of fraternally oriented digraphs having no symmetrical arcs has been obtained by Gavril and Urrutia [213], who also proved that triangulated graphs and circular arc graphs are all fraternally orientable graphs. An orientation is *transitive* if $(x, y) \in E(\vec{G})$ and $(y, z) \in E(\vec{G})$ implies $(x, z) \in E(\vec{G})$. It is obvious that a graph has an acyclic transitive fraternal orientation in which every vertex has indegree at most $k - 1$ if and only if it is the closure of a rooted forest of height k. It follows that tree-depth and transitive fraternal orientation are closely related.

Also, a local version of a transitive and fraternal orientation is formalized by *transitive fraternal augmentations* (each augmentation step consists in adding the missing arcs while applying the fraternity and transitivity rules on the initial arcs) which will be the key concept defined in Sect. 7.3. Towards this end, we define in the next section a local version of tree-depth by means of centered colorings.

7.2 Low Tree-Depth Coloring and p-Centered Colorings

We introduce p-centered colorings, as a local approximation of centered-colorings:

Definition 7.3. A p-*centered coloring* of a graph G is a vertex coloring such that, for any (induced) connected subgraph H, either some color c(H) appears exactly once in H, or H gets at least p colors.

Immediately from the definition we get that any p-centered coloring of a graph G uses at least td(G) colors when p is sufficiently large. This we formulate as follows.

Lemma 7.1. *Let* G, G_0 *be graphs, let* $p = td(G_0)$*, let* c *be a* q*-centered coloring of* G *where* $q \geq p$*. Then every subgraph* G *isomorphic to* G_0 *gets at least* p *colors in the coloring of* G. □

This lemma implies that p-centered colorings induce low tree-depth colorings:

Corollary 7.1. *Let* p *be an integer, let* G *be a graph and let* c *be a* p*-centered coloring of* G*.*

Then $i < p$ *parts induce a subgraph of tree-depth at most* i*.*

Proof. Let G' be any subgraph of G induced by $i < p$ parts. Assume $td(G') > i$. According to (6.1) of Sect. 6.1, the deletion of one vertex decreases the tree-depth by at most one. Hence there exists an induced subgraph H of G' such that $td(H) = i + 1 \leq p$. According to Lemma 7.1 (choosing $G_0 = H$), H gets at least p colors, a contradiction. □

7.3 Transitive Fraternal Augmentation

We start this Section with the main definitions related to transitive fraternal augmentation.

Definition 7.4. Let $\vec{G}$ be a directed graph. A 1-*transitive fraternal augmentation* of $\vec{G}$ is a directed graph $\vec{H}$ with the same vertex set, including all the arcs of $\vec{G}$ and such that, for all vertices x, y, z,

- If (x, z) and (z, y) are arcs of $\vec{G}$ then (x, y) is an arc of $\vec{H}$ (*transitivity*),
- If (x, z) and (y, z) are arcs of $\vec{G}$ then (x, y) or (y, x) is an arc of $\vec{H}$ (*fraternity*).

A 1-transitive fraternal augmentation $\vec{H}$ of $\vec{G}$ is *tight* if for each arc (x, y) in $\vec{H}$ which is not in $\vec{G}$ there exists a vertex z so that (x, z) and at least one of $(z, y), (y, z)$ are arcs of $\vec{G}$.

A *transitive fraternal augmentation* of a directed graph $\vec{G}$ is a sequence $\vec{G} = \vec{G}_1 \subseteq \vec{G}_2 \subseteq \cdots \subseteq \vec{G}_i \subseteq \vec{G}_{i+1} \subseteq \cdots$, such that $\vec{G}_{i+1}$ is a 1-transitive fraternal augmentation of $\vec{G}_i$ for every $i \geq 1$. The transitive fraternal augmentation is *tight* if all the 1-transitive fraternal augmentations of the sequence are tight.

The relationship between transitive fraternal augmentation and low tree-depth coloring can be suggested by the simple example of the relation between 1-transitive fraternal augmentations and *star coloring* (i.e., χ_2-coloring): The star chromatic number of a graph G is the minimum over all the orientations $\vec{G}$ of G of the chromatic number of a tight 1-transitive fraternal augmentation of $\vec{G}$ (see Exercise 7.2).

The key lemma to generalize our approach to general ∇_r states that grads are relatively stable under tight 1-fraternal augmentations. More precisely:

Lemma 7.2. *Let $\vec{G}$ be a directed graph and let $\vec{H}$ be a tight 1-transitive fraternal augmentation of $\vec{G}$. Then*

$$\widetilde{\nabla}_r(H) \leq \left(16r\Delta^-(\vec{G})^2\widetilde{\nabla}_{2r+1/2}(G)\right)^2 \tag{7.2}$$

Proof. As schematically pictured on Fig. 7.3, the graph H is a subgraph of the graph obtained from G by first multiply the vertices by $\Delta^-(\vec{G})+1$ (i.e., take the lexicographic product of G by $K_{\Delta^-(\vec{G})+1}$) an then contract a star forest.

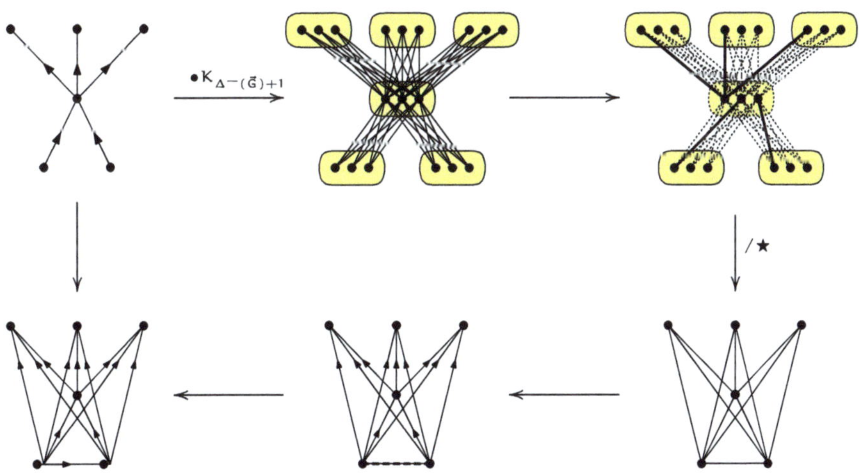

Fig. 7.3 Illustration for the proof of Lemma 7.2. Here, /★ schematically denotes the contraction of a star forest

Hence we have

$$\widetilde{\nabla}_r(H) \le \widetilde{\nabla}_r((G \bullet K_{\Delta^-(\vec{G})+1}) \triangledown 1/2)$$

Thus, according to Proposition 4.6 and Theorem 4.3:

$$\begin{aligned}\widetilde{\nabla}_r(H) &\le 8\big(\max((4r+1)\Delta^-(\vec{G})+1, (\Delta^-(\vec{G})+1)^2)\widetilde{\nabla}_{2r+1/2}(G)+\Delta^-(\vec{G})\big)^2 \\ &\le 32(2r+1)^2(\Delta^-(\vec{G})+1)^4(\widetilde{\nabla}_{2r+1/2}(G)+1)^2 \\ &\le \big(16r\Delta^-(\vec{G})^2\widetilde{\nabla}_{2r+1/2}(G)\big)^2\end{aligned}$$

□

By induction, and using the monotonicity of ∇_r, we get the following consequence:

Corollary 7.2. *There exist polynomials* $R_i(X, Y)$ $(i \ge 1)$, *such that every directed graph* $\vec{G}$ *has a transitive fraternal augmentation* $\vec{G} = \vec{G}_1 \subseteq \vec{G}_2 \subseteq \cdots \subseteq \vec{G}_i \subseteq \cdots$ *where*

$$\Delta^-(\vec{G}_i) \le R_i(\Delta^-(\vec{G}), \nabla_{2^{i+1}-1}(G)) \tag{7.3}$$

The transitive fraternal augmentation changes a given graph by locally creating many complete bipartite subgraphs. This will allow us to control the tree-depth of subgraphs. A transitive fraternal augmentation of a graph naturally induce a transitive fraternal augmentation of its subgraphs. This will allow us, by applying the following lemma to the subgraphs of a given graph G, to control the tree-depth of the subgraphs of G induced by p colors.

Lemma 7.3. *Let* $N(p, t) = 1 + (t-1)(2 + \lceil \log_2 p \rceil)$, *let* $\vec{G}$ *be a directed graph and let* $\vec{G} = \vec{G}_1 \subseteq \vec{G}_2 \subseteq \cdots \subseteq \vec{G}_i \subseteq \cdots$ *be a transitive fraternal augmentation of* $\vec{G}$.

Then $\vec{G}_{N(p,td(G))}$ *either includes an acyclically oriented clique of size* p *or it includes a rooted directed tree* $\vec{Y}$ *such that* $G \subseteq \mathrm{clos}(\vec{Y})$ *and* $\mathrm{clos}(\vec{Y}) \subseteq \vec{G}_{N(p,td(G))}$.

Proof. We fix the integer p and prove the lemma by induction on $t = td(\vec{G})$. The base case $t = 1$ corresponds to a graph without edges, for which the property is obvious. Assume the lemma has been proved for directed graphs with tree-depth at most t and let $\vec{G}$ be a directed graph with tree-depth $t+1$. As we may consider each connected component of $\vec{G}$ independently, we may assume that $\vec{G}$ is connected. Then there exists a vertex $s \in V(\vec{G})$ such that the connected components $\vec{H}_1, \ldots, \vec{H}_k$ of $G - s$ have tree-depth at most t. As $\vec{H}_i = \vec{G}_1[V(\vec{H}_i)] \subseteq \cdots \subseteq \vec{G}_j[V(\vec{H}_i)] \subseteq \cdots$ is a transitive fraternal augmentation of $\vec{H}_i$ we have, according to the induction hypothesis, that, for each $1 \le i \le k$, there exists in $\vec{H}_i$ either an acyclically oriented clique of size p or a rooted tree $\vec{Y}_i$ rooted at r_i such that $H_i \subseteq \mathrm{clos}(Y_i)$ and $\mathrm{clos}(\vec{Y}_i) \subseteq \vec{G}_{N(p,td(G))}[V(\vec{H}_i)]$. If the first case occurs for some i, then $\vec{G}$ includes an acyclically oriented clique of size p. Hence assume it does not and that for every i we have a rooted tree $\vec{Y}_i$ with the above properties. As $\vec{G}$ is connected, the vertex s has at least one neighbor x_i in $\vec{H}_i$ (for each $1 \le i \le k$). Let x be any neighbor of s in $\vec{H}_i$. If y is an ancestor of x in $\vec{Y}_i$, (y, x) is an arc of $\vec{G}_{N(p,t)}$ hence s and y are adjacent in $\vec{G}_{N(p,t)+1}$. Moreover, if (x, s) is an arc of $\vec{G}_{N(p,t)}$ then (y, s) is an arc of $\vec{G}_{N(p,t)+1}$. Let D_i be the subset of $V(\vec{H}_i)$ of the vertices x such that (x, s) belongs to $\vec{G}_{N(p,t)}$ and of their ancestors in $\vec{Y}_i$ and let $D = \bigcup_{i=1}^{k} D_i$. Then D induces a clique in $\vec{G}_{N(p,t)+2}$. Thus there exists a directed Hamiltonian path $\vec{P}$ in $\vec{G}_{N(p,t)+2}[D]$, i.e., a directed path containing all the vertices of $\vec{G}_{N(p,t)+2}[D]$.

Let r be the start vertex of $\vec{P}$. Define $\pi : V(G) - r \to V(G)$ as follows:

- If $x \in D$, the $\pi(x)$ is the predecessor y of x in $\vec{P}$ (the arc (y, x) belongs to $\vec{G}_{N(p,t)+2}$);
- Otherwise, if $x = s$, $\pi(x)$ is the end vertex y of $\vec{P}$ (the arc (y, x) belongs to $\vec{G}_{N(p,t)+1}$);
- Otherwise, if $x = r_i$ then $\pi(x) = s$ (the arc (s, r_i) belongs to $\vec{G}_{N(p,t)+2}$);
- Otherwise, if the father of $x \in V(\vec{H}_i) \setminus D$ does not belong to D, then $\pi(x)$ is the father of x in $\vec{Y}_i$;
- Otherwise, if no descendant of x in $\vec{Y}_i$ has an arc coming from s in $\vec{G}_{N(p,t)+1}$, $\pi(x)$ is the father of x in $\vec{Y}_i$;
- Otherwise, $\pi(x) = s$ (the arc (s, x) belongs to $\vec{G}_{N(p,t)+2}$).

It is easily checked that this *father mapping* π actually defines a directed rooted tree $\vec{Y}$ in $\vec{G}_{N(p,t)+2}$ with root r and that $G \subseteq \mathrm{clos}(\vec{Y})$. Moreover, either $\vec{Y}$ has height at least p and $\vec{G}_{N(p,t)+2+\lceil \log_2 p \rceil}$ includes an acyclically oriented clique of size p or $\mathrm{clos}(\vec{Y}) \subseteq \vec{G}_{N(p,t)+2+\lceil \log_2 p \rceil}$. As $N(p, t+1) = N(p, t) + 2 + \lceil \log_2 p \rceil$, the induction follows. □

Lemma 7.4. *Let* p *be an integer, let* $\vec{G}$ *be a directed graph and let* $\vec{G} = \vec{G}_1 \subseteq \vec{G}_2 \subseteq \cdots \subseteq \vec{G}_i \subseteq \cdots$ *be a transitive fraternal augmentation of* $\vec{G}$. *Then either* $\vec{G}_{N(p,p)}$ *includes an acyclically oriented clique of size* p *or* $td(G) \le p - 1$ *and there exists in* $\vec{G}_{N(p,p)}$ *a rooted directed tree* Y *so that* $G \subseteq \mathrm{clos}(Y)$ *and* $\mathrm{clos}(\vec{Y}) \subseteq \vec{G}_{N(p,p)}$.

Proof. If $td(G) > p$ we may consider a connected subgraph of H of tree-depth p. According to Lemma 7.3, there will exists in $\vec{G}_{N(p,p)}[V(H)]$ an acyclically oriented clique of size p or a rooted directed tree $\vec{Y}$ so that $H \subseteq \mathrm{clos}(Y)$ and

$\mathrm{clos}(\vec{Y}) \subseteq \vec{G}_{N(p,p)}[V(H)]$. In the later case, if $\mathrm{td}(G) = p$ then the height of $\vec{Y}$ is at least $\mathrm{td}(H) = p$ and $\mathrm{clos}(\vec{Y})$ includes an acyclically oriented clique of size p. □

The previous two results translate immediately to low tree-depth colorings:

Corollary 7.3. *Let* $S(p) = 1 + (p-1)(2 + \lceil \log_2 p \rceil) = O(p \log_2 p)$.

For every graph G, *for every transitive fraternal augmentation* $\vec{G} = \vec{G}_1 \subseteq \vec{G}_2 \subseteq \cdots \subseteq \vec{G}_i \subseteq \cdots$ *of* G *and for every integer* p*:*

$$\chi_p(G) \le 2\Delta^-(\vec{G}_{S(p)}) + 1 \tag{7.4}$$

Hence we have (which completes the item (1) above):

Theorem 7.7 (Low tree-depth coloring). *For every integer* $p \ge 1$, *there exist a polynomial* R_p, *(different from the one of Corollary 7.2), such that for every graph* G *holds*

$$\chi_p(G) \le R_p(\nabla_0(G), \nabla_{(8p)^{p-1}}(G)).$$

Proof. This is a direct consequence of Corollary 7.3 and 7.2. □

But this is not the end of the augmentation story. In Sect. 7.5 we shall introduce a related probabilistic proof of [476] and in the next section we introduce a different strategy of computing augmentations. This strategy will give us the best dependence ($\chi_p(G)$ is bounded by a polynomial function of the top-grad of rank $2^{p-2} + \frac{1}{2}$). In this approach we compute fraternal augmentations first and then consider the coloring induced by an implicit transitive augmentation. The advantage of this approach is that the degrees of the vertices cannot augment more than a ratio bounded by a polynomial of the top-grads.

7.4 Fraternal Augmentations of Graphs

We present the core of our coloring algorithm as a relabelling procedure. This is of course a technical part of the algorithm.

Let V be a finite set and let k be an integer. A k*-fraternity function* is a function $w : V \times V \to \mathbb{N} \cup \{\infty\}$ such that for every $x, y \in V$ one of $w(x, y)$ and $w(y, x)$ (at least) is ∞ and such that for every $x \ne y \in V$:

- Either $\min(w(x,y), w(y,x)) = 1$,
- Or $\min(w(x,y), w(y,x)) = \min_{z \in V \setminus \{x,y\}} w(x,z) + w(y,z)$,
- Or $\min(w(x,y), w(y,x)) > k$
  and $\min_{z \in V \setminus \{x,y\}} w(x,z) + w(y,z) > k$.

Notice that a k-fraternity function is a k' fraternity function for $k' \leq k$. A *fraternity function* (or ∞*-fraternity function*) is a function which is a k-fraternity function for every k.

Corresponding to a fraternity function w, we define

- The directed graph $\vec{G}_i^w$ (for $i \geq 1$), whose vertex set is $\vec{G}^w$ is V and whose arcs are all the pairs (x,y) such that $w(x,y) = i$;
- The directed graph $\vec{G}_{\leq i}^w$ (for $i \geq 1$), whose vertex set is V and whose arcs are all the pairs (x,y) such that $w(x,y) \leq i$;
- The directed graph $\vec{G}^w$ whose vertex set is V and whose arcs are all the pairs (x,y) such that $w(x,y) \leq i$ (see Fig. 7.4).

Obviously

$$\vec{G}_1^w = \vec{G}_{\leq 1}^w \subseteq \cdots \subseteq \vec{G}_{\leq k}^w \subseteq \vec{G}^w.$$

We define the values $\Delta_i^-(w)$ by

$$\Delta_i^-(w) = \Delta^-(\vec{G}_i^w) = \max_{y \in V} |\{x \in V : w(x,y) = i\}|.$$

Let Γ_i^w be the graph whose vertex set is the disjoint union of V and a of set of cardinality at most $(i-1)|w^{-1}(i)|$, which is obtained from the empty graph by adding, for each pair $(x,y) \in V^2$ such that $w(x,y) = i$ an induced path of length i linking x and y (with no interior vertex in V), see Fig. 7.5. In other words, Γ_i^w is the $(i-1)$-subdivision of the graph G_i^w underlying $\vec{G}_i^w$. Hence

$$G_i^w \in \Gamma_i^w \,\widetilde{\nabla}\, (i-1)/2.$$

Let w be a k-fraternity function. We shall prove that for every integer $a \leq k$, the graph Γ_i^w may be injectively embedded into a blowing $\Gamma_1^W \bullet \overline{K}_{N(a)}$ of Γ_1^W (see Fig. 7.6).

Toward this end, we introduce the following notations:

- For a finite set X, the empty graph with vertex set X is denoted by E_X.
- For each integer $1 \leq a \leq k$, define the sets Ω_a and Z_a by

$$\Omega_a = \{(i,\alpha) : 1 \leq i < a \text{ and } 1 \leq \alpha \leq \Delta_{a-i}^-(w)\}$$

(ordered according to lexicographic order), and

$$Z_a = \{((i,\alpha),(j,\beta)) \in \Omega_a^2 : (i,\alpha) < (j,\beta) \text{ and } i+j = a\}.$$

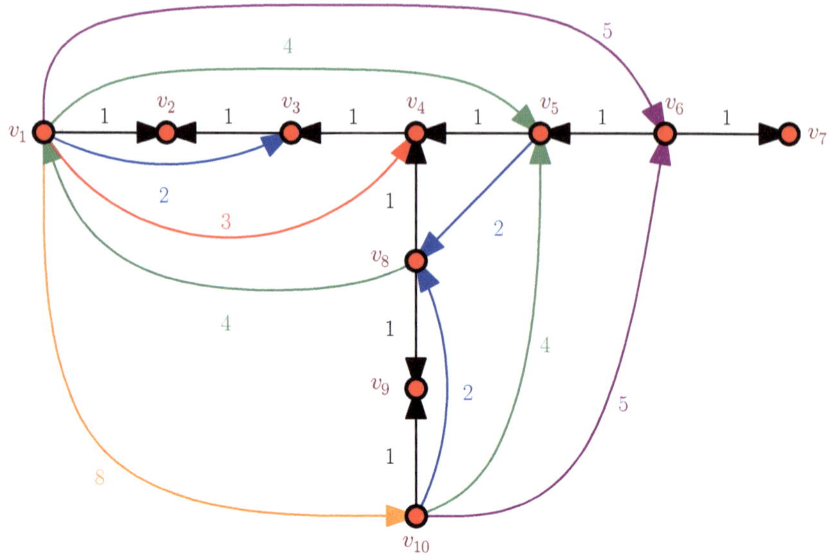

$$\left(w(v_i, v_j)\right)_{i,j} = \begin{pmatrix} \infty & 1 & 2 & 3 & 4 & 5 & \infty & \infty & \infty & 8 \\ \infty & \infty & \infty & \infty & \infty & \infty & \infty & \infty & \infty & \infty \\ \infty & 1 & \infty & \infty & \infty & \infty & \infty & \infty & \infty & \infty \\ \infty & \infty & 1 & \infty & \infty & \infty & \infty & \infty & \infty & \infty \\ \infty & \infty & \infty & 1 & \infty & \infty & \infty & 2 & \infty & \infty \\ \infty & \infty & \infty & \infty & 1 & \infty & 1 & \infty & \infty & \infty \\ \infty & \infty & \infty & \infty & \infty & \infty & \infty & \infty & \infty & \infty \\ 4 & \infty & \infty & 1 & \infty & \infty & \infty & \infty & 1 & \infty \\ \infty & \infty & \infty & \infty & \infty & \infty & \infty & \infty & \infty & \infty \\ \infty & \infty & \infty & \infty & 4 & 5 & \infty & 2 & 1 & \infty \end{pmatrix}$$

Fig. 7.4 The directed graph $\vec{G}^w$ defined by a fraternity function w. The value on arc (x, y) is $w(x, y)$; a pair (x, y) is an arc if $w(x, y) < \infty$

- We define inductively the sets R_i $(1 \leq i \leq k)$ by:

$$R_i = \begin{cases} \emptyset, & \text{if } i = 1; \\ Z_2, & \text{if } i = 2; \\ \left(\bigcup_{j=2}^{i-1} R_j \times [\Delta^-_{i-j}(w)]\right) \cup Z_i, & \text{otherwise.} \end{cases}$$

(All the unions above are disjoint unions, and $[N]$ denotes the set $\{1, \ldots, N\}$.)

We take time out for a lemma:

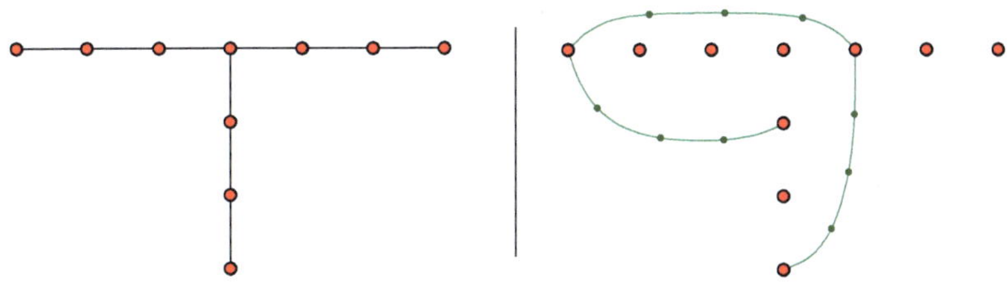

Fig. 7.5 The graphs Γ_1^w and Γ_4^w defined by the fraternity function shown Fig. 7.4

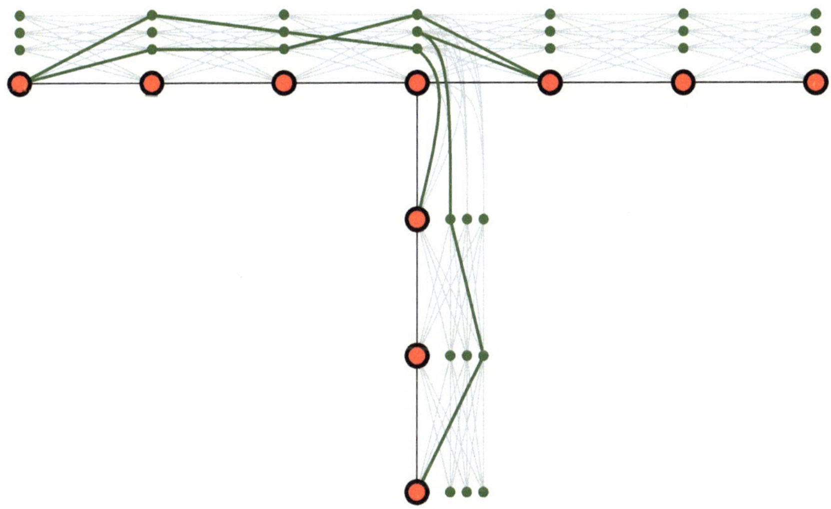

Fig. 7.6 Injective embedding of Γ_4^w into a blowing of Γ_1^w

Lemma 7.5. *For every integer* $1 \leq a \leq k$ *there exists an injective homomorphism* $f_a : \Gamma_a^w \to \Gamma_1^w \bullet E_{\{\epsilon\}\cup R_a}$ *such that for every* $v \in V$ *holds* $f_a(v) = (v, \epsilon)$.

Proof. The proof is by induction on a. If $a = 1$ the proof is straightforward. Assume that the properties holds for all $a \leq n$ (for some $1 \leq n < k$) and let $a = n + 1$.

Consider a coloring λ of the pairs (x, y) such that $w(x, y) < a$ such that

- If $w(x, z) = w(y, z)$ and $w(x, z) < a$ then $\lambda(x, z) \neq \lambda(y, z)$,
- If $w(x, y) < a$ then $1 \leq \lambda(x, y) \leq \Delta^-_{w(x,y)}$.

For $(i, \alpha) \in \Omega_a$, let $\phi_{i,\alpha}$ be an injective homomorphism of Γ_i^w to $\Gamma_1^w \bullet E_{\{\epsilon\}\cup R_i \times \{\alpha\}}$ such that $\phi_{i,\alpha}(v) = (v, \epsilon)$ for every $v \in V$ (such a homomorphism exists by induction hypothesis).

We now construct the injective homomorphism f_a: for every $v \in V$, define $f_a(v) = (v, \epsilon)$; for every pair $(x, y) \in w^{-1}(a)$, there exists $z \in V$ such that $w(x,y) = w(x,z) + w(y,z)$. Let $i = w(x,z), j = w(y,z), \alpha = w(x,z)$ and $\beta = \lambda(y,z)$. Because the involved graphs are not oriented, we may assume without loss of generality that $(i, \beta) < (j, \alpha)$ (the pairs cannot be equal according to the definition of λ). Let $P_{x,y} = (x, v_1, \dots, v_{a-1}, y)$ be the path of Γ_a^w linking x and y, let $P'_{x,z} = (x, s_1, \dots, s_{i-1}, z)$ be the path of Γ_i^w linking x and z, and let $P''_{z,y} = (z, t_1, \dots, t_{j-1}, y)$ be the path of Γ_j^w linking z and y. Then we define

$$f_a(v_l) = \begin{cases} \phi_{i,\beta}(s_i), & \text{if } 1 \le l < i, \\ (z, ((i,\beta),(j,\alpha))), & \text{if } l = i, \\ \phi_{j,\alpha}(t_{l+j-a}), & \text{otherwise.} \end{cases}$$

Then it is easily checked that f_a meets the requirements of the lemma. □

We shall now prove that it is possible to extend a k-fraternity function w into a $(k+1)$-fraternity function w' while bounding $\Delta^-_{k+1}(w')$ by some polynomial function of $\Delta^-_k(w)$ and $\widetilde{\nabla}_{k/2}(G_1^w)$. For this purpose, we associate to every k-fraternity function w the function $N_w : [k+1] \to \mathbb{R}$ inductively defined by as follows:

$$N_w(i) = \begin{cases} 0, & \text{if } i = 1; \\ \binom{\Delta^-_1(w)}{2}, & \text{if } i = 2; \\ \sum_{j=2}^{i-1} N_w(j)\Delta^-_{i-j}(w) & \\ \quad + \sum_{j=1}^{(i-1)/2} \Delta^-_j(w)\Delta^-_{i-j}(w), & \text{if } i \equiv 1 \pmod 2; \\ \sum_{j=2}^{i-1} N_w(j)\Delta^-_{i-j}(w) & \\ \quad + \sum_{j=1}^{i/2-1} \Delta^-_j(w)\Delta^-_{i-j}(w) + \binom{\Delta^-_{i/2}(w)}{2}, & \text{if } i \equiv 0 \pmod 2. \end{cases}$$

Notice that $N_w(i)$ is a polynomial in $\Delta^-_1(w), \dots, \Delta^-_{i-1}(w)$ of degree i. Precisely, for each monomial $\Delta^-_1(w)^{a_1}\Delta^-_2(w)^{a_2}\dots\Delta^-_{i-1}(w)^{a_{i-1}}$ appearing in $N_w(i)$ we have $\sum_{j=1}^{i-1} j\, a_j \le i$.

The function N_w is governing the blowing procedure:

Lemma 7.6. *Let V be a finite set, let $k \ge 1$ be an integer, let w be a k-fraternity function and let $G = G_1^w$.*

There exists a $(k+1)$-fraternity function w' such that

$$\forall (x,y) \in V^2, \quad w(x,y) \le k \implies w'(x,y) = w(x,y)$$
$$\Delta^-_{k+1}(w') \le \widetilde{\nabla}_{k/2}(G \bullet \overline{K}_{1+N_w(k+1)})$$

Proof. Consider any linear order on V. Let w_1 be the $(k+1)$-fraternity function defined from the truncation of w:

- $w_1(x,y) = w(x,y)$ if $w(x,y) \le k$ (i.e., w_1 extends w);
- $w_1(x,y) = k+1$ if $x < y$, $\min(w(x,y), w(y,x)) > k$, and there exists $z \in V$ such that $w(x,z) + w(y,z) = k+1$;
- $w_1(x,y) = \infty$, otherwise.

The function w_1 is clearly a $(k+1)$-fraternity function.

According to Lemma 7.5, $\Gamma^{w_1}_{k+1} \in G \bullet \overline{K}_{1+N_w(k+1)}$. As $G^{w_1}_{k+1} \in \Gamma^{w_1}_{k+1} \,\widetilde{\nabla}\, k/2$ we deduce that

$$\nabla_0(G^{w_1}_{k+1}) = \widetilde{\nabla}_{k/2}(\Gamma^{w_1}_{k+1}) \le \widetilde{\nabla}_{k/2}(G \bullet \overline{K}_{1+N_w(k+1)}).$$

Hence there exists an orientation $\vec{H}$ of $G^{w_1}_{k+1}$ such that $\Delta^-(\vec{H}) \le \widetilde{\nabla}_{k/2}(G \bullet \overline{K}_{1+N_w(k+1)})$.

Let w' be the $(k+1)$-fraternity function defined by $w_1(x,y) = w(x,y)$ if $w(x,y) \le k$, $w_1(x,y) = k+1$ if (x,y) is an arc of $\vec{H}$, and such that $w_1(x,y) = \infty$ otherwise. Then the conclusion of the lemma holds. □

We shall now see that fraternity functions are related to p-centered colorings.

Let p be an integer, let $k = 2^{p-1} + 2$, let w be a k-fraternity function on a set V, and let $G = G^w_1$. Let $c : V \to \mathbb{N}$ be a coloring such that $c(x) \ne c(y)$ if there exists in $\vec{G}^w_{\le k}$ a directed path from x to y of length at most p. Intuitively, the coloring c will handle the transitivity part of the augmentation.

Lemma 7.7. *Let P be a p-colored path in G and let V_P be the vertex set of P.*

Then the length of P is at most $2^p - 2$, and there exists a vertex $s \in V_P$ such that every other vertex $v \in V_P$ may be reached from s by a directed path of $\vec{G}^w_{\le k}[V_P]$.

Proof. Let P be a p-colored path in G. By considering a subpath, we may assume that the length of P is at most $2^p - 1$ (as there exists a p-colored path of length $L > 2^p - 2$ if and only if there exists a p-colored path of length exactly $2^p - 1$).

Let $\phi : \mathbb{N} \to \mathbb{N}$ be defined by $\phi(i) = \lceil \log_2(i) \rceil$.

Let L be the length of P. We defined inductively oriented paths $\vec{P}_i = (x_{i,1}, \ldots, x_{i,L_i})$ in $\vec{G}^w_{\le k}[V_P]$ (i.e., in the subgraph of $\vec{G}^w_{\le k}$ induced by the vertex set of P) of length $L_i = L + 1 - i$ such that

- Every vertex of V_P out of $\vec{P}_i$ can be reached from a vertex of $\vec{P}_i$ by a directed path of $\vec{G}^w_{\leq k}[V_P]$;
- For every adjacent vertices u, v of $\vec{P}_i$ such that $w(u,v) < \infty$ there exists in $\vec{G}^w_{\leq k}[V_P]$ a directed path of length $\phi(w(u,v))$ starting from v which intersects $\vec{P}_i$ only at v.

Let $\vec{P}_1 = (x_{1,1}, \dots, x_{1,L_1})$ be the oriented path corresponding to P and assume $\vec{P}_i$ has been constructed, for some $i \geq 1$.

If there exists an internal vertex $x_{i,t}$ of P_i which is a sink and $w(x_{i,t-1}, x_{i,t}) + w(x_{i,t+1}, x_{i,t}) \leq k$ then we define

$$\vec{P}_{i+1} = (x_{i+1,1}, \dots, x_{i+1,L_{i+1}}),$$

where $L_{i+1} = L_i - 1$, $x_{i+1,j} = x_{i,j}$ if $j < t$, and $x_{i+1,j} = x_{i,j+1}$ if $j \geq t$ (see Fig. 7.7).

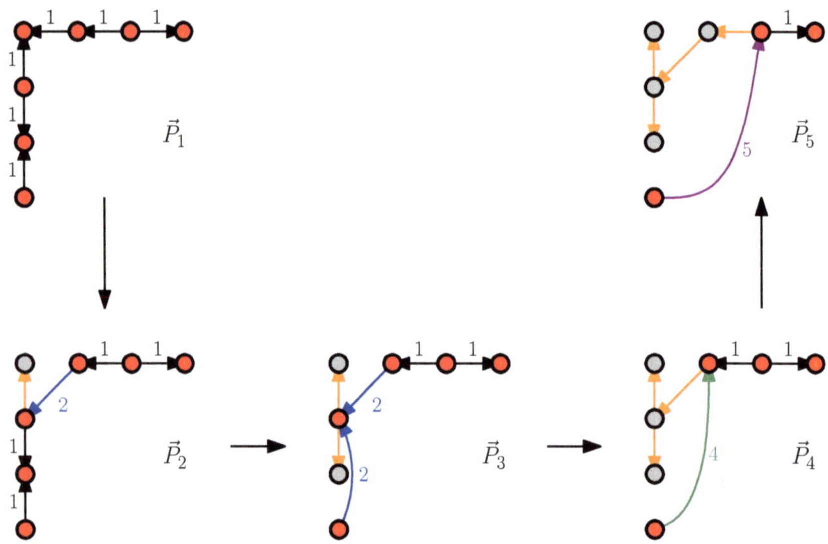

Fig. 7.7 Construction of the sequence of paths $\vec{P}_1, \vec{P}_2, \dots$

To check that this directed path has the required properties we only need to consider vertices $u = x_{i+1,t-1} = x_{i,t-1}$ and $v = x_{i+1,t} = x_{i,t+1}$. Up to a relabeling, we may assume that $w(u,v) < \infty$. By induction, there exists a directed path $\vec{P}'$ in $\vec{G}^w_{\leq k}[V_P]$ starting from $x_{i,t}$ and intersecting $\vec{P}_i$ only at $x_{i,t}$, whose length is at least

$$\begin{aligned}\|\vec{P}'\| &= \max(\phi(w(x_{i,t-1}, x_{i,t})), \phi(w(x_{i,t+1}, x_{i,t})))\\ &\geq \phi(w(x_{i,t-1}, x_{i,t}) + w(x_{i,t+1}, x_{i,t})) - 1\\ &\geq \phi(w(u,v)) - 1.\end{aligned}$$

We deduce that there exists a directed path in $\vec{G}^{w}_{\leq k}[V_P]$ starting from v and intersecting $\vec{P}_{i+1}$ only at v whose length is $\phi(w(u,v))$ (this is $(v, x_{i,t})$ followed by $\vec{P}'$).

By iterating this process as long as possible, we end with an directed path $\vec{P}_n = (x_{n,1}, \dots, x_{n,L_n})$ which will make evident that the property of the lemma holds.

Assume for contradiction that $\vec{P}_n$ has an internal vertex $x_{n,t}$ which is a sink. Then $w(x_{n,t-1}, x_{n,t}) + w(x_{n,t+1}, x_{n,t}) > 2^p - 2$ thus one of $w(x_{n,t-1}, x_{n,t})$ and $w(x_{n,t+1}, x_{n,t})$ (say $w(x_{n,t-1}, x_{n,t})$) is at least $2^{p-2}+1$. Thus there exists a directed path in $\vec{G}^{w}_{\leq k}[V_P]$ starting from $x_{n,t}$ and intersecting $\vec{P}_n$ only at $x_{n,t}$, whose length is $\phi(w(x_{n,t-1}, x_{n,t})) \geq \phi(2^{p-2}+1) = p-1$. Hence there exists a directed path in $\vec{G}^{w}_{\leq k}[V_P]$ starting from $x_{n,t-1}$ whose length is p. This path cannot be p-colored according to our assumption on the coloring, hence a contradiction.

It follows that there exists in $\vec{P}_n$ a vertex s which is linked to all the other vertices of $\vec{P}_n$ by a directed path in $\vec{P}_n$ hence to all the other vertices of V_P by a directed path of $\vec{G}^{w}_{\leq k}[V_P]$ (of length at most $p-1$, according to the coloration assumption).

The removal of s splits the path P into two $(p-1)$-colored sub-paths, one of them having length at least $\lceil L/2 \rceil$. By induction, it follows that $L \leq 2^p - 1$. □

Lemma 7.8. *The coloring c is a* $(p+1)$*-centered coloring of* G*. Hence* $\chi_p(G) \leq |c(V)|$.

Proof. Consider any induced subgraph $G[A]$ of G and assume that A contains at most p colors.

We define a function s which maps paths of $G[A]$ to A in such a way that for every path P, the vertex $s(P)$ is linked to all the other vertices of P by a directed path in $\vec{G}^{w}_{k}[A]$.

Define a relation $\prec_A$ on A by $u \prec_A v$ if there exists v' with the same color as v and a path $P_{v,v'}$ linking v to v' such that there exists a directed path in $\vec{G}_q$ from u to $s(P_{v,v'})$. This relation is antisymmetric: otherwise, we would find in $\vec{G}_q$ a directed path from u to $s(P_{v,v'})$, one from $s(P_{v,v'})$ to v, one from v to $s(P_{u,u'})$ and one from $s(P_{u,u'})$ to u' thus there would exist in $\vec{G}_q$ a directed path from u to u', contradicting the hypothesis that u and u' are colored the same. Also, the relation is transitive (similar proof) hence it defines a partial order. As $u \prec_A v$ implies that there exists in $\vec{G}_q$ a directed path from u to v, we deduce that the partial order $\prec_A$ has no chain of length strictly greater than p.

Let r be a minimal element of $\prec_A$. Then no other vertex in A has the same color as r for otherwise there would exist a vertex r' colored the same as r and a path P linking r and r' (as $G[A]$) is connected hence we would get $s(P) \prec_A r$, contradicting the minimality of r.

It follows that the coloring if $(p+1)$-centered. □

After all these preliminaries (and admittedly technical) results we arrive to the following main result of this section:

Theorem 7.8 (Low tree-depth coloring). *Let $\vec{G}$ be a directed graph. Define recursively*

$$A_1 = 0$$
$$B_1 = \Delta^-(\vec{G})$$

and inductively, for $i > 1$:

$$A_i = \sum_{j=2}^{i-1} A_j B_{i-j} + \frac{1}{2}\sum_{j=1}^{i-1} B_j B_{i-j}$$
$$B_i = \max((i-1)A_i + 1, (A_i+1)^2)\widetilde{\nabla}_{(i-1)/2}(G) + A_i$$

Then, for every integer $p \geq 2$:

$$\chi_p(G) \leq 1 + 2\sum_{i=1}^{2^{p-1}+2} B_i.$$

Hence $\chi_p(G)$ is bounded by a polynomial $P_p(\widetilde{\nabla}_{2^{p-2}+1/2}(G))$, where P_p has degree about 2^{2^p}.

Actually, $\widetilde{\nabla}_{(p-1)/2}(G)$ and $\min\{\Delta^-_{\leq p}(w) : G_1^w \cong G\}$ are polynomially equivalent for each fixed p. The proof of Theorem 7.8 follows from the computation formulated in the following two lemmas.

For a path P in $G^w_{\leq k}$ we define the *w-length* of $P = (x_1, \dots, x_q)$ by

$$\|P\|_w = \sum_{i=1}^{q-1} \min(w(x_i, x_{i+1}), w(x_{i+1}, x_i))$$

Lemma 7.9. *Let w be a k-fraternity function and let P be a path of w-length at most k of $G^w_{\leq k}$ linking vertices x and y. Then there exists $z \in V(P)$ such that either $z \neq x$ and $w(z,x) \leq k$ or $z \neq y$ and $w(z,y) \leq k$.*

Proof. We prove the lemma by induction on the length of P. If $\|P\| = 1$, then one of $w(x,y)$ and $w(y,x)$ is at most k hence we let $z = x$ or $z = y$. Assume that the lemma holds if $\|P\| = i$ where $1 \leq i < k$ and assume $\|P\| = i+1$. Let $P = (x = v_1, v_2, \ldots, v_i, v_{i+1} = y)$. If $w(v_2, v_1) \leq k$ then let $z = v_2$. Otherwise, if $w(v_i, v_{i+1}) \leq k$ then let $z = v_i$. Otherwise, let j be the smallest integer such that $w(v_{j+1}, v_j) \leq k$ (hence $1 < j \leq i$). By minimality we have $w(v_{j-1}, v_j) \leq k$ hence

$$\min(w(v_{j-1}, v_{j+1}), w(v_{j+1}, v_{j-1})) \leq w(v_{j-1}, v_j) + w(v_{j+1}, v_j) \leq \|P\|_w \leq k.$$

It follows that $\{v_{j-1}, v_{j+1}\}$ is an edge of $G^w_{\leq k}$. Consider the path

$$P' = (v_1, \ldots, v_{j-1}, v_{j+1}, \ldots, v_{i+1}).$$

Then $\|P'\| = i$ and $\|P'\|_w \leq \|P\|_w \leq k$. By induction there exists $z \in V(P') \subset V(P)$ such that either $z \neq x$ and $w(z,x) \leq k$ or $z \neq y$ and $w(z,y) \leq k$. □

Lemma 7.10. *For every k-fraternity function w we have:*

$$\Delta^-_{\leq k}(w) \geq \widetilde{\nabla}_{(k-1)/2}(G^w_1).$$

Proof. Consider a $\leq k$-subdivision S of a graph H in G^w_1 such that $\nabla_0(H) = \widetilde{\nabla}_{(k-1)/2}(H)$. Let $\vec{H}$, be an orientation of H computed as follows: Each edge $\{u,v\}$ of H corresponds to a path P_{uv} of length at most k linking u to v in S hence in G. As $\|P_{uv}\|_w = \|P\| \leq k$, there exists, according to Lemma 7.9 a vertex z $V(P_{uv})$ such that either $z \neq u$ and $w(z,u) \leq k$ or $z \neq v$ and $w(z,v) \leq k$. In the first case we orient $\{u,v\}$ from v to u in $\vec{H}$ and from u to v otherwise. As the branches are internally vertex disjoints the indegree of a vertex u in $\vec{H}$ will be at most equal to the number of vertices z such that $w(z,u) \leq k$. Hence

$$\Delta^-_{\leq k}(w) \geq \Delta^-(\vec{H}) \geq \nabla_0(H) = \widetilde{\nabla}_{(k-1)/2}(G).$$

□

Theorem 7.8 gives us yet another characterization of nowhere dense classes:

Theorem 7.9. *Let $\mathcal{C}$ be an infinite class of graphs. Define*

$$f(\mathcal{C}) = \lim_{p\to\infty} \limsup_{G\in\mathcal{C}} \frac{\log \chi_p(G)}{\log|G|}.$$

Then, the class $\mathcal{C}$ is

- *A nowhere dense class if and only if $f(\mathcal{C}) = 0$,*
- *A somewhere dense class if and only if $f(\mathcal{C}) \geq 1/2$.*

Proof. For $p-2$ subdivisions of large complete graphs, we have $\chi_p(G) \approx \sqrt{|G|/p}$ hence $f(\mathcal{C}) \geq 1/2$ if $\mathcal{C}$ is somewhere dense.

According to Theorem 7.8, $\chi_p(G) \leq P_p(\widetilde{\nabla}_{2^{p-2}+1/2}(G))$ (for some polynomial P_p independent of G) Hence if G is not edgeless we have:

$$\log \chi_p(G) = O(\log \widetilde{\nabla}_{2^{p-2}+1/2}(G)),$$

Thus

$$\lim_{p\to\infty} \limsup_{G\in\mathcal{C}} \frac{\log \chi_p(G)}{\log|G|} \leq \overline{\ell\mathrm{dens}}(\mathcal{C}^{\widetilde{\nabla}}) - 1.$$

As for every graph G (of order at least 1) and every positive integer p we have $\chi_p(G) \geq 1$, it follows that $f(\mathcal{C}) = 0$ if $\mathcal{C}$ is nowhere dense. □

The transitive fraternal augmentation and the fraternal augmentation processes have many advantages. Particularly, as we shall see in Chap. 18, transitive fraternal augmentation leads to low complexity algorithms for very diverse problems. On the one hand these augmentations have the advantage of allowing any initial indegree bounded orientation; on the other hand, they have the disadvantage that they cannot preserve in the augmentation the acyclicity of an initial orientation while keeping the indegrees bounded (see Exercise 7.4).

7.5 The Weak-Coloring Approach

In [476], Xuding Zhu proposed an alternate approach to proving existence of low tree-depth colorings, by using generalized coloring numbers. He then showed that these are polynomially equivalent to grads. In this section we review Zhu's approach [476]. The notations are those introduced in Sect. 4.9.

Theorem 7.10. *If* G *is a graph with* $\mathrm{wcol}_{2^{p-2}}(G) \leq m$, *then* G *has a* p*-centered coloring using at most* m *colors.*

Proof. Let $L = v_1 v_2 \ldots v_n$ be an ordering of the vertices of G with

$$\max_{v \in V(G)} |Q_{2^{p-2}}(G_L, v)| \leq m - 1.$$

Color the vertices of G greedily, using the order L, so that the color assigned to v is distinct from colors assigned to vertices in $Q_{2^{p-2}}(G_L, v)$. As $|Q_{2^{p-2}}(G_L, v)| \leq m - 1$, m colors suffice. We claim that such a coloring is a p-centered coloring. Let H be a connected subgraph of G. Let v be the minimum vertex of H with respect to L. If the color $c(v)$ of v appears exactly once in H, then we are done.

Assume $c(v)$ is used more than once in H. We shall prove that H uses at least p colors. Let $u \neq v$ be a vertex of H with $c(u) = c(v)$, and let $P_0 = (v = v_0, v_1, v_2, \ldots, v_= u)$ be a path in H connecting v and u. We must have $q > 2^{p-2}$, for otherwise v is weakly 2^{p-2}-accessible from u, i.e., $v \in$, $Q_{2^{p-2}}(G_L, v)$ and we should have $c(u) \neq c(v)$.

Let $u_0 = v$, and let $P_1 = (v_1, \ldots, v_{2^{p-2}})$ be the subpath of P_0. Observe that no vertex of P_1 uses color $c(u_0)$ and P_1 contains 2^{p-2} vertices. Assume $0 \leq j \leq p-2$, and a vertex u_j of P_j, and a subpath P_{j+1} of P_j are chosen such that the following hold:

- No vertex of P_{j+1} uses the color of $c(u_j)$.
- P_{j+1} contains at least 2^{p-j-2} vertices.

Let u_{j+1} be the minimum vertex of P_{j+1} with respect to L. If $j \leq p-3$, then let P_{j+2} be the largest component of $P_{j+1} - u_{j+1}$. Then u_{j+1} is weakly 2^{p-2}-accessible from each vertex of P_{j+1}, and hence no vertex of P_{j+2} uses the color $c(u_{j+1})$. Moreover P_{j+2} is a path containing at least 2^{p-j-3} vertices. So we can repeat the process until $j = p-2$, and we obtain vertices $u_0, u_1, \ldots, u_{p-1}$. By the choices of these vertices, their colors are distinct. So H uses at least p colors. □

It follows that $\chi_p(G)$ is bounded by $\mathrm{wcol}_{2^{p-1}}(G)$ (as a $(p+1)$-centered coloring is a χ_p-coloring, see Corollary 7.1). Next we prove (without relying on the previous two sections) that wcol_p is polynomially equivalent to $\nabla_{(p-1)/2}$. This will follow from the following two lemmas:

Lemma 7.11. *For any graph* G*, for any integer* k*,*

$$\nabla_{(k-1)/2}(G) + 1 \leq \mathrm{wcol}_k(G). \tag{7.5}$$

Proof. Consider a linear order L on the vertex set of G such that

$$\max_{v \in V(G)} |Q_k(G_L, v)| = \mathrm{wcol}_k(G) - 1.$$

Let $\mathcal{P} = \{V_1, V_2, \ldots, V_n\}$ be a witness of $\nabla_{(k-1)/2}$. Let H be the subgraph obtained from G by replacing each $G[V_i]$ by a tree T_i such that for each $x \in V_i$, $\mathrm{dist}_{T_i}(v_i, x) = \mathrm{dist}_{G[V_i]}(v_i, x)$, and that there is between V_i and V_j one edge $\{x, y\}$ with $\mathrm{dist}_{T_i}(v_i, x) + \mathrm{dist}_{T_j}(v_j, y) \leq k-1$, if such an edge exists. Then we have

$$\frac{|E(H/\mathcal{P})|}{|\mathcal{P}|} = \nabla_{(k-1)/2}(H) = \nabla_{(k-1)/2}(G).$$

For V_i, V_j adjacent in $H/\mathcal{P}$, orient the edge $\{V_i, V_j\}$ from V_i to V_j (in $H/\mathcal{P}$) if the minimum vertex $m_{i,j}$ of the (unique) $v_i v_j$-path in $H[V_i \cup V_j]$ (with respect to the linear order L) belongs to V_i. In such a case, $m_{i,j}$ is weakly k-accessible from v_j. Moreover, as V_i and $V_{i'}$ are disjoint sets, if $i \neq i'$ then $m_{i,j} \neq m_{i',j}$. It follows that V_j has indegree at most $|Q_k(G_L, v_j)|$ in $H/\mathcal{P}$, and hence $\nabla_{(k-1)/2}(G) + 1 \leq \mathrm{wcol}_k(G)$. □

Theorem 7.11. *Suppose* G *is a graph and* k *is a positive integer. Let* $p = (k-1)/2$ *and* $\nabla_p(G) \leq m$*. Then*

$$\mathrm{col}_k(G) \leq 1 + q_k, \tag{7.6}$$

where q_k *is defined as* $q_1 = 2m$ *and for* $i \geq 1$*,* $q_{i+1} = q_1 q_i^{2i^2}$*.*

Proof. The proof of this result of [476] is similar to the proof of Theorem 4 in [274]. Suppose G is a graph, k is a positive integer, put $p = (k-1)/2$ and $\nabla_p(G) \leq m$. We shall construct a linear ordering L on the vertices of G so that for each vertex u, $|R_k(G_L), u)| \leq q_k$ (where q_k is defined in the statement of the lemma).

The linear ordering $L = x_1 x_2 \ldots x_n$ is defined recursively. Suppose that we have constructed the subsequence $x_{i+1} x_{i+2} \ldots x_n$ of L. (If $i = n$, then the sequence is empty.) Let $M = \{x_{i+1}, x_{i+2}, \dots, x_n\}$ and let $U = V(G) \setminus M$. Let Ω be the probability space of all graphs of the form $H = (U, F)$, where F is defined as follows: For each pair $\{u, v\}$ of vertices in U for which there is a uv-path P of length at most k with all interior vertices in M, choose one such path P_{uv}. For $z \in M$, let

$$S_z = \{uv : \{u, v\} \subseteq U, z \in V(P_{uv})\}.$$

Label each $z \in M$ with a random element chosen from S_z. If $S_z = \emptyset$, then z is unlabeled. Let F be the set of pairs uv such that either $uv \in E(G)$ or for all $z \in V(P_{uv})$, z is labeled uv. So if uv is an edge of G, then $uv \in F$ with probability 1. Otherwise,

$$\Pr(uv \in F) = \prod_{z \in M \cap V(P_{uv})} \frac{1}{|S_z|}.$$

Let $E[d_H(x)]$ be the expected value of the degree of vertex x in H. Choose $x_i \in U$ so that $E[d_H(x_i)]$ is minimum.

Next, we prove that the expected minimal degree of H is at most $2m$. Toward this end, let H be a random graph defined as above. For each $\{u, v\} \in F \setminus E(G)$, partition the vertices in P_{uv} into two parts $A(uv, u), A(uv, v)$ by puting $z \in A(uv, u)$ if $\mathrm{dist}_{P_{uv}}(u, z) < \mathrm{dist}_{P_{uv}}(v, z)$, and $z \in A(uv, v)$ if $\mathrm{dist}_{P_{uv}}(v, z) < \mathrm{dist}_{P_{uv}}(u, z)$. If z is in the middle of P_{uv}, then arbitrarily put z in $A(uv, u)$ and $A(uv, v)$. Put $V_u = \bigcup_{v \in N_H(u)} A(uv, u)$. Then all sets $V_u, u \in U$ are disjoint subsets of $V(G)$, and each induces a connected subgraph of radius at most $k/2$. Let $\mathcal{P} = \{V_u : u \in U\}$. Then $\rho(\mathcal{P}) \leq k/2$ and H is a subgraph of $G/\mathcal{P}$. Thus the minimum degree of H satisfies

$$\delta(H) \leq 2|E(G/\mathcal{P})|/|\mathcal{P}| \leq 2m.$$

Therefore $E[d_H(x_i)] \leq q_1$.

We now prove by induction on $i \leq k$ that $|R_i(G_L, y)| \leq q_i$ for all vertices $y \in U$. Fix the time when y was added to the final sequence of L. For $i = 1$, this is true as if $x \in R_1(G_L, y)$, then $x \in U$ and $\{x, y\} \in F$ with probability 1, and thus $E[d_H(y)] \leq q_1$. Assume now that $|R_i(G_L, y)| \leq q_i$ holds for all vertices $y \in U$ and all $i \leq t$ and consider the case that $i = t + 1$. For each $z \in M, x \in R_{t+1}(G_L, y)$ and $xy \in S_z$, both x, y are in $Q_t(G_L, z)$. According to Proposition 4.8, we have:

$$\max_{v \in V(G)} |Q_k(G_L, v)| \leq \max_{v \in V(G)} |R_k(G_L, v)|^k,$$

and by the induction hypothesis,

$$|S_z| \leq |Q_t(G_L, z)| \leq (q_t)^{2t}.$$

Therefore,

$$q_1 \geq E[d_H(y)] \geq \sum_{x \in R_{t+1}(G_L, y)} \Pr(xy \in F)$$
$$= \sum_{x \in R_{t+1}(G_L, y)} \prod_{z \in M \cap V(P_{xy})} \frac{1}{|S_z|} \geq |R_{t+1}(G_L, y)|(q_t)^{-2t^2}.$$

So finally

$$|R_{t+1}(G_L, y)| \leq q_1 (q_t)^{2t^2} = q_{t+1}.$$

□

One important advantage of this approach (apart from its elegance and shortness) is that it allows to construct an acyclic transitive fraternal augmentations for each value of p. One disadvantage is that the proof is non constructive and that it does not allow to extend a partially defined the linear order (in a way similar to the use of an arbitrary initial indegree bounded orientation for transitive fraternal augmentations or transitive augmentations), see Exercise 7.4.

Exercises

7.1. Let $l(G)$ be the minimum length of a monochromatic path in a 2-coloring of the vertices of G. Prove that the maximum of $l(G)$ over all outerplanar graphs G of order n is at least $c\log(n-1)$ for some positive constant c.

Deduce that there is no constant C such that every outerplanar graph admits a vertex partition into two parts, each of them inducing a subgraph of tree-depth at most C.

7.2. (1) Consider any star coloring of a graph G and let N be the number of colors used by the coloring. Prove that there exists an orientation $\vec{G}$ of G such that every tight 1-transitive fraternal augmentation of $\vec{G}$ has chromatic number at most N;

(2) Deduce that there exist planar subcubic graphs of arbitrary high girth and star chromatic number 4;

(3) Consider any orientation $\vec{G}$ of a graph G and a tight 1-transitive fraternal augmentation H of G. Prove that $\chi_s(G) \leq \chi(H)$.

(4) Deduce that the star chromatic number of a graph G is the minimum over all the orientations $\vec{G}$ of G of the chromatic number of a tight 1-transitive fraternal augmentation of $\vec{G}$.

7.3. Define the tree-depth of a **countable** graph G has the minimum height of a rooted forest Y such that $G \subseteq \mathrm{Clos}(Y)$.

Prove that for every countable graph G it holds

$$\mathrm{td}(G) = \sup_{A \subseteq V(G) \text{ finite}} \mathrm{td}(G[A]).$$

For integer p, define $\chi_p(G)$ (for countable G) as the minimum N such that $V(G)$ can be N-colored in such a way that every subset of $i \leq p$ colors induce a subgraph with tree-depth at most i. Prove that

$$\chi_p(G) = \sup_{A \subseteq V(G) \text{ finite}} \chi_p(G[A]).$$

7.4. Show that acyclicity and bounded indegrees cannot be both preserved by a (transitive) fraternal augmentation and hence that the acyclicity ensured by the weak-coloring approach of Sect. 7.5 cannot be combined with the flexibility of the augmentation processes used in Sects. 7.3 and 7.4.

7.5. Consider the closure of a highly branching tree of height 3. Let ϕ be an integral graph function (i.e., $\phi(G)$ is an integer for every graph G). Assume that for any integer k and for any proper minor closed class $\mathcal{K}$ there exists an integer $N(\mathcal{K}, k)$ such that every graph $G \in \mathcal{K}$ has a partition into $\leq N(\mathcal{K}, k)$

parts with the property that any subgraph $H \subseteq G$ gets at least $\min(k, \phi(H))$ colors.

Prove that $\phi(H) \leq \mathrm{td}(H)$.

Chapter 8
Independence

Eyes wide open at independence.

We know that the dichotomy nowhere dense vs. somewhere dense can be expressed by means of decompositions, maximal cliques, and coloring numbers. We now give in a certain sense a dual characterization by independence (or stability).

8.1 How Wide is a Class?

Recall that the *distance* $\mathrm{dist}_G(x, y)$ in a graph G between two vertices x and y is the minimum length of a path linking x and y (or ∞ if x and y do not belong to the same connected component of G). Let $G = (V, E)$ be a graph and let d be an integer. Recall that the d-*neighborhood* $N_d^G(u)$ of a vertex $u \in V$ (sometimes called d-*ball*) is the subset of vertices of G at distance at most d from u in G: $N_d^G(u) = \{v \in V : \mathrm{dist}_G(u, v) \le d\}$.

Let $r \ge 1$ be an integer. A subset A of vertices of a graph G is r-*independent* if the distance between any two distinct elements of A is strictly greater than r. We denote by $\alpha_r(G)$ the maximum size of an r-independent set of G. Thus $\alpha_1(G)$ is the usual independence number $\alpha(G)$ of G. The notion of r-independence is closely related to the notion of an r-scattered set which originated in mathematical logic: a subset A of vertices of G is d-*scattered* if $N_d^G(u) \cap N_d^G(v) = \emptyset$ for every two distinct vertices $u, v \in A$. Thus A is d-scattered if and only if it is $2d$-independent. It is possible to say that this whole chapter is motivated by mathematical logic.

J. Nešetřil and P.O. de Mendez, *Sparsity*, Algorithms and Combinatorics 28, DOI 10.1007/978-3-642-27875-4_8, © Springer-Verlag Berlin Heidelberg 2012

The notion of r-independence leads to the notion of *wide*, *almost wide* and *quasi wide* classes. We find it useful to study these classes by means of the following function $\Phi_{\mathcal{C}}$ defined for any class of graphs. It is essential for our approach that we also define the uniform version $\overline{\Phi}_{\mathcal{C}}$ of this function.

Function $\Phi_{\mathcal{C}}$

This function has domain $\mathbb{N}$ and range $\mathbb{N} \cup \{\infty\}$ and $\Phi_{\mathcal{C}}(d)$ is defined for $d \geq 1$ as the minimum s such that the class $\mathcal{C}$ satisfies the following property:

There exists a function $f : \mathbb{N} \to \mathbb{N}$ such that for every integer m, every graph $G \in \mathcal{C}$ with order at least $f(m)$ contains a subset S of size at most s so that $G - S$ has a d-independent set of size m (see Fig. 8.1).

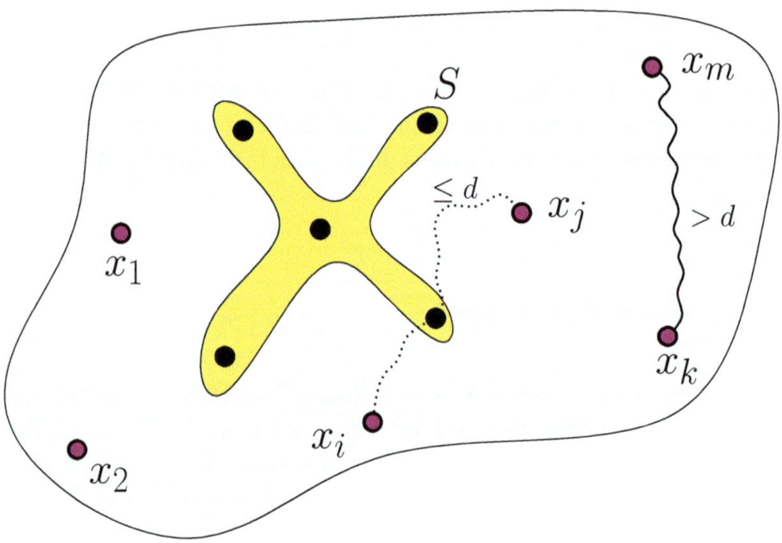

Fig. 8.1 If $|G| \geq f(m)$ then there exists S (with $|S| \leq \Phi_{\mathcal{C}}(d)$) and $x_1, \ldots, x_m$ such that $\mathrm{dist}_{G-S}(x_i, x_j) \geq d$ (i.e. every x_i–x_j path of G with length $< d$ intersects S)

We put $\Phi_{\mathcal{C}}(d) = \infty$ if $\mathcal{C}$ does not satisfy the above property for any value of s. Moreover, we define $\Phi_{\mathcal{C}}(0) = 0$.

Function $\overline{\Phi}_{\mathcal{C}}$

This function has domain $\mathbb{N}$ and range $\mathbb{N} \cup \{\infty\}$ and $\overline{\Phi}_{\mathcal{C}}(d)$ is defined for $d \geq 1$ as the minimum s such that $\mathcal{C}$ satisfies the following property:

There exists a function $f : \mathbb{N} \to \mathbb{N}$ such that for every integer m, every graph $G \in \mathcal{C}$ and every subset A of vertices of G of size at least $f(m)$, the graph G contains a subset S of size at most s so that A includes a d-independent set of size m of $G - S$ (see Fig. 8.2).

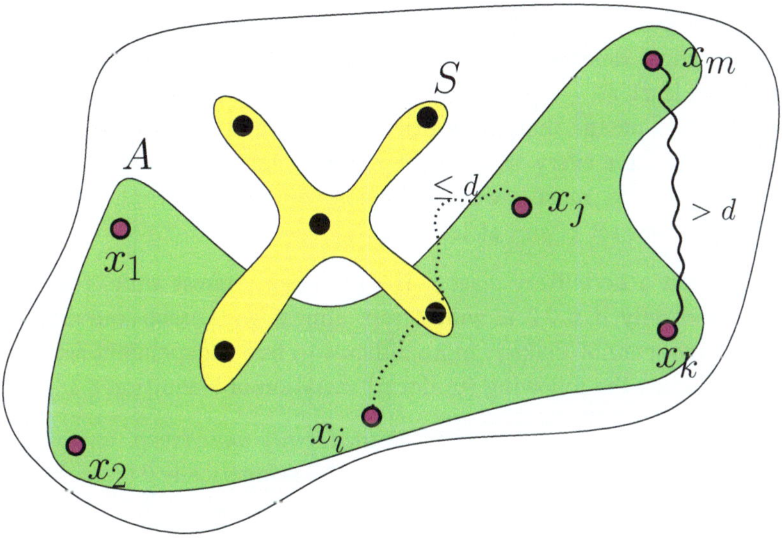

Fig. 8.2 If $|A| \geq f(m)$ then there exists S (with $|S| \leq \overline{\Phi}_{\mathcal{C}}(d)$) and $x_1, \ldots, x_m \in A$ such that $\mathrm{dist}_{G-S}(x_i, x_j) \geq d$ (i.e. every x_i–x_j path of G with length $< d$ intersects S)

We put $\overline{\Phi}_{\mathcal{C}}(d) = \infty$ if $\mathcal{C}$ does not satisfy the above property for any value of s. Moreover, we define $\overline{\Phi}_{\mathcal{C}}(0) = 0$.

Notice that obviously $\overline{\Phi}_{\mathcal{C}}(d) \geq \Phi_{\mathcal{C}}(d)$ for every class $\mathcal{C}$ and for every integer d. We view $\overline{\Phi}_{\mathcal{C}}(d)$ as a uniform version of $\Phi_{\mathcal{C}}(d)$. Obviously it is possible for a class $\mathcal{C}$ that $\Phi_{\mathcal{C}}(d)$ is finite while its uniform version $\overline{\Phi}_{\mathcal{C}}(d)$ is infinite.

Definition 8.1. A class of graphs $\mathcal{C}$ is *wide* (resp. *almost wide*, resp. *quasi-wide*) if $\Phi_{\mathcal{C}}$ is identically 0 (resp. bounded, resp. finite) [42, 109]:

$$\begin{aligned}
\mathcal{C} \text{ is wide} &\iff \forall d \in \mathbb{N} : \Phi_{\mathcal{C}}(d) = 0\\
\mathcal{C} \text{ is almost wide} &\iff \sup_{d \in \mathbb{N}} \Phi_{\mathcal{C}}(d) < \infty\\
\mathcal{C} \text{ is quasi-wide} &\iff \forall d \in \mathbb{N} : \Phi_{\mathcal{C}}(d) < \infty
\end{aligned}$$

What do these definitions mean? For a better understanding, let us reformulate them.

- $\Phi_{\mathcal{C}}(d) = 0$ means that for every m, every sufficiently large graph in $\mathcal{C}$ contains a d-independent set of size at least m;
- $\sup_{d \in \mathbb{N}} \Phi_{\mathcal{C}}(d) = 0$ means that there exists a constant s such that for every d and m, every sufficiently large graph in $\mathcal{C}$ contains a subset of size at most s whose deletion results in a graph containing a d-independent set of size at least m.
- $\Phi_{\mathcal{C}}(d) < \infty$ means that there exists a constant $s = s(d)$ (depending on d) such that for every m, every sufficiently large graph in $\mathcal{C}$ contains a subset of size at most s whose deletion results in a graph containing a d-independent set of size at least m;

Notice that a hereditary class $\mathcal{C}$ is wide (resp. almost wide, resp. quasi-wide) if and only if $\mathcal{C} \triangledown 0$ is wide (resp. almost wide, resp. quasi-wide) as deleting edges cannot make it more difficult to find independent sets.

We introduce the following (uniform) variation of Definition 8.1.

Definition 8.2. A class of graphs $\mathcal{C}$ is *uniformly wide* (resp. *uniformly almost wide*, resp. *uniformly quasi-wide*) if $\overline{\Phi}_{\mathcal{C}}$ is identically 0 (resp. bounded, resp. finite):

$$\begin{aligned}
\mathcal{C} \text{ is uniformly wide} &\iff \forall d \in \mathbb{N} : \overline{\Phi}_{\mathcal{C}}(d) = 0\\
\mathcal{C} \text{ is uniformly almost wide} &\iff \sup_{d \in \mathbb{N}} \overline{\Phi}_{\mathcal{C}}(d) < \infty\\
\mathcal{C} \text{ is uniformly quasi-wide} &\iff \forall d \in \mathbb{N} : \overline{\Phi}_{\mathcal{C}}(d) < \infty
\end{aligned}$$

Notice that a class $\mathcal{C}$ is uniformly wide (resp. uniformly almost wide, resp. uniformly quasi-wide) if and only if $\mathcal{C} \triangledown 0$ is uniformly wide (resp. uniformly almost wide, resp. uniformly quasi-wide): the uniform properties are monotone as deleting edges or vertices cannot make it more difficult to find independent sets.

Based on a construction of Kreidler and Seese [289], Atserias et al. [42] proved that if a class excludes a graph minor then it is almost wide. Classes locally excluding a minor have been shown to be quasi-wide by Dawar et al. [111]. In Sects. 8.4 and 8.5 we explicitly characterize these classes in graph theoretic terms. Particularly we show that a hereditary class is quasi wide if and only if it is nowhere dense (Theorem 8.2).

8.2 Wide Classes

Recall that $\Delta(G)$ denotes the maximal degree of a graph G and that $\Delta(\mathcal{C})$ denotes the supremum of all $\Delta(G)$ for $G \in \mathcal{C}$. Thus $\Delta(\mathcal{C}) = \infty$ just means that the graphs in $\mathcal{C}$ have arbitrary large degrees.

Characterizing wide classes is not difficult. Advancing this, we take time for two easy lemmas.

Lemma 8.1. *Let $\mathcal{C}$ be a hereditary class of graphs. If $\Delta(\mathcal{C}) = \infty$ then $\Phi_{\mathcal{C}}(2) > 0$.*

Proof. Assume for contradiction that $\mathcal{C}$ satisfies $\Phi_{\mathcal{C}}(2) = 0$. Then there exists a function $f : \mathbb{N} \to \mathbb{N}$ such that every graph $G \in \mathcal{C}$ with order at least $f(2)$ has a 2-independent set of size 2. As $\Delta(\mathcal{C}) = \infty$, the class $\mathcal{C}$ contains a graph G with maximum degree at least $f(m)$. As $\mathcal{C}$ is hereditary it contains a graph of order at least $f(m) + 1$ with a universal vertex (that is: a vertex adjacent to all the other vertices). Although this graph has order greater than $f(m)$, it contains no 2-independent set of size 2. □

Lemma 8.2. *Let $G = (V, E)$ be a graph and let d, m be integers. If $A \subseteq V$ has size at least $(\Delta(G)^d + 1)m$ then A includes a d-independent set of size at least m.*

Proof. Notice that G^d (the graph with vertex set $V(G)$ in which two vertices are adjacent if their distance in G is at most d) has maximum degree at most $\Delta(G)\sum_{i=0}^{d-1}(\Delta(G)-1)^i \leq \Delta(G)^d$ (hence chromatic number at most $\Delta(G)^d+1$) and that any independent set of G^d is a d-independent set of G. As at least one color class of G^d intersects A on a subset of size at least $|A|/\chi(G^d)$ the lemma follows. □

As a consequence of previous lemmas we deduce our first characterization theorem:

Theorem 8.1. *Let $\mathcal{C}$ be a hereditary class of graphs. Then the following are equivalent:*

- $\Phi_{\mathcal{C}}(2) = 0$,
- $\overline{\Phi}_{\mathcal{C}}(2) = 0$,
- $\Delta(\mathcal{C}) < \infty$,
- *$\mathcal{C}$ is wide,*
- *$\mathcal{C}$ is uniformly wide.*

Proof. The theorem follows from the following implications (where the non-obvious implications follow from the two above Lemmas).

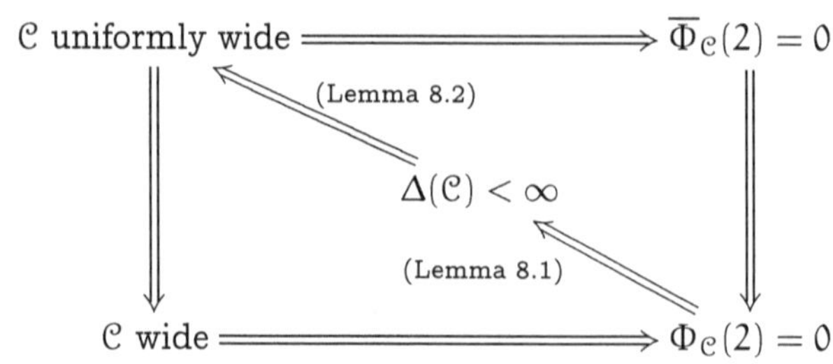

□

Although the hypothesis that $\mathcal{C}$ is hereditary is not necessary to prove that $\mathcal{C}$ is uniformly wide if and only if $\Delta(\mathcal{C})$ is finite (because the property of being uniformly wide is hereditary in nature), this assumption is necessary in order to prove that a wide class has bounded maximum degree (see Exercise 8.1).

8.3 Finding d-Independent Sets in Graphs

The characterization of almost wide and quasi wide classes is not as simple as for wide classes. What we are looking for are obstructions to these two properties. This problem has a Ramsey theory flavor. In this section we (elaborately) prove the possible types of such obstruction sets. This is then combined in Sect. 8.4 to obtain a characterization of quasi wide classes (Theorem 8.2) and in Sect. 8.5 to obtain a characterization of almost wide classes (Theorem 8.4).

8.3.1 Finding 1-Independent Sets in Graphs

The following is a restatement of the Ramsey theorem for graphs. It implies the existence of 1-independent sets in large graphs without large cliques:

Lemma 8.3. *Let G be a graph and let c, n be integers. Let A be a subset of at least $R(c, n)$ vertices of G. Then either G contains a K_c or A includes an independent set of size n.* □

8.3.2 Finding a 2-Independent Set in a 1-Independent Set

Define the following variant of Ramsey numbers:

$$R^\star(p, q, n) = R(\overbrace{q, q, \ldots, q}^{\binom{n-1}{2} \text{ times}}, p).$$

Then we have the following:

Lemma 8.4. *Let* $G = (A \cup B, E)$ *be a bipartite graph and let* p, q, n *be integers. If* $|A| \geq R^\star(p, q, n)$ *then at least one of the following properties holds:*

- *A includes a set* A' *of size* p *such that no two vertices in* A' *have a common neighbor;*
- *A includes the principal (i.e. branching) vertices of a* 1*-subdivision of the complete graph* K_q *(*1*-subdivision means that all the edges are subdivided by exactly one vertex);*
- *B includes a vertex of degree at least* n*.*

Proof. Assume that B includes no vertex of degree at least n. Let $k = |B|$ and let $b_1, b_2, \ldots, b_k$ be the vertices in B in some arbitrary order. Let Γ be the complete graph with vertex set A, whose edges are colored using $\binom{n-1}{2} + 1$ colors and defined as follows: To begin, let Γ be the empty graph with vertex set A. Then we add the edges in $k+1$ steps. At step $i \leq k$ we add to Γ all the edges (which have not been previously added) between the neighbors of b_i, coloring them with integers between 1 and $\binom{n-1}{2}$ in such a way that no two edges added at this step get the same color. This is possible as the degree of b_i is at most $n - 1$. At step $k + 1$, we add all the missing edges and assign to them the color $\binom{n-1}{2} + 1$.

As $|A| \geq R^\star(p, q, n)$, there exists in Γ a monochromatic clique of size q with color in $\{1, \ldots, \binom{n-1}{2}\}$ or a monochromatic clique of size p with color $\binom{n-1}{2} + 1$. If the edges of the clique have color $\binom{n-1}{2} + 1$, its vertices define a subset A' of size p, such that no two vertices in A' have a common neighbor. Otherwise, all the edges of the monochromatic clique of size q have been added at different $\binom{q}{2}$ steps (as they got the same color), hence G includes a 1-subdivision of K_q having its principal vertices in A. □

We shall need a stronger statement by refining the third possibility, namely: "B includes a vertex of degree at least n". Define inductively the number $\Theta(m, a, b, s)$ by:

$$\Theta(m, a, b, s) = \begin{cases} R^\star(m, a, b), & \text{if } s = 0; \\ R^\star(m, a, \Theta(m, a, b, s-1)), & \text{otherwise.} \end{cases}$$

Lemma 8.5. *Let* m, a, b, s *be integers. Let* $G = (A \cup B, E)$ *be a bipartite graph such that* $|A| \geq \Theta(m, a, b, s)$*. Then at least one of the following properties holds (cf. Fig. 8.3):*

- *There exist subsets* $A' \subseteq A$ *and* $B' \subseteq B$ *with* $|A'| = m$ *and* $|B'| = s$ *such that only the vertices in* B' *can have more than one neighbor in* A'*;*
- A *includes all principal vertices of a* 1*-subdivision of* K_a*;*
- B *includes the* $s+1$ *vertices of the complete bipartite graph* $K_{s+1,b}$.

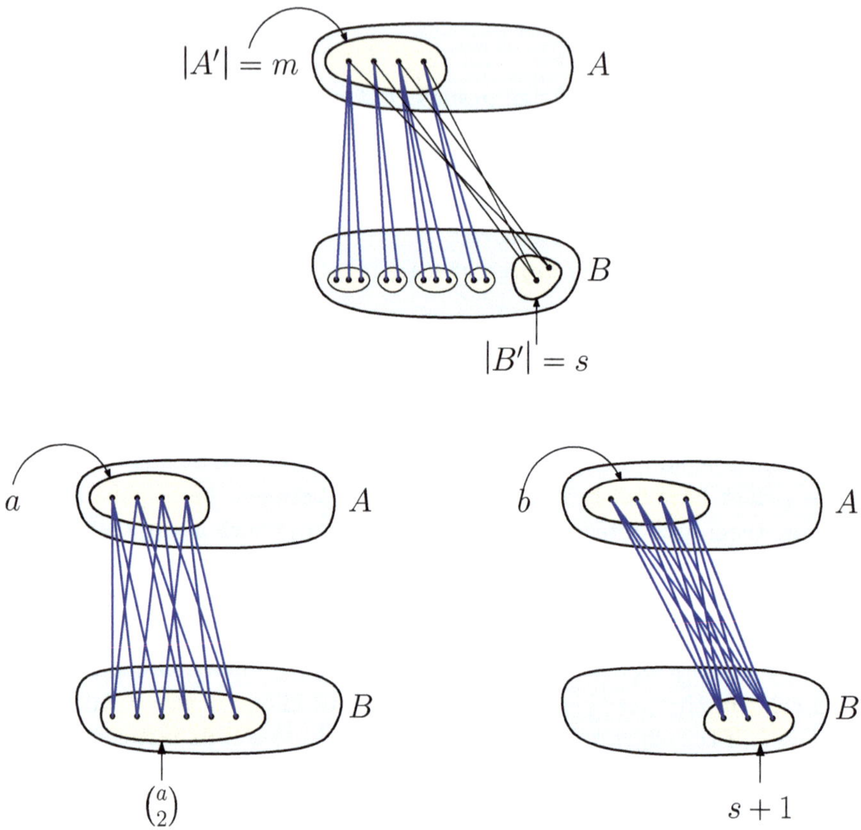

Fig. 8.3 Unavoidable structures of Lemma 8.5

Proof. We proceed by induction on s.

Assume $s = 0$. Let $p = m, q = a$ and $n = b$. According to Lemma 8.4, either A includes a subset A' of size m such that no two vertices of A' have a common neighbor or G includes a 1-subdivision of K_a with principal vertices in A or B includes a vertex of degree at least b hence G includes a star $K_{1,b}$ with the center of the star in B.

Assume that $s > 0$ and that the result has been proved for $s-1$. Let $p = m, q = a, n = \Theta(m, a, b, s-1)$. According to Lemma 8.4, either A

includes a subset A' of size m such that no two vertices of A' have a common neighbor or G includes a 1-subdivision of K_a with principal vertices in A or B includes a vertex of degree at least $\Theta(m, a, b, s-1)$.

In the two first cases we are done. Thus assume that G contains a vertex $v \in B$ of degree at least $\Theta(m, a, b, s-1)$. Let H be the subgraph of G induced by the neighborhood X of v and the set Y of the vertices in $B - v$ having at least a neighbor in common with v. Then $|X| \geq \Theta(m, a, b, s-1)$. By induction, either there exist $A' \subseteq X \subseteq A$ and $Y' \subseteq Y$ with $|A'| = m$, $|Y'| = s-1$ and only the vertices in Y' can have in H more than one neighbor in A' (hence only the vertices of $B' = Y' \cup \{v\}$ can have in G more than one neighbor in A') or H includes a complete bipartite graph $K_{s,b}$ with the s vertices in Y (thus G includes a complete bipartite graph $K_{s+1,b}$ with the $s+1$ vertices in B as v is adjacent to all the vertices in X) or H (hence G) contains a 1-subdivision of the complete graph K_a with principal vertices in $X \subseteq A$. □

Lemma 8.6. *Let G be a graph and let A be an independent set of G of order at least $\Theta(m, a, b, s)$. Then at least one of the following properties holds:*

- *There exists in G a subset of size at most s whose removal leaves in A a 2-independent set of size m;*
- *G includes a $\dot{K}_a$ or a $K_{s+1,b}$.*

Proof. Consider the bipartite graph $G' = (A \cup B, E')$ where B is the set of all the vertices of G adjacent to a least a vertex in A and E' is the subset of the edges of G linking a vertex in A to a vertex in B. The result is then a direct consequence of Lemma 8.5. □

8.3.3 Finding a $(2r+1)$-independent Set in a $2r$-independent Set

Lemma 8.7. *Let G be a graph and let c, n be integers. Let A be a $2r$-independent subset of G of size at least $R(c, n)$. Then either $K_c \in G \,\triangledown\, r$ or A includes a $(2r+1)$-independent set of size n.*

Proof. Consider the graph $H \in G \,\triangledown\, r$ obtained from G by contracting the r-neighborhoods of the vertices in A into a set A' identified to A (see Fig. 8.4). According to Lemma 8.3, either H contains a K_c (thus $K_c \in G \,\triangledown\, r$) or A' includes an independent set of size n of H, which corresponds to a $(2r+1)$-independent set of G included in A. □

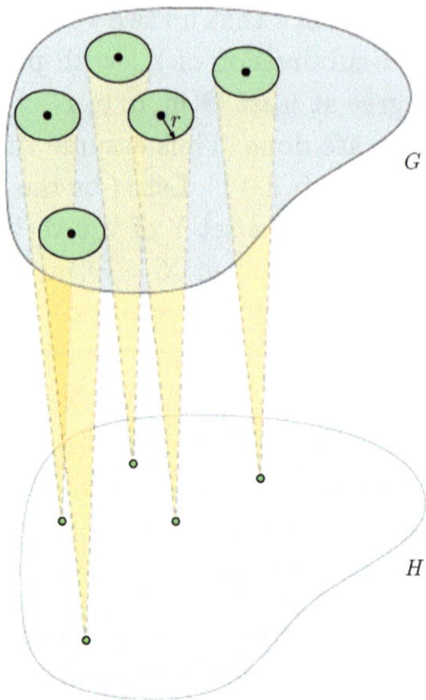

Fig. 8.4 The reduced graph H is obtained by contracting the r-neighborhoods of the vertices in A

8.3.4 Finding a $(2r+2)$*-Independent Set in a* $(2r+1)$*-Independent Set*

Lemma 8.8. *Let* G *be a graph and let* A *be a* $(2r+1)$*-independent set of* G *of order at least* $\Theta(m, a, b, s)$*. Then at least one of the following properties holds:*

- *There exists in* G *a subset of size at most* s *whose removal leaves in* A *a* $(2r+2)$*-independent set of size* m*;*
- $G \triangledown r$ *includes a* $\dot{K}_a$ *or a* $K_{s+1,b}$*.*

Proof. Consider the graph $H \in G \triangledown r$ obtained from G by contracting the r-neighborhoods of the vertices in A into a set A' identified to A (again see Fig. 8.4). According to Lemma 8.6, either H contains a $\dot{K}_a$ or a $K_{s+1,b}$ (thus $\dot{K}_a \in G \triangledown r$ or $K_{s+1,b} \in G \triangledown r$) or A' includes a 2-independent set of size n of H, which corresponds to a $(2r+2)$-independent set of G included in A. □

8.4 Quasi-Wide Classes

We are now ready to decide finiteness of $\Phi_{\mathcal{C}}(r)$ for a general monotone class $\mathcal{C}$.

Lemma 8.9. *Let $\mathcal{C}$ be a monotone class of graphs and let $r \geq 0$ be a half-integer. Assume $\omega(\mathcal{C} \,\widetilde{\nabla}\, r) = \infty$. Then $\Phi_{\mathcal{C}}(6r+1) = \infty$.*

Proof. Let $c = \omega(\mathcal{C} \,\widetilde{\nabla}\, r)$. Assume for contradiction that $s = \Phi_{\mathcal{C}}(6r+1) < \infty$ and put $m = rs(s-1) + \sqrt{2s} + 2$. Then there exists an integer N such that every graph G of order at least N has a subset S of at most s vertices, such that $G - S$ contains a $(6r+1)$-independent set I of size m.

Let G be a $\leq 2r$-subdivision of K_N in $\mathcal{C}$ (such a graph has to exist in $\mathcal{C}$ as $\mathcal{C}$ is monotone and $\omega(\mathcal{C} \,\widetilde{\nabla}\, r) = \infty$). Put $S_0 = \{v \in S : d(v) = 2\}$ and $S_1 = S \setminus S_0$. The deletion of S in G leaves one big connected component and several small ones. Each small connected component is included in a branch of G and does not included any non-subdivision vertex of G. Hence the sum of the orders of the small component is bounded by $2r\binom{s}{2} = rs(s-1)$. As $|I| \geq rs(s-1) + \sqrt{s} + 1$, there exists $I_0 \subseteq I$ of cardinality at least $\sqrt{2s} + 2$ which contains only vertices of the big connected component. For each $x \in I_0$ there exists a principal vertex $\tau(x)$ of $G - S$ such that $\mathrm{dist}_{G-S}(x, \tau(x)) \leq 2r$. Put $J_0 = \{\tau(x) : x \in I_0$. Obviously, no two distinct vertices $x, y \in I_0$ may be such that $\tau(x) = \tau(y)$ for otherwise $\mathrm{dist}_{G-S}(x,y) \leq \mathrm{dist}_{G-S}(x,\tau(x)) + \mathrm{dist}_{G-S}(y,\tau(y) \leq 4r$. Hence $|J_0| = |I_0|$. As at most s branches of G have an internal vertex in S, it is possible to color the principal vertices in $G - S$ by $\lfloor\sqrt{2s}\rfloor$ colors in such a way that if two principal vertices x and y are colored the same then the branch of G linking them does not meet S. As $|J_0| = |I_0| \geq \sqrt{2s} + 2$, there exists in J_0 two vertices $\tau(x)$ and $\tau(y)$ with the same color, hence such that the branch linking them does not meet S. Thus $\mathrm{dist}_{G-S}(x,y) \leq \mathrm{dist}_{G-S}(x,\tau(x)) + \mathrm{dist}_{G-S}(\tau(x),\tau(y)) + \mathrm{dist}_{G-S}(\tau(y),y) \leq 6r+1$. Hence a contradiction (see schematic Fig. 8.5 on page 192). □

The following characterization theorem follows:

Theorem 8.2. *Let $\mathcal{C}$ be a hereditary class of graphs. The following conditions are equivalent:*

- *$\mathcal{C}$ is quasi-wide;*
- *$\mathcal{C}$ is uniformly quasi-wide;*
- *For every integer d, $\omega(\mathcal{C} \,\nabla\, d) < \infty$;*
- *For every integer d, $\omega(\mathcal{C} \,\widetilde{\nabla}\, d) < \infty$;*
- *$\mathcal{C}$ is a class of nowhere dense graphs.*

Proof. First remark that a hereditary class is quasi-wide if and only if its monotone closure is quasi-wide (as noticed after Definition 8.1), and that a class $\mathcal{C}$ is nowhere dense if and only if its monotone closure $\mathcal{C} \triangledown 0$ is nowhere dense (direct from definition). Also note that the equivalence of the three last items has already been established.

Assume $\mathcal{C} \triangledown 0$ is a class of nowhere dense graphs. According to Lemmas 8.3, 8.6, 8.7 and 8.8 then $\mathcal{C} \triangledown 0$ (hence $\mathcal{C}$) is uniformly quasi-wide hence quasi-wide. Conversely, if $\mathcal{C}$ (hence $\mathcal{C} \triangledown 0$) is not a class of nowhere dense graphs, then $\mathcal{C} \triangledown 0$ (hence $\mathcal{C}$) is not quasi-wide according to Lemma 8.9 thus $\mathcal{C}$ is also not uniformly quasi-wide. □

We can be more precise about the function $\overline{\Phi}_{\mathcal{C}}$. Toward this end, we introduce the following class parameter: For a class $\mathcal{C}$ we define

$$\omega'(\mathcal{C}) = \sup\{s : \forall n \in \mathbb{N}, K_{s,n} \in \mathcal{C} \triangledown 0\}.$$

We show that $\overline{\Phi}_{\mathcal{C}}$ and ω' are closely related (Proposition 8.1).

Lemma 8.10. *Let $\mathcal{C}$ be a monotone class of graphs and let $r \geq 0$ be a half-integer. Then $\overline{\Phi}_{\mathcal{C}}(8r+6) \geq \omega'(\mathcal{C} \triangledown r)$.*

Proof. Assume for contradiction that $s = \overline{\Phi}_{\mathcal{C}}(8r+6) < \omega'(\mathcal{C} \triangledown r)$. Put $m = 2(r+1)s(s+1)+2$. There exists an integer N such that for every graph $G \in \mathcal{C}$ and every subset A of at least N vertices of G, the graph G has a subset S of cardinality at most s and A includes a subset I of cardinality m such that I is $(8r+6)$-independent in $G - S$. Put $d = \lceil r \rceil + 1$ and $N_0 = N^{d^{s+1}}$.

As $\mathcal{C}$ is monotone and as $\omega'(\mathcal{C} \triangledown r) > s$ there exists in $\mathcal{C}$ a ramification G_0 of K_{s+1,N_0} witnessing $K_{s+1,N_0} \in \mathcal{C} \triangledown r$. Let $\{a_1, \dots, a_{s+1}\}$ and $\{b_1, \dots, b_{N_0}\}$ be the two parts of K_{s+1,N_0} and let $\{Y_1, \dots, Y_{s+1}, Y'_1, \dots, Y'_{N_0}\}$ be a corresponding K_{s+1,N_0}-decomposition of G_0. Orient every Y_i from its root, every Y'_j to its root and the edge between Y_i and Y'_j from Y_i to Y'_j. Each tree Y_i has radius at most $d = \lceil r \rceil$. Hence Y_1 contains a vertex y_1 of out-degree at least $N_1 = N_0^{1/d}$ thus the vertex y_1 can reach by internally vertex disjoint directed paths a subset $\mathcal{F}_1$ of $\{Y'_1, \dots, Y'_{N_0}\}$ of cardinality N_1. Similarly, Y_2 has a vertex y_2 which can reach by internally vertex disjoint directed paths a subset $\mathcal{F}_2$ of $\mathcal{F}_1$ of cardinality $N_2 = N_1^{1/d}$. Inductively, we obtain vertices $y_1 \in Y_1, y_2 \in Y_2, \dots y_{s+1} \in Y_{s+1}$ and a subset $\mathcal{F}$ of $\{Y'_1, \dots, Y'_{N_0}\}$ of cardinality at least $N = N_0^{1/d^{s+1}}$ such that each vertex y_i can reach the Y'_j in $\mathcal{F}$ by internally vertex-disjoint directed paths. Without loss of generality, we assume that $\mathcal{F} = \{Y'_1, \dots, Y'_N\}$ and we consider the subgraph of G induced by $y_1, \dots, y_{s+1}, Y'_1, \dots, Y'_N$ and the directed paths linking each y_i to each Y'_j.

This graph G has order at least N hence there exists a subset S of cardinality at most s such that $G - S$ contains a $(8r+6)$-independent set I of cardinality m. Each Y'_j has order at most $\lceil r \rceil(s+1)$. The deletion of the vertices in S leaves some small connected components and a unique big connected component. The total order of the small components is bounded by rs^2. Hence there exists a subset $I_0 \subseteq I$ of cardinality at least $m - rs^2$ of vertices belonging in the big connected component which forms a $(8r+6)$-independent set in $G - S$. There exist a vertex y_i (say y_1) whose bush does not intersect S (as $|S| < s+1$). Moreover, this vertex obviously belongs to the big connected component. For each $x \in I_0$, there exists a vertex $\tau(x)$ in one of the Y'_j such that $\mathrm{dist}_{G-S}(x, \tau(x)) \leq d$. Of course, two distinct vertices of I_0 have different τ-values for otherwise their distance in $G - S$ would be at most $2d$. Moreover, at most s trees Y'_j may intersect S hence at most $\lceil r \rceil(s+1)s$ vertices in I_0 may have their τ-value in some Y'_j intersecting S. Thus there exists $I_1 \subseteq I_0$ of cardinality at least $m - 2(r+1)s(s+1) \geq 2$ such that $\tau(x)$ does not belong to a Y'_j intersecting S for every $x \in I_1$. Let x, y be two distinct elements of I_1. Then:

$$\begin{aligned}
\mathrm{dist}_{G-S}(x,y) &\leq \mathrm{dist}_{G-S}(x,\tau(x)) + \mathrm{dist}_{G-S}(\tau(x), y_1) \\
&\quad + \mathrm{dist}_{G-S}(y_1, \tau(y)) + \mathrm{dist}_{G-S}(\tau(y), y) \\
&\leq d + (2r+d) + (2r+d) + d \\
&\leq 8r + 6.
\end{aligned}$$

Hence we are led to a contradiction (see schematic Fig. 8.6 on page 193). □

Lemma 8.11. *Let* $G = (V, E)$ *be a graph, let* $d \geq 1, m, s$ *be integers and let* A *be a subset of* V.

Assume there is $S \subseteq V$ *and* $A' \subseteq A$ *such that* A' *is* $2d$*-independent in* $G - S$ *and* $|A'| \geq m2^s$.

Then there exists $C \subseteq S$ *and* $A'' \subseteq A'$ *such that* A'' *is* $2d$*-independent in* $G - C$, $|A''| = m$ *and* $K_{|C|,m} \in G \,\nabla\, (d-1)$.

Proof. For $v \in A'$, let L_v be a minimal subset of S such that $N_d^{G-L_v}(v) \cap (S \setminus L_v) = \emptyset$. Such a set obviously exists (it can be S). As S has only 2^s distinct subsets, there exists a subset $A'' \subset A'$ of size m such that $L_x = L_y$ for every $x, y \in A''$ (call this set C). Let x, y be any two distinct elements of A''. We have $N_d^{G-C}(x) \cap N_d^{G-C}(y) = \emptyset$ for otherwise there would exist an x–y path of length at most $2d$ avoiding C but not S because A'' is $2d$-independent in $G - S$ thus some element of $S \setminus C$ would belong to L_x or L_y. Thus A'' is $2d$-independent in $G - C$. Moreover, by the minimality of L_x, every vertex $v \in L_x$ is such that $N_d^{G-(C-v)}(x) \cap (S \setminus (C - v)) \neq \emptyset$ and more precisely $v \in N_d^{G-(C-v)}(x)$. It follows that for every $x \in A''$ there exists a tree Y_x of depth at most d, whose leaves are exactly the vertices in C and such that the Y_x's are pairwise internally vertex disjoint. By contracting the Y_x's we obtain the desired $K_{|C|,m}$ in $G \triangledown (d-1)$. □

We deduce the following inequalities:

Proposition 8.1. *Let $\mathcal{C}$ be a class of graphs and let $r \geq 0$ be an integer. Then*

$$\overline{\Phi}_{\mathcal{C}}(8r+6) \geq \omega'(\mathcal{C} \triangledown r) \geq \overline{\Phi}_{\mathcal{C}}(2r+1)$$

8.5 Almost Wide Classes

The following characterizations follow from Theorem 8.2.

Theorem 8.3. *Let $\mathcal{C}$ be a class with bounded expansion. Then $\overline{\Phi}_d(\mathcal{C}) \leq \nabla_{\lfloor d/2 \rfloor - 1}(\mathcal{C})$.*

Proof. According to Theorem 8.2, $\mathcal{C}$ is uniformly quasi-wide, hence $\overline{\Phi}_{\mathcal{C}}(d) < \infty$ for every d. According to Lemma 8.11,

$$\overline{\Phi}_{\mathcal{C}}(d) \leq \omega'(\mathcal{C} \triangledown (\lfloor d/2 \rfloor - 1))$$

Then the result follows as

$$\begin{aligned}\nabla_{\lfloor d/2 \rfloor - 1}(\mathcal{C}) &= \nabla_0(\mathcal{C} \triangledown (\lfloor d/2 \rfloor - 1)) \\ &\geq \sup_{n \in \mathbb{N}} \nabla_0(K_{\omega'(\mathcal{C} \triangledown (\lfloor d/2 \rfloor - 1)),n}) \\ &= \omega'(\mathcal{C} \triangledown (\lfloor d/2 \rfloor - 1)).\end{aligned}$$

□

We have the following characterization of hereditary almost wide classes of graphs:

Theorem 8.4. *Let $\mathcal{C}$ be a hereditary class of graphs. Then the following are equivalent:*

- *$\mathcal{C}$ is almost wide;*
- *$\mathcal{C}$ is uniformly almost wide;*
- *There are $s \in \mathbb{N}$ and $t : \mathbb{N} \to \mathbb{N}$ such that $K_{s,t(r)} \notin \mathcal{C} \triangledown r$ (for all $r \in \mathbb{N}$).*

Proof. If $\mathcal{C}$ is almost wide then the two next items follow from Proposition 8.1. If $\mathcal{C}$ is such that each $\mathcal{C} \triangledown r$ excludes some $K_{s,t(r)}$, then it is uniformly quasi-wide according to Theorem 8.2 and the bounding of $\overline{\Phi}_{\mathcal{C}}(d)$ then follows from Proposition 8.1. □

Remark 8.1. Uniformly almost wide classes do not need to be topologically closed and may not even have bounded local expansion: Consider the class $\mathcal{C}$ of all graphs G satisfying $\Delta(G) \leq \text{girth}(G)$ (not necessarily 2-connected). Then $\mathcal{C}$ is uniformly almost wide although it does not have a bounded average degree: As the class is hereditary, it is sufficient to prove that $\mathcal{C}$ is almost wide. Let d and m be integers. If a graph $G \in \mathcal{C}$ has diameter at least $D = dm$ then G includes a d-independent set of size m. Otherwise, if G includes a cycle, then this cycle has length $\text{girth}(G) \leq 2D$, hence $\Delta(G) \leq 2D$ and G has at most about $(2D)^D$ vertices. Otherwise, if G is acyclic, it is a forest, and the deletion of one vertex is sufficient to get a big d-independent set. Hence $\mathcal{C}$ is almost wide. Also, let $\mathcal{D} = \{G + K_1 : G \in \mathcal{C}\}$ ($G + K_1$ is graph obtained from G by adding a vertex adjacent to all the other vertices). Obviously, $\mathcal{D}$ is also uniformly almost wide but does not have a bounded local expansion.

We may be more precise when $\mathcal{C}$ is actually minor closed:

Theorem 8.5. *Let $\mathcal{C}$ be a minor closed class of graphs and let s be an integer. Then the following are equivalent:*

- *$\mathcal{C}$ is almost wide and $\Phi_{\mathcal{C}}(d) < s$ for every integer $d \geq 2$;*
- *$\mathcal{C}$ is uniformly almost wide and $\overline{\Phi}_{\mathcal{C}}(d) < s$ for every integer $d \geq 2$;*
- *$\mathcal{C}$ excludes some graph $K_{s,t}$ (for a $t \geq 1$).*

Proof. If $K_{s,t}$ belongs to $\mathcal{C}$ for every $t \in \mathbb{N}$ then $\overline{\Phi}_{\mathcal{C}}(d) \geq \Phi_{\mathcal{C}}(d) \geq \Phi_{\mathcal{C}}(2) \geq s$. Otherwise, according to Proposition 8.1 we have $s > \overline{\Phi}_{\mathcal{C}}(d) \geq \Phi_{\mathcal{C}}(d)$. □

Example 8.1. For a surface Σ, let $\mathcal{C}_\Sigma$ be the class of the graphs which embed on Σ. It has been proved in [42] that $\mathcal{C}_\Sigma$ is almost wide for every surface Σ and that $\Phi_{\mathcal{C}_\Sigma}(d)$ is at most equal to the order of the smallest clique which does not embed on Σ. Again, according to Theorem 8.2, the class $\mathcal{C}_\Sigma$ is uniformly quasi wide. Hence by Proposition 8.1, we deduce that $\Phi_{\mathcal{C}_\Sigma}(d) = \overline{\Phi}_{\mathcal{C}_\Sigma}(d) = 2$ for every integer d, as every $K_{2,n}$ embed on any surface but not every $K_{3,n}$ does.

8.6 A Nice (Asymmetric) Application

In Chap. 6, we have proved that a class of graphs $\mathcal{C}$ has bounded tree-depth if and only if it is degenerate and graphs in $\mathcal{C}$ exclude some path P_n as an induced subgraph (Proposition 6.4). A natural question is whether the requirement that $\mathcal{C}$ is degenerate could be replaced by another one. The uniformly quasi-wide property seems to be unrelated to degeneracy (see Exercise 8.2). However, it appears that the uniformly quasi-wide property is sufficient to characterize bounded tree-depth classes by the existence of a forbidden induced path:

Proposition 8.2. *Let $\mathcal{C}$ be a class of graphs. Then $\mathcal{C}$ has bounded tree-depth if and only if $\mathcal{C}$ is uniformly quasi-wide and graphs in $\mathcal{C}$ exclude some path P_k as an induced subgraph.*

Proof. One way is obvious as the class of graphs with tree-depth at most t is minor closed (hence is uniformly quasi-wide) and actually excludes P_{2^t-1} as a subgraph. So assume $\mathcal{C}$ is a quasi-wide class so that graphs in $\mathcal{C}$ exclude some path P_k as an induced subgraph. Without loss of generality, we may assume that $\mathcal{C}$ is infinite (for otherwise the result is straightforward). There exists some $s(k)$ such that for every m there exists $N(k,m)$, so that if $G \in \mathcal{C}$ and A is a subset of vertices of G of size at least $N(k,m)$ then there exists a subset S of vertices of size at most $s(k)$ such that A includes a subset A' of size at least m which is k-independent in $G-S$. Let us prove by contradiction that the tree-depths of the graphs in $\mathcal{C}$ are at most $N(k-1, s(k-1)+2)$. Otherwise, there exists $G \in \mathcal{C}$ including $P_{N(k-1,s(k-1)+2)}$ as a subgraph. Let A be the vertex set of this subgraph. Deleting $s(k-1)$ vertices cannot disconnect the path into more than $s(k-1)+1$ connected components hence A' has to include at least two vertices x and y belonging to a same connected component of $G-S$. As the distance from x to y is at least k in $G-S$, there exists in $G-S$ (hence in G) an induced path from x to y having length at least k, a contradiction. □

As consequence of this Proposition and Corollary 6.4 we get:

Corollary 8.1. *Every large graph either has a non-trivial symmetry or it contains a long induced path or it contains a shallow subdivision of a large complete graph.*

More precisely:

Theorem 8.6. *For every choice of positive integers* k, c *there exists* $N = N(k, c)$ *such that for every graph* G *of order at least* N *one of the following holds:*

(1) G *includes* P_{k+1} *as an induced subgraph;*
(2) G *includes* K_c *as a minor at depth* $(k-1)/2$*;*
(3) G *has a non trivial involutive automorphism.*

Proof. Assume P_{k+1} is not an induced subgraph of G and $K_c \notin G \triangledown (k-1)/2$. As $K_c \notin G \triangledown (k-1)/2$ there exist $s = s(k-1, c)$ and $N_0 = N_0(k-1, c)$ such that if A is a subset of at least N_0 vertices of G then there exists a subset S of at most s vertices of G such that A includes a $(k-1)$-independent set of $G - S$ of cardinality at least $s + 2$. Assume for contradiction that G includes a path P of length P_{N_0} and let A be the vertex set of P. After the deletion of some s vertices in G, the set A would include a $(k-1)$ independent set I of $G - S$ of cardinality $s + 2$ hence at least two vertices of I would belong to the same connected component (a path cannot be cut in more than $s + 1$ connected components by deleting s vertices), thus $G - S$ would include two vertices at distance at least k, contradicting the hypothesis that G includes no induced path of length k. As G includes no path of length N_0, the graph G has tree-depth at most N_0. According to Theorem 6.5, if the order of G is greater than $\mathcal{F}(N_0)$ then G has a non-trivial involutive automorphism. □

Recall that a graph is *asymmetric* if the identity is its unique automorphism. A graph G is *strongly minimal asymmetric* if it is asymmetric while none of its proper subgraphs with at least one edge is asymmetric. Corollary 8.1 implies that the number of strongly minimal asymmetric graphs is finite. Nešetřil conjectured that the number of *minimal asymmetric* graphs (i.e. those asymmetric graphs not containing an induced subgraph of order at least two which is asymmetric) is finite. It has been proved that there are only finitely many minimal asymmetric graphs containing an induced subgraph isomorphic to P_l for $l > 4$ [366] and the case where $l = 4$ was solved in [204].

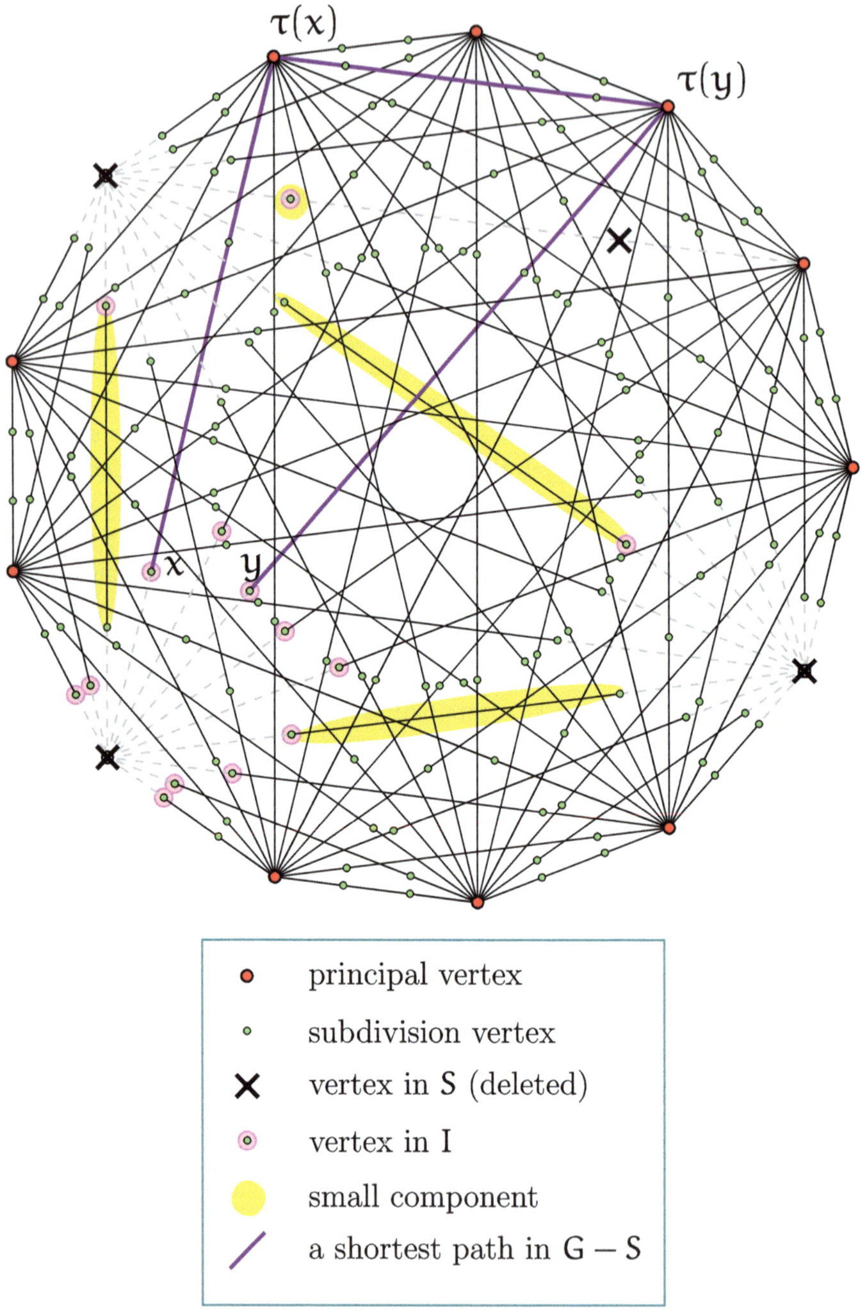

Fig. 8.5 Illustration of the proof of Lemma 8.9

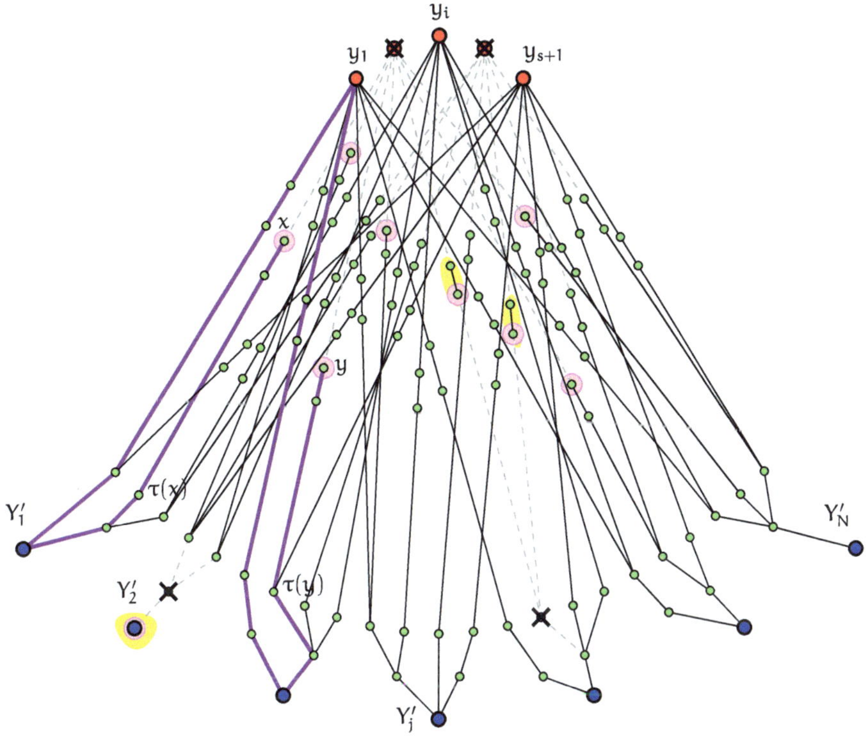

Fig. 8.6 Illustration of the proof of Lemma 8.10 (symbols have the same meaning as in Fig. 8.5)

Exercises

8.1. Give an example of a non-hereditary class of graphs which is wide but does not have bounded maximum degree.

8.2. (1) Prove that there exist classes of graphs which are degenerate but not uniformly quasi wide;
(2) Prove that there exist classes of graphs which are uniformly quasi-wide but not degenerate.

8.3. Prove that there exists a unique minimal asymmetric tree.

Chapter 9
First-Order Constraint Satisfaction Problems, Limits and Homomorphism Dualities

If you are not what you should not be,
you might well be as you should be.

9.1 Introduction

An important part of the *classical model theory* studies properties of abstract mathematical structures (finite or not) expressible in first-order logic [257]. In the setting of *finite model theory*, which developed more recently, one studies first-order logic (and its various extensions) just on finite structures [141, 303]. Both theories share many similarities but also display important and profound differences. This will be illustrated in this chapter by means of several examples. We start with a generalization of the coloring problem.

Suppose we want to check whether a graph has a particular property, for example admitting a decomposition (interpreted as a coloring) of a certain kind, satisfying certain local properties (such as containing no monochromatic edge, demanding that every path of length three gets at least three colors etc.). Such a task appears in the theory of scheduling and other applied areas. Typical examples are *Constraint Satisfaction Problems* (CSPs), which are formulated here as H-coloring problems.

J. Nešetřil and P.O. de Mendez, *Sparsity*, Algorithms and Combinatorics 28,
DOI 10.1007/978-3-642-27875-4_9,

Given a fixed relational structure **H**, the **H**-*coloring problem* is the decision problem to determine whether an input relational structure **G** admits a homomorphism to **H** (does $\mathbf{G} \to \mathbf{H}$?).

Coloring problems tend to be very difficult problems to solve exactly, but also to enumerate and even to approximate. In this direction, it has been proved by Hell and Nešetřil [252] that in the context of finite undirected graphs the H-coloring problem is NP-complete unless H is bipartite, in which case the H-coloring problem trivially belongs to P. So no non-trivial H-coloring problem is easy. Another early result by Schaefer [430] characterized complexities of Boolean constraint satisfaction problems. These results led Feder and Vardi [174, 175] to formulate the celebrated *Dichotomy Conjecture* which asserts that, for every constraint language over an arbitrary finite domain, either the constraint satisfaction problem is in P or it is NP-complete. It was soon noticed that this conjecture is equivalent to the existence of a dichotomy for **H**-coloring problems (for general relational structures **H**). This is one of the reasons why in this chapter we mainly consider the general case of relational structures. (To make this change more visible we denote the structures by bold face capital letters. The Dichotomy Conjecture is presently an open problem. It is even open for the class oriented graphs (and it suffices to establish dichotomy for this special case.) There are many results and suggested methods but the problem is hard, see e.g. [254].

On the other side of complexity spectrum are **H**-coloring problems that may be expressed in first-order logic, thus allowing fast checking (at most polynomial time). By this we mean the following:

Let **H** be a structure. The **H**-coloring is said to be *first-order definable* (on the finite) if there exists a first-order formula ϕ such that for every structure **G** it holds

$$\mathbf{G} \to \mathbf{H} \qquad \iff \qquad \mathbf{G} \models \phi.$$

The **H**-coloring problems which are first-order definable are related (in fact equivalent) to finite dualities. They are treated in the next section using logical tools which we introduce now and which are interesting in their own.

This is related to so-called preservation theorems, which form a classical area of mathematical logic. Results of this type are closely related to results in this chapter. Actually, we devote the whole Chap. 10 to preservation theorems. Here we state only one of the deepest results in this area, which is the following [425].

Theorem 9.1. *Let ϕ be a first order formula of quantifier rank n. Then,*

$$\mathbf{G} \to \mathbf{H} \text{ and } \mathbf{G} \models \phi \quad \Longrightarrow \quad \mathbf{H} \models \phi$$

holds for all finite $\mathbf{G}, \mathbf{H} \in \mathrm{Rel}(\sigma)$ *if and only if* ϕ *is equivalent for finite structures in* $\mathrm{Rel}(\sigma)$ *to an existential-positive first-order formula* ψ *of quantifier rank* $\rho(n)$ *(for some explicit function* $\rho : \mathbb{N} \to \mathbb{N}$*).*

It follows from Theorem 9.1 (and this is its combinatorial meaning) that for finite structures, the only $\mathbf{H}$-coloring problems which are expressible in first-order logic are those for which there exists a finite family $\mathcal{F}$ of finite structures with the property that for every $\mathbf{G}$:

$$\exists \mathbf{F} \in \mathcal{F} \quad \mathbf{F} \to \mathbf{G} \qquad \Longleftrightarrow \qquad \mathbf{G} \nrightarrow \mathbf{H}. \tag{9.1}$$

This found nice interpretation in the context of homomorphism dualities which will be introduced in the next section (see also Sect. 3.7 where we handled some special cases).

9.2 Homomorphism Dualities and the Functor U

9.2.1 Finite Dualities

Let us introduce finite dualities in the context of σ-structures.

Let $(\mathcal{F}, \mathcal{D})$ be two finite families of finite σ-structures. Assume that the following statement holds for every finite σ-structure $\mathbf{G}$:

$$\exists \mathbf{F} \in \mathcal{F} \quad \mathbf{F} \to \mathbf{G} \qquad \Longleftrightarrow \qquad \forall \mathbf{D} \in \mathcal{D} \quad \mathbf{G} \nrightarrow \mathbf{D}. \tag{9.2}$$

This statement is called a *finite homomorphism duality.* The pair $(\mathcal{F}, \mathcal{D})$ is called a *dual pair.* We also say that $\mathcal{D}$ has *finite duality.*

The notion of finite duality captures the fact that the existence of a homomorphism into a given set $\mathcal{D}$ of structures (called *duals* or *templates*) can be alternatively expressed dually by the non-existence of homomorphisms from a given set $\mathcal{F}$ of (forbidden) structures. In other words, the finite set $\mathcal{F}$ is a complete set of obstructions for one of the $\mathbf{D}$-coloring, for $\mathbf{D} \in \mathcal{D}$. Here is one example for directed graphs:

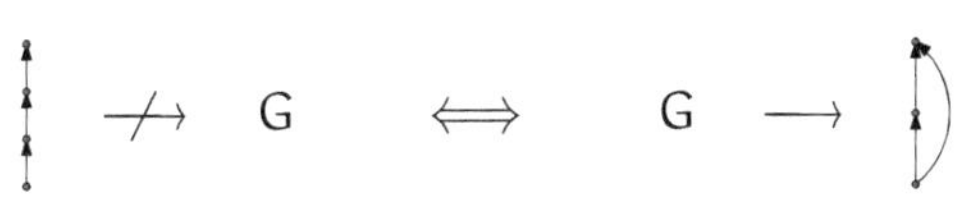

Recall that in Theorem 3.13 we stated that for every directed graph $\vec{G}$ it holds:

$$\vec{P}_{k+1} \nrightarrow \vec{G} \quad \Longleftrightarrow \quad \vec{G} \rightarrow \vec{T}_k.$$

Singleton dualities are finite dualities where $\mathcal{F}$ and $\mathcal{D}$ are one element sets (i.e. singletons). Another example of a singleton duality is

Singleton dualities have been characterized by Nešetřil and Pultr [362] for undirected graphs, by Komárek [279] for directed graphs and by Nešetřil and Tardif [370] for general finite structures:

Theorem 9.2. *For any signature σ and any finite set $\mathcal{F}$ of σ-structures the following two statements are equivalent:*

1. *There exists* **D** *such that $\mathcal{F}$ and* **D** *form a finite duality,*
2. *$\mathcal{F}$ is homomorphically equivalent to a set of finite (relational) trees. (A relational tree may be defined via its incidence graph, see Sect. 3.8.3.)*

A direct consequence of Theorem 9.1 is:

Corollary 9.1. *Let* **H** *be a structure. Then, the* **H***-coloring is first-order definable if and only if there is a family $\mathcal{F}$ of structures such that $(\mathcal{F}, \mathbf{H})$ is a finite homomorphism duality.*

Notice that Atserias [38, 39] derived this theorem independently of Theorem 9.1. Combining the above results yields the following.

Theorem 9.3. *Let* **D** *be a σ-structure. Then the following are equivalent.*

1. **D***-coloring is first-order definable;*
2. **D** *has finite duality;*
3. **D** *has a complete set of obstructions consisting of a finite set of trees.*

By passing let us note that the seemingly more general setting when the set $\mathcal{D}$ consists from more graphs can be treated similarly: the obstruction set $\mathcal{F}$ is consisting from *relational forests*, see [180]. We do not include elegant proofs of these results, see e.g. [253, 369]. Larose et al. [297] proved that it is decidable (and even just NP-complete) to determine whether a constraint satisfaction problem is first-order definable. We can do better if we restrict input structures to cores: the problem of deciding, for a core **A**, whether

A-coloring is first-order decidable may be done in polynomial time (without requiring an actual certificate that the input **A** is a core).

9.2.2 Tree Dualities and the Functor U

Let us take a more general view. Finite homomorphism duality finds for a **A**-coloring a complete finite set of obstructions $\mathcal{F}$. (These sections of this chapter have algebraic flavour and thus we use **A**, **B**, **C** for relational structures.) In the other words **A**-coloring is defined by a (very) simple set of obstructions. What happens if we allow other (possibly infinite) sets of simple obstructions such as trees, cycles, graphs with bounded tree depth or graphs with bounded tree width? In the case of a set of obstructions with tree-width bounded by an integer k, we speak about k-*tree-width duality* or, if the actual value of k is not important, *bounded tree-width duality*. Examples of such dualities are abundant (think for example of two-coloring of undirected graphs, which is an example of a two-tree-width duality) and they were studied in many papers. Particularly the bounded tree width duality was studied in [255] and it coincides with problems which have a *Datalog* description [175] (see [89] for more informations on Datalog). Such problems have a polynomial algorithm for **A**-coloring (see also [371]). Recently it has been proved [53] that it is decidable whether a CSP has a bounded tree width duality.

A particular case is when for **A**-coloring there exists a complete set of obstructions consisting from trees only (i.e. in case of a one-tree-width duality). Then the characterization is particularly elegant using the following construction (defined in [175]): Given a structure **A** (with universe A and relations $R_1, \ldots, R_m$) we define the structure $\mathbf{U}(\mathbf{A})$ with same signature, whose universe U is the set of all nonempty subsets of A, and for $i = 1, \ldots, m$, $R_i(\mathbf{U}(\mathbf{A}))$ is the set of all r_i-tuples $(X_1, \ldots, X_{r_i})$ such that for all $j \in \{1, \ldots, r_i\}$ and for every choice $x_j \in X_j$ there exist $x_k \in X_k$ for all $k \in \{1, \ldots, r_i\} \setminus \{j\}$ such that $(x_1, \ldots, x_{r_i}) \in R_i(\mathbf{A})$.

The mapping $\mathbf{A} \mapsto \mathbf{U}(\mathbf{A})$ has nice algebraic properties, which are expressed by the following.

Proposition 9.1. *The mapping* **U** *has the following properties:*

(a) *The mapping* **U** *is functorial.*
Precisely, for every homomorphism $f : \mathbf{A} \to \mathbf{B}$ *define* $\mathbf{U}(f) : \mathbf{U}(\mathbf{A}) \to \mathbf{U}(\mathbf{B})$ *by*

$$\mathbf{U}(f)(X) = \{f(x) : x \in X\}.$$

Then $\mathbf{U}(f)$ *is a homomorphism of* $\mathbf{U}(\mathbf{A})$ *to* $\mathbf{U}(\mathbf{B})$.

(b) *Every* $\mathbf{A}$ *is homomorphic to* $\mathbf{U}(\mathbf{A})$.
Precisely, the mapping $\eta_{\mathbf{A}}$ *which maps the element* x *of* $\mathbf{A}$ *to the element* $\{x\}$ *of* $\mathbf{U}(\mathbf{A})$ *is a homomorphism of* $\mathbf{A}$ *to* $\mathbf{U}(\mathbf{A})$*. Actually, we have the following commutative diagram (which expresses that* η *is a natural transformation from the identical functor* Id *to* $\mathbf{U}$*):*

$$\begin{array}{ccc} \mathbf{A} & \xrightarrow{f} & \mathbf{B} \\ \downarrow{\scriptstyle \eta_{\mathbf{A}}} & & \downarrow{\scriptstyle \eta_{\mathbf{B}}} \\ \mathbf{U}(\mathbf{A}) & \xrightarrow{\mathbf{U}(f)} & \mathbf{U}(\mathbf{B}) \end{array}$$

(c) *The functor* $\mathbf{U}$ *is idempotent.*
Precisely, for every structure $\mathbf{A}$ *it holds*

$$\mathbf{U}(\mathbf{U}(\mathbf{A})) \rightleftarrows \mathbf{U}(\mathbf{A}).$$

Proof. (a) Let $(X_1, \dots, X_{r_i}) \in R_i(\mathbf{U}(\mathbf{A}))$. Then (by definition)

$$\forall j \in [r_i]\ \forall x_j \in X_j\ \forall k \in [r_i] \setminus \{j\}\ \exists x_k \in X_k :\ (x_1, \dots, x_{r_i}) \in R_i(\mathbf{A}).$$

Hence

$$\forall j \in [r_i]\ \forall x_j \in X_j\ \forall k \in [r_i] \setminus \{j\}\ \exists x_k \in X_k :\ (f(x_1), \dots, f(x_{r_i})) \in R_i(\mathbf{B}),$$

that is:

$$\forall j \in [r_i]\ \forall u_j \in f(X_j)\ \forall k \in [r_i] \setminus \{j\}\ \exists u_k \in f(X_k) :\ (u_1, \dots, u_{r_i}) \in R_i(\mathbf{B}).$$

It follows that $\mathrm{U}(f)$ is a homomorphism of $\mathbf{U}(\mathbf{A})$ to $\mathbf{U}(\mathbf{B})$. Moreover, $\mathbf{U}(\mathrm{Id}_{\mathbf{A}}) = \mathrm{Id}_{\mathbf{U}(\mathbf{A})}$ and if $f : \mathbf{A} \to \mathbf{B}$ and $g : \mathbf{B} \to \mathbf{C}$ are homomorphisms, then $\mathbf{U}(g \circ f) = \mathbf{U}(g) \circ \mathbf{U}(f)$ hence $\mathbf{U}$ is a functor.

(b) Let $(u_1, \dots, u_{r_i}) \in R_i(\mathbf{A})$. Then obviously

$$(\{u_1\}, \dots, \{u_{r_i}\}) \in R_i(\mathbf{U}(\mathbf{A})).$$

(c) According to (b), $\mathbf{A} \to \mathbf{U}(\mathbf{A})$ hence, according to (a), $\mathbf{U}(\mathbf{A}) \to \mathbf{U}(\mathbf{U}(\mathbf{A}))$. Consider the mapping F from the universe of $\mathbf{U}(\mathbf{U}(\mathbf{A}))$ (nonempty families of nonempty subsets of elements of $\mathbf{A}$) to the one of $\mathbf{U}(\mathbf{A})$ (nonempty subsets of elements of $\mathbf{A}$) defined by $F(\mathcal{X}) = \bigcup \mathcal{X}$.
Let $(\mathcal{X}_1, \dots, \mathcal{X}_{r_i}) \in R_i(\mathbf{U}(\mathbf{U}(\mathbf{A})))$. Let $x_j \in F(\mathcal{X}_j)$. Then there exists $X_j \in \mathcal{X}_j$ such that $x_j \in X_j$. As $(\mathcal{X}_1, \dots, \mathcal{X}_{r_i}) \in R_i(\mathbf{U}(\mathbf{U}(\mathbf{A})))$, there exists for each $k \in [r_i] \setminus \{j\}$ some $X_k \in \mathcal{X}_k$ such that $(X_1, \dots, X_{r_i}) \in R_i(\mathbf{U}(\mathbf{A}))$. Hence, as $x_j \in X_j$, there exists for each $k \in [r_i] \setminus \{j\}$ some $x_k \in X_k \subseteq F(\mathcal{X}_k)$ such that $(x_1, \dots, x_{r_i}) \in R_i(\mathbf{A})$. As this holds for every $x_j \in F(\mathcal{X}_j)$, it

follows that $F(\mathcal{X}_1, \ldots, \mathcal{X}_{r_i}) \in R_i(\mathbf{U}(\mathbf{A}))$ hence F is a homomorphism of $\mathbf{U}(\mathbf{U}(\mathbf{A}))$ to $\mathbf{U}(\mathbf{A})$.

□

One of the main aspect of this mapping is indeed that it captures the property of two structures to be distinguishable by testing existence of homomorphisms from finite trees:

Lemma 9.1. *Let* $\mathbf{T}$ *be a tree. Assume* ϕ *is a homomorphism of* $\mathbf{T}$ *to* $\mathbf{U}(\mathbf{A})$, $x_0 \in T$ *and* $a_0 \in \phi(x_0)$.

Then there exist a homomorphism $\widetilde{\phi}$ *of* $\mathbf{T}$ *to* $\mathbf{A}$ *such that* $\widetilde{\phi}(x) \in \phi(x)$ *for every* $x \in \mathbf{T}$.

Proof. We proceed by induction over the order of $\mathbf{F}$. If $|\mathbf{T}| = 1$, the result is straightforward. Assume that the statement holds for every tree $\mathbf{T}$ of order at most n_0 for some $n_0 \geq 1$, and let $\mathbf{T}$ be a tree of order $n_0 + 1$.

Let $x_0 \in T$ and let $x_1, \ldots, x_k$ be the elements of $\mathbf{T}$ adjacent to x_0 (that is the elements belonging to a common relation with x_0). As $\mathbf{T}$ is a relational tree, the relational forest $\mathbf{T} - x_0$ obtained by deleting x_0 (and all the relations including x_0) is a union of relational trees, each of those include exactly one of the x_i's (for $1 \leq i \leq k$). Denote by $\mathbf{T}_i$ the component which contains x_i. According to the definition of $\mathbf{U}(\mathbf{A})$, (and because $\mathbf{T}$ is a tree) it is possible to choose in each $\phi(x_i)$ and element a_i such that for every $i_1, \ldots, i_p \in \{0, 1, \ldots, k\}$, if $(x_{i_1}, \ldots, x_{i_p}) \in R_j$ in $\mathbf{T}$ then $(a_{i_1}, \ldots, a_{i_p}) \in R_j$ in $\mathbf{A}$. By induction, there exist for every $1 \leq i \leq k$ a homomorphism $\widetilde{\phi}_i : \mathbf{T}_i \to \mathbf{A}$ such that $\widetilde{\phi}(x_i) = a_i$ and $\widetilde{\phi}(x) \in \phi(x)$ for every $x \in \mathbf{T}_i$. We define $\widetilde{\phi}$ by $\widetilde{\phi}(x_0) = a_0$ and $\widetilde{\phi}(v) = \widetilde{\phi}_i(v)$ whenever $v \in \mathbf{T}_i$. Then $\widetilde{\phi}$ meets the conditions of the lemma. □

Corollary 9.2. *Let* $\mathbf{T}$ *be a tree. Then* $\mathbf{T}$ *admits a homomorphism to* $\mathbf{A}$ *if and only if* $\mathbf{T}$ *admits a homomorphism to* $\mathbf{U}(\mathbf{A})$.

Proof. If $\mathbf{T} \to \mathbf{A}$ then $\mathbf{T} \to \mathbf{U}(\mathbf{A})$, as $\mathbf{A} \to \mathbf{U}(\mathbf{A})$ (by Proposition 9.1–(b)).

Conversely, if $\mathbf{T} \to \mathbf{U}(\mathbf{A})$ then $\mathbf{T} \to \mathbf{A}$ by Lemma 9.1.

It is then not surprising that the functor U characterizes tree dualities (Feder and Vardi [175]): □

Theorem 9.4. *A* σ*-structure* $\mathbf{A}$ *has tree duality if and only if there exists a homomorphism of* $\mathbf{U}(\mathbf{A})$ *to* $\mathbf{A}$.

It is easy to see that the existence of a homomorphism $\mathbf{U}(\mathbf{A}) \to \mathbf{A}$ is equivalent to the fact that a particular algorithm—the so-called *consistency check algorithm*—succeeds. The existence of a tree duality can be expressed in terms of consistency check [255] (see Exercise 9.10 where we sketch the proof in the case of graphs).

Consequently, determining whether a given structure $\mathbf{A}$ has tree duality is decidable. Tree dualities were studied in several papers. Let us mention at least the recent paper by Foniok and Tardif [181].

We conclude this section with the following alternate characterization of structures which cannot be distinguished by existence of homomorphisms from finite trees:

Lemma 9.2. *Let $\mathbf{A}, \mathbf{B}$ be structures. The following conditions are equivalent:*

1. *$\mathbf{A}$ and $\mathbf{B}$ cannot be distinguished by existence of homomorphisms from finite trees:*
$$\forall \text{ tree } \mathbf{T}: \quad (\mathbf{T} \to \mathbf{A}) \iff (\mathbf{T} \to \mathbf{B});$$
2. *$\mathbf{U}(\mathbf{A})$ is homomorphically equivalent to $\mathbf{U}(\mathbf{B})$:*
$$\mathbf{U}(\mathbf{A}) \rightleftarrows \mathbf{U}(\mathbf{B}).$$

Proof. Assume that (1) holds. According to Proposition 9.1–(c) we have $\mathbf{U}(\mathbf{U}(\mathbf{A})) \to \mathbf{U}(\mathbf{A})$ hence, by Theorem 9.4, $\mathbf{U}(\mathbf{A})$ has a tree duality. It follows that for every $\mathbf{B}$ we have $\mathbf{B} \to \mathbf{U}(\mathbf{A})$ if and only if for every tree $\mathbf{T}$ it holds

$$\mathbf{T} \to \mathbf{B} \quad \Longrightarrow \quad \mathbf{T} \to \mathbf{U}(\mathbf{A}).$$

According to Corollary 9.2, we have

$$\mathbf{T} \to \mathbf{U}(\mathbf{A}) \quad \iff \quad \mathbf{T} \to \mathbf{A}.$$

Hence we deduce $\mathbf{B} \to \mathbf{U}(\mathbf{A})$ thus, by Proposition 9.1–(a) and (c), $\mathbf{U}(\mathbf{B}) \to \mathbf{U}(\mathbf{U}(\mathbf{A})) \to \mathbf{U}(\mathbf{A})$. Similarly, we have $\mathbf{U}(\mathbf{A}) \to \mathbf{U}(\mathbf{B})$ hence (2) holds.

Conversely, assume that (2) holds. For every tree $\mathbf{T}$ we have:

$$\begin{aligned} \mathbf{T} \to \mathbf{A} &\iff \mathbf{T} \to \mathbf{U}(\mathbf{A}) && \text{(by Corollary 9.2)} \\ &\iff \mathbf{T} \to \mathbf{U}(\mathbf{B}) && \text{(by (2))} \\ &\iff \mathbf{T} \to \mathbf{B} && \text{(by Corollary 9.2).} \end{aligned}$$

Hence (1) holds. □

9.3 Metrics on the Homomorphism Order

We would like to characterize more general setting of dualities than above: the restricted dualities (i.e. dualities restricted to a class) will be treated in the next chapter. Towards this end we introduce in this chapter metric theory for the homomorphism order and related limits.

9.3.1 Partially Ordered Sets

We review here standard notions and notations defined on partially ordered sets which are relevant to this chapter. Let $(X, \leq)$ be a (finite or infinite) *partially ordered set* (or *poset*).

- A *lower set* (*downset*, or *initial segment*) is a subset D with the property that
$$\forall x \in D \text{ and } y \leq x \quad \Rightarrow \quad y \in D;$$
- An *ideal* is a lower set I with the additional property
$$\forall x, y \in I, \ \exists z \in I: \quad x \leq z \text{ and } y \leq z;$$
- The *principal ideal* ${}_{\downarrow}p$ of an element p is the smallest ideal that contains p:
$${}_{\downarrow}p = \{x : \ x \leq p\}.$$

Dual to these notions are the one of

- *Upper set* (or *upset*),
- *Filter*,
- *Principal filter* $p^{\uparrow} = \{x : x \geq p\}$.

The poset $(X, \leq)$ is a *lattice* if any two elements x and y have a unique supremum $x \vee y$ (their *join*) and an infimum $x \wedge y$ (their *meet*). A lattice is

- *Distributive* if
$$a \wedge (b \vee c) = (a \wedge b) \vee (a \wedge c)$$
holds for every $a, b, c \in X$;
- *Complete* if all its subsets (finite or not) have both a join and a meet.

(An example of distributive lattice has been introduced in Exercise 3.3.)

Of particular interest are ideals whose complements are filters. An ideal with this property is a *prime ideal.* If $(X, \leq)$ is a lattice, then an ideal I is prime if and only if

$$\forall x, y \in X: \quad (x \wedge y \in I) \Rightarrow (x \in I) \text{ or } (y \in I).$$

The notion dual to prime ideal is *prime filter*.

For a subset A of X, one defines

- The *set of upper bounds* A^u of A by
$$A^u = \{x: \forall y \in A \; x \geq y\};$$

- The *set of lower bounds* A^ℓ of A by
$$A^\ell = \{x: \forall y \in A \; x \leq y\}.$$

These constructions are of particular importance. For instance, the *Dedekind-MacNeille completion* of $(X, \leq)$, which is the smallest complete lattice that contains $(X, \leq)$ is defined by means of the inclusion order on the subsets A of X such that $(A^u)^\ell = A$. An element $x \in X$ then embeds in the completion of $(X, \leq)$ as its principal ideal, as

$$(({}_\downarrow x)^u)^\ell = {}_\downarrow x.$$

Notice that for every subset A, the set A^u is a filter and the set A^ℓ is an ideal.

Another classical completion of $(X, \leq)$ is the set of its downsets ordered by inclusion. The poset $(X, \leq)$ is also embedded in this lattice by sending each element x to ${}_\downarrow x$. This completion is a distributive lattice, which is used in Birkhoff's representation theorem. However, it may be much bigger than Dedekind-MacNeille completion.

9.3.2 The Homomorphism Order of σ-Structures

At several places of this book (starting with the Prolegomena) we dealt with the notion of homomorphism order. We now (and in the next sections) review the algebraical and metrical side of it in a greater detail (for general structures).

The existence of a homomorphism of a σ-structure **A** to a σ-structure **B** naturally defines a quasi-order on Rel(σ). The derived equivalence relation is the *homomorphism equivalence* $\mathbf{A} \rightleftarrows \mathbf{B}$. The quotient of Rel(σ) by homomorphism equivalence is denoted by $[\mathrm{Rel}(\sigma)]$. We also denote by $[\mathbf{A}]$ the equivalence class of the structure $\mathbf{A}$ (a natural representative of which is the core of $\mathbf{A}$, see Sect. 3.7) and by $[\mathcal{C}]$ the set $\{[\mathbf{A}], \mathbf{A} \in \mathcal{C}\}$. The set $[\mathrm{Rel}(\sigma)]$ is ordered by the *homomorphism order* $\leq_h$ defined by

$$[\mathbf{A}] \leq_h [\mathbf{B}] \iff \mathbf{A} \to \mathbf{B}.$$

In previous parts of this book we already considered the homomorphism order of graphs. Notice that $([\mathrm{Rel}(\sigma)], \leq_h)$ has a structure of distributive lattice, with meet $[\mathbf{A}] \vee [\mathbf{B}] = [\mathbf{A} + \mathbf{B}]$ and join $[\mathbf{A}] \wedge [\mathbf{B}] = [\mathbf{A} \times \mathbf{B}]$.

We denote by $(\to \mathbf{A})$ the set of all $\mathbf{F} \in \mathrm{Rel}(\sigma)$ having a homomorphism to $\mathbf{A}$ and we denote by $(\mathbf{A} \to)$ the class of all $\mathbf{F} \in \mathrm{Rel}(\sigma)$ that admit a homomorphism of $\mathbf{A}$.

We have the following correspondences between $\mathrm{Rel}(\sigma)$ and $[\mathrm{Rel}(\sigma)]$:

$\mathbf{A} \to \mathbf{B}$	$\xrightarrow{[\,]}$	$[\mathbf{A}] \leq_h [\mathbf{B}]$
$(\to \mathbf{A})$	$\to$	${}_{\downarrow}[\mathbf{A}]$
$(\mathbf{A} \to)$	$\to$	$[\mathbf{A}]^{\uparrow}$
$\mathbf{A} + \mathbf{B}$	$\to$	$[\mathbf{A}] \vee [\mathbf{B}]$
$\mathbf{A} \times \mathbf{B}$	$\to$	$[\mathbf{A}] \wedge [\mathbf{B}]$

Also we have:

$$[(\to \mathbf{A})] = {}_{\downarrow}[\mathbf{A}] \qquad [(\mathbf{A} \to)] = [\mathbf{A}]^{\uparrow}$$

$$(\to \mathbf{A}) \cap (\mathbf{A} \to) = [\mathbf{A}] \qquad {}_{\downarrow}[\mathbf{A}] \cap [\mathbf{A}]^{\uparrow} = \{[\mathbf{A}]\}$$

For an ideal $\mathcal{I}$ and a filter $\mathcal{F}$ we define

$$\mathcal{I}^{\star} = \mathcal{I}^{u} = \{[\mathbf{A}],\ \forall [\mathbf{B}] \in \mathcal{I}:\ [\mathbf{A}] \geq_h [\mathbf{B}]\} = \bigcap_{[\mathbf{B}] \in \mathcal{I}} [\mathbf{B}]^{\uparrow}$$

$$\mathcal{F}^{\star} = \mathcal{F}^{\ell} = \{[\mathbf{A}],\ \forall [\mathbf{B}] \in \mathcal{F}:\ [\mathbf{A}] \leq_h [\mathbf{B}]\} = \bigcap_{[\mathbf{B}] \in \mathcal{F}} {}_{\downarrow}[\mathbf{B}]$$

Notice that $\mathcal{I}^{\star}$ is a filter and that $\mathcal{F}^{\star}$ is an ideal and that for every $\mathbf{A}$: $({}_{\downarrow}[\mathbf{A}])^{\star} = [\mathbf{A}]^{\uparrow}$ and $([\mathbf{A}]^{\uparrow})^{\star} = {}_{\downarrow}[\mathbf{A}]$.

9.3.3 Connectivity and Multiplicativity

A structure $\mathbf{A} \in \mathrm{Rel}(\sigma)$ is *connected* if its Gaifman graph (or equivalently, its incidence graph) is connected. We denote by $\mathrm{Rel}_{\mathrm{con}}(\sigma)$ the set of all connected σ-structures. A *connected component* of a structure $\mathbf{A}$ is each substructure induced by all the vertices of $\mathbf{A}$ in a connected component of the Gaifman graph of $\mathbf{A}$. Remark that if $\mathbf{A}$ is connected, then $\mathbf{A} \to \mathbf{B} + \mathbf{C}$ if and only if $\mathbf{A} \to \mathbf{B}$ or $\mathbf{A} \to \mathbf{C}$. By extension, we say that $[\mathbf{A}]$ is *connected* if $[\mathbf{A}]$ contains a connected structure. Similarly, a structure $\mathbf{A}$ is *multiplicative* if, for every $\mathbf{B}, \mathbf{C}$, we have $\mathbf{B} \times \mathbf{C} \to \mathbf{A}$ if and only if $\mathbf{B} \to \mathbf{A}$ or $\mathbf{C} \to \mathbf{A}$. Also, $[\mathbf{A}]$ is *multiplicative* if $[\mathbf{A}]$ contains a multiplicative structure.

The following properties are straightforward:

Lemma 9.3. *A class* $[\mathbf{A}] \in [\mathrm{Rel}(\sigma)]$ *is:*

(a) Connected if and only if

$$[\mathbf{A}] \leq_h [\mathbf{B}] \vee [\mathbf{C}] \quad \iff \quad [\mathbf{A}] \leq_h [\mathbf{B}] \text{ or } [\mathbf{A}] \leq_h [\mathbf{C}]$$

holds for every $[\mathbf{B}], [\mathbf{C}] \in [\mathrm{Rel}(\sigma)]$.

(b) Not connected if and only if there exist $\mathbf{B}$ *and* $\mathbf{C}$ *such that* $\mathbf{B} \not\rightleftarrows \mathbf{C}$ *(i.e.* $\mathbf{B} \nrightarrow \mathbf{C} \nrightarrow \mathbf{B}$*) and* $\mathbf{A} \rightleftarrows \mathbf{B} + \mathbf{C}$.

(c) Multiplicative if and only if

$$[\mathbf{B}] \wedge [\mathbf{C}] \leq_h [\mathbf{A}] \quad \iff \quad [\mathbf{B}] \leq_h [\mathbf{A}] \text{ or } [\mathbf{C}] \leq_h [\mathbf{A}]$$

holds for every $[\mathbf{B}], [\mathbf{C}] \in [\mathrm{Rel}(\sigma)]$.

(d) Non multiplicative if and only if there exist $\mathbf{B}$ *and* $\mathbf{C}$ *such that* $\mathbf{B} \not\rightleftarrows \mathbf{C}$ *and* $\mathbf{A} \rightleftarrows \mathbf{B} \times \mathbf{C}$.

For every σ*-structures* $\mathbf{A}$ *and* $\mathbf{B}$ *the following implications hold:*

(e) If $\mathbf{A}$ *and* $\mathbf{B}$ *are not connected, then* $\mathbf{A} + \mathbf{B}$ *is not connected;*

(f) If $\mathbf{A}$ *and* $\mathbf{B}$ *are not multiplicative, then* $\mathbf{A} \times \mathbf{B}$ *is not multiplicative.*

Proof. Items (a) to (c) are straightforward.

(d) Assume $\mathbf{B} \not\rightleftarrows \mathbf{C}$ and $\mathbf{A} \rightleftarrows \mathbf{B} \times \mathbf{C}$. Then $\mathbf{B} \times \mathbf{C} \to \mathbf{A}$ although $\mathbf{B} \nrightarrow \mathbf{A}$ and $\mathbf{C} \nrightarrow \mathbf{A}$.

Conversely, assume that $\mathbf{A}$ is non multiplicative. Then there exist $\mathbf{F}, \mathbf{D}$ such that $\mathbf{F} \nrightarrow \mathbf{A}$, $\mathbf{D} \nrightarrow \mathbf{A}$ and $\mathbf{F} \times \mathbf{D} \to \mathbf{A}$. Assume $\mathbf{F}_1, \dots, \mathbf{F}_p$ are the connected components of $\mathbf{F}$. Then there exists $1 \leq i \leq p$ such that $\mathbf{F}_i \nrightarrow \mathbf{A}$ (for otherwise $\mathbf{F} \to \mathbf{A}$). Moreover, $\mathbf{F}_i \times \mathbf{D} \to \mathbf{F} \times \mathbf{D} \to \mathbf{A}$. Hence we may request that both $\mathbf{F}$ and $\mathbf{D}$ are connected. Notice that $\mathbf{F} \nrightarrow \mathbf{D}$ for otherwise $\mathbf{F} \to \mathbf{F} \times \mathbf{D} \to \mathbf{A}$. As $\mathbf{F}$ is connected and $\mathbf{F} \nrightarrow \mathbf{A}$ we deduce $\mathbf{F} \nrightarrow \mathbf{D} + \mathbf{A}$ thus $\mathbf{F} + \mathbf{A} \nrightarrow \mathbf{D} + \mathbf{A}$. Similarly we have

$\mathbf{D}+\mathbf{A} \nrightarrow \mathbf{F}+\mathbf{A}$. Let $\mathbf{B}=\mathbf{F}+\mathbf{A}$ and $\mathbf{C}=\mathbf{D}+\mathbf{A}$. Then $\mathbf{B} \mathrel{\substack{\nrightarrow \\ \nleftarrow}} \mathbf{C}$ and $\mathbf{B}\times\mathbf{C}=(\mathbf{F}+\mathbf{D})\times(\mathbf{F}+\mathbf{A}) \rightleftarrows \mathbf{F}\times\mathbf{D}+\mathbf{A} \rightleftarrows \mathbf{A}$.

(e) Is straightforward.

(f) Assume $\mathbf{C}_1$ and $\mathbf{C}_2$ are non-multiplicative and let $\mathbf{A}_1 \mathrel{\substack{\nrightarrow \\ \nleftarrow}} \mathbf{B}_1$ and $\mathbf{A}_2 \mathrel{\substack{\nrightarrow \\ \nleftarrow}} \mathbf{B}_2$ be such that $\mathbf{C}_1 \rightleftarrows \mathbf{A}_1\times\mathbf{B}_1$ and $\mathbf{C}_2 \rightleftarrows \mathbf{A}_2\times\mathbf{B}_2$.

- Notice that $\mathbf{A}_1 \nrightarrow \mathbf{A}_2\times\mathbf{B}_1\times\mathbf{B}_2$ for otherwise $\mathbf{A}_1 \to \mathbf{B}_1$. If $\mathbf{A}_2\times\mathbf{B}_1\times\mathbf{B}_2 \nrightarrow \mathbf{A}_1$ then $\mathbf{C}_1\times\mathbf{C}_2=\mathbf{A}_1\times(\mathbf{A}_2\times\mathbf{B}_1\times\mathbf{B}_2)$ is non-multiplicative according to (d).
- Otherwise, $\mathbf{A}_2\times\mathbf{B}_1\times\mathbf{B}_2 \to \mathbf{A}_1$ thus $\mathbf{C}_1\times\mathbf{C}_2 \rightleftarrows \mathbf{A}_2\times\mathbf{B}_1\times\mathbf{B}_2$. Notice that $\mathbf{A}_2 \nrightarrow \mathbf{B}_1\times\mathbf{B}_2$ for otherwise $\mathbf{A}_2 \to \mathbf{B}_2$. If $\mathbf{B}_1\times\mathbf{B}_2 \nrightarrow \mathbf{A}_2$ then $\mathbf{C}_1\times\mathbf{C}_2 \rightleftarrows \mathbf{A}_2\times(\mathbf{B}_1\times\mathbf{B}_2)$ is non-multiplicative according to (d).
- Otherwise, $\mathbf{B}_1\times\mathbf{B}_2 \to \mathbf{A}_2$ thus $\mathbf{C}_1\times\mathbf{C}_2 \rightleftarrows \mathbf{B}_1\times\mathbf{B}_2$. We have $\mathbf{B}_1 \nrightarrow \mathbf{B}_2$ for otherwise $\mathbf{B}_1 \to \mathbf{B}_1\times\mathbf{B}_2 \rightleftarrows \mathbf{C}_1\times\mathbf{C}_2 \to \mathbf{C}_1$. Similarly, $\mathbf{B}_2 \nrightarrow \mathbf{B}_1$. Hence $\mathbf{C}_1\times\mathbf{C}_2 \rightleftarrows \mathbf{B}_1\times\mathbf{B}_2$ is non-multiplicative according to (d).

□

Hedetniemi conjecture [88, 249] (sometimes called *product conjecture*) asks whether every complete graphs is multiplicative (as undirected graphs). It is usually formulated as follows:

Conjecture 9.1 (Hedetniemi conjecture). For every graphs G and H it holds

$$\chi(G\times H)=\min(\chi(G),\chi(H)).$$

This conjecture can be reformulated [298] by means of retracts:

If K_n is a retract of $G\times H$ then K_n is a retract of G or H.

Surprisingly, this later statement is known to hold for connected graphs G and H, see [297].

The product conjecture seem to be very hard and not much is known even for its weaker forms. It is still possible that there exists a constant k_0 such that for every n there is a pair G_n, H_n of graphs each with the chromatic number n such that the product $G_n\times H_n$ has the chromatic number at most K_0. (In fact k_0 may be as low as 9 (see [388]). Quite surprisingly, recently Zhu verified the analog of the product conjecture for fractional chromatic numbers, see [477]. For further reading about Hedetniemi conjecture, see surveys [454, 475].

9.3.4 Left and Right Distances for the Homomorphism Order

For a finite or infinite subset $\mathcal{A} \subseteq \mathrm{Rel}(\sigma)$, we define the *weight* $w(\mathcal{A})$ of $\mathcal{A}$ by:

$$w(\mathcal{A}) = \begin{cases} 0, & \text{if } \mathcal{A} = \emptyset, \\ 2^{-\min\{|\mathbf{A}|,\ \mathbf{A}\in\mathcal{A}\}}, & \text{otherwise.} \end{cases} \tag{9.3}$$

This naturally defines an ultrametric d on the powerset $2^{\mathrm{Rel}(\sigma)}$ of $\mathrm{Rel}(\sigma)$ by:

$$\mathrm{d}(\mathcal{A}, \mathcal{B}) = w\big((\mathcal{A} \setminus \mathcal{B}) \cup (\mathcal{B} \setminus \mathcal{A})\big) \tag{9.4}$$

The correspondences $[\mathbf{A}] \mapsto (\to \mathbf{A})$ and $[\mathbf{A}] \mapsto (\mathbf{A} \to)$ are injective mappings from $[\mathrm{Rel}(\sigma)]$ to $2^{\mathrm{Rel}(\sigma)}$. They naturally define two distances, the *left distance* dist_L and the *right distance* dist_R on $[\mathrm{Rel}(\sigma)]$ by

$$\mathrm{dist}_L([\mathbf{A}_1], [\mathbf{A}_2]) = \mathrm{d}((\to \mathbf{A}_1), (\to \mathbf{A}_2)) \tag{9.5}$$

and

$$\mathrm{dist}_R([\mathbf{A}_1], [\mathbf{A}_2]) = \mathrm{d}((\mathbf{A}_1 \to), (\mathbf{A}_2 \to)). \tag{9.6}$$

Also, one defines the *full distance* dist on $[\mathrm{Rel}(\sigma)]$ by

$$\mathrm{dist}([\mathbf{A}_1], [\mathbf{A}_2]) = \max(\mathrm{dist}_L([\mathbf{A}_1], [\mathbf{A}_2]), \mathrm{dist}_R([\mathbf{A}_1], [\mathbf{A}_2])). \tag{9.7}$$

We remark that the left distance can be computed by considering connected test structures only:

Lemma 9.4. *For every $[\mathbf{A}], [\mathbf{B}]$ in $[\mathrm{Rel}(\sigma)]$:*

$$\mathrm{dist}_L([\mathbf{A}], [\mathbf{B}]) = \mathrm{d}((\to \mathbf{A}) \cap \mathrm{Rel}_{\mathrm{con}}(\sigma), (\to \mathbf{B}) \cap \mathrm{Rel}_{\mathrm{con}}(\sigma)) \tag{9.8}$$

Proof. Obviously, $\mathrm{dist}_L([\mathbf{A}], [\mathbf{B}]) \geq \mathrm{d}((\to \mathbf{A}) \cap \mathrm{Rel}_{\mathrm{con}}(\sigma), (\to \mathbf{B}) \cap \mathrm{Rel}_{\mathrm{con}}(\sigma))$. Let $\mathrm{d} = \mathrm{dist}_L([\mathbf{A}], [\mathbf{B}])$, let $\mathbf{F} \in \mathrm{Rel}(\sigma)$ be such that $2^{-|\mathbf{F}|} > \mathrm{d}$ and let $\mathbf{F}_1, \dots, \mathbf{F}_p$ be the connected components of $\mathbf{F}$. Then

$$\begin{aligned} \mathbf{F} \to \mathbf{A} \quad & \iff \quad \forall 1 \leq i \leq p,\ \mathbf{F}_i \to \mathbf{A} \\ & \iff \quad \forall 1 \leq i \leq p,\ \mathbf{F}_i \to \mathbf{B} \qquad (\text{as } 2^{-|\mathbf{F}_i|} \geq 2^{-|\mathbf{F}|} > \mathrm{d}) \\ & \iff \quad \mathbf{F} \to \mathbf{B}. \end{aligned}$$

Hence $\mathrm{dist}_L([\mathbf{A}], [\mathbf{B}]) \leq \mathrm{d}((\to \mathbf{A}) \cap \mathrm{Rel}_{\mathrm{con}}(\sigma), (\to \mathbf{B}) \cap \mathrm{Rel}_{\mathrm{con}}(\sigma))$. □

Denote by $\overline{[\mathrm{Rel}(\sigma)]}_L$, $\overline{[\mathrm{Rel}(\sigma)]}_R$ and $\overline{[\mathrm{Rel}(\sigma)]}$ the completions of the metric spaces $([\mathrm{Rel}(\sigma)], \mathrm{dist}_L)$, $([\mathrm{Rel}(\sigma)], \mathrm{dist}_R)$ and $([\mathrm{Rel}(\sigma)], \mathrm{dist})$. Notice that these completions are compact (as they are totally bounded). We shall come back to these spaces and investigate these completions in more detail later on. In the next section we deal with the approximation of complex structures by simple ones.

9.3.5 Density

9.3.5.1 Density and Ambivalence of the Homomorphism Order

The homomorphism order has several spectacular properties already mentioned in Sect. 3.7: it is universal (for all countable posets), and it is dense (with a few exceptions). The later property is the subject of this section.

We make use of the following result which is often called *sparse incomparability lemma* [253, 365]. In the setting of this book we call such results *ambivalence theorems.* First we deal with graphs as a typical case.

Theorem 9.5. *For every* $\epsilon > 0$ *and for every non-discrete graph* G *there exists a graph* G' *with the following properties:*

(a) $\mathrm{dist}_L(K_2, G') < \epsilon$,
(b) $\mathrm{dist}_R(G', G) < \epsilon$.

Proof. Let us start by rewriting the above metrical conditions: The meaning of (a) is simply that G has odd-girth greater than $\log_2 \epsilon^{-1}$, as $F \nrightarrow K_2$ if and only if F contains an odd cycle. On the other hand, $\mathrm{dist}_R(G', G) < \epsilon$ if for every graph F of order $|F| < \log_2 \epsilon^{-1}$ holds ($G' \to F \iff G \to F$). Particularly, if $\mathrm{dist}_R(G', G) < 2^{-\chi(G)}$ then $\chi(G') = \chi(G)$ and this in turn can be used to define G'. Put $n = |G|$ and $l = \lceil \log_2 \epsilon^{-1} \rceil$. Let H be a graph with the following properties:

(i) The graph H has chromatic number $\chi(H) > l^n$,
(ii) The graph H contains no odd cycle of length smaller than l.

The existence of a graph H with these properties is not trivial, but follows from Theorem 3.2. However, the condition which we need here is weaker than the girth condition of Theorem 3.2, as we only need the graph H to have high chromatic number and odd-girth, and the result is easier to prove. This can be achieved, for example, by iterating l times the construction of the so-called shift graphs, see [253] and Exercise 9.8 for a detailed account.

Now put $G' = G \times H$. Obviously the odd-girth of G' is greater than l and thus $\mathrm{dist}_L(K_2, G') < \epsilon$. Let F be a graph of order $|F| < l$. By composition of homomorphisms, $G \to F$ implies $G' \to F$. Now assume that f is a homomorphism of $G' = G \times H$ to F. For each $x \in V(H)$ consider the *fiber map* $f_x : V(G) \to V(F)$ defined by $f_x(y) = f(x, y)$. As the chromatic number of H is very large, there exist an edge $\{x, y\}$ of H such that $f_x = f_y$. It is easily seen that then f_x is a homomorphism of G to F. □

This result generalizes to relational structures (with a nearly identical proof). This mainly needs a proper generalization of the notion of odd-girth: Given a σ-structure $\mathbf{A} = (A, (R_i, i \in I))$ we consider its incidence multigraph $\mathrm{Inc}(\mathbf{A})$ (see Sect. 3.8.3). The *odd-girth* is the smallest odd integer $l > 1$ such that $\mathrm{Inc}(\mathbf{A})$ contains a cycle of length $2l$. (Note that the cycles of length 2 in $\mathrm{Inc}(\mathbf{A})$—which correspond to relations with a same element appearing more than once—are ruled out by the condition $l > 1$.) The product of σ-structures was already defined in Sect. 3.8.1.

Although in the undirected case the girth bound was sufficient to deduce that every small graph having a homomorphism to G would have a homomorphism to any graph G' such that $K_2 \to G'$, things are more complex in the general case of σ-structures and the functor $\mathbf{U}$ will be invoked again.

According to Theorem 9.2, each tree has a dual, that is: for every tree $\mathbf{T}$ there exists $\mathbf{D}(\mathbf{T})$, the *dual* of $\mathbf{T}$, such that for every $\mathbf{A}$ it holds

$$\mathbf{T} \nrightarrow \mathbf{A} \iff \mathbf{A} \to \mathbf{D}(\mathbf{T}). \tag{9.9}$$

Consequently, if $\mathbf{A}_1 \to \mathbf{A}_2$ and if $\mathbf{A}_1$ and $\mathbf{A}_2$ are distinguished by a tree $\mathbf{T}$ (without loss of generality $\mathbf{T} \nrightarrow \mathbf{A}_1$ and $\mathbf{T} \to \mathbf{A}_2$, that is $\mathbf{A}_1 \to \mathbf{D}(\mathbf{T})$ and $\mathbf{A}_2 \nrightarrow \mathbf{D}(\mathbf{T})$) then for every $\mathbf{B}$ such that $\mathbf{A}_1 \to \mathbf{B} \to \mathbf{A}_2$ holds either $\mathrm{dist}_L(\mathbf{A}_1, \mathbf{B}) \geq 2^{-|\mathbf{T}|}$ or $\mathrm{dist}_R(\mathbf{B}, \mathbf{A}_2) \geq 2^{-|\mathbf{D}(\mathbf{T})|}$. However, the separability by trees is the only obstacle.

Using these notions we can state and prove the generalization of Theorem 9.5 for σ-structures:

Theorem 9.6. *For every $\epsilon > 0$ and for every σ-structure $\mathbf{A}$ there exists a σ-structure $\mathbf{A}'$ with the following properties:*

(a) $\mathbf{A}'$ *has odd-girth at least* $\lceil \log_2 \epsilon^{-1} \rceil$,
(b) $\mathrm{dist}_R(\mathbf{A}, \mathbf{A}') < \epsilon$.

Proof. Denote by A be the universe of $\mathbf{A}$. Put $l = \lceil \log_2 \epsilon^{-1} \rceil$ and $t = l^{|A|}$.

It is easy to check if $\mathbf{A} \times \mathbf{B}$ has odd-girth at most l then both $\mathbf{A}$ and $\mathbf{B}$ have odd-girth at most l. So, analogously with the proof of Theorem 9.5 we will put $\mathbf{A}' = \mathbf{A} \times \mathbf{H}$, where $\mathbf{H}$ is a σ-structure with both large odd-girth and a property which will be slightly stronger than having a high chromatic number. Precisely, the properties we shall request for $\mathbf{H}$ are the following:

1. $\mathbf{H}$ has odd-girth at least l,
2. For every partition $X_1 \cup \cdots \cup X_t$ of the universe H of $\mathbf{H}$, one of the classes contains a tuple of every relation in σ.

Let $r = \sum r_i$ be the sum of the arities of the relations in σ. Let σ' be a signature with a single relation of arity r. We first construct a σ'-structure $\mathbf{H}_0$ with odd-girth greater than l and chromatic number at least t. Then we split each tuple $(x_1, \ldots, x_r)$ into tuples of all the arities of relations in σ: $(x_1, \ldots, x_{r_1})$, $(x_{r_1+1}, \ldots, x_{r_1+r_2})$, ... The σ-structure $\mathbf{H}$ obtained this way has properties 1. and 2.

The remaining of the proof of the theorem follows the same lines as the proof of Theorem 9.5. □

Theorem 9.7 (Local ambivalence theorem). *Let* $\mathbf{A}_1 \to \mathbf{A}_2$ *and let* k *be a positive integer. Assume there exists no tree* $\mathbf{T}$ *with* $\max(|\mathbf{T}|, |D(\mathbf{T})|) \leq k$ *such that* $\mathbf{T} \to \mathbf{A}_2$ *but* $\mathbf{T} \nrightarrow \mathbf{A}_1$.

Then there exists $\mathbf{C}$ *such that:*

- $\mathbf{A}_1 \to \mathbf{C} \to \mathbf{A}_2$,
- $\mathrm{dist}_L([\mathbf{A}_1], [\mathbf{C}]) < 2^{-k}$,
- $\mathrm{dist}_R([\mathbf{A}_2], [\mathbf{C}]) < 2^{-k}$.

Proof. According to Theorem 9.6 there exists $\mathbf{B}$ such that $\mathbf{B} \to \mathbf{A}_2$, the odd-girth of $\mathbf{B}$ is greater than k, and $\mathrm{dist}_R([\mathbf{A}_2], [\mathbf{B}]) < 2^{-k}$.

Let $\mathbf{C} = \mathbf{A}_1 + \mathbf{B}$. Then $\mathbf{A}_1 \to \mathbf{C} \to \mathbf{A}_2$ and $\mathrm{dist}_R([\mathbf{A}_2], [\mathbf{C}]) < 2^{-k}$ (as $\mathbf{A}_1 \to \mathbf{A}_2$).

Assume for contradiction that $\mathrm{dist}_L([\mathbf{C}], [\mathbf{A}_1]) \geq 2^{-k}$. Then, according to Lemma 9.4, there exists a connected $\mathbf{Z}$ of order at most k such that $\mathbf{Z} \nrightarrow \mathbf{A}_1$ but $\mathbf{Z} \to \mathbf{C}$ (hence $\mathbf{Z} \to \mathbf{B}$). As the odd-girth of $\mathbf{B}$ is greater than $|\mathbf{Z}|$ we infer that the homomorphic image of $\mathbf{Z}$ in $\mathbf{B}$ is a tree $\mathbf{T}$ and thus $\mathbf{Z} \to \mathbf{T} \to \mathbf{B}$. However, $\mathbf{T} \nrightarrow \mathbf{A}_1$ as for otherwise $\mathbf{Z} \to \mathbf{T} \to \mathbf{A}_1$. But $\mathbf{T} \to \mathbf{B}$ implies $\mathbf{B} \nrightarrow D(\mathbf{T})$ hence $\mathbf{A}_2 \nrightarrow D(\mathbf{T})$ (as $|D(\mathbf{T})| \leq k$) hence $\mathbf{T} \to \mathbf{A}_2$. As $\mathbf{T} \nrightarrow \mathbf{A}_1$ and $\mathbf{T} \to \mathbf{A}_2$ we are led to a contradiction. □

Recall the definition of the functor $\mathbf{U}(\mathbf{A})$ given in Sect. 9.2.2 and define $\mathbf{U}([\mathbf{A}]) = [\mathbf{U}(\mathbf{A})]$ (this makes sense as $\mathbf{A} \rightleftarrows \mathbf{B}$ implies $\mathbf{U}(\mathbf{A}) \rightleftarrows \mathbf{U}(\mathbf{B})$). Then $\mathbf{U}$ is a *closure operator* of the poset $([\mathrm{Rel}(\sigma)], \leq_h)$. (Note that this closure operator arises as the lower adjoint of a Galois connection between the poset of the subposet of the classes $[\mathbf{A}]$ with tree dualities and the whole poset $[\mathrm{Rel}(\sigma)]$.)

As a corollary of Lemmas 9.2 and 9.7 we get:

Theorem 9.8 (Ambivalence theorem). *Let* $[\mathbf{A}_1] \leq_h [\mathbf{A}_2]$. *The following conditions are equivalent:*

1. $\mathbf{U}([\mathbf{A}_1])] = \mathbf{U}([\mathbf{A}_2])$.
2. *for every* $\epsilon > 0$ *there exists* $\mathbf{C}_\epsilon$ *such that:*
   - $[\mathbf{A}_1] \leq_h [\mathbf{C}_\epsilon] \leq_h [\mathbf{A}_2]$,
   - $\mathrm{dist}_L([\mathbf{A}_1], [\mathbf{C}_\epsilon]) < \epsilon$,
   - $\mathrm{dist}_R([\mathbf{A}_2], [\mathbf{C}_\epsilon]) < \epsilon$.

9.4 Left Limits and Countable Structures

9.4.1 Left Limits

A sequence $([\mathbf{G}_i])$ is a *Left Cauchy Sequence* if it is a Cauchy sequence according to the left distance dist_L. Recall that the completion $\overline{[\mathrm{Rel}(\sigma)]}_L$ of $([\mathrm{Rel}(\sigma)], \mathrm{dist}_L)$ is a compact space. If $([\mathbf{G}_i])$ is a Left Cauchy Sequence we will denote its limit by left $\lim_{i\to\infty}[\mathbf{G}_i]$. For limits we reserve blackboard fonts letters $\mathbb{L}, \mathbb{A}, \mathbb{B}$.

Sometimes, a sequence of graph may converge, with respect to dist_L to a graph which does not belong to the sequence. For instance, the sequence of odd cycles converges to $[K_2]$, that is:

$$\text{left} \lim_{i\to\infty}[C_{2i+1}] = [K_2].$$

However we shall see below that in many cases the left limit corresponds to an infinite graph. We extend the homomorphism order relation to left limits by

$$\begin{aligned}\text{left} \lim_i[\mathbf{G}_i] \leq_h \text{left} \lim_i[\mathbf{H}_i] \quad &\iff \quad \lim_{i\to\infty} w((\to \mathbf{G}_i) \setminus (\to \mathbf{H}_i)) = 0 \\ &\iff \quad \lim_{i\to\infty} 2^{-\min\{|\mathbf{A}|,\ \mathbf{A}\to\mathbf{G}_i \text{ and } \mathbf{A}\nrightarrow\mathbf{H}_i\}} = 0\end{aligned}$$

Notice that also this relation does not depend of the converging sequences and that it extends the homomorphism order (identified with constant sequences). We extend the operators ${}_{\downarrow}[\mathbf{A}]$ and $[\mathbf{A}]^{\uparrow}$ to left limits by:

$$
\begin{aligned}
{}_{\downarrow}\mathbb{L}_L &= \{[\mathbf{A}] \in [\mathrm{Rel}(\sigma)] : \ [\mathbf{A}] \leq_h \mathbb{L}_L\} \\
\mathbb{L}_L{}^{\uparrow} &= \{[\mathbf{A}] \in [\mathrm{Rel}(\sigma)] : \ [\mathbf{A}] \geq_h \mathbb{L}_L\}
\end{aligned}
$$

Notice that ${}_{\downarrow}\mathbb{L}_L$ and $\mathbb{L}_L{}^{\uparrow}$ are subsets of $[\mathrm{Rel}(\sigma)]$ and are not the lower sets and upper sets defined by $\mathbb{L}_L$ in the poset $(\overline{[\mathrm{Rel}(\sigma)]}_L, \leq_h)$.

We shall now prove that the two basic operations $\vee$ and $\wedge$ (meet and join of the homomorphism order) are continuous for the topology induced by dist_L. Precisely, we have:

Lemma 9.5. *Let* $[G_1], [G_2], [H_1], [H_2]$ *be elements of* $[\mathrm{Rel}(\sigma)]$. *Then*

$$
\begin{aligned}
&\mathrm{dist}_L([G_1] \vee [H_1], [G_2] \vee [H_2]) \leq \max(\mathrm{dist}_L([G_1], [G_2]), \mathrm{dist}_L([H_1], [H_2])) \\
&\mathrm{dist}_L([G_1] \wedge [H_1], [G_2] \wedge [H_2]) \leq \max(\mathrm{dist}_L([G_1], [G_2]), \mathrm{dist}_L([H_1], [H_2])).
\end{aligned}
$$

Proof. Let $T \in [\mathrm{Rel}_{\mathrm{con}}(\sigma)]$ be such that $t = |T|$ satisfies

$$2^{-t} > \max(\mathrm{dist}_L(G_1, G_2), \mathrm{dist}_L(H_1, H_2)).$$

Then

$$
\begin{aligned}
T \to G_1 + H_1 \quad &\iff \quad T \to G_1 \text{ or } T \to H_1 \\
&\iff \quad T \to G_2 \text{ or } T \to H_2 \\
&\iff \quad T \to G_2 + H_2
\end{aligned}
$$

According to Lemma 9.4 we deduce that $\mathrm{dist}_L([G_1] \vee [H_1], [G_2] \vee [H_2]) < 2^{-t}$. Also, for any $T \in [\mathrm{Rel}(\sigma)]$ we have

$$
\begin{aligned}
T \to G_1 \times H_1 \quad &\iff \quad T \to G_1 \text{ and } T \to H_1 \\
&\iff \quad T \to G_2 \text{ and } T \to H_2 \\
&\iff \quad T \to G_2 \times H_2
\end{aligned}
$$

Thus $\mathrm{dist}_L([G_1] \wedge [H_1], [G_2] \wedge [H_2]) < 2^{-t}$. □

Lemma 9.6. *Every ideal* $\mathcal{I}$ *of* $[\mathrm{Rel}(\sigma)]$ *has a supremum in* $\overline{[\mathrm{Rel}(\sigma)]}_L$ *which is the limit of an increasing sequence* $[\mathbf{G}_1] \leq_h [\mathbf{G}_2] \leq_h \cdots \leq_h [\mathbf{G}_t] \leq_h \ldots$ *of elements of* $\mathcal{I}$.

Proof. For an integer t, denote by $\mathcal{I}(t)$ the subset of $\mathcal{I}$ with structures of order at most t. Let $\mathbf{G}_t = \sum \mathcal{I}(t)$. By construction, $([\mathbf{G}_t])$ is a Left Cauchy

Sequence. Let $\mathbb{I}_L$ be its left limit. As $(\to \mathbf{G}_1) \subseteq (\to \mathbf{G}_2) \subseteq \cdots \subseteq (\to \mathbf{G}_t) \subseteq \ldots$ we have $\mathbf{H} \leq_h \mathbb{I}_L$ for every $\mathbf{H} \in \mathcal{I}$.

Now assume that $[\mathbf{H}] \leq_h$ left $\lim_t [\mathbf{G}'_t]$ for every $\mathbf{H} \in \mathcal{I}$ and some Left Cauchy Sequence $([\mathbf{G}'_t])$. Then for every t we have $[\mathbf{G}_t] \leq_h$ left $\lim_t [\mathbf{G}'_t]$ hence left $\lim_t [\mathbf{G}_t] \leq_h$ left $\lim_t [\mathbf{G}'_t]$. □

As a corollary, we obtain that the metric completion $\overline{[\mathrm{Rel}(\sigma)]}_L$ of $[\mathrm{Rel}(\sigma)]$ coincides with the ideal completion of $[\mathrm{Rel}(\sigma)]$:

Corollary 9.3. *The mapping* $\mathbb{L}_L \mapsto {\downarrow}\mathbb{L}_L$ *is a bijection between* $\overline{[\mathrm{Rel}(\sigma)]}_L$ *and ideals of* $[\mathrm{Rel}(\sigma)]$.

Proof. Assume $([\mathbf{G}_t])$ is a Left Cauchy Sequence and let $\mathcal{I} = {\downarrow}(\text{left} \lim_i [\mathbf{G}_i])$. According to Lemma 9.6, $\mathcal{I}$ has a supremum which is a left limit of some sequence $([\mathbf{H}_t])$. Obviously, $\lim_{t\to\infty} \mathrm{dist}_L([\mathbf{G}_t], [\mathbf{H}_t]) = 0$ hence the two limits are equal. The special case of the empty structure corresponds to the empty ideal. □

It follows that every left limit $\mathbb{L}$ is represented by a countable structure, the disjoint union $\sum \mathcal{I}$ of all the members of the ideal $\mathcal{I}$ corresponding to $\mathbb{L}$. There is more to this than meets the eye and this relates to both classical and contemporary model theory. We can only be brief here:

The *age* of a countable σ-structure $\mathbf{A}$ is the set of all finite σ-structures that (isomorphically) embed in $\mathbf{A}$. An important property of the age of countable structures is the following (which has been proved by Fraïssé, see [257]): a non-empty countable class $\mathcal{C} \subseteq \mathrm{Rel}(\sigma)$ is the age of some countable σ-structure if and only if $\mathcal{C}$ is *hereditary* (i.e. closed by taking induced substructures) and has the following *joint embedding property*: if $\mathbf{A}, \mathbf{B} \in \mathcal{C}$ then there is $\mathbf{C} \in \mathcal{C}$ such that $\mathbf{A}$ and $\mathbf{B}$ are embeddable in $\mathcal{C}$. Notice that the countable structure is generally not unique.

Let $\mathcal{I}$ be an ideal of $[\mathrm{Rel}(\sigma)]$. Put $\mathcal{I}^+ = \{\mathbf{A} \in \mathrm{Rel}(\sigma) : \ [\mathbf{A}] \in \mathcal{I}\}$. If $\mathcal{I}^+$ is non-empty, it is the age of some countable σ-structure.

Conversely, if $\mathcal{C} \subseteq \mathrm{Rel}(\sigma)$ is the age of a countable structure then the set (the *core age* of $\mathcal{C}$)

$$ {\downarrow}\mathcal{C} = \{[\mathbf{A}] \in [\mathrm{Rel}(\sigma)] : \ \exists \mathbf{B} \in \mathcal{C}, \ \mathbf{A} \to \mathbf{B}\} $$

is clearly an ideal (thanks to the joint embedding property) hence it defines a left limit $\mathbb{L}$, which can be represented by the sum $\sum {\downarrow}\mathcal{C}$.

Homomorphism equivalence is easy for finite structures: every finite structure contains, up to isomorphism, a unique core (i.e. a unique minimal retract). For infinite structures this is usually not the case. In turn the representation of limits in $\overline{[\mathrm{Rel}(\sigma)]_{\mathrm{L}}}$ is not unique: As above we can represent it by the disjoint union of the cores in the corresponding ideal $\mathcal{I}$. Sometimes, the ideal $\mathcal{I}$ gives rise to an ultra-homogeneous structure. This is characterized by another Fraïssé theorem (see [257]) (The ideal has to be a so-called *amalgamation class.*) Yet another possiblity is that the ideal $\mathcal{I}$ is induced as the age of an ω-categorical structure. In our setting, as our classes are homomorphism monotone, this happens if and only if the ideal $\mathcal{I}$ is given by a set of forbidden homomorphisms, i.e. $\mathcal{I} = \mathcal{F} \nrightarrow$, where $\mathcal{F}$ is a set of finite connected structures. This follows from a result of Cherlin et al. [95]. A more combinatorial approach can be found in [343]. This paper also generalizes the duality to limits: For $\mathcal{I} = \mathcal{F} \nrightarrow$ where $\mathcal{F}$ is a finite set of trees, the limit is represented by a finite graph.

Recall that we have, according to the extension of the partial order $\leq_{\mathrm{h}}$ to $\overline{[\mathrm{Rel}(\sigma)]_{\mathrm{L}}}$ (see Sect. 9.3.1 for definitions):

$$\begin{aligned} \mathbb{A}_{\mathrm{L}} \leq_{\mathrm{h}} \mathbb{B}_{\mathrm{L}} &\iff {}_{\downarrow}\mathbb{A}_{\mathrm{L}} \subseteq {}_{\downarrow}\mathbb{B}_{\mathrm{L}} \\ &\iff \mathbb{A}_{\mathrm{L}}{}^{\uparrow} \supseteq \mathbb{B}_{\mathrm{L}}{}^{\uparrow} \end{aligned}$$

and even

$$\mathbb{L}_{\mathrm{L}}{}^{\uparrow} = ({}_{\downarrow}\mathbb{L}_{\mathrm{L}})^{\star}.$$

It is now not difficult to prove that $(\overline{[\mathrm{Rel}(\sigma)]_{\mathrm{L}}}, \leq_{\mathrm{h}})$ is a complete lattice and that meet and joins are "compatible" with the topology of $\overline{[\mathrm{Rel}(\sigma)]_{\mathrm{L}}}$.

Lemma 9.7. *The poset* $(\overline{[\mathrm{Rel}(\sigma)]_{\mathrm{L}}}, \leq_{\mathrm{h}})$ *is a complete lattice.*

Proof. Let $\mathcal{F} \subsetneq \overline{[\mathrm{Rel}(\sigma)]_{\mathrm{L}}}$. According to the compactness of $\overline{[\mathrm{Rel}(\sigma)]_{\mathrm{L}}}$ (see Exercise 9.4), there exists a function $\mathrm{f} : \mathcal{F} \times \mathbb{N} \to \mathrm{Rel}(\sigma)$ such that for every $\mathbb{L} \in \mathcal{F}$ we have

$$\mathrm{dist}_{\mathrm{L}}(\mathbb{L}, [\mathrm{f}(\mathbb{L}, \mathrm{t})]) < 2^{-\mathrm{t}} \qquad \text{and} \qquad |\mathrm{f}(\mathbb{L}, \mathrm{t})| \leq \ell(\mathrm{t}).$$

For integer t, let $\mathcal{F}_{\mathrm{t}} = \{[\mathrm{f}(\mathbb{L}, \mathrm{t})],\ \mathbb{L} \in \mathcal{F}\}$. Obviously, $\mathcal{F}_{\mathrm{t}}$ is a finite set (for each t). Define

$$\mathbf{I}_{\mathrm{t}} = \prod_{\mathbb{L} \in \mathcal{F}'} \mathrm{f}(\mathbb{L}, \mathrm{t}), \qquad \text{and} \qquad \mathbf{S}_{\mathrm{t}} = \sum_{\mathbb{L} \in \mathcal{F}'} \mathrm{f}(\mathbb{L}, \mathrm{t})$$

(where $\mathcal{F}'$ is any minimal subset of $\mathcal{F}$ with $\mathrm{f}(\mathcal{F}', \mathrm{t}) = \mathrm{f}(\mathcal{F}, \mathrm{t})$).

By construction, for every $\mathbf{F} \in \mathrm{Rel}(\sigma)$ and every $\mathrm{t} \geq |\mathrm{F}|$, we have

$$\begin{aligned} \mathbf{F} \to \mathbf{I}_{\mathrm{t}} &\iff \mathbf{F} \to \mathbf{I}_{|\mathbf{F}|}, \\ \mathbf{F} \to \mathbf{S}_{\mathrm{t}} &\iff \mathbf{F} \to \mathbf{S}_{|\mathbf{F}|}. \end{aligned}$$

Hence both $(\mathbf{I}_t)$ and $(\mathbf{S}_t)$ converge. Let

$$\mathbb{I} = \operatorname{left}\lim_{t\to\infty} \mathbf{I}_t \quad \text{and} \quad \mathbb{S} = \operatorname{left}\lim_{t\to\infty} \mathbf{S}_t.$$

It is easily checked that

$$\mathbb{I} = \bigwedge \mathcal{F} \qquad \text{and} \qquad \mathbb{S} = \bigvee \mathcal{F}.$$

□

It is not difficult to see that the functor $\mathbf{U}$ can be extended by continuity to $[\mathrm{Rel}(\sigma)]_L$.

Lemma 9.8. *For every* $\mathbf{A}, \mathbf{B} \in \mathrm{Rel}(\sigma)$ *we have*

$$\mathrm{dist}_L([\mathbf{U}(\mathbf{A})], [\mathbf{U}(\mathbf{B})]) \leq \mathrm{dist}_L([\mathbf{A}], [\mathbf{B}]).$$

Proof. Let k be an integer and let $\mathbf{A}, \mathbf{B} \in \mathrm{Rel}(\sigma)$ be such that $\mathrm{dist}_L([\mathbf{A}], [\mathbf{B}]) < 2^{-k}$.

Let $\mathbf{F} \in \mathrm{Rel}(\sigma)$ be such that $|\mathbf{F}| \leq k$ and let $\mathcal{X}$ be the set of all homomorphic images of trees contained in $\mathbf{F}$. Of course, every structure $\mathbf{X} \in \mathcal{X}$ is such that $|\mathbf{X}| \leq |\mathbf{F}| \leq k$.

Then we have the equivalences:

$$\begin{aligned}
\mathbf{F} \to \mathbf{U}(\mathbf{A}) &\iff \forall \mathbf{T} \in \mathrm{Tree}(\sigma), \quad (\mathbf{T} \to \mathbf{F}) \Longrightarrow (\mathbf{T} \to \mathbf{A}) \\
&\iff \forall \mathbf{X} \in \mathcal{X},\ \mathbf{X} \to \mathbf{A} \\
&\iff \forall \mathbf{X} \in \mathcal{X},\ \mathbf{X} \to \mathbf{A} \quad (\text{as } \mathrm{dist}_L(\mathbf{A}, \mathbf{B}) < 2^{-k}) \\
&\iff \forall \mathbf{T} \in \mathrm{Tree}(\sigma), \quad (\mathbf{T} \to \mathbf{F}) \Longrightarrow (\mathbf{T} \to \mathbf{B}) \\
&\iff \mathbf{F} \to \mathbf{U}(\mathbf{B})
\end{aligned}$$

Thus $\mathrm{dist}_L([\mathbf{U}(\mathbf{A})], [\mathbf{U}(\mathbf{B})]) < 2^{-k}$. □

As a consequence, we have the following dual limit formulation of the functor $\mathbf{U}$:

Corollary 9.4. *Let* $\mathbf{A} \in \mathrm{Rel}(\sigma)$*. Then*

$$[\mathbf{U}(\mathbf{A})] = \operatorname{left}\lim_{t\to\infty} \bigwedge \{[\mathbf{D}(\mathbf{T})] :\ \mathbf{T} \in \mathrm{Tree}(\sigma), |\mathbf{T}| \leq t,\ \textit{and } \mathbf{T} \nrightarrow \mathbf{A}\}.$$

Thanks to the continuity of $\mathbf{U}$ we define the mapping $\mathbb{U} : \overline{[\mathrm{Rel}(\sigma)]_L} \to \overline{[\mathrm{Rel}(\sigma)]_L}$ by

$$\mathbb{U}(\operatorname{left}\lim_{t\to\infty}[\mathbf{A}_t]) = \operatorname{left}\lim_{t\to\infty}[\mathbf{U}(\mathbf{A}_t)].$$

As $\mathbf{U}$ is a functor, we easily deduce:

$$\mathbb{L}_1 \leq_h \mathbb{L}_2 \quad \Rightarrow \quad \mathbb{U}(\mathbb{L}_1) \leq_h \mathbb{U}(\mathbb{L}_2).$$

The limit structure leads to some interesting structural results, for example about the set of multiplicative (resp. of non-multiplicative) elements of $[\mathrm{Rel}(\sigma)]$. We know that for finite models these sets are very difficult to describe, even in the simple case of undirected graphs. However from the point of view of graph limits we get at least some global information.

As noticed earlier, connected and multiplicative elements are of special importance. We now give some properties of these subsets of elements (actually, of their complements) in the space $\overline{[\mathrm{Rel}(\sigma)]}_L$.

Lemma 9.9. *The set of non-multiplicative elements of* $[\mathrm{Rel}(\sigma)]$ *is open and dense in* $\overline{[\mathrm{Rel}(\sigma)]}_L$.

Proof. To prove that the set of non-multiplicative elements of $[\mathrm{Rel}(\sigma)]$ is open, we prove that if $[\mathbf{G}]$ is non-multiplicative then $[\mathbf{G}]$ has a neighborhood which contains no multiplicative element.

If $[\mathbf{G}]$ is non multiplicative then there exist $\mathbf{A}$ and $\mathbf{B}$ such that $\mathbf{A} \nrightarrow \mathbf{G}$, $\mathbf{B} \nrightarrow \mathbf{G}$ and $\mathbf{A} \times \mathbf{B} \rightarrow \mathbf{G}$. Assume $\mathrm{dist}_L([\mathbf{G}], [\mathbf{H}]) < 2^{-|\mathbf{A}|\cdot|\mathbf{B}|}$. Then $\mathbf{A} \times \mathbf{B} \rightarrow \mathbf{H}$, $\mathbf{A} \nrightarrow \mathbf{H}$ and $\mathbf{B} \nrightarrow \mathbf{H}$. It follows that $\mathbf{H}$ is non-multiplicative.
To prove that the set of non-multiplicative elements of $[\mathrm{Rel}(\sigma)]$ is dense, we prove that for every multiplicative $[\mathbf{M}]$ and every $\epsilon > 0$ there exists non-multiplicative $[\mathbf{N}]$ such that $\mathbf{M} \rightarrow \mathbf{N}$ and $\mathrm{dist}_L([\mathbf{M}], [\mathbf{N}]) < \epsilon$.

Let $[\mathbf{M}]$ be multiplicative and let $\epsilon > 0$ be a positive real. There exist $\mathbf{G}$ and $\mathbf{H}$ such that $\mathbf{G} \nrightarrow \mathbf{H} \nrightarrow \mathbf{G}$, $\mathbf{M} \mathrel{\substack{\rightarrow \\ \not\leftarrow}} \mathbf{G}$ and $\mathbf{M} \mathrel{\substack{\rightarrow \\ \not\leftarrow}} \mathbf{H}$. According to Theorem 9.8 there exist $\mathbf{G}'$ such that $\mathbf{M} \mathrel{\substack{\rightarrow \\ \not\leftarrow}} \mathbf{G}' \mathrel{\substack{\rightarrow \\ \not\leftarrow}} \mathbf{G}$, $\mathrm{dist}_L(\mathbf{M}, \mathbf{G}') < \epsilon$ and $\mathrm{dist}_R(\mathbf{G}', \mathbf{G}) < 2^{-|\mathbf{H}|}$. Then $\mathbf{H} \nrightarrow \mathbf{G}'$ for otherwise $\mathbf{H} \rightarrow \mathbf{G}' \rightarrow \mathbf{G}$ and $\mathbf{G}' \nrightarrow \mathbf{H}$ as $\mathbf{G} \nrightarrow \mathbf{H}$ and $\mathrm{dist}_R(\mathbf{G}', \mathbf{G}) < 2^{-|\mathbf{H}|}$. According to Lemma 9.3–(d), $\mathbf{N} = \mathbf{G}' \times \mathbf{H}$ is non-multiplicative and, according to Lemma 9.5, $\mathrm{dist}_L([\mathbf{M}], [\mathbf{N}]) \leq \mathrm{dist}_L([\mathbf{M}], [\mathbf{G}']) < \epsilon$.

□

On the other hand, one can also prove that the set of non-connected elements of $[\mathrm{Rel}(\sigma)]$ is dense (see Exercise 9.5).

9.5 Right Limits and Full Limits

9.5.1 The Right Distance

A sequence $([\mathbf{G}_i])$ is a *Right Cauchy Sequence* if it is a Cauchy sequence according to the distance dist_R (introduced in Sect. 9.3.4). The completion

$\overline{[\mathrm{Rel}(\sigma)]}_R$ of $([\mathrm{Rel}(\sigma)], \mathrm{dist}_R)$ is again a compact space, according to Heine-Borel theorem. If $([\mathbf{G}_i])$ is a Right Cauchy Sequence we will denote its limit by right $\lim_{i\to\infty}[\mathbf{G}_i]$.

For instance, the sequence of odd cycles converges, but this time (as opposed to left limits) we cannot associate a graph to this limit.

We extend the homomorphism order to right limits by

$$\text{right}\lim_i[\mathbf{G}_i] \leq_h \text{left}\lim_i[\mathbf{H}_i] \iff \lim_{i\to\infty} w((\mathbf{H}_i \to)\setminus(\mathbf{G}_i \to)) = 0 \quad (9.10)$$

Notice that this relation does not depend of the converging sequences and that it extends the homomorphism order (identified with constant sequences).

Lemma 9.10. *Every filter $\mathcal{F}$ of $[\mathrm{Rel}(\sigma)]$ has an infimum in $\overline{[\mathrm{Rel}(\sigma)]}_R$ which is the limit of a decreasing sequence $[\mathbf{G}_1] \geq_h [\mathbf{G}_2] \geq_h \cdots \geq_h [\mathbf{G}_t] \geq_h \ldots$ of elements of $\mathcal{F}$.*

Proof. For an integer t, denote by $\mathcal{F}(t)$ the subset of $\mathcal{F}$ with structures of order at most t. Let $\mathbf{G}_t = \prod \mathcal{F}(t)$. By construction, $([\mathbf{G}_t])$ is a Right Cauchy Sequence. Let $\mathbb{F}_L$ be its right limit. As $([\mathbf{G}_1]\to) \subseteq ([\mathbf{G}_2]\to) \subseteq \cdots \subseteq ([\mathbf{G}_t]\to) \subseteq \ldots$ we have $\mathbb{I}_R \leq_h [\mathbf{H}]$ for every $[\mathbf{H}] \in \mathcal{F}$.

Now assume that right $\lim_t[\mathbf{G}'_t] \leq_h [\mathbf{H}]$ for every $[\mathbf{H}] \in \mathcal{F}$ and some Right Cauchy Sequence $([\mathbf{G}'_t])$. Then for every t we have right $\lim_t[\mathbf{G}'_t] \leq [\mathbf{G}_t]$ hence right $\lim_t[\mathbf{G}'_t] \leq_h$ right $\lim_t[\mathbf{G}_t]$. □

Corollary 9.5. *The mapping $\mathbb{L}_R \mapsto \mathbb{L}_R{}^{\uparrow}$ is a bijection between $\overline{[\mathrm{Rel}(\sigma)]}_R$ and filters of $[\mathrm{Rel}(\sigma)]$.*

Proof. Assume $([\mathbf{G}_t])$ is a Right Cauchy Sequence and let $\mathcal{F} = (\text{right}\lim_i[\mathbf{G}_i])^{\uparrow}$. According to Lemma 9.10, $\mathcal{F}$ has an infimum which is a right limit of some sequence $([\mathbf{H}_t])$. Obviously, $\lim_{t\to\infty}\mathrm{dist}_R([\mathbf{G}_t],[\mathbf{H}_t]) = 0$ hence the two limits are equal. The special case of the empty structure corresponds to the empty filter. □

$$\mathbb{A}_R \leq_h \mathbb{B}_R \iff \mathbb{A}_R{}^{\uparrow} \supseteq \mathbb{B}_R{}^{\uparrow}.$$

9.5.2 Full Distance

Let dist be the distance between homomorphism equivalence classes defined by:

$$\operatorname{dist}([\mathbf{G}],[\mathbf{H}]) = \max(\operatorname{dist}_L([\mathbf{G}],[\mathbf{H}]), \operatorname{dist}_R([\mathbf{G}],[\mathbf{H}]))$$

Notice that

$$\operatorname{dist}([\mathbf{G}_1],[\mathbf{G}_3]) \leq \max(\operatorname{dist}([\mathbf{G}_1],[\mathbf{G}_2]), \operatorname{dist}([\mathbf{G}_2],[\mathbf{G}_3]))$$

hence $([\mathrm{Rel}(\sigma)], \operatorname{dist})$ is an ultrametric space. The completion $\overline{[\mathrm{Rel}(\sigma)]}$ of $([\mathrm{Rel}(\sigma)], \operatorname{dist})$ is a compact space. If $([\mathbf{G}_i])$ is a Cauchy sequence for dist, we will denote by its limit by $\lim_{i\to\infty}[\mathbf{G}_i]$. Limits will be denoted as $\mathbb{A}, \mathbb{B}, \mathbb{L}$, etc. For a sequence $([\mathbf{G}_i])$ to convergence it is necessary and sufficient that the sequence converges for both the distances dist_L and dist_R.

We extend the homomorphism order to limits by defining $\lim_i[\mathbf{G}_i] \leq_h \lim_i[\mathbf{H}_i]$ if

$$\lim_{i\to\infty} \max(w((\to G_i) \setminus (\to H_i)), w((H_i \to) \setminus (G_i \to))) = 0.$$

Hence we have, for limit objects $\mathbb{A}$ and $\mathbb{B}$:

$$\mathbb{A} \leq_h \mathbb{B} \iff ({}_{\downarrow}\mathbb{A} \subseteq {}_{\downarrow}\mathbb{B}) \wedge (\mathbb{A}^{\uparrow} \supseteq \mathbb{B}^{\uparrow}).$$

Notice that this relation does not depend of the converging sequences and that it extends the homomorphism order on $[\mathrm{Rel}(\sigma)]$ (identified with constant sequences).

We deduce the following characterization of limits.

Theorem 9.9. *The mapping $\mathbb{L} \mapsto ({}_{\downarrow}\mathbb{L}, \mathbb{L}^{\uparrow})$ is a bijection from $\overline{[\mathrm{Rel}(\sigma)]}$ to the set of pairs $(\mathcal{I}, \mathcal{F})$ such that:*

1. *The set $\mathcal{I}$ is an ideal of $[\mathrm{Rel}(\sigma)]$,*
2. *The set $\mathcal{F}$ is a filter of $[\mathrm{Rel}(\sigma)]$,*
3. *Every element of $\mathcal{I}$ is smaller than every element of $\mathcal{F}$, that is: $\mathcal{I} \subseteq \mathcal{F}^\star$,*
4. *Every tree smaller than every element of $\mathcal{F}$ belongs to $\mathcal{I}$, that is: $\mathcal{F}^\star \cap [\mathrm{Tree}(\sigma)] \subseteq \mathcal{I}$.*

Proof. Assume (1) to (4) hold. For integer t define

$$I_t = \sum\{\mathbf{A} : |\mathbf{A}| \leq t \text{ and } [\mathbf{A}] \in \mathcal{I}\}$$
$$F_t = \prod\{\mathbf{A} : |\mathbf{A}| \leq t \text{ and } [\mathbf{A}] \in \mathcal{F}\}.$$

Let $\mathbf{T}$ be a tree with $\max(|\mathbf{T}|, |D(\mathbf{T})|) \leq t$. If $\mathbf{T} \to \mathbf{F}_t$ then $\mathbf{T} \to \mathbf{F}_{t'}$ for every $t' \geq t$ (by the definition of $\mathbf{F}_t$) hence $[\mathbf{T}] \in \mathcal{F}^\star$. Then, by (4), $[\mathbf{T}] \in \mathcal{I}$ thus

$\mathbf{T} \nrightarrow \mathbf{I}_t$. It follows, according to Theorem 9.7, that there exists $[\mathbf{M}_t]$ such that

- $[\mathbf{I}_t] \leq_h [\mathbf{M}_t] \leq_h [\mathbf{F}_t]$,
- $\mathrm{dist}_L([\mathbf{I}_t], [\mathbf{M}_t]) < 2^{-t}$,
- $\mathrm{dist}_R([\mathbf{F}_t], [\mathbf{M}_t]) < 2^{-t}$.

By compactness, one extracts a converging subsequence of $[\mathbf{M}_j]$ with limit $\mathbb{L}$. By construction, $\mathcal{I} = {}_{\downarrow}\mathbb{L}$ and $\mathcal{F} = \mathbb{L}^{\uparrow}$ (hence $\mathbb{L}$ is unique).

Conversely, let $\mathbb{L}$ be a limit of a sequence $[\mathbf{L}_t]$. Let $\mathcal{I} = {}_{\downarrow}\mathbb{L}$ and $\mathcal{F} = \mathbb{L}^{\uparrow}$. Then $\mathcal{I}$ is an ideal, $\mathcal{F}$ is a filter and conditions (1) to (3) obviously hold. Let $[\mathbf{T}] \in \mathcal{F}^{\star} \cap [\mathrm{Tree}(\sigma)]$. Assume for contradiction that $[\mathbf{T}] \notin \mathcal{I}$. Then, there exists t_0 such that for every $t \geq t_0$ we have $\mathbf{L}_t \nrightarrow \mathbf{A}$. According to Lemma 9.2, $\mathbf{T}$ has a dual $\mathbf{D}(\mathbf{T})$. Thus $\mathbf{L}_t \to \mathbf{D}(\mathbf{T})$ holds for every $t \geq t_0$. Hence $[\mathbf{D}(\mathbf{T})] \in \mathcal{F}$ thus $\mathbf{T} \to \mathbf{D}(\mathbf{T})$, a contradiction. □

Let $\mathcal{I}$ be an ideal and let $\mathcal{F}$ be a filter, such that

$$\mathcal{F}^{\star} \cap [\mathrm{Tree}(\sigma)] \subseteq \mathcal{I} \subseteq \mathcal{F}^{\star}.$$

Then we denote the unique limit defined by $\mathcal{I}$ and $\mathcal{F}$ by

$$\mathbb{L} = [\,\mathcal{I}\,,\,\mathcal{F}\,].$$

Proposition 9.2. *There is a natural identification*

- *Of* $(\overline{[\mathrm{Rel}(\sigma)]}_R, \leq_h)$ *with the sub-poset of* $(\overline{[\mathrm{Rel}(\sigma)]}, \leq_h)$ *with elements of the form* $[\,\mathcal{F}^{\star}\,,\,\mathcal{F}\,]$,
- *Of* $(\overline{[\mathrm{Rel}(\sigma)]}_L, \leq_h)$ *with the sub-poset of* $(\overline{[\mathrm{Rel}(\sigma)]}, \leq_h)$ *with elements of the form* $[\,\mathcal{I}\,,\,\mathcal{I}^{\star}\,]$.

Proof. According to Corollary 9.5, there is a natural identification between right limits and filters of $[\mathrm{Rel}(\sigma)]$. For any filter $\mathcal{F}$ of $[\mathrm{Rel}(\sigma)]$, $\mathcal{F}^{\star}$ is an ideal and we have $\mathcal{F}^{\star} \cap \mathrm{Tree}(\sigma) \subseteq \mathcal{F}^{\star}$ hence, according to Theorem 9.9 there exists $[\,\mathcal{F}^{\star}\,,\,\mathcal{F}\,]$ in $\overline{[\mathrm{Rel}(\sigma)]}$. The mapping

$$\mathbb{L}_R \mapsto P_R(\mathbb{L}_R) = [\,(\mathbb{L}_R{}^{\uparrow})^{\star}\,,\,\mathbb{L}_R{}^{\uparrow}\,]$$

is clearly a bijection from $\overline{[\mathrm{Rel}(\sigma)]}_R$ to the subset of limits of the form $[\,\mathcal{F}^{\star}\,,\,\mathcal{F}\,]$. According to the definitions of the extensions of $\leq_h$ on $\overline{[\mathrm{Rel}(\sigma)]}_R$ and $\overline{[\mathrm{Rel}(\sigma)]}$, it is clear than P_R induces an homomorphism of $(\overline{[\mathrm{Rel}(\sigma)]}_R, \leq_h)$ to the sub-poset of $(\overline{[\mathrm{Rel}(\sigma)]}, \leq_h)$ induced by the limits of the form $[\,\mathcal{F}^{\star}\,,\,\mathcal{F}\,]$.

According to Corollary 9.3, there is a natural identification between left limits and ideals of $[\mathrm{Rel}(\sigma)]$. For any ideal $\mathcal{I}$ of $[\mathrm{Rel}(\sigma)]$, $\mathcal{I}^\star$ is a filter and we have $\mathcal{I} \subseteq \mathcal{I}^{\star\star}$. Assume for contradiction that $\mathcal{I}^{\star\star} \cap \mathrm{Tree}(\sigma) \not\subseteq \mathcal{I}$. Let $\mathbf{T}_0 \in \mathcal{I}^{\star\star} \cap \mathrm{Tree}(\sigma) \setminus \mathcal{I}$. According to Theorem 9.2, $\mathbf{T}_0$ has a dual $\mathbf{D}(\mathbf{T}_0)$. For every $[\mathbf{A}] \in \mathrm{Rel}(\sigma)$ we have $\mathbf{T}_0 \nrightarrow \mathbf{A}$ hence $\mathbf{A} \rightarrow \mathbf{D}(\mathbf{T}_0)$. Thus $\mathbf{D}(\mathbf{T}_0) \in \mathcal{I}^\star$. As $\mathbf{T}_0 \nrightarrow \mathbf{D}(\mathbf{T}_0)$ we get $\mathbf{T}_0 \notin \mathcal{I}^{\star\star}$, a contradiction. Thus $\mathcal{I}^{\star\star} \cap \mathrm{Tree}(\sigma) \subseteq \mathcal{I}$ and, according to Theorem 9.9, there exists $[\mathcal{I}, \mathcal{I}^\star]$ in $\overline{[\mathrm{Rel}(\sigma)]}$. The mapping

$$\mathbb{L}_\mathrm{L} \mapsto \mathrm{P}_\mathrm{L}(\mathbb{L}_\mathrm{L}) = [\,{}_\downarrow\mathbb{L}_\mathrm{L}\,,\, ({}_\downarrow\mathbb{L}_\mathrm{L})^\star\,]$$

is clearly a bijection from $\overline{[\mathrm{Rel}(\sigma)]}_\mathrm{L}$ to the subset of limits of the form $[\mathcal{I}, \mathcal{I}^\star]$. According to the definitions of the extensions of $\leq_\mathrm{h}$ on $\overline{[\mathrm{Rel}(\sigma)]}_\mathrm{L}$ and $\overline{[\mathrm{Rel}(\sigma)]}$, it is clear than P_L induces an homomorphism of $(\overline{[\mathrm{Rel}(\sigma)]}_\mathrm{L}, \leq_\mathrm{h})$ to the subposet of $(\overline{[\mathrm{Rel}(\sigma)]}, \leq_\mathrm{h})$ induced by the limits of the form $[\mathcal{I}, \mathcal{I}^\star]$. □

Also, the Ambivalence theorem (Theorem 9.8) has the following consequence (with the possibility to define a chimera):

Lemma 9.11. *Let* $[\mathbf{A}_1] \leq_\mathrm{h} [\mathbf{A}_2]$. *The following conditions are equivalent:*

1. $\mathbf{U}([\mathbf{A}_1])] = \mathbf{U}([\mathbf{A}_2])$.
2. $[\,{}_\downarrow[\mathbf{A}_1]\,,\, [\mathbf{A}_2]^\uparrow\,] \in \overline{[\mathrm{Rel}(\sigma)]}$. *This limit is called a* chimera.

Example of chimeras are given in Exercise 9.7

9.5.3 Full Dualities

Recall that singleton homomorphism duality is captured by the following scheme:

$$\mathsf{F} \nrightarrow \mathsf{G} \quad \iff \quad \mathsf{G} \longrightarrow \mathsf{H}$$

Unfortunately there exists only one duality in $\mathcal{G}raph$, namely the pair $(\mathsf{K}_2, \mathsf{K}_1)$ and although there are countably many dualities in general in $\mathrm{Rel}(\sigma)$, only tree-like structures have duals. However, if we consider limits, the set of dualities drastically increases:

A *full duality* is a pair $(\mathbb{F}, \mathbb{D})$ such that

$$\forall \mathbb{L} \in \overline{\mathrm{Rel}(\sigma)}: \qquad \mathbb{F} \not\leq_\mathrm{h} \mathbb{L} \iff \mathbb{L} \leq_\mathrm{h} \mathbb{D}.$$

We have the following easy result which has, as we shall see below, an interesting context.

Lemma 9.12. *Let* $(\mathbb{F}, \mathbb{D})$ *be a full duality. Then*

- *Either* $\mathbb{F} \in [\mathrm{Rel}(\sigma)]$ *and* $\mathbb{D} \in \overline{[\mathrm{Rel}(\sigma)]}_L$,
- *Or* $\mathbb{F} \in \overline{[\mathrm{Rel}(\sigma)]}_R$ *and* $\mathbb{D} \in [\mathrm{Rel}(\sigma)]$.

Proof. Let us prove first that either $\mathbb{F} \in [\mathrm{Rel}(\sigma)]$ or $\mathbb{D} \in [\mathrm{Rel}(\sigma)]$: As $\mathbb{D} \to \mathbb{D}$ we have $\mathbb{F} \nrightarrow \mathbb{D}$. This means that there exists a graph T such that either $T \to \mathbb{F}$ and $T \nrightarrow \mathbb{D}$—hence $\mathbb{F} \to T$ according to duality thus $T \rightleftarrows \mathbb{F}$, or $\mathbb{D} \to T$ and $\mathbb{F} \nrightarrow T$—hence $T \to \mathbb{D}$ by duality thus $T \rightleftarrows \mathbb{D}$.

Assume $\mathbb{F} = [\mathbf{F}] \in [\mathrm{Rel}(\sigma)]$. Let $\mathbb{L} = [{}_{\downarrow}\mathbb{D}, ({}_{\downarrow}\mathbb{D})^\star] \in \overline{[\mathrm{Rel}(\sigma)]}_L$. Thus $\mathbb{D} \leq_h \mathbb{L}$. As $[\mathbf{F}] \notin {}_{\downarrow}\mathbb{D}$ we have $[\mathbf{F}] \not\leq_h \mathbb{L}$. By duality, $\mathbb{L} \leq_h \mathbb{D}$ hence $\mathbb{L} = \mathbb{D}$.

Assume $\mathbb{D} = [\mathbf{D}] \in [\mathrm{Rel}(\sigma)]$. Let $\mathbb{L} = [(\mathbb{F}^{\uparrow})^\star, \mathbb{F}^{\uparrow}] \in \overline{[\mathrm{Rel}(\sigma)]}_R$. Thus $\mathbb{L} \leq_h \mathbb{F}$. As $[\mathbf{D}] \notin \mathbb{F}^{\uparrow}$ we have $[\mathbf{D}] \not\geq_h \mathbb{L}$. By duality, $\mathbb{L} \geq_h \mathbb{F}$ hence $\mathbb{L} = \mathbb{F}$. □

Nešetřil and Shelah conjectured in [368] that every finite maximal antichain in the homomorphism order of countable graphs contains a finite graph. Finite maximal antichains in $\mathcal{G}\mathrm{raph}$ are in a correspondence with finite dualities. Thus the lemma settles the characterization of dualities for the completion $\overline{[\mathrm{Rel}(\sigma)]}$ of $[\mathrm{Rel}(\sigma)]$.

Theorem 9.10. *Every connected structure has a full dual, every multiplicative structure is a full dual, and these are the only full dualities.*

Proof. We already have proved that every duality pair contains a graph. Moreover, it is easily checked that if one of $\mathbb{F}, \mathbb{D}$ is a graph and if duality holds for graphs, then it also holds for limits:

- Let $(F, \mathbb{D})$ be a duality which holds for graphs. Let $\mathbb{L}$ be the limit of a Cauchy sequence (L_i). Then $F \nrightarrow \mathbb{L}$ is equivalent to $\exists i_0 \forall i \geq i_0\ F \nrightarrow L_i$, i.e. $\exists i_0 \forall i \geq i_0\ L_i \to \mathbb{D}$, what is equivalent to $\mathbb{L} \to \mathbb{D}$;
- Similarly, let $(\mathbb{F}, D)$ be a duality which holds for graphs. Let $\mathbb{L}$ be the limit of a Cauchy sequence (L_i). Then $\mathbb{L} \nrightarrow D$ is equivalent to $\exists i_0 \forall i \geq i_0\ L_i \nrightarrow D$, i.e. $\exists i_0 \forall i \geq i_0\ \mathbb{F} \to L_i$, what is equivalent to $\mathbb{D} \to \mathbb{L}$.

We have already proved that every graph has a dual hence we only have to characterize those full duality pairs $(\mathbb{F}, D)$ such that D is a graph. If D is multiplicative, let $\mathcal{S} = \{G : G \nrightarrow D\}$ and let $\mathcal{I}$ be the class of all graphs having a homomorphism to every graph in $\mathcal{S}$. The set $\mathcal{I}$ is an ideal by construction and, as D is multiplicative, $\mathcal{S}$ is a filter hence the pair $(\mathcal{I}, \mathcal{S})$ defines a limit object $\mathbb{F}$. For every graph G, we have $\mathbb{F} \longrightarrow G$ if and only if $G \in \mathcal{S}$, i.e.: if and only if $G \nrightarrow D$. It follows that every multiplicative graph is a full dual. Conversely, let $(\mathbb{F}, D)$ be a full duality. Let $\mathcal{I} = \mathcal{I}(\mathbb{F})$ and $\mathcal{S} = \mathcal{S}(\mathbb{F})$. According to the duality, $\mathbb{F} \longrightarrow G$ if and only if $G \nrightarrow D$ thus $\mathcal{S} = \{G : G \nrightarrow D\}$ is a filter, what means that D is multiplicative. □

Exercises

9.1. Prove that for every digraph $\vec{G}$ the following conditions are equivalent:

1. $\vec{G}$ is acyclically oriented,
2. $\mathcal{U}(\vec{G})$ is loopless,
3. $\mathcal{U}(\vec{G})$ is acyclically oriented.

Hence, if a loopless digraph has tree duality then it is acyclically oriented.
Also, the following conditions are equivalent:

1. $\vec{G}$ is homomorphically equivalent to some dipath,
2. $\vec{G}$ is balanced,
3. $\mathcal{U}(\vec{G})$ is balanced.

9.2. Let $(\mathbf{F}, \mathbf{D})$ and $(\mathbf{F}', \mathbf{D}')$ be finite dualities. Prove that $\mathbf{F} \to \mathbf{F}'$ if and only if $\mathbf{D} \to \mathbf{D}'$.

9.3. As a further comparison of connectedness and multiplicativity, let us consider now the problem of the decomposition. Although it is well known that every structure is the sum of finitely many connected structures (namely its connected components), things get more complex with the product.

- Prove that for every $\mathbf{A}$ there exists a unique decomposition

$$[\mathbf{A}] = [\mathbf{C}_1] \vee [\mathbf{C}_2] \vee \cdots \vee [\mathbf{C}_p]$$

  where $[\mathbf{C}_i]$'s are connected and distinct.
- Assume $[\mathbf{A}] = [\mathbf{M}_1] \wedge [\mathbf{M}_2] \wedge \cdots \wedge [\mathbf{M}_p]$, where the $[\mathbf{M}_i]$'s are multiplicative and distinct. Prove that this decomposition is unique.

9.4. Prove that there is a function $\ell : \mathbb{N} \to \mathbb{N}$ such that for every integer t and for every $\mathbb{L} \in \overline{[\mathrm{Rel}(\sigma)]}_L$ there exists $\mathbf{A} \in [\mathrm{Rel}(\sigma)]$ such that

$$\mathrm{dist}_L(\mathbb{L}, \mathbf{A}) < 2^{-t} \qquad \text{and} \qquad |\mathbf{A}| \leq \ell(t).$$

9.5. Prove that the set of non-connected elements of $[\mathrm{Rel}(\sigma)]$ is dense in $\overline{[\mathrm{Rel}(\sigma)]}_L$.

9.6. The aim of this exercise is to prove that existence of homomorphisms from every finite substructures of a countable structure $\mathbf{A}$ to an ω-categorical structure $\mathbf{B}$ implies the existence of a homomorphism from $\mathbf{A}$ to $\mathbf{B}$.

- Number the vertices of $\mathbf{A}$ as $v_1, \dots, v_n, \dots$ For $n \geq 1$, say that two homomorphisms $f, g : \mathbf{A}[v_1, \dots, v_n] \to \mathbf{B}$ are equivalent if there exits an automorphism h of $\mathbf{B}$ such that $f = g \circ h$;

- Prove that for every homomorphism $f : A[v_1, \ldots, v_n] \to \mathbf{B}$ there exists finitely many non-equivalent $f' : A[v_1, \ldots, v_{n+1}] \to \mathbf{B}$ extending f;
- Organize all the equivalence classes of these homomorphisms (for arbitrary n) as a tree and conclude using König's lemma that there exists a homomorphism from $\mathbf{A}$ to $\mathbf{B}$.

9.7. Let $G = K_2$ and $H = K_n$. There exists a chimera $\mathbb{E} = [\,{}_{\downarrow}[K_2]\,,\,[K_n]^{\uparrow}\,]$. Construct a sequence (G_i) of graphs with limits $\mathbb{E}$;
Let $\vec{G} = \vec{C}_3$ and $\vec{H} = \vec{C}_6$ be the circuits of order 3 and 6. Prove that there exists a chimera $\mathbb{L} = [\,{}_{\downarrow}[\vec{G}]\,,\,[\vec{H}]^{\uparrow}\,]$. Construct a sequence $(\vec{F}_i)$ of directed graphs which limits is $\mathbb{L}$.
Prove that the chimera $\mathbb{L} = [\,{}_{\downarrow}[\vec{G}]\,,\,[\vec{H}]^{\uparrow}\,]$ exists for

$\vec{G} =$ 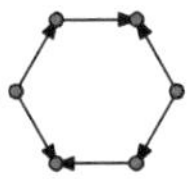 and $\vec{H} =$  .

9.8. Let G be a graph with vertex set $V = \{v_1, \ldots, v_n\}$ and edge set E. Denote by ∂G the graph with vertex set E and edge set F, where $\{\{v_i, v_j\}, \{v_k, v_l\}\} \in E \times E$ forms an edge of ∂G if $i < j = k < l$. The graph ∂G is the *shift graph* of G.

Prove that $\chi(\partial G) \geq \log \chi(G)$;
Prove that the odd-girth of ∂G is at least odd-girth$(G) + 2$.

9.9. Modify the construction in Exercise 9.8 for hypergraphs.

9.10. Let $\vec{G}, \vec{H}$ be oriented graphs.

Prove that $\vec{H}$ has (oriented) tree duality if and only if $\mathbf{U}(\vec{H}) \to \vec{H}$.
Hint: Use the procedure known as the *consistency check* algorithm, which in case of graphs takes the following form: To find a homomorphism $G \to H$ enumerate all the edges of G as $e_1, \ldots, e_m$. To every vertex $x \in V(G)$ we assign a set $\ell_i(x) \subset V(H)$, where $i = 0, 1, \ldots$. We put $\ell_0(x) = V(H)$. In the $i-$th step we consider the arc $e_j = (x, y)$, where $j \cong i \bmod m$ and we modify the labeling ℓ_{i-1} by removing from $\ell_{i-1}(y)$ those labels s for which there is no $r \in \ell_{i-1}(x)$ with $(r, s) \in E(\vec{H})$, and removing from $\ell_{i-1}(x)$ those labels r for which there is no $s \in \ell_{i-1}(y)$ with $(r, s) \in E(\vec{H})$. The resulting labeling is then consistent with e_j; it could however have become inconsistent with some previously treated edges. Nevertheless, after a sequence of at most $|\vec{G}|\,|\vec{H}|$ steps the procedure stabilizes with a labelling ℓ^*. If all sets $\ell^*(x)$ are non-empty then using $\mathbf{U}(H) \to H$ we get a homomorphism $G \to H$. It follows that the consistency check succeeds if and only if $\mathbf{U}(H) \longrightarrow H$. Finish the proof by proving an analogue of Lemma 9.1 for the consistency check: $r \notin \ell^*(x)$ if there is a rooted oriented tree $(\vec{T}, t)$ which is homomorphic to $(\vec{G}, x)$ but not to $(\vec{H}, r)$.

9.11. A *connected gap* in the homomorphism order $(\mathrm{Rel}(\sigma), <_{\mathrm{h}})$ is a pair $(\mathbf{A}, \mathbf{B})$, where $\mathbf{B}$ is connected, $\mathbf{A} <_{\mathrm{h}} \mathbf{B}$ but no $\mathbf{C}$ satisfies $\mathbf{A} <_{\mathrm{h}} \mathbf{C} <_{\mathrm{h}} \mathbf{B}$.

Prove that connected gaps and duality pairs are in 1–1 correspondence [370]. Hint: If $(\mathbf{F}, \mathbf{D})$ is a singleton duality pair then $(\mathbf{F} \times \mathbf{D}, \mathbf{F})$ is a gap. If $\mathbf{A}, \mathbf{B}$ is a connected gap then $\mathbf{B}, \mathbf{A}^{\mathbf{B}}$ is a singleton duality. See [370] for a the definition of the power $\mathbf{A}^{\mathbf{B}}$. (This can be defined by the validity of $(\mathbf{A} \times \mathbf{B}) \to \mathbf{C}$ if and only if $\mathbf{A} \to \mathbf{B}^{\mathbf{C}}$; this fact is sufficient for this exercise.)

Chapter 10
Preservation Theorems

Preserve: to prepare so as to resist to decomposition.

10.1 Introduction

We start with the following observation: If a formula Φ expresses a $\mathbf{H}$-coloring problem (i.e. $\mathbf{G} \models \Phi$ if and only if $\mathbf{G} \to \mathbf{H}$) the negated formula $\neg\Phi$ is preserved by homomorphisms:

$$\mathbf{G} \models \neg\Phi \quad \text{and} \quad \mathbf{G} \to \mathbf{G}' \qquad \Longrightarrow \qquad \mathbf{G}' \models \neg\Phi. \tag{10.1}$$

(This is easy to see: otherwise $\mathbf{G} \to \mathbf{G}' \to \mathbf{H}$, a contradiction.)

The complementary class $(\nrightarrow \mathbf{H})$ seems to have a very special form. (For example, note that every graph is a homomorphic image of a matching.) Examples of classes closed under homomorphisms are classes of the form $(\mathbf{A} \to)$ (introduced in Sect. 9.3.2) and this suggests that they can be described by a sentence with a specific syntactic form. Indeed classical preservation theorems are statements of the form [224]: "A class $\mathcal{C}$ of structures, defined by a first-order sentence, is preserved under some specified algebraic operation if and only if it is definable by a first-order sentence of a certain syntactic form". Several such theorems have been proved. They connect syntactic and semantic properties of first-order formulas. We list the following (see e.g. [257]):

- The *Łoś-Tarski theorem*, which asserts that a first-order formula is preserved under extensions on all structures if, and only if, it is logically equivalent to an existential formula;

J. Nešetřil and P.O. de Mendez, *Sparsity*, Algorithms and Combinatorics 28,
DOI 10.1007/978-3-642-27875-4_10,

- *Lyndon's theorem*, which asserts that a first-order formula is preserved under surjective homomorphisms on all structures if, and only if, it is logically equivalent to a positive formula [319];
- The *Homomorphism Preservation Theorem*, which asserts that a first-order formula is preserved under homomorphisms on all structures if, and only if, it is logically equivalent to an existential-positive formula.

The terms "all structures", which means finite and infinite structures, is crucial in the statement of these theorems. It has been proved that the two first theorems fail when relativized to the finite: there exists a first-order formula that is preserved under extensions on finite structures, but is not equivalent in the finite to an existential formula [10, 238, 452] (see Problem 10.1) and there exists a first-order formula that is preserved under surjective homomorphisms on finite structures, but is not equivalent in the finite to a positive formula [5, 445]. The case of the homomorphism preservation theorem remained open for a long time until Rossman showed [425] that the homomorphism preservation theorem actually holds when relativized to the finite. Rossman's result was stated in Chap. 9 as Theorem 9.1.

In this chapter we investigate preservation theorems in greater details and particularly we are interested in determining for which class of structures the homomorphism preservation holds. We proceed in two main directions: infinite and finite.

The homomorphism preservation holds when unrestricted (so for the class of all finite and infinite structures) and we shall show that it also holds for special classes of (at most) countable structures. This is related to left limits and to a study of approximations and homomorphisms, which leads to a new setting of Theorem 9.1.

In the second direction, we deal with special classes of finite structures and we shall see that the homomorphism preservation theorem holds for nearly all naturally defined sparse classes of finite structures.

10.2 Primitive Positive Theories and Left Limits

We shall now consider an alternative view of left limits in the context of first-order logic.

In the following, we consider first-order logic and a fixed finite signature σ of relations. Within the first-order language defined by σ, we shall sometimes consider *formulas* $\phi(x_1, \dots, x_n)$ with *free variables* $x_1, \dots, x_n$, but most of the time we will consider *sentence*, that is formulas without free variables.

The structures we consider here are finite or infinite; if they are finite, it will be specified.

For $n \in \mathbb{N}$ we denote by FO^n the set of all sentences with quantifier rank at most n and we define $\mathrm{FO} = \bigcup_n \mathrm{FO}^n$.

A formula (or a sentence) is *primitive positive* if it is constructed using conjunction $\wedge$ and existential quantification $\exists$ only. For $n \in \mathbb{N}$ we denote by P^n the subset of FO^n of primitive positive sentences with quantifier rank at most n. We also define $\mathrm{P} = \bigcup_n \mathrm{P}^n$.

As noted by Chandra and Merlin [90], the study of primitive positive formulas is intimately connected to homomorphisms. This is in particular due to the following correspondence between P and $\mathrm{Rel}(\sigma)$: to each sentence $\phi \in \mathrm{P}$ corresponds a finite structure $\mathbf{A}_\phi \in \mathrm{Rel}(\sigma)$ in such a way that for every structure $\mathbf{M}$ (finite or not) it holds:

$$\mathbf{M} \models \phi \quad \iff \quad \mathbf{A}_\phi \to \mathbf{M}.$$

Such $\mathbf{A}_\phi$ is easy to construct as conjunction of tuples of $\mathbf{A}$ preceded by existential quantifications of all the variables. Converse correspondence also holds. This is made more precise by the following construction:

Let $\mathbf{A}$ be a connected structure with tree-depth t. First assume that $\mathbf{A}$ is connected. We consider a canonical rooted plane tree Y such that the Gaifman graph of $\mathbf{A}$ is included in the closure of Y (see Exercise 17.1 for hints on how such a canonical tree can be computed). By construction, every relation of $\mathbf{A}$ is included in a chain of Y. The sentence $\vartheta(\mathbf{A})$ is constructed from a traversal of Y as follows: The first time we reach a vertex v_i we write down "$\exists x_a($" where a is the height of v_i and the conjunction of all relations involving v_i and its ancestors in Y (replacing each v_i by the corresponding x_a); when we backtrack from one element v_i to its father we close the parenthesis opened when reaching v_i; and when we move from one element to its next brother in Y we write $\wedge$ (see Fig. 10.1).

If $\mathbf{A}$ is not connected the $\vartheta(\mathbf{A})$ is defined as the conjunction of the sentences $\vartheta(\mathbf{A}_i)$ where the $\mathbf{A}_i$'s are the connected components of $\mathbf{A}$. By construction, the quantifier rank of $\vartheta(\mathbf{A})$ is exactly t.

Conversely, for a primitive positive sentence ϕ with quantifier rank t, we build a rooted forest Y and construct in a natural way the set of relations between these elements from the atoms present in ϕ. The so-obtained structure $\mathbf{M}(\phi)$ has tree-depth at most t, as its Gaifman graph is included in the closure of Y. Hence we have:

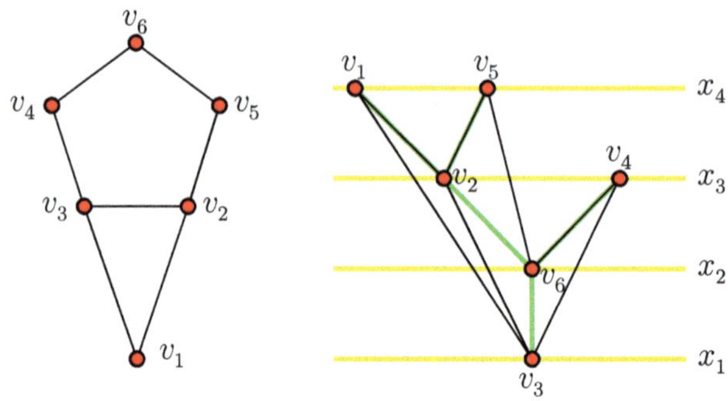

F $\quad\quad$ $F \subseteq \mathrm{Clos}(Y)$

$$\begin{array}{lllll} \vartheta(F): & \exists x_1\,(& \exists x_2\,(& \exists x_3\,(& R(x_3,x_1) \\ & & & \wedge\exists x_4\,(& R(x_4,x_1) \wedge R(x_4,x_2)) \\ & & & \wedge\exists x_4\,(& R(x_4,x_2) \wedge R(x_4,x_3))) \\ & & \wedge\exists x_3\,(& & R(x_3,x_1) \wedge R(x_3,x_2)))) \end{array}$$

Fig. 10.1 computation of $\vartheta(\mathsf{F})$ for an undirected graph F

$$\begin{aligned} \mathrm{qrank}(\vartheta(\mathbf{A})) &= \mathrm{td}(\mathbf{A}) \\ \mathrm{qcount}(\vartheta(\mathbf{A})) &= |\mathbf{A}| \\ \mathrm{td}(\mathbf{M}(\phi)) &\leq \mathrm{qrank}(\phi) \\ |\mathbf{M}(\phi)| &\leq \mathrm{qcount}(\phi) \quad . \\ \vartheta(\mathbf{A}_1 + \mathbf{A}_2) &= \vartheta(\mathbf{A}_1) \wedge \vartheta(\mathbf{A}_2) \\ \mathbf{M}(\vartheta(\mathbf{A})) &\cong \mathbf{A}. \end{aligned}$$

This also leads to the following:

Proposition 10.1. *There two mappings* $\vartheta : \mathrm{Rel}(\sigma) \to \mathsf{P}$ *and* $\mathbf{M} : \mathsf{P} \to \mathrm{Rel}(\sigma)$ *are such that for every* $\mathbf{A}, \mathbf{B} \in \mathrm{Rel}(\sigma)$ *and every* $\phi, \psi \in \mathsf{P}$ *it holds:*

$$\begin{aligned} \mathbf{A} \to \mathbf{B} \quad &\iff \quad \mathbf{B} \models \vartheta(\mathbf{A}) \\ \phi \vdash \psi \quad &\iff \quad \mathbf{M}(\phi) \models \psi \\ \mathbf{B} \models \phi \quad &\iff \quad \mathbf{M}(\phi) \to \mathbf{B} \\ &\iff \quad \vartheta(\mathbf{B}) \vdash \phi \end{aligned}$$

This somewhat dual approach to homomorphism problems will be further extended to theories. Recall that a (first-order) *theory* T is a set of (first-order) sentences. A theory T is *consistent* if it does not contain a contradiction, in the sense that there is no sentence ϕ such that both ϕ and its negation are provable from the sentences in T. If T is a theory and ϕ, ψ are sentences we say that T (resp. ϕ) *entails* ψ if $T \cup \{\neg\psi\}$ (resp. $\{\phi, \neg\psi\}$) is not consistent, and we denote this by $T \vdash \psi$ (resp. $\phi \vdash \psi$).

It follows that Proposition 10.1 establishes an order-reversing isomorphism between the homomorphism quasi-orders $(\mathrm{Rel}(\sigma), \leq_h)$ and $(P, \vdash)$. It follows that two finite structures $\mathbf{A}$ and $\mathbf{B}$ are homomorphically equivalent if and only if $\vartheta(\mathbf{A})$ and $\vartheta(\mathbf{B})$ are logically equivalent. Also note that for a finite structure $\mathbf{A}$ the minimum tree-depth of a finite structure $\mathbf{A}' \rightleftarrows \mathbf{A}$ equals the minimum quantifier rank of a primitive positive sentence logically equivalent to $\vartheta(\mathbf{A})$.

The above connection extends when considering sets of primitive positive sentences. A *PP-theory* is a subset of P. Every PP-theory may be naturally identified with a subset of $\mathrm{Rel}(\sigma)$ by the mapping $\phi \mapsto \mathbf{M}(\phi)$. Note that every PP-theory is consistent. An PP-theory T is *closed* if for every primitive positive sentence ϕ, if $T \vdash \phi$ then $\phi \in T$. The class of all closed PP-theories is denoted by $\mathfrak{P}$. Then, for a primitive positive sentence ϕ, the closed PP-theory consisting of all primitive positive sentences which are logical consequences of ϕ corresponds to the set of structures having a homomorphism to $\mathbf{M}(\phi)$, that is to $(\rightarrow \mathbf{M})$. Closed PP-theories are easy to characterize:

Proposition 10.2. *A PP-theory T is closed if and only if the two following conditions holds:*

1. $\forall \phi \in T, \forall \psi \in P, \quad \phi \vdash \psi \ \Rightarrow \ \psi \in T,$
2. $\forall \phi_1, \phi_2 \in T, \phi_1 \wedge \phi_2 \in T.$

Proof. If T is a closed PP-theory, the two conditions obviously hold. Conversely, assume that the two conditions hold. If $\psi \in P$ and $T \vdash \psi$ then there exists a finite subset $T' \subseteq T$ such that $T' \vdash \psi$ (proofs are finite). By applying the second condition iteratively, $\bigwedge_{\phi \in T'} \phi \in T$. As $\bigwedge_{\phi \in T'} \phi \in T \vdash \psi$, the first condition implies $\psi \in T$. It follows that T is a closed PP-theory. □

The identification of PP-theories and classes of finite structures extends further.

Proposition 10.3. *The mapping*

$$\mathsf{T} \in \mathfrak{P} \mapsto \mathcal{I}(\mathsf{T}) = \{\mathbf{M}(\phi) : \phi \in \mathsf{T}\}$$

is a bijection of $\mathfrak{P}$ *and the class of ideals of* $[\mathrm{Rel}(\sigma)]$, *whose inverse is defined by*

$$\overline{\vartheta}(\mathcal{I}) = \bigcup_{\mathbf{A} \in \mathcal{I}} \{\phi \in \mathsf{P} : \vartheta(\mathbf{A}) \vdash \phi\}.$$

Proof. Indeed, if $\mathbf{A} \to \mathbf{M}(\phi) \in \mathcal{I}(\mathsf{T})$ and $\mathbf{A}$ is finite then $\phi \vdash \vartheta(\mathbf{A})$, thus $\mathbf{A} \in \mathcal{I}(\mathsf{T})$. If $\mathbf{M}(\phi_1), \mathbf{M}(\phi_2) \in \mathcal{I}(\mathsf{T})$ then $\phi_1, \phi_2 \in \mathsf{T}$ hence $\mathsf{T} \vdash \phi_1 \wedge \phi_2$. As $\mathbf{M}(\phi_1 \wedge \phi_2) \rightleftarrows \mathbf{M}(\phi_1) + \mathbf{M}(\phi_2)$ we deduce that $\mathcal{I}(\mathsf{T})$ is an ideal.
Conversely, assume $\mathcal{I}$ is an ideal of $[\mathrm{Rel}(\sigma)]$. Assume $\phi \in \overline{\vartheta}(\mathcal{I})$ and $\phi \vdash \psi$. Then obviously $\psi \in \mathsf{T}(\mathcal{I})$. Assume $\phi_1, \phi_2 \in \overline{\vartheta}(\mathcal{I})$. Then there exist $\mathbf{A}_1, \mathbf{A}_2 \in \mathcal{I}$ such that for $i = 1, 2$ it holds $\vartheta(\mathbf{A}_i) \vdash \phi_i$ hence $\mathbf{M}(\phi_i) \to \mathbf{A}_i$. As $\mathcal{I}$ is an ideal we have $\mathbf{A}_1 + \mathbf{A}_2 \in \mathcal{I}$. As $\vartheta(\mathbf{A}_1 + \mathbf{A}_2) = \vartheta(\mathbf{A}_1) \wedge \vartheta(\mathbf{A}_2)$ we have $\vartheta(\mathbf{A}_1 + \mathbf{A}_2) \vdash \phi_1 \wedge \phi_2$ thus $\phi_1 \wedge \phi_2 \in \overline{\vartheta}(\mathcal{I})$. According to Proposition 10.2, it follows that $\overline{\vartheta}(\mathcal{I})$ is closed.

As obviously $\mathcal{I}(\overline{\vartheta}(\mathcal{I})) = \mathcal{I}$ we deduce that $\mathsf{T} \mapsto \mathcal{I}(\mathsf{T})$ is a bijection of $\mathfrak{P}$ and the class of ideals of $[\mathrm{Rel}(\sigma)]$. □

Thus the class $\mathfrak{P}$ of closed PP-theories is in bijection with the completion $\overline{[\mathrm{Rel}(\sigma)]}_{\mathrm{L}}$ (given by all left limits of $[\mathrm{Rel}(\sigma)]$ introduced in Sect 9.4).

Let $\mathfrak{P}_{\mathrm{F}}$ be the image of $[\mathrm{Rel}(\sigma)]$ by $\overline{\vartheta}$.

Proposition 10.4. *Let* $\mathsf{T} \in \mathfrak{P}$. *Then the following conditions are equivalent:*

(i) $\mathsf{T} \in \mathfrak{P}_{\mathrm{F}}$,
(ii) $\exists \mathbf{A} \in \mathrm{Rel}(\sigma)$, $\mathsf{T} = \overline{\vartheta}([\mathbf{A}])$ *(such* $\mathbf{A}$ *is uniquely determined up to homomorphism equivalence),*
(iii) $\exists \phi_0 \in \mathsf{T}\ \forall \phi \in \mathsf{T}\ \phi_0 \vdash \phi$ *(such* ϕ_0 *is uniquely determined up to logical equivalence),*
(iv) $\mathcal{I}(\mathsf{T})$ *is a principal ideal of* $[\mathrm{Rel}(\sigma)]$.

Proof. From (i) it follows that there exists $\mathbf{A} \in \mathrm{Rel}(\sigma)$ such that $\mathsf{T} = \overline{\vartheta}([\mathbf{A}])$. Assume $\mathsf{T} = \overline{\vartheta}([\mathbf{B}])$. Then $\vartheta(\mathbf{A}) \vdash \mathbf{B}$ and $\vartheta(\mathbf{B}) \vdash \mathbf{B}$, that is: $\mathbf{A} \rightleftarrows \mathbf{B}$ hence (ii) holds.
(ii) and (iii) are equivalent, by putting $\phi_0 = \vartheta(\mathbf{A})$ or $\mathbf{A} = \mathbf{M}(\phi_0)$.
From (ii) follows that $\mathcal{I}(\mathsf{T}) = (\to \mathbf{A})$ hence (iv).

From (iv) follows (i) as $\mathsf{T} = \overline{\vartheta}((\to \mathbf{A})) = \overline{\vartheta}([\mathbf{A}])$.

□

10.3 Theories and Countable Structures

A theory T is *satisfiable* if it has a *model*, that is a structure $\mathbf{M}$ which satisfies all the sentences in T (this is denoted by $\mathbf{M} \models \mathsf{T}$). We denote by Mod(T) the class of the models of the theory T. Conversely, for any structure $\mathbf{A}$ we denote by Th($\mathbf{A}$) the *theory* of $\mathbf{A}$, that is the class of all the sentences satisfied by $\mathbf{A}$. A theory is *complete* if it is a maximal consistent set of sentences. By *Gödel completeness theorem*, every consistent theory is satisfiable, thus has at least a model. A class $\mathfrak{C}$ of structures is an *elementary class* if there is a theory T such that $\mathfrak{C} = \mathrm{Mod}(\mathsf{T})$. It is known that the homomorphism preservation theorem holds for every elementary class (see Exercise 2 in Sect. 5.5 in [258]). According to downward *Löwenheim-Skolem theorem*, every infinite first-order structure has a countable elementary substructure. Recall that a substructure $\mathbf{A}$ of a structure $\mathbf{B}$ is an *elementary substructure* of $\mathbf{B}$ if for every first-order formula $\phi(x_1, \dots, x_n)$ with free variables $x_1, \dots, x_n$, and for every tuple $(a_1, \dots, a_n)$ of elements of $\mathbf{A}$ it holds

$$\mathbf{A} \models \phi(a_1, \dots, a_n) \quad \iff \quad \mathbf{B} \models \phi(a_1, \dots, a_n).$$

It follows that every consistent theory has at least a model that is at most countable.

Here, as in the whole chapter, we refer to [257, 299, 303, 325] for a background on logic and model theory information.

We denote by $\mathfrak{T}$ the class of all consistent theories. For $\mathsf{T} \in \mathfrak{T}$, the *deductive closure* of T is the theory $\mathsf{T}^{\vdash} = \{\phi : \ \mathsf{T} \vdash \phi\}$ of all the sentences which are logical consequences of those in T. We denote by $\mathfrak{T}_C$ the subclass of all complete theories. Recall that a theory T is *complete* if it is consistent and maximal for this property; this means that for every sentence $\phi \in \mathrm{FO}$, either $\phi \in \mathsf{T}$ or $\neg\phi \in \mathsf{T}$. Every theory $\mathsf{T} \in \mathfrak{T}_C$ has a model but all the countable models of T do not need to be isomorphic. However, all the models of T are elementarily equivalent. Hence $\mathfrak{T}_C$ can be viewed has the quotient $\mathfrak{Rel}(\sigma)/\equiv$ of the class $\mathfrak{Rel}(\sigma)$ of all σ-structures by the elementary equivalence relation $\equiv$. Although elementary equivalence reduces to isomorphism on finite structures, this relation is quite a complicated equivalence for infinite structures and the language of theories provides a convenient setting.

Of particular interest will be the following two subclasses of $\mathfrak{T}_C$:

- The class $\mathfrak{T}_F$ of all complete theories that have a finite model.

- The class $\mathfrak{T}_{FMP}$ of all complete theories T with the *finite model property*, i.e. such that every sentence $\phi \in T$ has a finite model. (Notice that this is equivalent to the statement that every finite $T' \subseteq T$ has a finite model.)

Obviously we have

$$\mathfrak{T}_F \subseteq \mathfrak{T}_{FMP} \subseteq \mathfrak{T}_C \subseteq \mathfrak{T}.$$

For an example of a sentence ϕ of quantifier rank 5 with no finite model see Exercise 10.2. For an example of a consistent theory T without a finite model, but such that every finite subset T' of T has a finite model, see Exercise 10.3.

The mapping Th from the class $\mathrm{Rel}(\sigma)$ of finite structures to the class $\mathfrak{T}_F$ is clearly a bijection, as a complete theory T with a finite model has no other model.

In the complete analogy to left distances of finite σ-structures we define a distance on $\mathfrak{T}$. A natural ultrametric can be defined on $\mathfrak{T}$, which expresses that two theories are close if they contain the same sentences of small quantifier rank. Formally, for two theories $T_1, T_2 \in \mathfrak{T}$, the *first-order distance* $\mathrm{dist}_{FO}(T_1, T_2)$ is 0 if $T_1 = T_2$, and otherwise it is 2^{-t}, where t is the minimum quantifier rank of a formula belonging to exactly one of T_1 and T_2.

Proposition 10.5. *With the metric* dist_{FO}*:*

- $\mathfrak{T}$ *is compact,*
- $\mathfrak{T}_C$ *is compact,*
- $\mathfrak{T}_{FMP}$ *is compact,*
- $\mathfrak{T}_F$ *is dense in* $\mathfrak{T}_{FMP}$.

Proof. The first two items are direct consequences of the compactness theorem. The two next items will follow from the proof that $\mathfrak{T}_{FMP}$ is the set of all limits of sequences of elements of $\mathfrak{T}_F$ which converge in $\mathfrak{T}_C$.

Assume (T_i) is a sequence of elements of $\mathfrak{T}_F$ converging to some $T \in \mathfrak{T}_C \setminus \mathfrak{T}_F$. Let $\phi \in T$. For every i such that $\mathrm{dist}_{FO}(T_i, T) < 2^{-\mathrm{qrank}(\phi)}$, the finite model of T_i is a model of ϕ.

Conversely, assume $T \in \mathfrak{T}_{FMP}$. Then for every $n \in \mathbb{N}$, $T \cap FO^n$ has a finite model $\mathbf{A}_n$. As both $\mathrm{Th}(A_n)$ and T are complete theories, we have $\mathrm{Th}(A_n) \cap FO^n = T \cap FO^n$ hence T is the limit of the (possibly constant) sequence $(\mathrm{Th}(A_n))$.

□

10.4 Primitive Positive Theories Again

It follows that $\mathfrak{P}_F$ is the trace of $\mathfrak{T}_F$ on P:

$$\mathfrak{P}_F = \{T \cap P : T \in \mathfrak{T}_F\}.$$

However, note that $\mathfrak{P}_F$ is not the class of all closed PP-theories with a finite model (see Exercise 10.4).

We shall prove that the metric space $(\mathfrak{P}, \mathrm{dist}_{FO})$, which is constructed from ideals similarly as $\overline{[\mathrm{Rel}(\sigma)]}_L$ is constructed from $[\mathrm{Rel}(\sigma)]$, is the completion of the metric space $(\mathfrak{P}_F, \mathrm{dist}_{FO})$. The proof is based on the following approximation result.

Proposition 10.6. *Let* $T \in \mathfrak{P}$ *and let* $n \in \mathbb{N}$. *Then there exists* $\phi_n \in I$ *such that*

1. $\mathrm{qrank}(\phi_n) \leq n$,
2. $\forall \psi \in T,\ (\mathrm{qrank}(\psi) \leq n) \Rightarrow (\phi_n \vdash \psi)$.

Proof. Let $\mathcal{C}_n = \{\mathbf{M}(\phi) : \phi \in T \text{ and } \mathrm{qrank}(\phi) \leq n\}$. According to Proposition 10.1, the class $\mathcal{C}_n$ is the union of homomorphism classes of structures with tree-depth at most n. According to Corollary 6.6 there exists N such that $\mathcal{C}_n$ is the union of the homomorphism equivalence classes of N structures $\mathbf{C}_1, \dots, \mathbf{C}_N$. As T is complete, $\vartheta(\mathbf{C}_i) \in T$ for every $1 \leq i \leq N$. Define $\phi_n = \bigwedge_{i=1}^{N} \vartheta(\mathbf{C}_i)$. Then, by construction, for every $\psi \in T$ with quantifier rank at most n it holds $\phi_n \vdash \psi$. □

Corollary 10.1. *The metric space* $(\mathfrak{P}, \mathrm{dist}_{FO})$ *is the completion of the space* $(\mathfrak{P}_F, \mathrm{dist}_{FO})$.

Proof. If (T_n) is a Cauchy sequence of elements of $\mathfrak{T}_F$ then for every integer n there exists $f(n)$ such that $T_i \cap P^n = T_{f(n)} \cap P^n$ for every $i > f(n)$. Define $T = \bigcup_n T_{f(n)} \cap P^n$. Then T is clearly the limit of (T_n).
Let $T \in \mathfrak{P}$. For every integer n there exists, according to Proposition 10.6, a sentence ϕ_0 such that every sentence $\psi \in T \cap P^n$ is such that $\phi_0 \vdash \psi$. Let $T_n = \{\psi : \phi_0 \vdash \psi\} = \mathrm{Th}(\mathbf{M}(\phi_0)) \cap P$. Then (T_n) converges to T.

□

The mapping $\overline{\vartheta}$ can be used to define an alternative distance on $[\mathrm{Rel}(\sigma)]$, by making use of the equality of the minimum tree-depth of structures in $[\mathbf{A}]$ and the minimum quantifier rank of the primitive positive sentences logically equivalent to $\vartheta(\mathbf{A})$. Define

$$\mathrm{dist}_{\mathrm{td}}([\mathbf{A}],[\mathbf{B}]) = \mathrm{dist}_{\mathrm{FO}}(\overline{\vartheta}([\mathbf{A}]),\overline{\vartheta}([\mathbf{B}])).$$

We call this metric the *tree-depth distance*, as it can be alternatively be defined using tree-depth as we shall prove now.

Proposition 10.7. *For every* $\mathbf{A},\mathbf{B} \in \mathrm{Rel}(\sigma)$, $\mathrm{dist}_{\mathrm{td}}([\mathbf{A}],[\mathbf{B}])$ *is* 0 *if* $[\mathbf{A}] = [\mathbf{B}]$ *and (otherwise) it is* 2^{-t}, *where* t *is the minimum tree-depth of a* $\mathbf{F} \in \mathrm{Rel}(\sigma)$ *such that* $(\mathbf{F} \to \mathbf{A}) \not\Leftrightarrow (\mathbf{F} \to \mathbf{B})\}$.

Proof. If $[\mathbf{A}] = [\mathbf{B}]$ then $\mathrm{dist}_{\mathrm{td}}([\mathbf{A}],[\mathbf{B}])$ is obviously 0. Assume $[\mathbf{A}] \neq [\mathbf{B}]$ hence $\overline{\vartheta}([\mathbf{A}]) \neq \overline{\vartheta}([\mathbf{B}])$. Let $2^{-n} = \mathrm{dist}_{\mathrm{FO}}(\overline{\vartheta}([\mathbf{A}]),\overline{\vartheta}([\mathbf{B}]))$. Without loss of generality, we can assume that there exists $\phi \in \overline{\vartheta}([\mathbf{A}]) \setminus \overline{\vartheta}([\mathbf{B}])$ such that $\mathrm{qrank}(\phi) = n$. By the definition of $\overline{\vartheta}$ we have $\vartheta(\mathbf{A}) \vdash \phi$ and $\vartheta(\mathbf{B}) \not\vdash \phi$ hence, according to Proposition 10.1, $\mathbf{M}(\phi) \to \mathbf{A}$, $\mathbf{M}(\phi) \nrightarrow \mathbf{B}$ and $\mathrm{td}(\mathbf{M}(\phi)) \leq n$. Conversely, assume $\mathbf{F} \to \mathbf{A}$ and $\mathbf{F} \nrightarrow \mathbf{B}$. Then $\vartheta(\mathbf{F}) \in \overline{\vartheta}([\mathbf{A}]) \setminus \overline{\vartheta}([\mathbf{B}])$ and $\mathrm{qrank}(\vartheta(\mathbf{F})) \leq \mathrm{td}(\mathbf{F})$.
Assume $\mathbf{F} \to \mathbf{A}$ and $\mathbf{F} \nrightarrow \mathbf{B}$. Then the same property holds for $\mathrm{Core}(\mathbf{F})$ instead of $\mathbf{F}$. As $\mathrm{Core}(\mathbf{F})$ is a substructure of $\mathbf{F}$, its order and tree-depth are not greater than those of $\mathbf{F}$. Moreover, according to Corollary 6.5 we have

$$\mathrm{td}(\mathrm{Core}(\mathbf{F})) \leq |\mathrm{Core}(\mathbf{F})| \leq F(\mathrm{td}(\mathrm{Core}(\mathbf{F}))).$$

$$\mathrm{dist}_{\mathrm{td}}([\mathbf{A}],[\mathbf{B}]) \leq \mathrm{dist}_{\mathrm{L}}([\mathbf{A}],[\mathbf{B}]) \leq 2^{-F(-\log_2 \mathrm{dist}_{\mathrm{td}}([\mathbf{A}],[\mathbf{B}]))}.$$

□

Corollary 10.2. *(1)* $([\mathrm{Rel}(\sigma)],\mathrm{dist}_{\mathrm{td}})$ *and* $([\mathrm{Rel}(\sigma)],\mathrm{dist}_{\mathrm{L}})$ *are homeomorphic;*
(2) $(\mathfrak{P}_{\mathrm{F}},\mathrm{dist}_{\mathrm{FO}})$ *and* $([\mathrm{Rel}(\sigma)],\mathrm{dist}_{\mathrm{L}})$ *are homeomorphic;*
(3) $(\mathfrak{P},\mathrm{dist}_{\mathrm{FO}})$ *and* $(\overline{[\mathrm{Rel}(\sigma)]},\mathrm{dist}_{\mathrm{L}})$ *are homeomorphic;*

Proof. (1) It follows directly from Proposition 10.7 that the metric spaces $([\mathrm{Rel}(\sigma)],\mathrm{dist}_{\mathrm{td}})$ and $([\mathrm{Rel}(\sigma)],\mathrm{dist}_{\mathrm{L}})$ are homeomorphic.
(2) As $\overline{\vartheta}$ is an isometry by the construction of $\mathrm{dist}_{\mathrm{FO}})$ on $[\mathrm{Rel}(\sigma)]$, the metric spaces $(\mathfrak{P}_{\mathrm{F}},\mathrm{dist}_{\mathrm{FO}})$ and $([\mathrm{Rel}(\sigma)],\mathrm{dist}_{\mathrm{L}})$ are homeomorphic.
(3) By considering the completions of the above spaces, the metric spaces $(\mathfrak{P},\mathrm{dist}_{\mathrm{FO}})$ and $(\overline{[\mathrm{Rel}(\sigma)]},\mathrm{dist}_{\mathrm{L}})$ are homeomorphic.

□

The class $\mathfrak{P}$ itself can be viewed even as the trace of $\mathfrak{T}$ on P:

Proposition 10.8. *The class $\mathfrak{P}$ is the trace of $\mathfrak{T}_C$ on P:*

$$\mathfrak{P} = \{T \cap P : \ T \in \mathfrak{T}_C\}.$$

Proof. If $T \in \mathfrak{P}$ then for every $\phi \in P \setminus T$ we have $T \vdash \neg\phi$ (as T is complete). It follows that $T' = P \cup \{\neg\phi, \ \phi \in P \setminus T\}$ is consistent hence has a model $\mathbf{A}$. As $\mathrm{Th}(\mathbf{A}) \cap P = T$ we deduce $T \in \{T \cap P : \ T \in \mathfrak{T}_C\}$.
Conversely, if $T = \mathrm{Th}(\mathbf{A}) \cap P$ then $T \cap P$ is a closed PP-theory as T is complete. □

It follows that $\mathfrak{P} = \mathfrak{T}_C / \sim_P$, where $\sim_P$ is the equivalence classes defined by

$$T_1 \sim_P T_2 \quad \Longleftrightarrow \quad T_1 \cap P = T_2 \cap P.$$

A global picture of the metric subspaces of $(\mathfrak{T}, \mathrm{dist}_{FO})$ is sketched in Fig. 10.2.

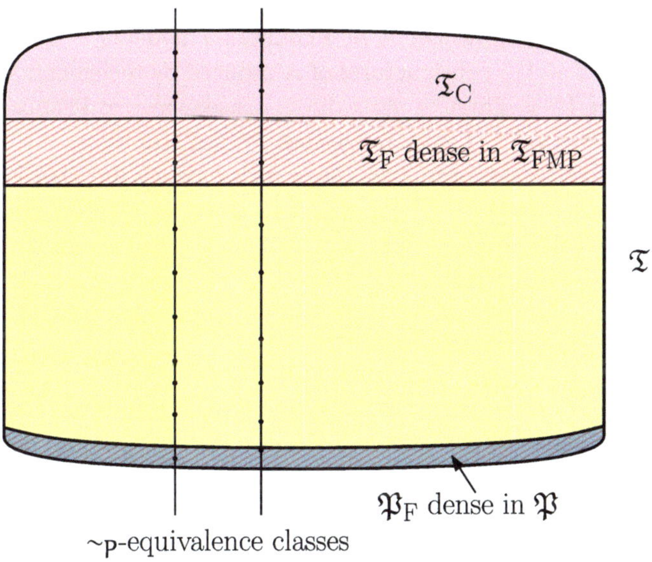

Fig. 10.2 Compact subspaces of the compact space $\mathfrak{T}$: $\mathfrak{P}$ and $\mathfrak{T}_{FMP} \subset \mathfrak{T}_C$. Density of $\mathbf{T}_F$ in $\mathfrak{T}_{FMP}$ and of $\mathbf{P}_F$ in $\mathfrak{P}$. The $\sim_P$ equivalence classes

10.5 Quotient Metric Spaces

It follows from Proposition 10.8 that yet another metric can be defined on $\mathfrak{P}$, by considering the quotient of the metric space $(\mathfrak{T}_C, \mathrm{dist}_{FO})$ by $\sim_P$. Let us denote the quotient metric by d_{FO}. Then, for every $T_1, T_2 \in \mathfrak{P}$, it holds:

$$d_{FO}(T_1, T_2) = \inf\{dist_{FO}(T_1', T_2') : T_1' \cap P = T_1, T_2' \cap P = T_2,\ T_1', T_2' \in \mathfrak{T}_C\}.$$

Notice that obviously, for every $T_1, T_2 \in \mathfrak{P}$, it holds:

$$d_{FO}(T_1, T_2) \geq dist_{FO}(T_1, T_2).$$

Indeed, $d_{FO}(T_1, T_2) < 2^{-n}$ means that there exist $T_1', T_2' \in \mathfrak{T}_C$ such that $T_1' \cap P = T_1, T_2' = T_2$ and for every $\phi \in FO^n$ it holds $(\phi \in T_1') \iff (\phi \in T_2')$. Hence, for every $\phi \in P^n$ it holds $(\phi \in T_1) \iff (\phi \in T_2)$ thus $dist_{FO}(T_1, T_2) < 2^{-n}$.

The remaining of this section will be devoted to the proof that the equality holds. This is not obvious and its proof will use one of the main theorems of [425].

Lemma 10.1. *Every closed PP-theory is the trace on* P *of a complete theory in* $\mathfrak{T}_{FMP}$*:*

For every $T \in \mathfrak{P}$ *there exists* $T' \in \mathfrak{T}_{FMP}$ *such that* $T = T' \cap P$.

Proof. Let $T \in \mathfrak{T}_C \setminus \mathfrak{T}_{FMP}$. Let $\mathbf{A}$ be a countable model of T. Let $\mathbf{A}_n$ be the disjoint union of all the substructures of $\mathbf{A}$ induced by n elements. Let $\mathbf{A}^+ = \bigcup_n \mathbf{A}_n$ and let $T^+ = Th(\mathbf{A}^+)$. As a direct consequence of Proposition 10.1, we have $T^+ \cap P = T \cap P$. Let us prove that $T^+ \in \mathfrak{T}_{FMP}$:

Let $\phi \in Th(\mathbf{A}^+)$. According to *Gaifman's locality theorem* [205], there exists a first-order sentence ϕ' equivalent to ϕ which is a Boolean combination of *basic local sentences*, that is of sentences ϕ_a' of the form

$$\exists x_1 \dots \exists x_{k_a} \Bigg(\bigwedge_{1 \leq i < j \leq k_a} dist(x_i, x_j) > 2r \wedge \bigwedge_{1 \leq i \leq k_a} \psi(x_i) \Bigg),$$

where $\psi(x)$ is r-*local* (i.e. such that $\mathbf{B} \models \psi(b)$ if and only if $\mathbf{B}[N_r(b)] \models \psi(b)$). Let $k = \max k_a$. For each ϕ_a', we have $\mathbf{A}^+ \models \phi_a'$ if and only if there exists a union of at most k connected components of $\mathbf{A}^+$ which satisfies ϕ_a', that is if and only if there exists some n_a such that $\bigcup_{i \leq n} \mathbf{A}_i \models \phi_a'$ for all $n \geq n_a$. Let N be the maximum of the n_a such that $\mathbf{A}^+ \models \phi_a'$. Then for every a and every $N \geq n$ we have $(\mathbf{A}^+ \models \phi_a') \iff (\bigcup_{i \leq n} \mathbf{A}_i \models \phi_a')$ hence $\bigcup_{i \leq n} \mathbf{A}_i \models \phi$. We deduce that $(Th(\bigcup_{i \leq n} \mathbf{A}_i))$ converges to T^+. Hence $T^+ \in \mathfrak{T}_{FMP}$, as T^+ has obviously an infinite model. □

Corollary 10.3. $\mathfrak{P}_F$ *is dense in* $(\mathfrak{P}, d_{FO})$.

Proof. Let $T \in \mathfrak{P}$. According to Lemma 10.1 there exists $T' \in \mathfrak{T}_{FMP}$ such that $T = T' \cap P$. According to Proposition 10.5, $\mathfrak{T}_{FMP}$ is the closure of $\mathfrak{T}_F$ for the topology induced by dist_{FO} on $\mathfrak{T}_C$. Hence there exists a sequence (T_i) of elements of $\mathfrak{T}_F$ such that $\mathrm{dist}_{FO}(T_i, T') \to 0$. As $(\mathfrak{P}, d_{FO}) = (\mathfrak{T}_C, \mathrm{dist}_{FO})/\sim_P$ we deduce that $d_{FO}(T_i \cap P, T) = d_{FO}(T_i \cap P, T' \cap P) \to 0$. As $T_i \cap P \in \mathfrak{P}_F$ we deduce that every $T \in \mathfrak{P}$ is a limit point of $\mathfrak{P}_F$. □

We recall Rossman's equirank homomorphism preservation theorem [425].

Theorem 10.1. *If* $(\mathbf{F} \to \mathbf{A}) \iff (\mathbf{F} \to \mathbf{B})$ *holds for every (finite) structure* $\mathbf{F}$ *with tree-depth at most* n, *then there exist* $\mathbf{A}' \rightleftarrows \mathbf{A}$ *and* $\mathbf{B}' \rightleftarrows \mathbf{B}$ *such that* $\mathbf{A}' \equiv^n \mathbf{B}'$.

We define the pseudo-metric $\mathrm{dist}_\equiv$ on structures by

$$\mathrm{dist}_\equiv(\mathbf{A}, \mathbf{B}) = \mathrm{dist}_{FO}(\mathrm{Th}(\mathbf{A}), \mathrm{Th}(\mathbf{B})).$$

In other words, $\mathrm{dist}_\equiv(\mathbf{A}, \mathbf{B}) = 0$ if $\mathbf{A} \equiv \mathbf{B}$. Otherwise, t is the minimum t such that $\mathbf{A} \not\equiv^t \mathbf{B}$.

From Proposition 10.7 and Theorem 10.1 we get immediately:

Corollary 10.4. *For every structures* $\mathbf{A}, \mathbf{B}$ *it holds*

$$\mathrm{dist}_{td}([\mathbf{A}], [\mathbf{B}]) = \inf\{\mathrm{dist}_\equiv(\mathbf{A}', \mathbf{B}'),\ \mathbf{A}' \in [\mathbf{A}], \mathbf{B}' \in [\mathbf{B}]\}.$$

The next fact is also immediate.

Lemma 10.2. *For every finite structure* $\mathbf{A}$ *and every (possibly non finite) structure* $\mathbf{B}$, *it holds*

$$(\mathbf{A} \rightleftarrows \mathbf{B}) \text{ if and only if } (\forall \mathbf{F} \in \mathrm{Rel}(\sigma), (\mathbf{F} \to \mathbf{A}) \iff (\mathbf{F} \to \mathbf{B})).$$

Theorem 10.2. *The metric spaces* $(\mathfrak{P}, \mathrm{dist}_{FO})$ *and* $(\mathfrak{P}, d_{FO})$ *are isometric.*

Proof. Let $T_1, T_2 \in \mathfrak{P}_F$, let $\mathbf{A}_1 = \overline{\vartheta}^{-1}(T_1)$ and $\mathbf{A}_2 = \overline{\vartheta}^{-1}(T_2)$.

By definition of dist_{td}, Corollary 10.4 and Lemma 10.2 we get:

$$\begin{aligned}\mathrm{dist}_{FO}(T_1, T_2) &= \mathrm{dist}_{td}(\mathbf{A}_1, \mathbf{A}_2)\\ &= \inf\{\mathrm{dist}_\equiv(\mathbf{A}', \mathbf{B}'),\ \mathbf{A}' \in [\mathbf{A}_1], \mathbf{B}' \in [\mathbf{A}_2]\}\\ &= \inf\{\mathrm{dist}_{FO}(T_1', T_2'),\ T_1' \cap P = T_1, T_2 \cap P = T_2,\ T_1', T_2' \in \mathfrak{T}_C\}\\ &= d_{FO}(T_1, T_2)\end{aligned}$$

As $\mathfrak{P}_F$ is dense in both $(\mathfrak{P}, \mathrm{dist}_{FO})$ (by Corollary 10.1) and $(\mathfrak{P}, d_{FO})$ (by Corollary 10.3) we deduce that $(\mathfrak{P}, \mathrm{dist}_{FO})$ and $(\mathfrak{P}, d_{FO})$ are isometric. □

10.6 The Topological Preservation Theorem

In this section, we prove the following general preservation theorem which relates metric properties to preservation theorems.

Theorem 10.3 (Topological Preservation Theorem). *Let* $\mathfrak{T}_0 \subseteq \mathfrak{T}_C$ *and let* $\mathfrak{X}_0 = \{T \cap X : T \in \mathfrak{T}_0\}$. *Assume that the two following condition hold:*

1. Uniform approximation property *for* $\mathfrak{X}_0$*: there exists* $C : \mathbb{N} \to \mathbb{N}$ *such that for every* $T \in \mathfrak{X}_0$ *and every* $n \in \mathbb{N}$ *there exists* $T' \in \mathfrak{X}_0$ *such that*

$$T \cap X^n \subseteq T' \subseteq T \quad \textit{and} \quad \mathrm{qrank}(T') \leq C(n);$$

2. *The quotient space* $(\mathfrak{T}_0, \mathrm{dist}_{FO})/\sim_X$ *is uniformly homeomorphic to the metric space* $(\mathfrak{X}_0, \mathrm{dist}_{FO})$.

Then, for every sentence $\phi \in FO$*, the following conditions are equivalent:*

1. $\forall T_1, T_2 \in \mathfrak{T}_0$ *it holds*

$$(\phi \in T_1) \wedge (T_1 \cap X \subseteq T_2 \cap X) \quad \Longrightarrow \quad \phi \in T_2;$$

2. *There exist a sentence* ψ*, which is a conjunction of disjunction of sentences in* X*, such that for every* $T \in \mathfrak{T}_0$ *it holds*

$$\phi \in T \quad \Longleftrightarrow \quad \psi \in T.$$

In the next section we will make use of this general theorem to prove specific homomorphism preservation theorems. Before proceeding to its proof, we introduce some definitions and notations.

The *quantifier rank* $\mathrm{qrank}(T)$ of a consistent theory $T \in \mathfrak{T}$ is the minimum over all theories T_0 with $T_0 \subseteq T \subseteq T_0^{\vdash}$ of the maximum quantifier rank of a formula in T_0. In particular, if T is closed under conjunctions then $\mathrm{qrank}(T) = \min\{\mathrm{qrank}(\phi) : \phi \in T, \phi \vdash T\}$.

Let $X \subseteq FO$ by a fragment of first-order logic. For integer n we define $X^n = X \cap FO^n$. We denote by $\sim_X$ the equivalence relation on $\mathfrak{T}_C$ defined by

$$T_1 \sim_X T_2 \qquad \Longleftrightarrow \qquad T_1 \cap X = T_2 \cap X.$$

Also, we define by $\mathfrak{X} = \{T \cap X : T \in \mathfrak{T}_C\}$.

Proof. Let ϕ be a first-order sentence with quantifier rank n.

It is clear that if there exist a sentence ψ, which is a conjunction of disjunction of sentences in X, such that for every $T \in \mathfrak{T}_0$ it holds $\phi \in T \iff \psi \in T$, then $\forall T_1, T_2 \in \mathfrak{X}_0$ it holds

$$(\phi \in T_1) \wedge (T_1 \cap X \subseteq T_2 \cap X) \quad \Longrightarrow \quad \phi \in T_2.$$

We shall thus restrict our attention to the converse implication.

As the quotient space $(\mathfrak{T}_0, \mathrm{dist}_{FO})/\sim_X$ is uniformly homeomorphic to the metric space $(\mathfrak{X}_0, \mathrm{dist}_{FO})$, there exists $g(n)$ such that for every $T_1, T_2 \in \mathfrak{T}_0$ such that $\mathrm{dist}_{FO}(T_1 \cap X, T_2 \cap X) < 2^{-g(n)}$ there exist $T_1', T_2' \in \mathfrak{T}_0$ such that $T_1' \sim_X T_1$, $T_2' \sim_X T_2$, and $\mathrm{dist}_{FO}(T_1', T_2') < 2^{-n}$.

Let $T \in \mathfrak{T}_0$. According to the uniform approximation property for $\mathfrak{X}_0$ there exists $T' \in \mathfrak{T}_0$ such that

$$T \cap X^{g(n)} \subseteq T' \cap X \subseteq T \cap X \quad \text{and} \quad \mathrm{qrank}(T' \cap X) \leq C(g(n)).$$

According to the definition of $\mathrm{qrank}(T' \cap X)$, there exists T_0 such that the maximum quantifier rank of a sentence in T_0 is $C(g(n))$ and $T_0 \subseteq T' \cap X \subseteq T_0^{\vdash}$, thus $T' \cap X = T_0^{\vdash} \cap X$ as T' is complete. By choosing inclusion minimal theory T_0 with this property, we can assume that T_0 is finite and has order bounded by some function $N(g(n))$ of $g(n)$. As there are only finitely many non-equivalent choices of theories of order at most $N(g(n))$ and maximum quantifier rank at most $C(g(n))$ there exists a finite set $\mathcal{T}$ of pairs $(\hat{T}, \tilde{T})$ of theories such that for every $T \in \mathfrak{T}_0$ there exists a pair $(\hat{T}, \tilde{T}) \in \mathcal{T}$ with the following properties:

- $\tilde{T} \subseteq X^{C(g(n))}$, $\hat{T} \in \mathfrak{T}_0$;
- $|\tilde{T}| \leq N(g(n))$;
- $\hat{T} \cap X = \tilde{T}^{\vdash} \cap X$;
- $T \cap X^{g(n)} \subseteq \hat{T} \cap X \subseteq T \cap X$.

It follows that $T \cap X^{g(n)} = \hat{T} \cap X^{g(n)}$, that is: $\mathrm{dist}_{FO}(T \cap X, \hat{T} \cap X) < 2^{-g(n)}$.

By the uniform homeomorphism of $(\mathfrak{T}_0, \mathrm{dist}_{FO})/\sim_X$ and $(\mathfrak{X}_0, \mathrm{dist}_{FO})$, we deduce from $\mathrm{dist}_{FO}(T \cap X, \hat{T} \cap X) < 2^{-g(n)}$ that there exists $T', \hat{T}' \in \mathfrak{T}_0$ such that $T' \sim_X T$, $\hat{T}' \sim_X \hat{T}$, and $T' \cap FO^n = \hat{T}' \cap FO^n$. It follows that if $\phi \in T$ then $\phi \in T'$ (as $T' \sim_X T$), $\phi \in \hat{T}'$ (as $\hat{T}' \cap FO^n = \hat{T} \cap FO^n$, and $\phi \in \hat{T}$ (as $\hat{T} \sim_X \hat{T}'$). Conversely, if $\phi \in \hat{T}$ then $\phi \in T$.

Define

$$\psi = \bigvee_{(\hat{T},\tilde{T})\in\mathcal{F}:\phi\in\hat{T}} \bigwedge \tilde{T}.$$

Assume $T_0 \in \mathfrak{T}_0$ and $\psi \in T_0$. Then there exists $(\hat{T},\tilde{T}) \in \mathcal{F}$ such that $\phi \in \hat{T}$ and $\bigwedge \tilde{T} \in T_0$. By the completeness of T_0, it follows that $\hat{T} \cap X \subseteq T_0 \cap X$. Hence, as also $\phi \in \hat{T}$, we deduce $\phi \in T_0$.
Conversely, assume $\phi \in T_0$. By construction, there exists $(\hat{T},\tilde{T}) \in \mathcal{F}$ such that $T \cap X^{g(n)} \subseteq \hat{T} \cap X \subseteq T_0 \cap X$ hence $\mathrm{dist}_{FO}(T_0 \cap X, \hat{T} \cap X) < 2^{-g(n)}$. Thus (as shown above) we deduce from $\phi \in T_0$ that $\phi \in \hat{T}$. It follows that $\tilde{T} \vdash \psi$ hence (as $\tilde{T} \subseteq \hat{T} \cap X \subseteq T_0$) it holds $\psi \in T_0$.

□

10.7 Homomorphism Preservation Theorems

We now consider the case where $X = P$, that is the case of primitive positive sentences, and where $\mathfrak{T}_0 = \{\mathrm{Th}(\mathbf{A}), \mathbf{A} \in \mathcal{C}\}$ and $\mathfrak{P}_0 = \{\mathrm{Th}(\mathbf{A}) \cap P, \mathbf{A} \in \mathcal{C}\}$, where $\mathcal{C}$ is a class of σ-structures. In this setting, the natural relativization of the homomorphism preservation theorem for the class $\mathcal{C}$ is the following:

Homomorphism Preservation Theorem for a class $\mathcal{C}$:
A first order formula ϕ is *preserved under homomorphisms* on $\mathcal{C}$, that is such that:

$$\forall \mathbf{A},\mathbf{B} \in \mathcal{C}: \qquad (\mathbf{A} \models \phi) \wedge (\mathbf{A} \to \mathbf{B}) \implies \mathbf{B} \models \phi,$$

if and only if it is equivalent on $\mathcal{C}$ to an existential-positive first-order formula.

Notice that one direction clearly holds for every class $\mathcal{C}$: if ϕ is an existential-positive first-order formula, it is preserved under homomorphisms. We shall now consider some contexts in which relativized homormophism theorems could derive from Topological Preservation Theorem 10.3.

A connection with concepts used in Sect. 10.6 follows from the following observation.

Lemma 10.3. *If the class $\mathcal{C}$ is hereditary then the uniform approximation property (introduced in Theorem 10.3) holds for $\mathfrak{P}_0$.*

Proof. For every $\mathbf{A} \in \mathfrak{T}_0$ and every $n \in \mathbb{N}$ consider the finite family $\mathcal{F}$ of the core structures $\mathbf{C}$ with tree-depth at most n such that $\mathbf{C} \to \mathbf{A}$. Then there exists a homomorphism $f : \sum \mathcal{F} \to \mathbf{A}$. Let $\mathbf{A}'$ be the substructure induced by the image of $\sum \mathcal{F}$ by f. Then $|\mathbf{A}'|$ is bounded by a function $C(n)$ of n and $\mathbf{A}' \in \mathcal{C}$. Moreover

$$\mathrm{Th}(\mathbf{A}) \cap P^n \subseteq \mathrm{Th}(\mathbf{A}') \cap P \subseteq \mathrm{Th}(\mathbf{A}) \cap P$$

and

$$\mathrm{qrank}(\mathrm{Th}(\mathbf{A}')) \leq \mathrm{qrank}(\vartheta(\mathbf{A}')) \leq \mathrm{td}(\mathbf{A}') \leq |\mathbf{A}'| \leq C(n).$$

□

We now consider hereditary classes structures without infinite connected components and show that the Topological Preservation Theorem gives a sufficient condition for the relativized homomorphism to hold.

Lemma 10.4. *Let* $\mathbf{A}, \mathbf{B}$ *be countable* σ*-structures without infinite connected components. Then it holds*

$$\mathrm{Th}(\mathbf{A}) \cap P \subseteq \mathrm{Th}(\mathbf{B}) \cap P \qquad \iff \qquad \mathbf{A} \to \mathbf{B}.$$

Proof. Let $\mathbf{A}_1, \dots, \mathbf{A}_n, \dots$ (resp. $\mathbf{B}_1, \dots, \mathbf{B}_n, \dots$) be the connected components of $\mathbf{A}$ (resp. $\mathbf{B}$). If $\mathrm{Th}(\mathbf{A}) \cap P \subseteq \mathrm{Th}(\mathbf{B})$ then for every integer i we deduce from $\vartheta(\mathbf{A}_i) \in \mathrm{Th}(\mathbf{A})$ that $\vartheta(\mathbf{A}_i) \in \mathrm{Th}(\mathbf{B})$ hence there exists a homomorphism $f_i : \mathbf{A}_i \to \mathbf{B}$. Define $f : \mathbf{A} \to \mathbf{B}$ by $f|\mathbf{A}_i = f_i$. Then f is clearly a homomorphism.

Conversely, if $\mathbf{A} \to \mathbf{B}$ then for every $\phi \in P$ it holds $(\mathbf{M}(\phi) \to \mathbf{A}) \Longrightarrow (\mathbf{M}(\phi) \to \mathbf{B})$, that is $(\phi \in \mathrm{Th}(\mathbf{A})) \Longrightarrow (\phi \in \mathrm{Th}(\mathbf{B}))$. □

Theorem 10.4. *Let* $\mathcal{C}$ *by a hereditary class of countable* σ*-structures without infinite connected components. If the quotient space* $(\mathcal{C}, \mathrm{dist}_{\equiv})/\rightleftarrows$ *is uniformly homeomorphic to* $([\mathcal{C}], \mathrm{dist}_{\mathrm{td}})$ *(or* $([\mathcal{C}], \mathrm{dist}_{\mathrm{L}})$*) then the homomorphism preservation theorem holds for* $\mathcal{C}$.

Proof. Let $\mathfrak{T}_0 = \{\mathrm{Th}(\mathbf{A}),\ \mathbf{A} \in \mathcal{C}\}$, $X = P$ and $\mathfrak{X}_0 = \{\mathrm{Th}(\mathbf{A}) \cap P,\ \mathbf{A} \in \mathcal{C}\}$. According to Lemma 10.3, the uniform approximation property holds for $\mathfrak{X}_0$.

According to Lemma 10.4, for every $\mathbf{A}, \mathbf{B} \in \mathcal{C}$ it holds

$$\begin{aligned} \mathrm{Th}(\mathbf{A}) \sim_X \mathrm{Th}(\mathbf{B}) &\qquad \iff \qquad \mathbf{A} \rightleftarrows \mathbf{B} \\ \mathrm{Th}(\mathbf{A}) \cap X \subseteq \mathrm{Th}(\mathbf{B}) \cap X &\qquad \iff \qquad \mathbf{A} \to \mathbf{B} \end{aligned}$$

Moreover, for every $\mathbf{A}, \mathbf{B} \in \mathcal{C}$ it holds

$$\mathrm{dist}_{\mathrm{FO}}(\mathrm{Th}(\mathbf{A}), \mathrm{Th}(\mathbf{B})) = \mathrm{dist}_{\equiv}(\mathbf{A}, \mathbf{B})$$
$$\mathrm{dist}_{\mathrm{FO}}(\mathrm{Th}(\mathbf{A}) \cap X, \mathrm{Th}(\mathbf{B}) \cap X) = \mathrm{dist}_{\mathrm{td}}([\mathbf{A}], [\mathbf{B}]).$$

According to Theorem 10.3, it follows that for every first-order sentence ϕ, the following conditions are equivalent:

1. $\forall T_1, T_2 \in \mathfrak{T}_0$ it holds
$$(\phi \in T_1) \wedge (T_1 \cap X \subseteq T_2 \cap X) \quad \Longrightarrow \quad \phi \in T_2,$$
that is: $\forall \mathbf{A}, \mathbf{B} \in \mathcal{C}$ it holds
$$(\mathbf{A} \models \phi) \wedge (\mathbf{A} \to \mathbf{B}) \quad \Longrightarrow \quad \mathbf{B} \models \phi;$$
2. There exist a sentence ψ, which is a conjunction of disjunction of sentences in X, such that for every $T \in \mathfrak{T}_0$ it holds
$$\phi \in T \quad \Longleftrightarrow \quad \psi \in T,$$
that is: there exists an existential-positive first-order sentence ψ such that for every $\mathbf{A} \in \mathcal{C}$ it holds
$$\mathbf{A} \models \phi \quad \Longleftrightarrow \quad \mathbf{A} \models \psi.$$

In other words, the homomorphism preservation theorem holds for $\mathcal{C}$.

Notice that one can consider $\mathrm{dist}_{\mathrm{L}}$ instead of $\mathrm{dist}_{\mathrm{td}}$ as $(\mathfrak{P}, \mathrm{dist}_{\mathrm{L}})$ is uniformly homeomorphic to $(\mathfrak{P}, \mathrm{dist}_{\mathrm{td}})$ (see Corollary 10.2). □

The existence of a uniform homeomorphism between the metric spaces $(\mathrm{Rel}(\sigma), \mathrm{dist}_{\equiv})$ and $([\mathrm{Rel}(\sigma)], \mathrm{dist}_{\mathrm{td}})$ follows from one of the main results of [425]:

Theorem 10.5. *For all finite structures* $\mathbf{A}$ *and* $\mathbf{B}$*, if* $(\mathbf{F} \to \mathbf{A}) \iff (\mathbf{F} \to \mathbf{B})$ *holds for every structure* $\mathbf{F}$ *with tree-depth at most* $\rho(n)$*, then there there exist finite structures* $\mathbf{A}' \rightleftarrows \mathbf{A}$ *and* $\mathbf{B}' \rightleftarrows \mathbf{B}$ *such that* $\mathbf{A}' \equiv^n \mathbf{B}'$.

From this, we deduce not only that the homomorphism preservation theorem holds when relativized to finite structures, but also that it holds when relativised to countable structures without infinite connected components.

Theorem 10.6. *The homormorphism preservation theorem holds for the following classes of σ-structures:*

- *The class* Rel(σ) *of all finite σ-structures,*
- *The class* $\mathfrak{A}$ *of all countable σ-structures without infinite connected components.*

Proof. According to Theorem 10.5, for every two finite σ-structures $\mathbf{A}, \mathbf{B}$ such that $\mathrm{dist_{td}}(\mathbf{A}, \mathbf{B}) < 2^{-\rho(n)}$ there exist two finite σ-structures $\mathbf{A}', \mathbf{B}'$ such that $\mathbf{A} \rightleftarrows \mathbf{A}' \equiv^n \mathbf{B}' \rightleftarrows \mathbf{A}$, that is:

$$\inf\{\mathrm{dist}_{\equiv}(\mathbf{A}', \mathbf{B}') : \mathbf{A}' \rightleftarrows \mathbf{A}, \mathbf{B}' \rightleftarrows \mathbf{B},\ \mathbf{A}', \mathbf{B}' \in \mathrm{Rel}(\sigma)\} < 2^{-n}.$$

In other words, $(\mathrm{Rel}(\sigma), \mathrm{dist}_{\equiv})/\rightleftarrows$ and $([\mathrm{Rel}(\sigma)], \mathrm{dist_{td}})$ are uniformly homeomorphic. Thus Theorem 10.4 applies.
The distance $\mathrm{dist}_{\equiv}$ extends by continuity on $\mathfrak{A}$ (see proof of Lemma 10.1) and the distance $\mathrm{dist_{td}}$ extends by continuity on $[\mathfrak{A}]$. It follows that $(\mathfrak{A}, \mathrm{dist}_{\equiv})/\rightleftarrows$ is uniformly homeomorphic to $([\mathfrak{A}], \mathrm{dist_{td}})$. Thus Theorem 10.4 applies.

□

We now consider more complex countable structures. A complete theory $T \in \mathfrak{T}_C$ is *ω-categorical* if it has an infinite model and any two models of size $\aleph_0$ are isomorphic. An *ω-categorical* structure is a structure $\mathbf{M}$ of size $\aleph_0$ whose theory is ω-categorical.

A central result on ω-categoricity is the Ryll-Nardzewski Theorem [428], (which includes parts due to Svenonius [447] and Engeler [154])

Theorem 10.7. *Let* $\mathbf{M}$ *be a countably infinite first order structure in a countable language. Then the following are equivalent.*

(i) $\mathbf{M}$ *is ω-categorical;*
(ii) $\mathrm{Aut}(\mathbf{M})$ *acts* oligomorphically *on* $\mathbf{M}$ *(meaning that every cartesian product* $\mathbf{M}^n$ *of* $\mathbf{M}$ *has finitely many orbits under the action of* $\mathrm{Aut}(\mathbf{M})$*);*
(iii) For each $n > 0$*, there are finitely many formulas* $\phi(x_1, \ldots, x_n)$ *up to* $\mathrm{Th}(\mathbf{M})$*-provable equivalence.*

For a hereditary class $\mathcal{C}$ of structures which are either finite or ω-categorical we can derive from the Topological Preservation Theorem a sufficient condition for the homomorphism preservation theorem to hold when relativized to $\mathcal{C}$.

Theorem 10.8. *Let $\mathcal{C}$ by a hereditary class of σ-structures which are either finite or ω-categorical. If the quotient space $(\mathcal{C}, \mathrm{dist}_{\equiv})/\rightleftarrows$ is uniformly homeomorphic to $([\mathcal{C}], \mathrm{dist}_{\mathrm{td}})$ (or $([\mathcal{C}], \mathrm{dist}_{\mathrm{L}})$) then the homomorphism preservation theorem holds for $\mathcal{C}$.*

Proof. Let $\mathfrak{T}_0 = \{\mathrm{Th}(\mathbf{A}),\ \mathbf{A} \in \mathcal{C}\}$, $\mathrm{X} = \mathrm{P}$ and $\mathfrak{X}_0 = \{\mathrm{Th}(\mathbf{A}) \cap \mathrm{P},\ \mathbf{A} \in \mathcal{C}\}$. According to Lemma 10.3, the uniform approximation property holds for $\mathfrak{X}_0$.

It is a folklore result that a countable structure $\mathbf{A}$ is homomorphic to an ω-categorical structure $\mathbf{B}$ if and only if every finite substructure of $\mathbf{A}$ is homomorphic to $\mathbf{B}$ (see Exercise 9.6). Thus it holds

$$\begin{aligned} \mathrm{Th}(\mathbf{A}) \sim_{\mathrm{X}} \mathrm{Th}(\mathbf{B}) &\iff \mathbf{A} \rightleftarrows \mathbf{B} \\ \mathrm{Th}(\mathbf{A}) \cap \mathrm{X} \subseteq \mathrm{Th}(\mathbf{B}) \cap \mathrm{X} &\iff \mathbf{A} \to \mathbf{B} \end{aligned}$$

Moreover, for every $\mathbf{A}, \mathbf{B} \in \mathcal{C}$ it holds

$$\begin{aligned} \mathrm{dist}_{\mathrm{FO}}(\mathrm{Th}(\mathbf{A}), \mathrm{Th}(\mathbf{B})) &= \mathrm{dist}_{\equiv}(\mathbf{A}, \mathbf{B}) \\ \mathrm{dist}_{\mathrm{FO}}(\mathrm{Th}(\mathbf{A}) \cap \mathrm{X}, \mathrm{Th}(\mathbf{B}) \cap \mathrm{X}) &= \mathrm{dist}_{\mathrm{td}}([\mathbf{A}], [\mathbf{B}]). \end{aligned}$$

According to Theorem 10.3, as in the proof of Theorem 10.4 we deduce that the homomorphism preservation theorem holds for $\mathcal{C}$. □

Perhaps these last results are a convincing argument for a study of completions of the homomorphism order. Of course, by now there are multiple evidences for the importance of the study of "limits" and of convergence in combinatorial setting, see for example [358].

10.8 Homomorphism Preservation Theorems for Finite Structures

We now consider classes of finite structures. In this context Atserias, Dawar and Kolaitis defined classes of graphs called *wide*, *almost wide* and *quasi-wide* (cf. [109] for instance). It has been proved in [40] that the extension preservation theorem holds in any class $\mathcal{C}$ that is wide, hereditary (i.e. closed under taking substructures) and closed under disjoint unions, that is hereditary classes with bounded degree which are closed under disjoint unions. Also, it has been proved in [41, 42] that the homomorphism preservation theorem holds in any class $\mathcal{C}$ that is almost wide, hereditary and closed under disjoint unions. Almost wide classes of graphs include classes of graphs which exclude a minor [289].

Dawar and Malod announced that the homomorphism preservation theorem holds in any hereditary quasi-wide class that is closed under disjoint unions, see [110].

Theorem 10.9. *Let $\mathcal{C}$ be a class of structures such that* Gaifman($\mathcal{C}$) *is a hereditary addable quasi-wide class of graphs. Then the homomorphism preservation theorem holds for $\mathcal{C}$.*

The proof of this result is a strengthening of the result in [41] for minor closed classes and proceeds via Gaifman locality theorem (which we already encountered in Sect. 10.5), and it is very different from the previous proofs of this chapter.

Recall that we proved in Chap. 8 (Theorem 8.2) that a hereditary class of graphs is quasi-wide if and only if it is nowhere dense.

As a corollary we get:

Corollary 10.5. *Let $\mathcal{C}$ be a class of structures such that* Gaifman($\mathcal{C}$) *is a hereditary addable nowhere dense class of graphs.*

Let $\mathbf{H}$ *be a structure. Then, the* $\mathbf{H}$*-coloring is first-order definable on* $\mathcal{C}$ *if and only if there is a finite family* $\mathcal{F}$ *of structures such that* $(\mathcal{F}, \mathbf{H})$ *is a restricted homomorphism duality for* $\mathcal{C}$.

We can also use interpretation to transport relativizations of the homomorphism preservation theorem. We refer the reader to Sect. 3.8.5 for definition and basic properties of interpretations. This will cover the relativization to somewhere dense classes.

Let $\mathcal{L}$ and $\mathcal{L}'$ be two languages and let T be a theory in $\mathcal{L}$. Let I be an interpretation of $\mathcal{L}'$ in $\mathcal{L}$. The interpretation I is a *functorial interpretation* if, for every models $\mathbf{A}$ and $\mathbf{B}$ of T such that $\mathbf{A} \to \mathbf{B}$ it holds $\mathrm{I}(\mathbf{A}) \to \mathrm{I}(\mathbf{B})$.

Lemma 10.5. *Let $\mathcal{C}$ be a hereditary class of σ-structures on which the homomorphism preservation theorem holds and let I be a functorial interpretation mapping $\mathcal{C}$ to a class $\mathcal{C}'$ of σ'-structures.*

Then the homomorphism preservation theorem holds for $\mathcal{C}'$.

Proof. Let Φ be a first-order formula of the language $\mathcal{L}(\sigma')$ preserved under homomorphisms on $\mathcal{C}'$, that is such that:

$$\forall \mathbf{A}', \mathbf{B}' \in \mathcal{C}' : \qquad (\mathbf{A}' \models \Phi) \wedge (\mathbf{A}' \to \mathbf{B}') \quad \Longrightarrow \quad (\mathbf{B}' \models \Phi).$$

According to Lemma 3.3, there exists a closed first order formula Ψ of $\mathcal{L}(\sigma)$ such that for every $\mathbf{A} \in \mathcal{C}$ we have

$$\mathbf{A} \models \Psi \quad \iff \quad \mathrm{I}(\mathbf{A}) \models \Phi.$$

Let $\mathbf{A}, \mathbf{B} \in \mathcal{C}$. Assume $\mathbf{A} \models \mathbf{B}$ and $\mathbf{A} \models \Psi$. Then $\mathrm{I}(\mathbf{A}) \to \mathrm{I}(\mathbf{B})$ (as I is a functorial interpretation) and $\mathrm{I}(\mathbf{A}) \models \Phi$. It follows that Ψ is preserved under homomorphism on $\mathcal{C}$.

As the homomorphism preservation theorem holds for $\mathcal{C}$, there exists an existential first-order property which is equivalent to Ψ for structures in $\mathcal{C}$. In other words, there is a finite family $\mathcal{F}$ of σ-structures such that

$$\forall \mathbf{A} \in \mathcal{C}: \qquad (\mathbf{A} \models \Psi) \quad \iff \quad (\exists \mathbf{F} \in \mathcal{F}:\ \mathbf{F} \to \mathbf{A}).$$

Moreover, as $\mathcal{C}$ is hereditary, we can assume $\mathcal{F} \subseteq \mathcal{C}$ (by considering all the possible images of the structures in $\mathcal{F}$). Hence every $\mathbf{F} \in \mathcal{C}$ is such that $\mathbf{F} \models \Psi$ thus $\mathrm{I}(\mathbf{F}) \models \Phi$.

Let $\mathbf{A}'$ in $\mathcal{C}'$. If there exists $\mathbf{F} \in \mathcal{F}$ such that $\mathrm{I}(\mathbf{F}) \to \mathbf{A}'$ then $\mathbf{A}' \models \Phi$ (as $\mathrm{I}(\mathbf{F}) \models \Phi$ and Φ is preserved under homomorphisms on $\mathcal{C}$). Conversely, assume $\mathbf{A}' \models \Phi$. By assumption, there exists a structure $\mathbf{A} \in \mathcal{C}$ such that $\mathbf{A}' = \mathrm{I}(\mathbf{A})$. Then $\mathrm{I}(\mathbf{A}) \models \Phi$ thus $\mathbf{A} \models \Psi$. It follows that there exists $\mathbf{F} \in \mathcal{F}$ such that $\mathbf{F} \to \mathbf{A}$ hence $\mathrm{I}(\mathbf{F}) \to \mathbf{A}'$.

Altogether, Φ is equivalent on $\mathcal{C}'$ with an existential first order property. Precisely:

$$\forall \mathbf{A}' \in \mathcal{C}': \qquad (\mathbf{A}' \models \Phi) \quad \iff \quad (\exists \mathbf{F} \in \mathcal{F}:\ \mathrm{I}(\mathbf{F}) \to \mathbf{A}').$$

□

For instance, for a class of graphs $\mathcal{C}$ and an integer p, denote by $\mathrm{Sub}_p(\mathcal{C})$ the class of the exact p-subdivisions of the graphs in $\mathcal{C}$.

Lemma 10.6. *For every class of graphs $\mathcal{C}$ and every integer p there exists a functorial interpretation*

$$\mathrm{I} : \mathcal{C} \to \mathrm{Sub}_{2p}(\mathcal{C}).$$

Proof. Let τ be the signature with a unique binary relation symbol R (meaning adjacency) and let T be the first order theory of undirected graphs.

Define the $\mathcal{L}(\tau)$-formulas $(1 \leq i \leq p)$:

$$\zeta_0(x_0,\dots,x_p) = \bigwedge_{j=1}^{p}(x_0 \simeq x_i)$$

$$\zeta_i(x_0,\dots,x_p) = R(x_0,x_p) \wedge \bigwedge_{j=1}^{p-i}(x_0 \simeq x_j) \wedge \bigwedge_{j=p-i+1}^{p-1}(x_j \simeq x_p)$$

The following tuple (p, U, E, F_R) defines the requested functorial interpretation I.

$$U[v_0,\dots,v_p] = \bigvee_{i=0}^{p} \zeta_i(v_0,\dots,v_p);$$

$$E[u_0,\dots,u_p,v_0,\dots,v_p] = \bigvee_{i=0}^{p}(u_i \simeq v_i);$$

$$\begin{aligned}F_R[u_0,\dots,u_p,v_0,\dots,v_p] &= \zeta_0(u_0,\dots,u_p) \wedge \zeta_1(v_0,\dots,v_p) \wedge (u_0 \simeq v_0)\\ &\vee \bigvee_{i=1}^{p-1} \zeta_i(u_0,\dots,u_p) \wedge \zeta_{i+1}(v_0,\dots,v_p) \wedge (u_0 \simeq v_0) \wedge (u_p \simeq v_p)\\ &\vee \zeta_p(u_0,\dots,u_p) \wedge \zeta_p(v_0,\dots,v_p) \wedge (u_0 \simeq v_p) \wedge (u_p \simeq v_0)\\ &\vee \bigvee_{i=1}^{p-1} \zeta_{i+1}(u_0,\dots,u_p) \wedge \zeta_i(v_0,\dots,v_p) \wedge (u_0 \simeq v_0) \wedge (u_p \simeq v_p)\\ &\vee \zeta_1(u_0,\dots,u_p) \wedge \zeta_0(v_0,\dots,v_p) \wedge (u_0 \simeq v_0)\end{aligned}$$

□

For a positive integer p and a graph G denote by $\mathrm{Sub}_{2p}(G)$ the graph obtained from G by subdividing every edge 2p times.

Corollary 10.6. *If the homomorphism preservation theorem holds for a hereditary class of graphs* $\mathcal{C}$, *it also holds for the class* $\mathrm{Sub}_p(\mathcal{C})$ *of all* p*-subdivisions of the graphs in* $\mathcal{C}$.

We deduce this extension of Theorem 9.1 to the class of p-subdivided graphs:

Corollary 10.7. *For every integer* p*, the homomorphism preservation theorem holds for* $\mathrm{Sub}_p(\mathcal{G}raph)$.

We will make use of this relativization later on.

Combining the above results wee see that the homomorphism preservation theorem holds for all (hereditray addable) nowhere dense classes and that every monotone somewhere dense class contains a class $\mathrm{Sub}_{\mathrm{p}}(\mathcal{G}raph)$, for which the homomorphism preservation holds.

Exercises

10.1. Prove that Łoś-Tarski theorem—which asserts that a first-order formula is preserved under extensions on all structures if, and only if, it is logically equivalent to an existential formula—does not holds when relativized to finite structures.

We follow the proof of Gurevich [239]: We consider a relational structure with two binary relations $<$ and S (plus "equality" which we always assume) and with two unary relations Min and Max.

- Prove that there exists a universal first-order formula α such that $\mathbf{A} \models \phi$ if
    1. $\mathbf{A}$ contains a unique element m such that $\mathbf{A} \models \mathrm{Min}(m)$;
    2. $\mathbf{A}$ contains a unique element M such that $\mathbf{A} \models \mathrm{Max}(M)$;
    3. $<$ is a linear order on $\mathbf{A}$ with minimum m and maximum M;
    4. S is compatible with the successor relation of $<$, that is: if xSy then y is the successor of x in $<$.
- Let β be a sentence expressing that for each $x < M$ there exists y with xSy, and let ϕ be the sentence $\alpha \rightarrow \beta$. Prove that if $\mathbf{A}$ is a substructure of a finite structure $\mathbf{B}$, $\mathbf{A} \models \phi$, and $\mathbf{B} \models \alpha$ then $\mathbf{A} = \mathbf{B}$;
- Deduce that ϕ is preserved under finite extensions;
- Suppose for contradiction that ϕ is equivalent to an existential sentence

$$\psi = \exists x_1 \ \ldots \ \exists x_k \ \Psi(x_1, \ldots, x_k)$$

where Ψ is quantifier free. Let $\mathbf{A}$ be a model of ϕ formed from the set of the first $k+3$ integers with natural linear order and successor relation. Fix witnesses $a_1, \ldots, a_k$ for Ψ and choose a non-initial and non-final element b different from all a_i. Prove that the structure $\mathbf{A}'$ obtained from $\mathbf{A}$ by discarding the relation $bS(b+1)$ satisfies Ψ but fails to satisfy ϕ.

10.2. The aim of this exercise is to give an example of a sentence ϕ with no finite model.

Define a first order sentence ϕ with quantifier rank 5 expressing that a graph has all its vertices of degree 2, but one which is of degree 1;
Check that the *ray*, which is the countable graph with vertex set $\mathbb{N}$ and edge set $\{\{i, i+1\},\ i \in \mathbb{N}\}$, is a model for ϕ and show that ϕ has other infinite models (and find infinitely many countable ones);
Prove that ϕ has no finite model.

10.3. Give an example of a consistent theory T without a finite model, but such that every finite subset T' of T has a finite model. Let $\mathcal{F}$ be the class of all triangle free 3-colorable graphs. Define the theory $T = \{\vartheta(G),\ G \in \mathcal{F}\} \cup \{\neg\vartheta K_3\}$.

Prove that T has no finite model (hint: use Theorem 9.6);
Prove that every finite subset of T has a model and deduce that T is consistent and thus has a model;
Find two countable models of T, including an ω-categorical one.

10.4. Show that $\mathfrak{P}_F$ (which is the trace on P of complete theories with a finite model) is different from the class of all closed PP-theories with a finite model. (Of course, every theory in $\mathfrak{P}_F$ is a closed PP-theory and has a finite model. But we show that the converse inclusion does not hold.)

Let $T_0 = \{\vartheta(F) : F \underset{\not\leftarrow}{\rightarrow} K_3\}$ and let T be the trace on P of the deductive closure of T_0, that is: $T = T_0^{\vdash} \cap P$. Prove that T has a finite model.
Using Exercise 10.3 prove that T has a triangle free infinite model hence $\vartheta(K_3) \notin T$.
Prove that every finite model of T contains a triangle hence satisfies $\vartheta(K_3)$.
Deduce that $T \notin \mathfrak{P}_F$.

Chapter 11
Restricted Homomorphism Dualities

From a restricted point of view, we gain in generality.

11.1 Introduction

In the case where the input structures are restricted to some class $\mathcal{C}$, some more "restricted" homomorphism dualities may appear. For instance, for every planar graphs we have [337] (where the graph on the right side is the *Clebsch graph*):

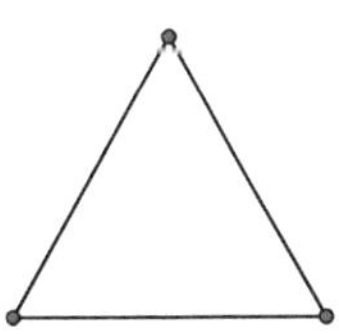

$$\nrightarrow \quad \mathsf{G} \quad \Longleftrightarrow \quad \mathsf{G} \quad \longrightarrow$$

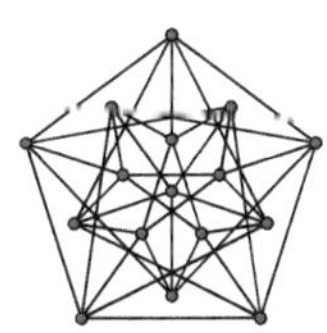

(and actually the same duality holds for K_5 minor-free graphs [338]).

This and other examples motivate the following: A *finite restricted homomorphism duality* for a class $\mathcal{C}$ (or a $\mathcal{C}$-*restricted duality*) is a pair $(\mathcal{F}, \mathcal{D})$ such that:

$$\forall \mathbf{G} \in \mathcal{C}: \qquad (\exists \mathbf{F} \in \mathcal{F} \quad \mathbf{F} \to \mathbf{G}) \quad \Longleftrightarrow \quad (\forall \mathbf{D} \in \mathcal{D} \quad \mathbf{G} \nrightarrow \mathbf{D}) \tag{11.1}$$

and we also add the requirement

$$\forall \mathbf{F} \in \mathcal{F},\ \forall \mathbf{D} \in \mathcal{D} \quad \mathbf{F} \nrightarrow \mathbf{D}. \tag{11.2}$$

J. Nešetřil and P.O. de Mendez, *Sparsity*, Algorithms and Combinatorics 28,
DOI 10.1007/978-3-642-27875-4_11, © Springer-Verlag Berlin Heidelberg 2012

In the case of finite restricted homomorphism dualities where both $\mathcal{F}$ and $\mathcal{D}$ are singletons, we will speak about *restricted dualities* for the sake of simplicity.

We say that a $\mathbf{D}$-coloring problem is *first-order definable on* $\mathcal{C}$ if there exists a first-order formula ϕ such that

$$\forall \mathbf{G} \in \mathcal{C}: \qquad \mathbf{G} \to \mathbf{D} \qquad \Longleftrightarrow \qquad \mathbf{G} \models \phi.$$

Obviously, if $(\mathcal{F}, \mathbf{D})$ is a finite restricted duality for a class $\mathcal{C}$, then $\mathbf{D}$-coloring is first-order definable on $\mathcal{C}$. It is a natural question whether the equivalence stated in Corollary 9.1 for finite structures could also apply in the restricted case. To address this question, we consider relativizations of the homomorphism preservation theorems to classes of structures.

11.2 Classes with All Restricted Dualities

We have seen that restricted dualities are abundant. As the extremal case we define the concept of "all restricted dualities": A class of σ-structures $\mathbf{A}$ has *all restricted dualities* if every connected σ-structure has a restricted dual for $\mathcal{C}$, that is: for every connected σ-structure $\mathbf{F}$ there exists a σ-structure $\mathbf{D}$ such that

$$\forall \mathbf{A} \in \mathcal{C}: \qquad (\mathbf{F} \to \mathbf{A}) \quad \Longleftrightarrow \quad (\mathbf{A} \nrightarrow \mathbf{D})$$

11.3 Characterization of Classes with All Restricted Dualities by Distances

The aim of this section is to give several characterizations of classes of structures having all restricted dualities. A bit surprisingly, the answer can be given in terms of distances: For a structure $\mathbf{A}$ and a real $\epsilon > 0$, define $\phi_L^\epsilon(\mathbf{A})$ as a structure of minimum order such that $\mathbf{A} \to \phi_L^\epsilon(\mathbf{A})$ and $\text{dist}_L([\mathbf{A}], [\phi_L^\epsilon(\mathbf{A})]) \leq \epsilon$ (we arbitrarily choose between those structures which have these properties, by using, for instance, some arbitrary linear order on $\text{Rel}(\sigma)$).

Lemma 11.1. *Let $\mathcal{C}$ be a bounded class of σ-structures. Then $\mathcal{C}$ has all restricted dualities if and only if for every $\epsilon > 0$ we have $\sup_{\mathbf{A} \in \mathcal{C}} |\phi_L^\epsilon(\mathbf{A})| < \infty$.*

Proof. Assume $\mathcal{C}$ has all restricted dualities, let $\epsilon > 0$ be a positive real and let $t \geq -\log_2 \epsilon$ be an integer. For a structure $\mathbf{A} \in \mathcal{C}$, let $\mathcal{F}_t(\mathbf{A})$ be the set of all connected cores $\mathbf{T}$ of order at most t such that $\mathbf{T} \nrightarrow \mathbf{A}$. If t is greater than the order of a bound of $\mathcal{C}$ then this set is not empty. For $\mathbf{T} \in \mathcal{F}_t(\mathbf{A})$, let $\mathbf{D_T}$ be the dual of $\mathbf{T}$ relative to $\mathcal{C}$ and let $\mathbf{A}'$ be the product of all the $\mathbf{D_T}$ for $\mathbf{T} \in \mathcal{F}_t(\mathbf{A})$. First notice that for every $\mathbf{T} \in \mathcal{F}_t(\mathbf{A})$ we have $\mathbf{T} \nrightarrow \mathbf{A}$ hence $\mathbf{A} \rightarrow \mathbf{D_T}$. It follows that $\mathbf{A} \rightarrow \mathbf{A}'$. Let $\mathbf{T}'$ be a connected structure of order at most t. Then $\mathbf{T}' \rightarrow \mathbf{A}$ implies $\mathbf{T}' \rightarrow \mathbf{A}'$ (as $\mathbf{A} \rightarrow \mathbf{A}'$) and $\mathbf{T}' \nrightarrow \mathbf{A}$ implies $\mathrm{Core}(\mathbf{T}') \in \mathcal{F}_t(\mathbf{A})$ hence $\mathbf{A}' \rightarrow \mathbf{D_{T'}}$ thus $\mathbf{T}' \nrightarrow \mathbf{A}'$ (as for otherwise $\mathbf{T}' \rightarrow \mathbf{D_{T'}}$). Thus $\mathrm{dist}_L([\mathbf{A}], [\mathbf{A}']) \leq \epsilon$ and $|\phi_L^\epsilon(\mathbf{A})| \leq |\mathbf{A}'| \leq C_\epsilon$ for some suitable finite constant C independent of $\mathbf{A}$ (for instance, one can choose C_ϵ to be the product of the orders of all the duals relative to $\mathcal{C}$ of connected cores of order at most t).

Conversely, assume that we have $\sup_{\mathbf{A}\in\mathcal{C}} |\phi_L^\epsilon(\mathbf{A})| < \infty$ for every $\epsilon > 0$ and let $\mathbf{F}$ be a connected σ-structure. Let $t \geq |\mathbf{F}|$, let $\epsilon = 2^{-t}$, let $\mathcal{D}$ be the set of all the $\phi_L^\epsilon(\mathbf{A})$ for $\mathbf{A} \in \mathcal{C}$ and $\mathbf{F} \nrightarrow \mathbf{A}$. As all the structures $\phi_L^\epsilon(\mathbf{A})$ have an order bounded by some constant C_ϵ, the set $\mathcal{D}$ is finite. Let $D_t(\mathbf{F})$ be the disjoint union of all the graphs in $\mathcal{D}$. First notice that $\mathbf{F} \nrightarrow D_t(\mathbf{F})$ as for otherwise $\mathbf{F}$ would have a homomorphism to some structure in $\mathcal{D}$ (as $\mathbf{F}$ is connected), that is to some $\phi_L^\epsilon(\mathbf{B})$ for a structure $\mathbf{B}$ such that $\mathbf{F} \nrightarrow \mathbf{B}$, what would contradict $\mathrm{dist}_L([\phi_L^\epsilon(\mathbf{B})], [\mathbf{B}]) \leq 2^{-|\mathbf{F}|}$. Also, if $\mathbf{F} \rightarrow \mathbf{A}$ then $\mathbf{A} \nrightarrow D_t(\mathbf{F})$ (for otherwise $\mathbf{F} \rightarrow D_t(\mathbf{F})$) and if $\mathbf{F} \nrightarrow \mathbf{A}$ then $\phi_L^\epsilon(\mathbf{A}) \in \mathcal{D}$ thus $\mathbf{A} \rightarrow D_t(\mathbf{F})$. Altogether, $D_t(\mathbf{F})$ is a dual of $\mathbf{F}$ relative to $\mathcal{C}$. □

To refine this result to the full distance, we make use of the following:

Lemma 11.2. *Assume* $\mathrm{dist}_L([\mathbf{A}], [\mathbf{B}]) < 2^{-t}$ *and* $\mathbf{A} \rightarrow \mathbf{B}$. *Let*

$$\mathbf{B}' = \mathbf{B} \times \prod\{\mathbf{T} : |\mathbf{T}| \leq t, \ \mathbf{T} \textit{ is a core and } \mathbf{A} \rightarrow \mathbf{T}\}.$$

Then $\mathrm{dist}([\mathbf{A}], [\mathbf{B}']) < 2^{-t}, \mathbf{A} \rightarrow \mathbf{B}'$ *and* $|\mathbf{B}'| \leq F(t)|\mathbf{B}|$.

Proof. First notice that $\mathbf{A} \rightarrow \mathbf{B}'$ has $\mathbf{A}$ has a homomorphism to each structure of the product defining $\mathbf{B}'$. Let $\mathbf{T}$ be a structure such that $|\mathbf{T}| \leq t$. Then

$$\begin{aligned}
&\mathbf{B}' \rightarrow \mathbf{T} \Longrightarrow \mathbf{A} \rightarrow \mathbf{T} && \text{(as } \mathbf{A} \rightarrow \mathbf{B}'\text{)}\\
&\mathbf{B} \rightarrow \mathbf{T} \Longrightarrow \mathbf{B}' \rightarrow \mathbf{T} && \text{(by construction)}\\
&\mathbf{T} \rightarrow \mathbf{B}' \Longrightarrow \mathbf{T} \rightarrow \mathbf{B} \Longrightarrow \mathbf{T} \rightarrow \mathbf{A} &&\\
&\mathbf{T} \rightarrow \mathbf{A} \Longrightarrow \mathbf{T} \rightarrow \mathbf{B}' && \text{(as } \mathbf{A} \rightarrow \mathbf{B}'\text{)}
\end{aligned}$$

□

For a graph $\mathbf{A}$ and a real $\epsilon > 0$, define $\phi^\epsilon(\mathbf{A})$ as a minimum order graph such that $\mathbf{A} \rightarrow \phi^\epsilon(\mathbf{A})$ and $\mathrm{dist}([\mathbf{A}], [\phi^\epsilon(\mathbf{A})]) \leq \epsilon$ (we arbitrarily choose between those graphs which have these properties, by using, for instance, some arbitrary linear order on $\mathrm{Rel}(\sigma)$).

Lemma 11.3. *There exists a function* $F : \mathbb{R} \to \mathbb{R}$ *such that for every* ϵ *and every structure* $\mathbf{A}$*,*

$$|\phi_L^\epsilon(\mathbf{A})| \leq |\phi^\epsilon(\mathbf{A})| \leq F(\epsilon)|\phi_L^\epsilon(\mathbf{A})|.$$

Proof. Obviously $|\phi^\epsilon(\mathbf{A})| \geq |\phi_L^\epsilon(\mathbf{A})|$. Let

$$\mathbf{B}(\mathbf{A}, \epsilon) = \phi_L^\epsilon(\mathbf{A}) \times \prod\{\mathbf{T} : \ |\mathbf{T}| \leq t, \ \mathbf{T} \text{ is a core and } \mathbf{A} \to \mathbf{T}\}.$$

According to Lemma 11.2, there exists a function F such that $|\mathbf{B}(\mathbf{A}, \epsilon)| \leq F(\epsilon)|\phi_L^\epsilon(\mathbf{A})|$, $\mathbf{A} \to \mathbf{B}(\mathbf{A}, \epsilon)$ and $\text{dist}([\mathbf{A}], [\mathbf{B}(\mathbf{A}, \epsilon)]) \leq \epsilon$ hence we conclude $|\phi^\epsilon(\mathbf{A})| \leq F(\epsilon)|\phi_L^\epsilon(\mathbf{A})|$. □

Theorem 11.1. *Let* $\mathcal{C}$ *be a bounded class of* σ*-structures. Then the following conditions are equivalent:*

1. *The* $\mathcal{C}$ *has all restricted dualities;*
2. *For every* $\epsilon > 0$ *we have* $\sup_{\mathbf{A} \in \mathcal{C}} |\phi_L^\epsilon(\mathbf{A})| < \infty$.
3. *For every* $\epsilon > 0$ *we have* $\sup_{\mathbf{A} \in \mathcal{C}} |\phi^\epsilon(\mathbf{A})| < \infty$.

Proof. The proof follows from Lemma 11.1 and 11.3. □

Note that this Theorem (and even Lemma 11.1 alone) allows to prove that bounded expansion classes have all restricted dualities (see Exercise 11.3).

11.4 Characterization of Classes with All Restricted Dualities by Local Homomorphisms

Definition 11.1. Let $\mathbf{A}, \mathbf{B}$ be structures and let $\mathcal{P}$ be a system of subsets of the universe A of $\mathbf{A}$. We say that $\mathbf{A}$ is $\mathcal{P}$-*locally homomorphic* to $\mathbf{B}$ and denoted by $\mathbf{A} \overset{\mathcal{P}}{\dashrightarrow} \mathbf{B}$ if for every subset $X \in \mathcal{P}$:

$$\mathbf{A}[X] \longrightarrow \mathbf{B}.$$

We shall deal mostly with systems of subsets of the following type: For a set of colors Γ, a coloring function $\gamma : A \longrightarrow \Gamma$ and a positive integer p, let $\mathcal{P}$ be the system $\{X \subset A : |\gamma(X)| \leq p\}$ (here we put $\gamma(X) = \{\gamma(x) : x \in X\}$). This system will be denoted by $\mathcal{P}_{(\gamma,p)}$. In this case we also say that $\mathbf{A}$ is (γ, p)-locally homomorphic to $\mathbf{B}$ (instead of $\mathcal{P}_{(\gamma,p)}$-locally homomorphic). This will be denoted by $\mathbf{A} \overset{(\gamma,p)}{\dashrightarrow} \mathbf{B}$. Of course, $\overset{(\gamma,p)}{\dashrightarrow}$ is not indicating the existence of a homomorphism.

Example 11.1. For $H = K_2$, a graph G and an injective coloring γ of the vertices of G, the graph G is (γ, p)-locally homomorphic to H if and only if the odd-girth of G is $> p$.

The following "truncated power construction" is a modification of a construction introduced in [349]:

Definition 11.2. Let $\mathbf{A}, \mathbf{B}$ be σ-structures with universes A and B and relations R_i of arity r_i ($i = 1, \ldots, m$). Let $1 \le p < |B|$ be an integer. For $v \in B$, define $\mathcal{I}_v = \{I : I \in \binom{B}{p} \text{ and } v \in I\}$, where $\binom{B}{p}$ stands for the subsets of B with cardinality p. For $v \in B$ denote by V_v the set of all mappings $\mathcal{I}_v \longrightarrow A$. We can write $V_v = A^{\mathcal{I}_v}$ and define the sets $W = \bigcup_{v \in B} V_v$ and the mapping $\alpha : W \to B$ by $\alpha(z) = v$ if $z \in V_v$.

The p-*truncated* $\mathbf{B}$-*power* $\mathbf{A}^{\Uparrow^{\mathbf{B}}_p}$ of $\mathbf{A}$ is the σ-structure whose universe is W and for $i = 1, \ldots, m$, $R_i(\mathbf{A}^{\Uparrow^{\mathbf{B}}_p})$ is the set of all r_i-tuples $(z_1, \ldots, z_{r_i})$ such that $(\alpha(z_1), \ldots, \alpha(z_{r_i})) \in R_i(\mathbf{B})$ and for every $I \in \bigcap_{j=1}^{m} \mathcal{I}_{\alpha(z)}$ holds $(z_1(I), \ldots, z_{r_i}(I)) \in R_i(\mathbf{A})$. The mapping α is called the *color projection* of $\mathbf{A}^{\Uparrow^{\mathbf{B}}_p}$.

Remark 11.1. Note that the universe of $\mathbf{A}^{\Uparrow^{\mathbf{B}}_p}$ has cardinality $|B| \cdot |A|^{\binom{|B|-1}{p-1}}$.

The use of the name "color projection" for the mapping α is justified by the following:

Lemma 11.4. *The color projection α of $\mathbf{A}^{\Uparrow^{\mathbf{B}}_p}$ is a homomorphism from $\mathbf{A}^{\Uparrow^{\mathbf{B}}_p}$ to $\mathbf{B}$:*

$$\alpha : \mathbf{A}^{\Uparrow^{\mathbf{B}}_p} \longrightarrow \mathbf{B}.$$

Proof. By definition we have

$$(z_1, \ldots, z_{r_i}) \in R_i(\mathbf{A}^{\Uparrow^{\mathbf{B}}_p}) \qquad \Longrightarrow \qquad (\alpha(z_1), \ldots, \alpha(z_{r_i})) \in R_i(\mathbf{B}).$$

□

Thus $\mathbf{A}^{\Uparrow^{\mathbf{B}}_p}$ is homomorphic to $\mathbf{B}$ but it is also locally homomorphic to $\mathbf{A}$:

Lemma 11.5. *Let α be the color projection of $\mathbf{A}^{\Uparrow^{\mathbf{B}}_p}$. Then the graph $\mathbf{A}^{\Uparrow^{\mathbf{B}}_p}$ is (α, p)-locally homomorphic to $\mathbf{A}$:*

$$\mathbf{A}^{\Uparrow^{\mathbf{B}}_p} \overset{(\alpha,p)}{\dashrightarrow} \mathbf{A}.$$

Proof. Let X be a subset of the universe W of $\mathbf{A}^{\Uparrow^{\mathbf{B}}_p}$ such that $|\alpha(X)| \le p$. Let I be some subset of B of cardinality p such that $\alpha(X) \subseteq I$. According to the definition of $\mathbf{A}^{\Uparrow^{\mathbf{B}}_{p}}$, for $i = 1, \dots, m$ we have $(z_1(I), \dots, z_{r_i}(I)) \in R_i(\mathbf{A})$ for every $(z_1, \dots, z_{r_i}) \in R_i(\mathbf{A}^{\Uparrow^{\mathbf{B}}_{p}}[X])$. It follows that the mapping $z \mapsto z(I)$ is a homomorphism of $\mathbf{A}^{\Uparrow^{\mathbf{B}}_{p}}[X]$ to $\mathbf{A}$. □

The following two propositions summarize the universality properties of $\mathbf{A}^{\Uparrow^{\mathbf{B}}_{p}}$ which will be needed in the sequel:

Lemma 11.6. *Let* $\mathbf{A}, \mathbf{B}, \mathbf{M}$ *be* σ*-structures with universes* A, B, M *and relations* R_i *of arity* r_i $(i = 1, \dots, m)$. *Let* $p \geq \max_i r_i$ *be an integer. Let* $\gamma : \mathbf{A} \longrightarrow \mathbf{B}$ *be a homomorphism and let* $\mathbf{A}$ *be* (γ, p)*-locally homomorphic to* $\mathbf{M}$. *Let* W *be the universe of* $\mathbf{M}^{\Uparrow^{\mathbf{B}}_{p}}$. *For* $I \in \binom{B}{p}$, *let* g_I *be a homomorphism of* $\mathbf{A}[\gamma^{-1}(I)]$ *to* $\mathbf{M}$ *and define* $h_I : W \to M$ *by* $h_I(z) = z(I)$.

Then there exists a homomorphism $f : \mathbf{A} \longrightarrow \mathbf{M}^{\Uparrow^{\mathbf{B}}_{p}}$ *such that* $\gamma = \alpha \circ f$ *and such that* $g_I = h_I \circ f_{|\gamma^{-1}(I)}$, *for any* $I \in \binom{B}{p}$ *(see Fig. 11.1). It follows that* $\mathbf{M}^{\Uparrow^{\mathbf{B}}_{p}}$ *is* (α, p)*-locally homomorphic to* $\mathbf{M}$ *(*α *is the color projection).*

Proof. For $I \in \binom{B}{p}$ put $\mathbf{A}_I = \mathbf{A}[\gamma^{-1}(I)]$. Then g_I is a homomorphism of $\mathbf{A}_I$ to $\mathbf{M}$. Define f as follows: Given $x \in A$ we define $f(x) \in V_{\gamma(x)}$ by the following formula $f(x)(I) = g_I(x)$ (see the above definition of $\mathbf{M}^{\Uparrow^{\mathbf{B}}_{p}}$). Obviously $\alpha \circ f = \gamma$ and $g_I = h_I \circ f_{|\gamma^{-1}(I)}$. We prove that f is a homomorphism.

Let $1 \leq i \leq m$ and let $(x_1, \dots, x_{r_i}) \in R_i(\mathbf{A})$. Then $\{\gamma(x_1), \dots, \gamma(x_{r_i})\} \in R_i(\mathbf{B})$ as $\gamma : \mathbf{A} \longrightarrow \mathbf{B}$. For every I that includes $\{\gamma(x_1) \dots, \gamma(x_{r_i})\}$ holds $(f(x_1)(I), \dots, f(x_{r_i})(I)) = (g_I(x_1), \dots, g_I(x_{r_i})) \in R_i(\mathbf{M})$. It follows that $(f(x_1), \dots, f(x_{r_i})) \in R_i(\mathbf{M}^{\Uparrow^{\mathbf{B}}_{p}})$ and thus f is a homomorphism. □

This lemma highlights a fundamental property of $\mathbf{A}^{\Uparrow^{\mathbf{B}}_{p}}$ which we will state as follows:

Lemma 11.7. *Let* $\mathbf{A}, \mathbf{B}, \mathbf{M}$ *be* σ*-structures with universes* A, B, M *and relations* R_i *of arity* r_i $(i = 1, \dots, m)$. *Let* $p \geq \max_i r_i$ *be an integer.*

Then there is a homomorphism $\mathbf{A} \longrightarrow \mathbf{M}^{\Uparrow^{\mathbf{B}}_{p}}$ *if and only if there exists a homomorphism* $\gamma : \mathbf{A} \longrightarrow \mathbf{B}$ *such that* $\mathbf{A}$ *is* (γ, p)*-locally homomorphic to* $\mathbf{M}$. *Schematically, this may be depicted as follows:*

$$\mathbf{A} \longrightarrow \mathbf{M}^{\Uparrow_{\mathrm{p}}^{\mathbf{B}}} \quad \Longleftrightarrow \quad \begin{array}{ccc} \mathbf{A} & \xrightarrow{\gamma} & \mathbf{B} \\ {\scriptstyle (\gamma,\mathrm{p})}\,\vdots & & \\ \downarrow & & \\ \mathbf{M} & & \end{array}$$

Proof. First, suppose that there is a homomorphism $f : \mathbf{A} \longrightarrow \mathbf{M}^{\Uparrow_{\mathrm{p}}^{\mathbf{B}}}$. Let α be the color projection of $M^{\Uparrow_{\mathrm{p}}^{\mathbf{B}}}$ to $\mathbf{B}$. Put $\gamma = \alpha \circ f$. We have $\gamma : \mathbf{A} \longrightarrow \mathbf{B}$. Let $X \subseteq A$. The condition $|\gamma(X)| \leq p$ is equivalent to the condition $|\alpha(f(X))| \leq p$. Hence the homomorphism $f : \mathbf{A} \longrightarrow \mathbf{M}^{\Uparrow_{\mathrm{p}}^{\mathbf{B}}}$ together with the (α, p)-local homomorphism of $\mathbf{M}^{\Uparrow_{\mathrm{p}}^{\mathbf{B}}}$ to $\mathbf{M}$ implies (γ, p)-local homomorphism from $\mathbf{A}$ to $\mathbf{M}$. The reverse implications follows from Lemma 11.6. □

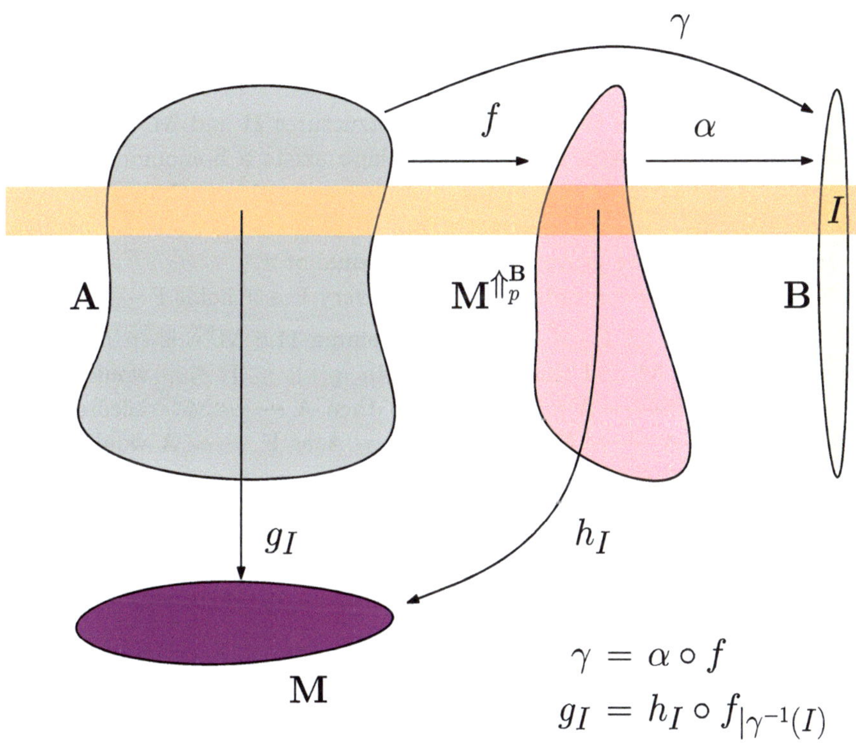

Fig. 11.1 Homorphisms composition

It is interesting to note that if we consider $\mathbf{M} = \mathbf{A}$ we get:

Corollary 11.1. $\mathbf{A} \longrightarrow \mathbf{A}^{\Uparrow_{\mathrm{p}}^{\mathbf{B}}} \iff \mathbf{A} \longrightarrow \mathbf{B}$. *In particular,* $\mathbf{A}$ *is homomorphism-equivalent to* $\mathbf{A}^{\Uparrow_{\mathrm{p}}^{\mathbf{A}}}$.

We now arrive to the following (implicit) characterization of classes with all restricted dualities.

Theorem 11.2. *A class of σ-structures $\mathcal{C}$ has all restricted dualities if and only if for every finite set $\mathcal{F}$ of connected σ-structures there exist σ-structures $\mathbf{B}$ and $\mathbf{M} \in \mathrm{Forb_h}(\mathcal{F})$ such that for every $\mathbf{A} \in \mathcal{C} \cap \mathrm{Forb_h}(\mathcal{F})$ there exists a homomorphism $\gamma : \mathbf{A} \longrightarrow \mathbf{B}$ for which $\mathbf{A}$ is (γ, p)-locally homomorphic to $\mathbf{M}$, where $p = \max\{r_i, 1 \le i \le m\} \cup \{|F|; \mathbf{F} \in \mathcal{F}\}$, where $r_1, \dots, r_m$ are the arities of the relations $R_1, \dots, R_m$ in σ.*

Proof. Let $\mathcal{F}$ be a finite set of connected σ-structures.

Assume $\mathcal{C}$ has all restricted dualities. Then $\mathcal{F}$ has a dual $\mathbf{D}_{\mathcal{F}}^{\mathcal{C}}$. Put $\mathbf{M} = \mathbf{B} = \mathbf{D}_{\mathcal{F}}^{\mathcal{C}}$.

Conversely, assume that there exist σ-structures $\mathbf{B}$ and $\mathbf{M} \in \mathrm{Forb_h}(\mathcal{F})$ such that for every $\mathbf{A} \in \mathcal{C} \cap \mathrm{Forb_h}(\mathcal{F})$ there exists a homomorphism $\gamma : \mathbf{A} \longrightarrow \mathbf{B}$ for which $\mathbf{A}$ is (γ, p)-locally homomorphic to $\mathbf{M}$, where $p = \max\{|F|; \mathbf{F} \in \mathcal{F}\}$.

In this situation we prove that $\mathbf{M}^{\Uparrow_p^{\mathbf{B}}}$ is a dual of $\mathcal{F}$.

First we prove by contradiction that for every $\mathbf{F} \in \mathcal{F}$ holds $\mathbf{F} \not\longrightarrow \mathbf{M}^{\Uparrow_p^{\mathbf{B}}}$. Suppose contrary, let $\mathbf{F} \xrightarrow{g} \mathbf{M}^{\Uparrow_p^{\mathbf{B}}}$. By Lemma 11.6 $\mathbf{M}^{\Uparrow_p^{\mathbf{B}}}$ is (p,p)-locally homomorphic to $\mathbf{M}$ and this together with $|g(F)| \le |F| \le p$ would imply $\mathbf{F} \longrightarrow \mathbf{M}$. If $\mathbf{F} \not\longrightarrow \mathbf{A}$ for every $\mathbf{F} \in \mathcal{F}$ then $\mathbf{A} \longrightarrow \mathbf{M}^{\Uparrow_p^{\mathbf{B}}}$ according to Lemma 11.7. If $\mathbf{A} \longrightarrow \mathbf{M}^{\Uparrow_p^{\mathbf{B}}}$ then $\mathbf{F} \not\longrightarrow \mathbf{A}$ as $\mathbf{F} \longrightarrow \mathbf{A}$ would imply $\mathbf{F} \longrightarrow \mathbf{M}^{\Uparrow_p^{\mathbf{B}}}$. $\square$

11.5 Restricted Dualities in Bounded Expansion Classes

Low tree-depth decompositions are suitable for the restricted dualities as they give us (a rather precise) control over subgraphs (of graphs in a bounded expansion class). Classes with bounded expansion are *a priori* good candidates if we look for classes with all restricted dualities. We have chosen to consider dualities from the point of view of relational structures. In the sequel of this chapter we say that a class of structures $\mathcal{C}$ has *bounded expansion* if the class $\mathrm{Gaifman}(\mathcal{C})$ of the Gaifman graphs of the structures in $\mathcal{C}$ has bounded expansion (i.e. if $\mathcal{C}$ has G-bounded expansion, see Sect. 5.8). We could have considered the class $\mathrm{Inc}(\mathcal{C})$ of the incidence graphs of the structures in $\mathcal{C}$ as well (hence I-bounded expansion classes). These two choices are equivalent as, according to Proposition 5.7, a class of relational structures has G-bounded expansion if and only if it has I-bounded expansion.

In Sect. 6.8, we proved that there is a function $\mathcal{F}$ such that every graph G is homomorphism-equivalent to an induced subgraph of G with order at most $\mathcal{F}(1, \mathrm{td}(G))$ (Corollary 6.5), and we know that the function $\mathcal{F}$ is extremely fast growing (of order the tower function). In order to use the decomposition techniques introduced in Chap. 7 for graphs on relational structures, we first prove a generalization of Corollary 6.5 to relational structures.

Let us first recall the inductive definition of the function $T(c, t)$ given in Sect. 6.8:

$$T(c,t) = \begin{cases} c, & \text{if } t = 1, \\ \sum_{i=1}^{T(c,t-1)+1} r_c(i), & \text{otherwise.} \end{cases}$$

Then we have the following reduction result for relational structures:

Lemma 11.8. *Let σ be a signature and let $r_1, \ldots, r_m$ be the signatures of the relational symbols $R_1, \ldots, R_m$ of σ.*

Then, for any $\mathbf{A} \in \mathrm{Rel}(\sigma)$, there exists $X \subseteq A$ such that

- $|A| \leq T(2^{1+\sum_{i=1}^{m} t^{r_i}}, \mathrm{td}(\mathrm{Gaifman}(\mathbf{A})))$,
- $\mathbf{A} \rightleftarrows \mathbf{A}[X]$.

Proof. Let $\mathbf{A} \in \mathrm{Rel}(\sigma)$, let $G = \mathrm{Gaifman}(\mathbf{A})$, and let $t = \mathrm{td}(G)$. Let Y be a rooted forest such that $G \subseteq \mathrm{Clos}(Y)$. Define $\pi : V(Y) \to V(Y) \cup \{\epsilon\}$ by

$$\pi(x) = \begin{cases} \text{the father of } x \text{ in } Y & \text{if } x \text{ is not a root of } Y, \\ \epsilon & \text{otherwise.} \end{cases}$$

Let $\mathcal{F}_i$ be the set of all mappings from $[r_i]$ to $\{0, \ldots, t-1\}$. Define $c : V(Y) \to \{0,1\} \times 2^{\mathcal{F}_1} \times \cdots \times 2^{\mathcal{F}_m}$ as follows: for $v \in V(Y), c(v) = (i(v), \mathcal{R}_1(v), \ldots, \mathcal{R}_m(v))$ where

- $i(v) = 1$ if $v \in V(G)$ and $i(v) = 0$ otherwise;
- For $i = 1, \ldots, m$, $\mathcal{R}_i(v)$ is the set of the functions $f : [r_i] \to \{0, \ldots, t-1\}$ such that $(\pi^{f(1)}(v), \ldots, \pi^{f(r_i)}(v)) \in R_i(\mathbf{A})$.

Let $z = 2^{1+\sum_{i=1}^{m} t^{r_i}}$. According to Lemma 6.10 there exists a subset $X_0 \subseteq V(Y)$ of cardinality at most $T(z, t)$ such that Y has a color preserving homomorphism to $Y[X_0]$. It follows clearly that $\mathbf{A}$ has a homomorphism to $\mathbf{A}[X_0 \cap A]$. We conclude the proof by putting $X = X_0 \cap A$. □

Theorem 11.3. *Let $\mathcal{F}$ be a finite set of connected σ-structures. Then, for every class of σ-structures $\mathcal{K}$ with bounded expansion there exists a σ-structure $\mathbf{D}_{\mathcal{F}}^{\mathcal{K}} \in \mathrm{Forb_h}(\mathcal{F})$ such that every σ-structure of $\mathcal{K} \cap \mathrm{Forb_h}(\mathcal{F})$ has a homomorphism to $\mathbf{D}_{\mathcal{F}}^{\mathcal{K}}$.*

In other words, every class of structures with bounded expansion has all restricted dualities.

Proof. Let $r_1, \ldots, r_m$ be the arities of the relational symbols $R_1, \ldots, R_m$ of σ. Let $p = \max\{r_i;\ 1 \le i \le m\} \cup \{|F|,\ \mathbf{F} \in \mathcal{F}\}$.

There exists an integer N, such that every graph $G \in \mathrm{Gaifman}(\mathcal{K})$ has a proper N-coloring in which every p colors induce a graph of tree depth at most p. Color the universe of each $\mathbf{A} \in \mathcal{K}$ accordingly and call $\gamma_{\mathbf{A}}$ the coloring for $\mathbf{A}$. Let $\mathbf{B}$ be the σ-structure with universe $[N]$ such that all r_i-tuples of distinct integers in $[N]$ belong to $R_i(\mathbf{B})$. Then, $\gamma_{\mathbf{A}}$ clearly defines a homomorphism from $\mathbf{A}$ to $\mathbf{B}$.

Let $I \in \binom{[N]}{p}$ and let $\mathbf{A}_I = \mathbf{A}[\gamma_{\mathbf{A}}^{-1}(I)]$. Then $\mathrm{td}(\mathrm{Gaifman}(\mathbf{A}_I)) \le p$.

Consequently to Lemma 11.8, there exists a finite set $\hat{\mathfrak{D}}_p^{\sigma}$ of σ-structures with Gaifman graphs of tree depth at most p, so that every σ-structures with Gaifman graph of tree-depth at most p is hom-equivalent to one structure in this set. Let $\mathbf{M}(\hat{\mathfrak{D}}_p^{\sigma}, \mathcal{F})$ be the disjoint union of the structures in $\hat{\mathfrak{D}}_p^{\sigma} \cap \mathrm{Forb_h}(\mathcal{F})$. In this situation we can use Theorem 11.2 and put $\mathbf{D}_{\mathcal{F}}^{\mathcal{K}} = \mathbf{M}(\hat{\mathfrak{D}}_p^{\sigma}, \mathcal{F})^{\Uparrow_p^{\mathbf{B}}}$. □

An alternative proof of this result (for graphs) using the characterization stated in Lemma 11.1 maybe found in Exercise 11.3.

11.6 Characterization of Classes with All Restricted Dualities by Reorientations

Let $\mathbf{A}, \mathbf{B} \in \mathrm{Rel}(\sigma)$. The structure $\mathbf{B}$ is a *weak reorientation* of $\mathbf{A}$ is

- $\forall (x_1, \ldots, x_{r_i}) \in R_i(\mathbf{A})\ \exists \sigma \in \mathfrak{S}_{r_i} :\ (x_{\sigma(1)}, \ldots, x_{\sigma(r_i)}) \in R_i(\mathbf{B})$;
- $\forall (y_1, \ldots, y_{r_i}) \in R_i(\mathbf{B})\ \exists \rho \in \mathfrak{S}_{r_i} :\ (y_{\rho(1)}, \ldots, y_{\rho(r_i)}) \in R_i(\mathbf{A})$.

Notice that this obviously defines an equivalence relation of $\mathrm{Rel}(\sigma)$, and that $R_i(\mathbf{A})$ and $R_i(\mathbf{B})$ can have different cardinality (see Fig. 11.2).

Let A be the universe of $\mathbf{A}$ and let $<$ be a linear order on A. Then $\mathbf{B}$ is the *linear $<$-reorientation* of $\mathbf{A}$ if

$$\forall (y_1, \ldots, y_{r_i}) \in R_i(\mathbf{B}) \quad y_1 < y_2 < \cdots < y_{r_i}.$$

(Notice that $\mathbf{A}$ has a unique linear $<$-reorientation for each linear order $<$.)

Fig. 11.2 Two directed graphs with different size, which are weak reorientations of each other

For a structure $\mathbf{A}$, a *circuit* of length p is a cycle $(x_1, b_1, x_2, b_2, \ldots, x_p, b_p)$ of $\mathrm{Inc}(\mathbf{A})$ (where $x_1, \ldots, x_p \in A$ and $b_1, \ldots, b_p$ are blocks of $\mathbf{A}$) such that (putting $x_{p+1} = x_1$ for the sake of simplicity) for $i = 1, \ldots, p$, x_i appears before x_{i+1} in the tuple of the block b_i. If $\mathbf{A}$ has no circuits, it is *acyclic*. It is easily checked that a structure $\mathbf{A}$ is acyclic if and only if there exists a linear order $<$ on A such that $\mathbf{A}$ is its own linear $<$-reorientation.

For a class $\mathcal{C} \subseteq \mathrm{Rel}(\sigma)$ we define

- The class $\mathcal{C}_{\mathrm{wor}}$ has the class of all weak reorientations of structures in $\mathcal{C}$;
- The class $\mathcal{C}_{\mathrm{acyc}}$ has the class of all acyclic weak reorientations of structures in $\mathcal{C}$.

Theorem 11.4. *Let $\mathcal{C} \subseteq \mathrm{Rel}(\sigma)$. The following properties are equivalent:*

1. *The class $\mathcal{C}$ has bounded expansion;*
2. *Therc class $\mathcal{C}_{\mathrm{wor}}$ has all restricted dualities;*
3. *For every integer p, there is $\mathbf{D}_p \in \mathrm{Rel}(\sigma)$ with no circuits of length at most p such that*

$$\forall \mathbf{A} \in \mathcal{C}_{\mathrm{acyc}}, \quad \mathbf{A} \to \mathbf{D}_p.$$

Proof. We prove the equivalence by means of three implications:

(1) $\Rightarrow$ (2) is a direct consequence of Theorem 11.3.
(2) $\Rightarrow$ (3) is straightforward (consider the product of the duals of all the minimal structures with a circuit of length at most p).
(3) $\Rightarrow$ (1) is proved by contradiction: Assume that (3) holds and that $\mathcal{C}$ does not have bounded expansion. According to Proposition 5.7 the class $\mathrm{Inc}(\mathcal{C})$ does not have bounded expansion. According to Proposition 5.5 there exists an integer p such that $\mathrm{Inc}(\mathcal{C}) \mathbin{\widetilde{\nabla}} p$ has unbounded chromatic number. Let N be

the order of $\mathbf{D}_{p+1}$. There exists in $\mathrm{Inc}(\mathcal{C})$ a graph G which contains the $\leq p$-subdivision S of a graph H having chromatic number strictly greater than N. We may further assume that the minimum degree of H is strictly greater than the maximum arity of a relational symbol in σ. Let $\mathbf{A} \in \mathcal{C}$ be such that G is isomorphic to the incidence graph of $\mathbf{A}$. By the assumptions on the minimal degree of H, the branching vertices of G correspond to vertices of $\mathbf{A}$. Consider a linear order $<$ on the universe A of $\mathbf{A}$ such that every branch of S will correspond to a monotone sequence. Consider the linear $<$-reorientation $\mathbf{B}$ of $\mathbf{A}$. According to (3), there exists a homomorphism $f : \mathbf{B} \to \mathbf{D}_{p+1}$. Moreover, the two endpoints of a branch of S cannot have the same image by f as then a circuit of length at most p would exist in $\mathbf{D}_{p+1}$. It follows that any two adjacent vertices in H are mapped by f to distinct vertices of $\mathbf{D}_{p+1}$ hence $\chi(H) \leq |\mathbf{D}_{p+1}|$, a contradiction.

□

For the sake of completeness, let us note a quite similar characterization of classes with bounded tree-depth:

Theorem 11.5. *Let $\mathcal{C} \subseteq \mathrm{Rel}(\sigma)$. The following properties are equivalent:*

1. *The class $\mathcal{C}$ has bounded tree-depth;*
2. *The class $\mathcal{C}_{\mathrm{acyc}}$ has an acyclic bound $\mathbf{D}$, that is: there exists an acyclic structure $\mathbf{D} \in \mathrm{Rel}(\sigma)$ such that $\mathbf{A} \to \mathbf{D}$ holds for every $\mathbf{A} \in \mathcal{C}_{\mathrm{acyc}}$.*

Proof. By definition, the class $\mathcal{C}$ has bounded tree-depth if and only if the class $\mathrm{Gaifman}(\mathcal{C})$ has bounded tree-depth.

Assume that Gaifman graphs of structures in $\mathcal{C}$ have tree-depth at most t. Let $\mathbf{D}$ be the structure with universe $\{1, \dots, t\}$, such that $(x_1, \dots, x_{r_i}) \in R_i(\mathbf{D})$ if $x_1 \leq x_2 \leq \dots \leq x_{r_i}$. Obviously, $\mathbf{D}$ is an acyclic bound for $\mathcal{C}_{\mathrm{acyc}}$.

Conversely, assume that there is an acyclic $\mathbf{D} \in \mathrm{Rel}(\sigma)$ such that every $\mathbf{A} \in \mathcal{C}_{\mathrm{acyc}}$ has a homomorphism to $\mathbf{D}$. Assume for contradiction that $\mathcal{C}$ has unbounded tree-depth. Then there exists in $\mathcal{C}$ a structure $\mathbf{A}$ such that $\mathrm{Gaifman}(\mathbf{A})$ contains a path of length strictly greater than $N = |\vec{D}|$. Let $<$ be a linear order on the universe A of $\mathbf{A}$. Let $\mathbf{B}$ be the linear $<$-reorientation of $\mathbf{A}$. As it is clear that $\mathbf{B}$ has no homomorphism to an acyclic structure of order at most N, we get a contradiction. □

11.7 Characterization of Classes with All Restricted Dualities by Subdivisions

In the case of graphs, we can consider subdivisions instead of considering orientations.

Theorem 11.6. *Let $\mathcal{C}$ be a class of undirected graphs closed under subdivisions.*[1]

Then the following properties are equivalent:

1. *The class $\mathcal{C}$ has bounded expansion;*
2. *The class $\mathcal{C}$ has all restricted dualities;*
3. *For every odd integer g there exists a non-bipartite graph H_g with odd-girth at least g such that every graph $G \in \mathcal{C}$ with odd-girth at least g has a homomorphism to H_g*

Proof. The proof follows from the next three implications:

(1) $\Rightarrow$ (2) is a direct consequence of Theorem 11.3.
(2) $\Rightarrow$ (3) is straightforward (consider for H_g a dual of C_g for $\mathcal{C}$).
(3) $\Rightarrow$ (1) is proved by contradiction: assume that (3) holds and that $\mathcal{C}$ does not have bounded expansion. According to Proposition 5.5 there exists an integer p such that $\mathcal{C} \mathbin{\widetilde{\nabla}} p$ has unbounded chromatic number. As $\mathcal{C}$ is topologically closed there exists an odd integer $g \geq p$ and a graph $G_0 \in \mathcal{C}$ such that G_0 is the $(g-1)$-subdivision of a graph H_0 with chromatic number $\chi(H_0) > |H_g|$. According to (3), there exists a homomorphism $f : G_0 \to H_g$. As $C_g \nrightarrow H_g$, the ends of a path of length g cannot have the same image by f. It follows that any two adjacent vertices in H_0 correspond to branching vertices of G_0 which are mapped by f to distinct vertices of H_g. It follows that $\chi(H_0) \leq |H_g|$, a contradiction.

□

11.8 First-Order Definable H-Colorings

Let us go back to our original motivation: which H-coloring problems are equivalent (when restricted to some class $\mathcal{C}$) to a first-order property?

According to Theorem 11.6, if $\mathcal{C}$ is a class of undirected graphs closed by subdivisions, then the following conditions are equivalent:

1. $\mathcal{C}$ has bounded expansion;
2. $\mathcal{C}$ has all restricted dualities;
3. for every odd integer g there exists a non-bipartite graph H_g with odd-girth at least g such that H_g coloring is equivalent, on $\mathcal{C}$, with the satisfaction of an existential positive first-order property.

[1] This should not be confused with "closed under topological minors": if a class $\mathcal{C}$ is closed under subdivision and contains a graph G, it contains all its subdivisions; if a class $\mathcal{C}$ is closed under topological minors and contains a subdivision of a graph G then it contains G.

Could it be possible that we could get rid of this "existential positive"? It is likely to be true if we add few technical conditions on $\mathcal{C}$:

Conjecture 11.1. Let $\mathcal{C}$ be a hereditary addable topologically closed class of graphs. The following conditions are equivalent:

1. The class $\mathcal{C}$ has bounded expansion;
2. The class $\mathcal{C}$ has all restricted dualities;
3. For every odd integer g there exists a non-bipartite graph H_g with odd-girth at least g such that H_g-coloring is first-order definable on $\mathcal{C}$, that is: there exists first-order formulas Φ_g such that

$$\forall g \in \mathbb{N} \quad \forall G \in \mathcal{C} \qquad (G \models \Phi_g) \iff (G \to H_g).$$

In view of the results proved in this chapter, we gather the evidences for the validity of Conjecture 11.1. First, we prove:

Lemma 11.9. *Let $\mathcal{C}$ be a hereditary class of graphs closed under subdivisions and disjoint unions.*

Assume that for every odd integer g there exists a non-bipartite graph H_g with odd-girth at least g and a first-order formula Φ_g such that

$$\forall g \in \mathbb{N}\, \forall G \in \mathcal{C} \qquad (G \models \Phi_g) \iff (G \to H_g).$$

Then:

1. *Either the class $\mathcal{C}$ has bounded expansion,*
2. *Or the class $\mathcal{C}$ is nowhere dense, does not have bounded expansion, and for every integer s there exists an integer l such that*

$$\sup\{\chi(G):\ G \in \mathcal{C} \,\widetilde{\nabla}\, s, \mathrm{girth}(G) \geq l\} < \infty.$$

Proof. If $\mathcal{C}$ is somewhere dense, then there exists an integer s such that $\mathcal{C} \,\widetilde{\nabla}\, s = \mathcal{G}raph$. As $\mathcal{C} \,\widetilde{\nabla}\, s \subseteq \mathcal{C} \,\widetilde{\nabla}\, (s+1)$. According to Corollary 10.6, the homomorphism preservation theorem holds for the class $\mathrm{Sub}_{2s}(\mathcal{G}raph)$ which is a subclass of $\mathcal{C}$ (as $\mathcal{C}$ is closed by subdivisions).

Assume for contradiction that H_{2s+1} and Φ_{2s+1} exist. As $\neg\Phi_{2s+1}$ is preserved by homomorphisms on $\mathcal{C}$ (hence on $\mathrm{Sub}_{2s}(\mathcal{G}raph)$) it is equivalent on $\mathrm{Sub}_{2s}(\mathcal{G}raph)$ with an existential first-order formula, that is: there exists a finite family $\mathcal{F}$ such that for every graph G it holds:

$$\forall F \in \mathcal{F}\ F \nrightarrow \mathrm{Sub}_{2s}(G) \iff \mathrm{Sub}_{2s}(G) \to H_{2s+1}.$$

Clearly, the graphs in $\mathcal{F}$ are non-bipartite. Let g be the maximum of girth of graphs in $\mathcal{F}$ and let G be a graph with chromatic number greater than $|H|$ and odd-girth greater than g. (see Sect. 3.4.) Then for every $F \in \mathcal{F}$ we have $F \nrightarrow \mathrm{Sub}_{2s}(G)$ hence $\mathrm{Sub}_{2s}(G) \to H$. However, has the odd-girth of H is strictly greater than $2s+1$ two branching vertices of $\mathrm{Sub}_{2s}(G)$ corresponding to adjacent vertices of G cannot be mapped to a same vertex. It follows that $|H| \geq \chi(G)$, a contradiction.

It follows that $\mathcal{C}$ is nowhere dense, $\neg\Phi_{2s+3}$ is equivalent on $\mathcal{C}$ with an existential positive first-order formula, that is: there exists a finite family $\mathcal{F}$ such that

$$\forall G \in \mathcal{C} \qquad (\forall F \in \mathcal{F}\ F \nrightarrow G) \iff (G \to H_{2s+3}).$$

If $\mathcal{C}$ includes only discrete graphs, then $\mathcal{C}$ has bounded expansion. Otherwise, as $\mathcal{C}$ is hereditary, $K_2 \in \mathcal{C}$. As $K_2 \to H_{2s+3}$ we deduce that the family $\mathcal{F}$ includes no bipartite graph. Let g be the maximum odd-girth of the graphs in $\mathcal{F}$. Assume for contradiction that the class $\mathcal{C}$ does not have bounded expansion and that (2) does not hold. Then there exists an integer s such that $\mathcal{C} \,\widetilde{\nabla}\, 2s$ contains graphs with both arbitrarily large girth and arbitrarily large chromatic number. Hence there exists a graph $S \in \mathcal{C} \,\widetilde{\nabla}\, 2s$ with odd-girth at least $g + 2$ and chromatic number at least $|H_{2s+3}| + 1$ and (as $\mathcal{C}$ is both hereditary and closed by subdivisions) the $2s$-subdivision G of S belongs to $\mathcal{C}$. Clearly, no $F \in \mathcal{F}$ maps to G because the odd-girth of G is strictly greater than g. Hence $G \to H_{2s+3}$. However, the endpoints of any $2s$-subdivided branch cannot be mapped by a homomorphism to a same vertex of H_{2s+3} as the odd-girth of H_{2s+3} is strictly greater than $2s + 1$. It follows that $|H_{2s+3}| \geq \chi(S)$, a contradiction with $\chi(S) \geq |H_{2s+3}| + 1$. □

It follows that Conjecture 11.1 would follow from the following conjecture:

Conjecture 11.2. For every monotone nowhere dense class without bounded expansion there exists an integer s such that the class includes s-subdivisions of graphs with arbitrarily large chromatic number and girth.

Observe that these two possibilities are clearly mutually exclusive. Conjecture 11.2 is related to an old conjecture already mentioned in Sect. 3.4:

Conjecture 11.3 (Erdős and Hajnal [164]). For all integers c, g there exists an integer f(c, g) such that every graph G of chromatic number at least f(c, g) contains a subgraph of chromatic number at least c and girth at least g..

Remark that the existence of graphs of both arbitrarily high chromatic number and high girth is a well known result of Erdős [163].

The following weakening of this difficult problem would actually easily imply Conjecture 11.2:

Conjecture 11.4. For all integers c, g there exist integers f(c, g) and s(g) such that every graph G of chromatic number at least f(c, g) contains a subgraph H such that

- Either $\chi(H) \geq c$ and $\mathrm{girth}(H) \geq g$,
- Or $K_c \in H \,\widetilde{\nabla}\, s(g)$.

Notice that if we do not ask the subdivision to be shallow (i.e. if we demand only that $K_c \in H \,\widetilde{\nabla}\, \infty$) then the conjecture of course holds.

As mentioned in Sect. 3.4, a conjecture with a similar flavor as Conjecture 11.3 has been proposed:

Conjecture 11.5 (Thomassen [457]). For all integers c, g there exists an integer f(c, g) such that every graph G of average degree at least f(c, g) contains a subgraph of average degree at least c and girth at least g.

Let us prove now that this conjecture also would imply Conjecture 11.2 (and hence Conjecture 11.1):

Proposition 11.1. *A positive answer to Conjecture 11.5 would imply a positive answer to Conjecture 11.2.*

Proof. Assume a positive answer to Conjecture 11.5.

Let $\mathcal{C}$ be a monotone nowhere dense class without bounded expansion. There exists an integer p such that $\mathcal{C} \,\widetilde{\nabla}\, p$ has unbounded average degree.

Let c, g be positive integers. There exists in $\mathcal{C}$ the $2p$-subdivisions of a graph G with average degree at least $56(c-1)^2 \frac{\log(c-1)}{\log c - \log(c-1)}$ and girth at least $2g$. According to Lemma 4.5, G contains as a subgraph the 1-subdivision of a graph H with girth at least g and chromatic number at least c.

Put $s = 4p + 2$. Then the class $\mathcal{C}$ includes s-subdivisions of graphs with arbitrarily large chromatic number and girth. □

11.9 Consequences and Related Problems

The fact that bounded expansion classes have all restricted dualities has many concrete consequences.

11.9.1 On Hadwiger Conjecture

We have the following corollary of Theorem 11.3:

Corollary 11.2. *Let $\mathcal{K}$ be a proper minor closed class of graphs. Let $\mathcal{F}$ be a finite set of connected graphs. Then there exists a finite graph $D_{\mathcal{F}}^{\mathcal{K}} \in$ $\mathrm{Forb}_h(\mathcal{F})$ such that every graph of $\mathcal{K} \cap \mathrm{Forb}_h(\mathcal{F})$ has a homomorphism to $D_{\mathcal{F}}^{\mathcal{K}}$.*

The celebrated Hadwiger Conjecture asserts that K_k is a minor of every graphs with chromatic number at least k. The largest k for which K_k is a minor of a graph G is the *Hadwiger number* of G and is denoted by $h(G)$. One can formulate the Hadwiger Conjecture as the existence of a maximum (in the homomorphism order) for every proper minor closed class [253, 347]. Let $h = \max\{h(G) : G \in \mathcal{K}\}$ be the Hadwiger number of the class $\mathcal{K}$. Then $K_{h+1} \notin \mathcal{K}$ and Corollary 11.2 gives at least a K_{h+1}-free bound of the class $\mathcal{K}$.

11.9.2 On Bounded Expansion Classes

Let $\mathcal{K}$ be the class of *all* graphs G which have bounded expansion with the expansion function f. Formally, $\mathcal{K} = \{G : \nabla_r(G) \leq f(r), r = 1, 2, \ldots\}$. Assume that p is minimal with $K_{p+1} \notin \mathcal{K}$. Then $\nabla_0(G) \leq p - 1$ for every $G \in \mathcal{K}$. Thus every $G \in \mathcal{K}$ is $(p-1)$-degenerate. If the function f is monotone then also $K_p \in \mathcal{K}$ and thus $\mathcal{K}$ has a maximum. Thus Hadwiger conjecture holds for those bounded expansion classes which are determined by a monotone expansion function.

Note also that for every constant expansion function the bounded expansion class is a proper minor closed class, while of course every proper minor closed class is contained in a bounded expansion class, [355].

11.9.3 On Distance Colorings – Powers and Exact Powers

We now explain a particular consequence of our main result in greater detail. Let G be a graph, p a positive integer. Denote by $G^{\natural p}$ the graph $(V(G), E^{\natural p})$ where $\{x, y\}$ is an edge of $E^{\natural p}$ if and only if there exists a path in G from x to y of length p. The graph $G^{\natural p}$ is called *exact* p*-power* of G. Clearly graphs $G^{\natural 2}$ and all graphs $G^{\natural p}, p$ even, may have unbounded chromatic number even for the case of trees (consider subdivided stars), and the only (obvious) bound is $\chi(G^{\natural p}) \leq \Delta(G)^p + 1$. Similarly, for every odd p there are 3-colorable graphs G for which the chromatic number $\chi(G^{\natural p})$ may be arbitrarily large (simply consider a large complete graph and subdivide every edge by $p - 1$ vertices). However for p odd and arbitrary proper minor closed class (and even for every class with bounded expansion) we have the following (perhaps surprising) result. Recall that *odd-girth* of a graph G is the length of the shortest odd cycle in G.

Theorem 11.7. *For every class* $\mathcal{K}$ *with bounded expansion and for every odd integer* $p \geq 1$*, there exists an integer* $N = N(\mathcal{K}, p)$ *such that all the graphs* $G^{\natural p}, G \in \mathcal{K}$ *and* $\text{odd-girth}(G) > p$ *have chromatic number* $\leq N$*: For every* $G \in \mathcal{K}$*,*

$$\text{odd-girth}(G) > p \Longrightarrow \chi(G^{\natural p}) \leq N$$

Proof. Theorem 11.7 follows immediately from Theorem 11.3. It suffices to consider $\mathcal{F} = \{C_p\}$. In this case every graph $D_{\mathcal{F}}^{\mathcal{K}} \in \text{Forb}_h(\mathcal{F})$ and every homomorphism $c : G \longrightarrow D_{\mathcal{F}}^{\mathcal{K}}$ gives a desired coloring by $N = |V(D_{\mathcal{F}}^{\mathcal{K}})|$ colors. □

With a little more care one can prove the following result about distance graphs: Let G be a graph, p a positive integer. Denote by $G^{[\natural p]}$ the graph $(V, E^{[\natural p]})$ where $\{x, y\}$ is an edge of $E^{[\natural p]}$ if and only if the distance of x and y in G is exactly p. The graph $G^{[\natural p]}$ is called *exact distance* graph.

Theorem 11.8. *For every class $\mathcal{K}$ with bounded expansion and for every odd integer $p \geq 1$, there exists an integer $N' = N(\mathcal{K}, p)$ such that all the graphs $G^{[\natural p]}, G \in \mathcal{K}$ have chromatic number at most* N.

Proof. According to Theorem 7.6 there exists an integer $X(p)$ such that every graph $G \in \mathcal{K}$ has a p-centered coloring ρ using a set X of a most $X(p)$ colors. Let $N_0 = X(p)$ and let $N = N_0 2^{N_0 2^{N_0}}$ and let $G \in \mathcal{K}$. For every subset $I \subseteq X$ of size at most p, let Y_I be a star forest of height at most p derived from the p-centered coloring, which is such that the subgraph G_I of G induced by the colors in I is a subgraph of the closure of Y_I. For every vertex v with $\rho(v) \in I$, denote by $G_I^{(v)}$ the connected component of G_I including v. For every color $c \in I$, notice that v has either 0 or 1 ancestor (with respect to Y_I) in $G_I^{(v)}$ with color c. We denote by d_{G_I} the distance in G_I and define the mapping $\pi_v : 2^X \times X \to \{0, 1\}$ by:

$$\pi_v(I, c) = \begin{cases} 1, & \text{if } c \in I, \rho(v) \in I, |I| \leq p, \\ & v \text{ has an ancestor } w \text{ of color } c \text{ in } G_I^{(v)}, \\ & \text{and } d_{G_I}(v, w) \equiv 1 \bmod 2; \\ 0, & \text{otherwise.} \end{cases}$$

Define the coloring ϕ of G by $\phi(v) = (\rho(v), \pi_v)$. This coloring uses at most N colors.

Assume there exists two vertices u, v at distance p in G, such that $\phi(u) = \phi(v)$. Let $P = (u = x_0, x_1, \dots, x_q = v)$ be a minimum distance path linking u and v (hence $q \leq p$). As $\phi(u) = \phi(v)$ we have $\rho(u) = \rho(v)$ hence P gets at most p colors in coloring ρ. Let $I = \rho(\{x_0, \dots, x_q\})$. According to the definition of a p-centered coloring, the path P includes a uniquely colored vertex z, which is a common ancestor of u and v in Y_I. As P is a minimum distance path, we have $d_{G_I}(u, v) = d_{G_I}(u, z) + d_{G_I}(z, v)$. As $\pi_u = \pi_v$ we deduce $\pi_u(I, \rho(z)) = \pi_v(I, \rho(z))$ hence $d_{G_I}(u, z) \equiv d_{G_I}(z, v) \bmod 2$. From this follows that $p = d_{G_I}(u, v)$ is even, a contradiction. □

Note that in Theorem 11.8 we cannot replace the condition in the definition of exact distance powers $G^{[\natural p]}$ by the existence of a path (or even induced path) of length p(see Figs. 11.3 and 11.4).

Theorem 11.8 implies, in particular, that for every odd integer p, there exists $N(p)$ such that for every planar graph G, the graph $G^{[\natural p]}$ has chromatic number at most $N(p)$. The value $N(p)$ seems to be difficult to estimate. Of

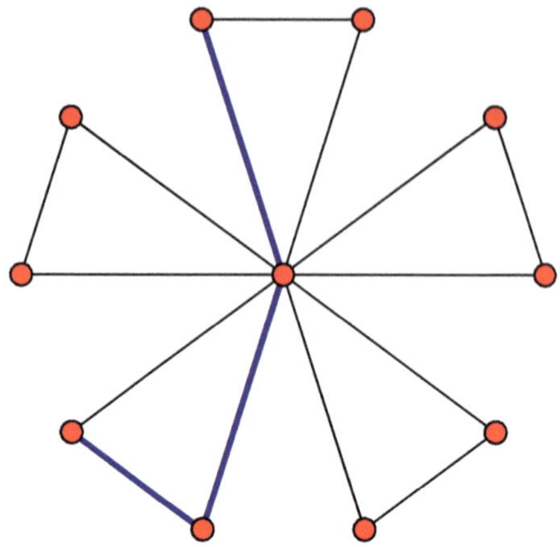

Fig. 11.3 For such graphs G, $G^{\natural 3}$ include arbitrarily large cliques

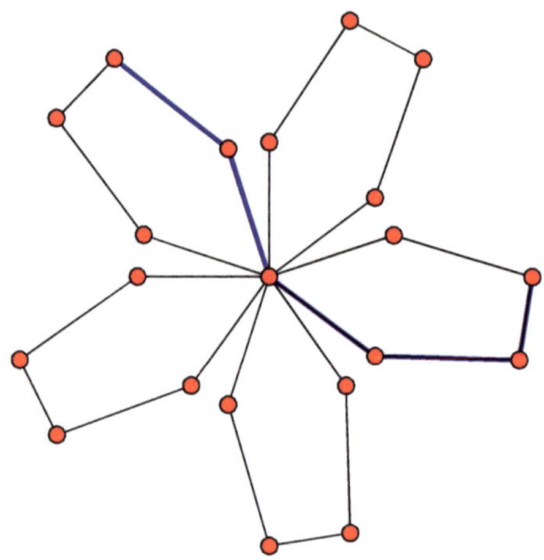

Fig. 11.4 For such graphs G, an arbitrarily large set of vertices exist, whose elements are pairwise connected by an induced path of length 5

course $N(1) = 4$,but for $p = 3$ we already only get huge upper bound! On the other side, we have $N(3) \geq 6$ (see Exercise 11.4).

More: although our upper bound for $N(p)$ grows very fast, we do not know if this function is actually unbounded. This motivates the following recent problem:

Problem 11.1 (van den Heuvel and Naserasr). Does there exist a constant C such that for every odd-integer p and any planar graph G it holds

$$\chi(G^{[\sharp p]}) \leq C?$$

Problem 11.2 (Thomassé). For a graph G, denote by G^{odd} the graph with vertex set $V(G)$ where two vertices are adjacent if they are at odd distance in G.

Does there exist a function f such that for every planar graph G it holds

$$\chi(G^{odd}) \leq f(\omega(G^{odd}))?$$

Note that for outerplanar graphs G, the graphs G^{odd} can have arbitrarily large clique number (see Exercise 11.5).

Exercises

11.1. Prove that if a class of graphs $\mathcal{C}$ has a restricted duality $\mathcal{F}, D$ then there exists another dual D' such that

$$\chi(D') = \max\{\chi(G) : \ G \in \mathcal{C} \cap \mathrm{Forb}_h(\mathcal{F})\}.$$

11.2. Thomassen proved [458] that a graph in the torus is 5-colorable if and only if it has no subgraph isomorphic to K_6, $C_3 \oplus C_5$, $K_2 \oplus H_7$ or C_{11}^3, where $G \oplus H$ denotes the *complete join* of G and H, that is the graph obtained from the disjoint union of G and H by making every vertex of G adjacent to every vertex of H.

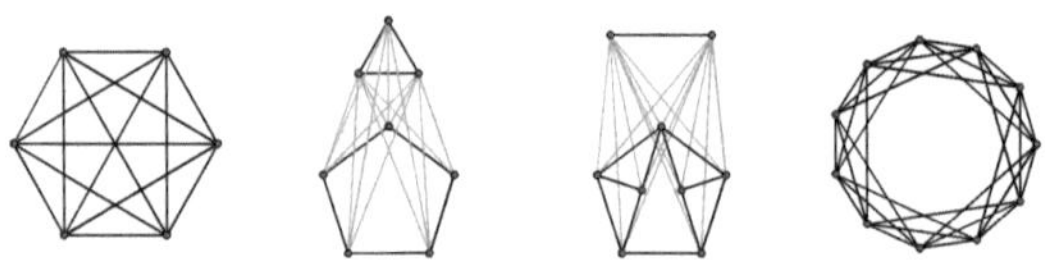

K_6 $\qquad$ $C_3 \oplus C_5$ $\qquad$ $K_2 \oplus H_7$ $\qquad$ C_{11}^3

Deduce the following restricted duality for toroidal graphs G:

$\not\to \ G \quad \Longleftrightarrow \quad G \ \longrightarrow$

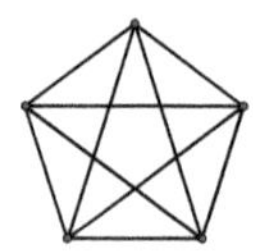

Other examples of such dualities are treated in [344].

11.3. Let G be a graph and let p be an integer. We consider a p-tree-depth coloring c of G by $N = \chi_p(G)$ colors.

Prove that for each $I \in \binom{[I]}{p}$ there exists a homomorphism $f_I : G_I \to G_I$ such that $|f_I(G_I)| \le F(p)$.
Let $x \sim y$ if $c(x) = c(y)$ and if $f_I(x) = f_I(y)$ holds for every $I \in \binom{[I]}{p}$. We define the graph $\hat{G}$ whose vertices are the equivalence classes $[x] \in V(G)/\sim$, whose edges are the pairs $\{[x], [y]\}$ such that for every $I \in \binom{[I]}{p}$ $\{f_I(x), f_I(y)\}$ is an edge of G. We also define a N-coloration of $\hat{G}$ by $\hat{c}([x]) = c(x)$. Check that $\hat{G}$ and $\hat{c}$ are well defined.
Prove that $x \mapsto [x]$ is a homomorphism $G \to \hat{G}$.
Prove that for every $I \in \binom{[N]}{p}$ the mapping $[x] \mapsto f_I(x)$ is a homomorphism $\hat{G}_I \to G_I$ (where $\hat{G}_I$ is the subgraph of $\hat{G}$ induced by colors in I).
Deduce that $|\hat{G}| \le F(p)^{N^p}$ and thus $|\Phi_L^{2^{-p}}(G)| \le F(p)^{N^p}$.
Conclude that bounded expansion classes have all restricted dualities (Hint: use Lemma 11.1). Thus the duals for dualities restricted to a bounded expansion class may be chosen in a "canonical" way.

11.4. Prove that the following planar graph G is such that $\chi(G^{[\sharp 3]}) \geq 6$.

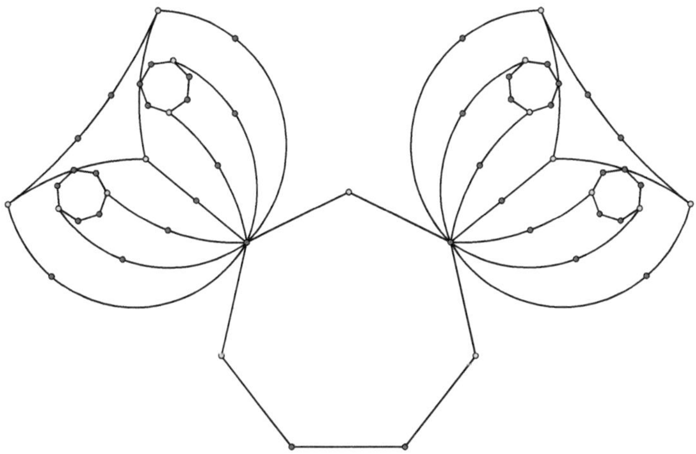

11.5. Prove that $\omega(G^{[\sharp 2p+1]}) \leq 4$ for G planar,
Prove that the clique number of the odd-distance graph of outerplanar graphs is unbounded.

Chapter 12
Counting

- How many leaves are there in a forest?
- Including the falling ones?

12.1 Introduction

In the previous chapters of this book we investigated the problem of the existence of particular structures with special properties such as the existence of special partitions, the value of special parameters (for example $\nabla_r(G)$) or the existence of special subsets.

The counting versions of our problems were not considered so far. However some counting was hidden. Indeed, computing the edge density of a graph of order n amounts to counting the number of K_2 in this graph (and dividing it by n) or, equivalently, counting the number of homomorphisms from K_2 to the graph and dividing this number by $2n$.

When considering a series of larger and larger graphs, counting the number of homomorphisms from small test graphs or counting the number of induced copies of a small pattern is the main tool in the study of a possible limit of the sequence. This is the case when one considers, on the one hand, the convergence criteria for dense graphs (as defined by Lovász et al. [78], linked to Szemerédi partitions) and, on the other hand, convergence of hyperfinite graphs (as defined by Elek and Lippner [151] linked to a finitarization of Farell-Varadarajan ergodic decomposition theorem). A main difference between these two approaches stands in the normalization needed to transform the number of induced copies of a fixed pattern into a "density", more precisely in the exponent of the order of the graph which is used to divide the

J. Nešetřil and P.O. de Mendez, *Sparsity*, Algorithms and Combinatorics 28, DOI 10.1007/978-3-642-27875-4_12, © Springer-Verlag Berlin Heidelberg 2012

number of copies. This exponent intuitively measures how independent the assignments of the vertices of the pattern graph may be, that is the "degree of freedom" of the pattern in the graphs of the class. The determination of this degree of freedom is the subject of this chapter. In this chapter we consider only graphs (although many results hold for finite relational structures).

Let us be more precise: in the dense case, it is natural to consider that each vertex of the pattern could be considered independently and thus to consider, for a small test graph F and a large graph G, the probability that a random map from $V(F)$ to $V(G)$ will be a homomorphism. (Notice that the number of induced copies of F in G may be easily derived from the number of homomorphisms from F to G). The considered density is thus

$$t(F, G) = \frac{\hom(F, G)}{|G|^{|F|}}, \tag{12.1}$$

(Recall that $\hom(F, G)$ stands for the number of homomorphisms from F to G).

Now consider the ultra-sparse case—for instance the case of bounded degree graphs excluding a minor [56]. A random map from a test graph F to a large ultra-sparse graph G is unlikely to be a homomorphism (except if F is edgeless). There are obviously only finitely many (rooted) isomorphism types of the balls $B_r(v)$ for $v \in V(G)$ (where we consider $B_r(v)$ as a graph rooted at v). It easily follows that the number of copies of F in G will be at most linear in the order of G and that the considered density should be

$$\mathrm{dens}(F, G) = \frac{(\#F \subseteq G)}{|G|}, \tag{12.2}$$

where $(\#F \subseteq G)$ stands for the number of induced copies of F in G.

When studying large graphs belonging to an infinite class of graphs $\mathcal{C}$ and fixing a small test graph F, the question arises to determine whether for a given infinite class $\mathcal{C}$ of finite graphs and a (small) test graph F there exists a "natural" exponent $f(F, \mathcal{C})$ such that the number of copies of F in any $G \in \mathcal{C}$ is bounded by $|G|^{f(F,\mathcal{C})}$, or even by $|G|^{f(F,\mathcal{C})+o(1)}$ (as $|G| \to \infty$). This motivates our study of the *asymptotic upper logarithmic density* of F in $\mathcal{C}$ defined as the limit

$$\limsup_{G \in \mathcal{C}} \frac{\log(\#F \subseteq G)}{\log |G|}.$$

Recall that, for any graph parameter f, we defined in Sect. 5.1.1

$$\limsup_{G \in \mathcal{C}} f(G) = \lim_{i \to \infty} \sup\{f(G) : \ G \in \mathcal{C} \text{ and } |G| \geq i\}.$$

In general, the asymptotic upper logarithmic density of F in $\mathcal{C}$ is not an integer. For instance, consider the class $\mathcal{C}_0$ of graphs of girth at least 5 (i.e. the class of C_3- and C_4-free graphs) and the class $\mathcal{C}_1$ consisting of subgraphs of the Cartesian product of a graph in $\mathcal{C}_0$ and the complete graph K_2 (see Fig. 12.1). Recall that the *Cartesian product* $G \square H$ of two graphs G and H is the graph with vertex set $V(G) \times V(H)$ in which two vertices (u, x) and (v, y) are adjacent if $u = v$ and $\{x, y\} \in E(H)$ or $x = y$ and $\{u, v\} \in E(G)$. Obviously, both $\mathcal{C}_0$ and $\mathcal{C}_1$ are addable (i.e. closed by disjoint union) and monotone (i.e. closed under subgraphs). According to a classical graph theory result of Erdős we have (see [162] and most books on Graph Theory [328]):

$$\limsup_{G \in \mathcal{C}_0} \frac{\log(\#K_2 \subseteq G)}{\log|G|} = \frac{3}{2}. \tag{12.3}$$

From which we deduce, for the class $\mathcal{C}_1$, that

$$\limsup_{G \in \mathcal{C}_1} \frac{\log(\#C_4 \subseteq G)}{\log|G|} = \frac{3}{2}. \tag{12.4}$$

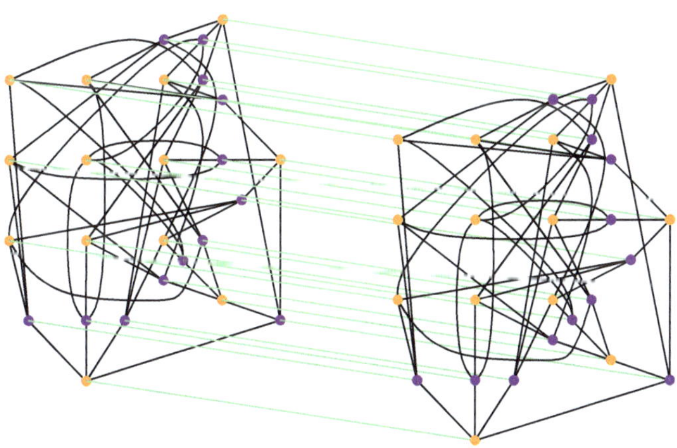

Fig. 12.1 Construction of a graph, all subgraphs of which have $O(n^{3/2})$ edges and $O(n^{3/2})$ C_4's

Of course, we suspect that the density of graphs in $\mathcal{C}$ will play an important role but the relation might be not obvious as if one considers the class $\mathcal{P}$ of planar graphs we have

$$\limsup_{G \in \mathcal{P}} \frac{\log(\#C_4 \subseteq G)}{\log|G|} = 2 > \frac{3}{2} = \limsup_{G \in \mathcal{C}_1} \frac{\log(\#C_4 \subseteq G)}{\log|G|}, \tag{12.5}$$

although

$$\limsup_{G \in \mathcal{P}} \frac{\log \|G\|}{\log |G|} = 1 < \frac{3}{2} = \limsup_{G \in \mathcal{C}_1} \frac{\log \|G\|}{\log |G|}. \tag{12.6}$$

This effect is actually related to the existence of shallow subdivisions of large dense graphs $\mathcal{C}_1$ and this is of course close to our setting.

In this setting, the trichotomy theorem may be expressed as follows:

Theorem (5.4). *Let $\mathcal{C}$ be an infinite class of graphs. Then the limit*

$$\mathrm{free}(K_2, \mathcal{C}) = \sup_r \limsup_{G \in \mathcal{C} \,\widetilde{\triangledown}\, r} \frac{\log \|G\|}{\log |G|}$$

is 2 if and only if $\mathcal{C}$ is somewhere dense and belongs to $\{-\infty, 0, 1\}$ otherwise (i.e. if $\mathcal{C}$ is nowhere dense).

In this chapter we generalize this theorem to a new characterization of nowhere dense classes via counting.

Theorem 12.1. *Let $\mathcal{C}$ be an infinite class of graphs, and let F be a fixed graph with at least one edge and stability number $\alpha(F)$. Then the limit*

$$\mathrm{free}(F, \mathcal{C}) = \sup_r \limsup_{G \in \mathcal{C} \,\widetilde{\triangledown}\, r} \frac{\log(\#F \subseteq G)}{\log |G|} \tag{12.7}$$

is $|F|$ if and only if $\mathcal{C}$ is somewhere dense, and otherwise (i.e. if $\mathcal{C}$ is nowhere dense) the limit belongs to $\{-\infty, 0, 1, \ldots, \alpha(F)\}$.

By analogy with similar situations we call $\mathrm{free}(F, \mathcal{C})$ the *degree of freedom* of F in $\mathcal{C}$. We will have a more precise result when restricting to nowhere dense classes. In this case, we don't have to consider shallow topological minors:

Theorem 12.2. *Let $\mathcal{C}$ be an infinite nowhere dense hereditary class of graphs, and let F be a fixed graph with at least one edge and stability number $\alpha(F)$. Then the limit*

$$\limsup_{G \in \mathcal{C}} \frac{\log(\#F \subseteq G)}{\log |G|} \tag{12.8}$$

belongs to $\{-\infty, 0, 1, \ldots, \alpha(F)\}$.

Note that the degree of freedom of F in $\mathcal{C}$ is the supremum of the asymptotic logarithmic densities of F in the classes $\mathcal{C} \,\widetilde{\triangledown}\, r$, and that generally free(F, $\mathcal{C}$) can be greater than the asymptotic logarithmic density of F in $\mathcal{C}$.

We shall actually prove a rigidity result: every sufficiently large graph G with many copies of F actually contains a large "regular" substructure with at least the same logarithmic density of copies of F. In order to state this structural result, we shall introduce in Sect. 12.2 the notion of a *generalized sunflower*. We will then proceed by a reduction from general graphs to graphs with bounded tree-depth and from graphs with bounded tree-depth to colored forests. Section 12.3 is devoted to the study of colored forests, Sect. 12.4 deals with graphs with bounded tree-depth, while Sect. 12.5 considers general graphs. Consequences for classes of graphs are stated in Sect. 12.6.

12.2 Generalized Sunflowers

One of the basic combinatorial results of Set Theory is the sunflower (or delta-system) lemma of Erdös and Rado [169]. Many extensions are known. Our result is yet another reincarnation of the original idea.

12.2.1 Generalized Sunflowers

Let F, G be graphs. A (k, F)-*sunflower* in G is a $(k+1)$-tuple $(C, \mathcal{F}_1, \dots, \mathcal{F}_k)$, such that $C \subseteq V(G), \mathcal{F}_i \subseteq \mathcal{P}(V(G))$, C and all the sets in the $\mathcal{F}_i$'s are pairwise disjoints and there exists a partition $(K, Y_1, \dots, Y_k)$ of $V(F)$ so that (see Fig. 12.2)

- There is no edge (of F) between vertices in Y_i and vertices in Y_j for $i \neq j$,
- There exists an isomorphism $\iota_0 : G[C] \to F[K]$ from the subgraph of G induced by C to the subgraph of F induced by K,
- For each $1 \leq i \leq k$ and each $X_i \in \mathcal{F}_i$ there exists an isomorphism ι_{X_i} : $G[X_i] \to F[Y_i]$,
- For any choice of $X_1 \in \mathcal{F}_1, \dots, X_k \in \mathcal{F}_k$ the mapping $\iota_{X_1,\dots,X_k}$ from $C \cup \bigcup_i X_i$ to $V(F)$ whose restriction to C is ι_0 and whose restriction to X_i is ι_{X_i} (for $1 \leq i \leq k$) is an isomorphism from $G[C \cup \bigcup_i X_i]$ to F.

For a better understanding we illustrate this construction for $k = 3$ and the Petersen graph on Fig. 12.2.

Some easy facts should be noticed about (k, F)-sunflowers:

Fact 1. *If a graph* G *includes a* (k, F)-*sunflower, then* $k \leq \alpha(F)$.

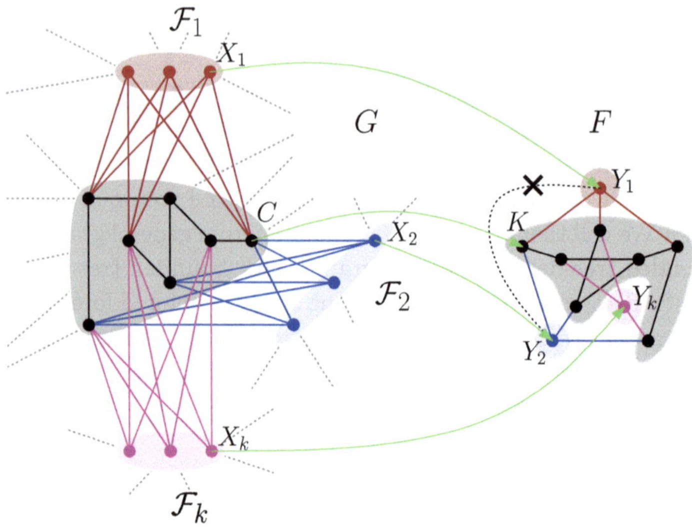

Fig. 12.2 A generalized (3, Petersen)-sunflower

Proof. Consider the partition $(K, Y_1, \ldots, Y_k)$ of $V(F)$ of the definition. As there is no edge between vertices in Y_i and vertices in Y_j for $i \neq j$, we can find in F an independent set of cardinality k, hence $k \leq \alpha(F)$. □

Fact 2. *A* (k, F)*-sunflower* $(C, \mathcal{F}_1, \ldots, \mathcal{F}_k)$ *includes at least* $\prod_{i=1}^{k} |\mathcal{F}_i|$ *copies of* F.

Proof. By definition of a (k, F)-sunflower, each choice of $X_1 \in \mathcal{F}_1, \ldots, X_k \in \mathcal{F}_k$ defines a copy of F. □

Fact 3. *If a graph* G *includes a* (k, F)*-sunflower* $(C, \mathcal{F}_1, \ldots, \mathcal{F}_k)$, *then* G *has an induced subgraph* G' *such that*

$$|G'| \geq k \min_i |\mathcal{F}_i|$$

$$(\#F \subseteq G) \geq \left(\frac{|G'| - |F|}{k}\right)^k.$$

Proof. This is a direct consequence of Fact 2. □

12.3 Counting Patterns of Bounded Height in a Colored Forest

Although colored forests with bounded height are very special and simple structures, their counting is not easy. In this section, we will not count copies of some fixed forest but rather the number of occurrences of more general patterns. This generalization will appear to be for free here and will ease the proofs in the subsequent sections. Still, this section is one of the technical parts of this book.

12.3.1 Patterns

Patterns will be colored forests. For the benefit of the reader we first recall some basic definitions. The considered color set will be $\mathbb{N}$. Let us recall some basic definitions given in Sect. 6.1: The *height* (or the *level*) of a vertex x in a rooted forest Y is the number of vertices of a path from the root (of the component of Y to which x belongs) to x and is noted $\mathrm{height}(x, Y)$. The *height* $h(Y)$ of a forest Y is the maximum height of the vertices of Y, that is one more than the maximum length of a path from a root. We denote by $<_Y$ the partial order on $V(Y)$ defined by putting $x <_Y y$ if there exists a tree path from a root of Y to y which includes x. A subset $A \subseteq V(Y)$ is an *antichain* if the elements of A are pairwise non-comparable.

The color of a vertex a will be denoted by $\gamma(a)$. For a vertex a linked to a root r of Y by a path $P = (r = x_1, x_2, \dots, x_k = a)$, we call the sequence $\Gamma_Y(a) = (\gamma(x_1), \dots, \gamma(x_k))$ the *color sequence* of a.

A *pattern* is formed by a rooted colored forest F and a mapping $\zeta : \bigcup_{s=0}^{\infty}(\mathbb{N}^s \times \mathbb{N}^s) \to \{0,1\}$. A mapping $f : F \to Y$ is ζ-*consistent* if

- f is an injective level preserving homomorphism,
- For every $a \in F$, $\zeta(\Gamma_F(a), \Gamma_Y(f(a))) = 1$.

Denote by $\sigma_\zeta(F, Y)$ the number of ζ-consistent mappings from F to Y. Notice that obviously $\sigma_\zeta(F, Y) \le |Y|^{|F|}$.

When F is a tree, we denote by $\mathrm{root}(F)$ the root of F and by $\pi_F : V(F) \setminus \{\mathrm{root}(F)\} \to V(F)$ the function mapping each vertex to its father in F. If F and Y are trees of height at least 2, with roots r_F and r_Y we denote by $\zeta/(r_F \to r_Y)$ the mapping ζ' defined by

$$\zeta'((a_1, \dots, a_k), (b_1, \dots, b_k)) = \zeta((\gamma(r_F), a_1, \dots, a_k), (\gamma(r_Y), b_1, \dots, b_k)).$$

From this definition it follows:

$$\sigma_\zeta(F, Y) = \sigma_{\zeta'}(F - r_F, Y - r_Y).$$

Example 12.1. Consider the rooted tree F formed by a directed chain (a, b, c) with all vertices colored 0 and $\zeta((0),(x)) = 1$ for every x, $\zeta((0,0),(x,y)) = 1$ if $x < y$ and $\zeta((0,0,0),(x,y,z)) = 1$ if (x,y,z) is an arithmetic progression.

Our functions correspond to the intuition of "color preserving mapping with a joker".

Example 12.2. The mapping ζ_0 defined by

$$\zeta_0((a_1, \ldots, a_s), (b_1, \ldots, b_s)) = 1$$

if for every $1 \le i \le s$ either $a_i = 0$ or $a_i = b_i$. This function allows us to consider mappings where non-zero colors are preserved and zero-colored vertices may be mapped to vertices of any color.

12.3.2 Blowing Patterns

Let F be a rooted forest, let $A = \{a_1, \ldots, a_k\}$ be an antichain of F and let $\rho : A \to \mathbb{N}$. We define the rooted forest $F \curlyvee_{(A,\rho)}$ as follows: The vertex set of $F \curlyvee_{(A,\rho)}$ is the set $V' \subset V \times \mathbb{N}$ defined as the union of $V'_0 = \{(x,0) : \forall a \in A\ x \not\geq_F a\}$ and the sets $V'_a = \{(x,i) : x \ge_F a, 1 \le i \le \rho(a)\}$ for $a \in A$; the father relation of $F \curlyvee_{(A,\rho)}$ is defined by letting (x,i) be the father of (y,j) if x is the father of y in F and either $i = j$ or $i = 0$ (Fig. 12.3).

The similarity between blowing patterns and generalized sunflowers is clear and we shall see that the extraction of a generalized sunflower will be actually related to the matching of a blown pattern.

12.3.3 Warm Up: Counting Patterns of Height 1

To start our inductive procedure, we first consider the case of patterns of height 1, which are nothing but sets of colored isolated vertices:

Lemma 12.1. *Let $0 < \epsilon < 1$, let F be colored rooted forest of height 1, and let k be an integer. For every rooted colored forest Y of order $n \ge |F|^{2|F|/\epsilon}$ including at least $n^{k+\epsilon}$ ζ-consistent images of F, there exists an antichain A of F and a mapping $\rho : A \to \mathbb{N}$ such that:*

- *The graph $F \curlyvee_{(A,\rho)}$ has (at least) a ζ-consistent image in Y,*
- *For every $a \in A$, we have $\rho(a) \ge n^{\epsilon/2|F|}$,*

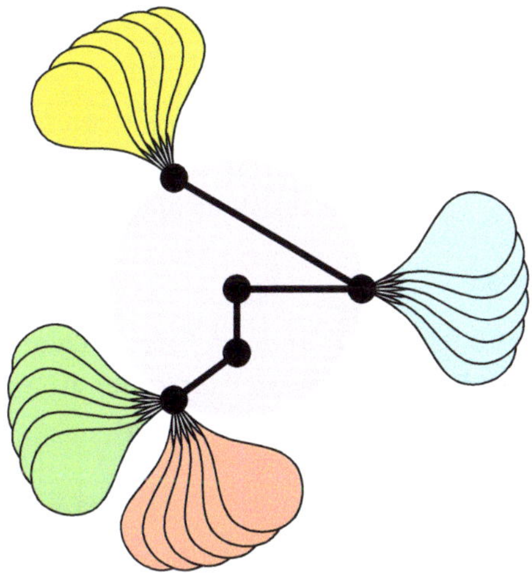

Fig. 12.3 Blowing a tree

- *The cardinality of* A *is at least* $k+1$.

Proof. Let the forest F consists of f_i roots of color c_i, $i = 1, \dots, k$ (of course $\sum f_i = |F|$). Put $\mu = \epsilon/(2|F|)$ and $N_0 = |F|^{3|F|/\epsilon}$.

Let $R_1, \dots, R_k$ be the sets of roots of Y such that for every $1 \le i \le k$ and for every $x \in R_i$ we have $\zeta(c_i, \gamma(x)) = 1$. Then obviously

$$\sigma_\zeta(F, Y) \le \prod_i |R_i|^{f_i}.$$

Let us partition $[k]$ into two sets X_1 and X_2 as follows: $i \in [k]$ belongs to X_1 if $|R_i| < |F| n^\mu$; otherwise, i belongs to X_2. Then

$$\begin{aligned} k + \epsilon &\le \frac{\log \sigma_\zeta(F, Y)}{\log n} \\ &\le \sum_{i \in X_2} f_i + \mu \sum_{i \in X_1} \left(f_i + \frac{\log|F|}{\mu \log n} \right) \\ &\le \sum_{i \in X_2} f_i + \mu|F| \left(1 + \frac{\log|F|}{\mu \log n} \right) \\ &< \sum_{i \in X_2} f_i + \epsilon. \end{aligned}$$

Hence

$$k+1 \leq \sum_{i \in X_2} f_i.$$

Now let A be the set of the roots with color f_i for $i \in X_2$, and for $i \in X_2$ let $\rho(r) = |R_i|/|F|$ for all roots r of F having color i. Then obviously $\rho(a) \geq n^{\mu}$ for every $a \in A$, $F \curlyvee_{(A,\rho)}$ has a ζ-consistent image in Y, and $|A| \geq k+1$. □

As a corollary, we get:

Corollary 12.1. *Let $0 < \epsilon < 1$ and let F be a colored rooted forest of height 1. Every rooted colored forest Y of order $n \geq |F|^{3|F|/\epsilon}$ includes a sub-forest Y' of order at least $n^{\epsilon/2|F|}$ such that*

$$\frac{\log \sigma_\zeta(F, Y')}{\log|Y'|} \geq \left\lceil \frac{\log \sigma_\zeta(F, Y)}{\log|Y|} - \epsilon \right\rceil - \epsilon. \tag{12.9}$$

In other words, the logarithmic density of ζ-consistent images can be shifted (in a non decreasing way) to the ϵ-neighborhood of an integer by consider a proper large sub-forest.

We shall now prove that this result generalizes to all patterns of any fixed height.

12.3.4 Counting Patterns of Height h

The remaining of this section will be devoted to the proof the following result:

Theorem 12.3. *Let $0 < \epsilon < 1$ and let (F, ζ) be a pattern, where F has height h. For positive integer t define the function*

$$\mu_t(x) = \frac{3}{2}(3|F|)^{-\frac{t(t+1)}{2}} x^t. \tag{12.10}$$

There exists $N = N_h(\epsilon)$ such that for every rooted colored forest Y of order $n \geq N$ there exists an antichain A of F and a mapping $\rho : A \to \mathbb{N}$ such that:

- *The graph $F \curlyvee_{(A,\rho)}$ has (at least) a ζ-consistent image in Y,*
- *For every $a \in A$, we have $\rho(a) \geq n^{\mu_t(\epsilon)}$ where $t = \mathrm{height}(a, F)$,*
- *The cardinality of A is at least*

$$\left\lceil \frac{\log \sigma_\zeta(F, Y)}{\log|Y|} - \epsilon \right\rceil.$$

Proof. First notice that if $\sigma_\zeta(F, Y) \leq n^\epsilon$ then we can put $A = \emptyset$. So let us assume in the following that $\sigma_\zeta(F, Y) > n^\epsilon$.

We shall proceed by induction on h, following steps analogous to those of Sect. 12.3.3:

1. We determine where the components of F have to be mapped to get a positive fraction of the ζ-consistent images while ensuring some regularity,
2. We partition the components of F depending on the type of images we get, and compute an upper bound on $\sigma_\zeta(F,Y)$ by looking independently to ζ-consistent images of the the components of F,
3. Using induction, we construct $F \curlyvee_{(A,\rho)}$ and a ζ-consistent image of it in Y.

According to Lemma 12.1, the property holds for patterns of height 1, thus assume the theorem has been proved for colored rooted forests of height at most $h-1 \geq 1$ and let F be a colored rooted forest of height h. Let $F_1,\dots,F_p$ be the connected components of F and let $Y_1,\dots,Y_q$ be the connected components of Y.

Step 1. Define $Z = \{0,1,\dots,|F|\} \cup \{\star\}$. Let $\epsilon_1 = \epsilon/(3|F|)$ and $\mu = \mu_0(\epsilon) = \epsilon/(2|F|)$. Define $\eta : [p] \times [q] \to Z \cup \{\emptyset\}$ by:

$$\eta(i,j) = \begin{cases} \emptyset & \text{if } \sigma_\zeta(F_i,Y_j) = 0, \\ \star & \text{otherwise if } |Y_j| < n^{\epsilon_1}, \\ \left\lceil \frac{\log \sigma_\zeta(F_i,Y_j)}{\log|Y_j|} - \epsilon_1 \right\rceil & \text{otherwise.} \end{cases}$$

Let $\mathcal{I}$ be the set of all the injective mappings $\phi : [p] \to [q]$ such that $\sigma_\zeta(F_i, Y_{\phi(i)}) > 0$ for every $i \in [p]$.

The *profile* P_ϕ of $\phi \in \mathcal{I}$ is the mapping from $[p]$ to Z defined by $P_\phi(i) = \eta(i,\phi(i))$.

For a mapping $c : [p] \to Z$ we denote by $\mathcal{I}_c$ the subset of $\mathcal{I}$ of the mappings with profile c, i.e. the mappings $\phi \in \mathcal{I}$ such that $P_\phi(i) = c(i)$ for every $i \in [p]$.

A simple computation shows that

$$\sigma_\zeta(F,Y) = \sum_{\phi\in\mathcal{I}} \prod_{i=1}^{p} \sigma_\zeta(F_i, Y_{\phi(i)}) = \sum_{c:[p]\to Z} \sum_{\phi\in\mathcal{I}_c} \prod_{i=1}^{p} \sigma_\zeta(F_i, Y_{\phi(i)}).$$

As there are at most $(|F|+2)^{|F|}$ different mapping $c : [p] \to Z$, we conclude that there exists a mapping $c : [p] \to Z$ such that

$$\sum_{\phi\in\mathcal{I}_c} \prod_{i=1}^{p} \sigma_\zeta(F_i, Y_{\phi(i)}) \geq \frac{1}{(|F|+2)^{|F|}} \sigma_\zeta(F,Y). \tag{12.11}$$

Step 2. For $i \in [p]$, let $T_c(i) = \{j : \eta(i,j) = \phi(i)\}$.

Then obviously

$$\sum_{\phi \in \mathcal{J}_c} \prod_{i=1}^{p} \sigma_\zeta(F_i, Y_{\phi(i)}) \leq \prod_{i=1}^{p} \sum_{j \in T_c(i)} \sigma_\zeta(F_i, Y_j).$$

Consider the following partition of $[p]$ into five parts:

- The part W_1 contains the $i \in [p]$ such that $c(i) = \star$ and $|T_c(i)| < |F|n^\mu$,
- The part W_2 contains the $i \in [p]$ such that $c(i) = \star$ and $|T_c(i)| \geq |F|n^\mu$,
- The part W_3 contains the $i \in [p]$ such that $c(i) = 0$ and $|T_c(i)| < |F|n^\mu$,
- The part W_4 contains the $i \in [p]$ such that $c(i) = 0$ and $|T_c(i)| \geq |F|n^\mu$,
- The part W_5 contains the $i \in [p]$ such that $c(i) \notin \{\star, 0\}$.

Let $\epsilon_3 = \mu + \log|F|/\log n$. We thus have the following bounds:

$$\text{If } i \in W_1 \text{ then } \sum_{j \in T_c(i)} \sigma_\zeta(F_i, Y_j) \leq n^{\epsilon_3} n^{\epsilon_1} = n^{\epsilon_1 + \epsilon_3}, \tag{12.12}$$

$$\text{if } i \in W_2 \text{ then } \sum_{j \in T_c(i)} \sigma_\zeta(F_i, Y_j) \leq n.n^{\epsilon_1} = n^{1+\epsilon_1}, \tag{12.13}$$

$$\text{if } i \in W_3 \text{ then } \sum_{j \in T_c(i)} \sigma_\zeta(F_i, Y_j) \leq n^{\epsilon_3} n^{\epsilon_1} = n^{\epsilon_1 + \epsilon_3} \tag{12.14}$$

$$\text{if } i \in W_4 \text{ then } \sum_{j \in T_c(i)} \sigma_\zeta(F_i, Y_j) \leq n.n^{\epsilon_1} = n^{1+\epsilon_1} \tag{12.15}$$

$$\text{if } i \in W_5 \text{ then } \sum_{j \in T_c(i)} \sigma_\zeta(F_i, Y_j) \leq n^{c(i)+\epsilon_1}. \tag{12.16}$$

Let $g(c) = |W_2| + |W_4| + \sum_{i \in W_5} c(i)$. Then

$$\prod_{i=1}^{p} \sum_{j \in T_c(i)} \sigma_\zeta(F_i, Y_j) \leq n^{g(c)+p\epsilon_1+(|W_1|+|W_3|)\epsilon_3}.$$

Hence we have

$$\left\lceil \frac{\log \sigma_\zeta(F, Y)}{\log |Y|} - \epsilon \right\rceil \leq g(c)$$

if

$$\epsilon \geq p\epsilon_1 + (|W_1| + |W_3|)\epsilon_3 + \frac{|F| \log(|F| + 2)}{\log n}.$$

what is the case in particular if

$$\frac{\log(|F| + 2)}{\log n} \leq \frac{\epsilon}{12}.$$

Step 3. We denote by r_i the root of F_i and by s_j the root of Y_j. For each pair (F_i, Y_j) such that Y_j contains a ζ-consistent image of F_i we define $\zeta_{i,j} = \zeta/(\mathrm{root}(F_i) \to \mathrm{root}(Y_j))$. Hence for any such pair we have $\zeta((\gamma(r_i)), (\gamma(s_j))) = 1$ and

$$\sigma_{\zeta_{i,j}}(F_i - r_i, Y_j - s_j) = \sigma_\zeta(F_i, Y_j)$$

Let $\phi_0 \in \mathcal{J}_c$. For $i \in W_1 \cup W_3$ let Y_i' be a ζ-consistent image of F_i in $Y_{\phi_0(i)}$. For $i \in W_5$, $Y_{\phi_0(i)}$ includes a ζ-consistent image of F_i. Hence

$$\left\lceil \frac{\log \sigma_{\zeta_{i,\phi_0(i)}}(F_i - r_i, Y_{\phi_0(i)} - s_{\phi_0(i)})}{\log|Y_{\phi_0(i)} - s_{\phi_0(i)}|} - \epsilon_1 \right\rceil \geq \left\lceil \frac{\log \sigma_\zeta(F_i, Y_{\phi_0(i)})}{\log|Y_{\phi_0(i)}|} - \epsilon_1 \right\rceil$$
$$\geq c(i).$$

As $|Y_{\phi_0(i)} - s_{\phi_0(i)}| \geq n^{\epsilon_1} \geq N_{h-1}(\epsilon_1)$ the induction holds and there exists an antichain A_i of $F_i - r_i$ and a mapping $\rho_i : A_i \to \mathbb{N}$ such that

- The graph $(F_i - r_i) \curlyvee (A_i, \rho_i)$ has (at least) a $\zeta_{i,\phi_0(i)}$-consistent image in $Y_{\phi_0(i)} - s_{\phi_0(i)}$ (hence $F_i \curlyvee (A_i, \rho_i)$ has (at least) a ζ-consistent image in $Y_{\phi_0(i)}$),
- For every $a \in A_i$, we have $\rho_i(a) \geq n^{\epsilon_1 \mu_t(\epsilon_1)}$ where $t = \mathrm{height}(a, F_i - r_i)$ (hence $\rho_i(a) \geq n^{\mu_{t'}(\epsilon)}$ where $t' = t + 1 = \mathrm{height}(a, F_i)$),
- The cardinality of A_i is at least

$$\left\lceil \frac{\log \sigma_{\zeta_{i,\phi_0(i)}}(F_i - r_i, Y_{\phi_0(i)} - s_{\phi_0(i)})}{\log|Y_{\phi_0(i)} - s_{\phi_0(i)}|} - \epsilon_1 \right\rceil \geq c(i).$$

For i in $W_2 \cup W_4$ we can choose disjoint subsets X_i of $T_c(i)$ of size $n' \geq n^{\epsilon_3}/|F| = n^\mu$. In each of the Y_j (for $j \in X_i$) we extract a $\zeta_{i,j}$-consistent image of F_i. □

12.4 Counting Subgraphs in Graphs with Bounded Tree Depth

For the benefit of the reader let us recall some basics of tree-depth definition (see Chap. 6). Let Y be a rooted forest. The vertex x is an *ancestor* of y in Y if x belongs to the path linking y and the root of the tree of Y to which y belongs. The *closure* $\mathrm{clos}(Y)$ of a rooted forest Y is the graph with vertex set $V(Y)$ and edge set $\{\{x, y\} : x \text{ is an ancestor of } y \text{ in } Y, x \neq y\}$. The *tree-depth* $\mathrm{td}(G)$ of a graph G is the minimum height of a rooted forest Y such that $G \subseteq \mathrm{clos}(Y)$. We shall need the following refinement.

Let G be a graph. A *td-representation* of G is a triple (Y, ν, γ) where:

- $\nu : V(G) \to V(Y)$ is an injective homomorphism of G to the closure of Y,
- Each vertex of Y is an ancestor of an image $\nu(x)$ of a vertex $x \in V(G)$,
- $\gamma : V(G) \to \{0,1\}^*$ is a bit string defined as follows (see Fig. 12.4):
  - $\gamma(x) = \epsilon$ (i.e. $\gamma(x)$ is empty) if x is not in the image of ν;
  - Otherwise, $\gamma(x) = (0)$ if x is a root;
  - Otherwise, $\gamma(x)$ is a bit string of length height$(x, Y) - 1$, whose ith bit is 1 if there exists adjacent u, v in G such that $\nu(u) = \pi_Y^i(x)$ and $\nu(v) = x$ (where π_Y maps each non-root vertex of Y to its father in Y).

We assume that the vertex set of Y is $\{1, 2, \ldots, |Y|\}$ (without loss of generality).

Some easy fact should be noticed about td-representations:

Fact 4. *Let $\Xi(G, p)$ denote the number of different td-representations (Y, ν, γ) of a graph G such that the height of Y is at most p. Then* $\Xi(G, p) \le (|G|p)^{|G|p}$.

Proof. Assume that (Y, ν, γ) is a td-representation of G. As the height of Y is at most p and as all the leaves of Y belongs to the image of ν, the forest Y has order at most $|G|p$. Thus there are at most $(|G|p)^{|G|p}$ choices for the pair (Y, ν) hence for the choice of a td-representation. □

Lemma 12.2. *Let F be a graph, let G be a graph, and let (Y, ν^Y, γ^Y) be a td-representation of G. Then the number of copies of F in G is the sum over all td-representations (T, ν^T, γ^T) of F such that* height$(T) \le$ height(Y) *of the ζ-consistent images of the pattern $(T, (\nu^T, \gamma^T))$ in $(Y, (\nu^Y, \gamma^Y))$, where ζ is defined by*

$$\zeta(((\ldots, (\nu_s, \gamma_s)), ((\ldots, (\nu'_s, \gamma'_s))) = \begin{cases} 1, & \text{if } \nu_s = 0 \text{ or } \nu'_s = 1 \text{ and } \forall i\ (\gamma_s)_i \le (\gamma'_s)_i \\ 0, & \text{otherwise.} \end{cases}$$

Proof. Copies of T are fully determined by the mapping of the leaves of F into Y. □

Lemma 12.2 allows us to reduce the counting in bounded tree-depth graphs to counting of a given pattern in trees. The following result is the main result of the section.

Theorem 12.4. *Let F be a graph, let $0 < \epsilon < 1$, and let t be a positive integer.*

Then there exists $N_t(\epsilon)$ such that every graph G of tree-depth at most t and order at least $N_t(\epsilon)$ contains a (k, F)-sunflower $(C, \mathcal{F}_1, \ldots, \mathcal{F}_k)$ where

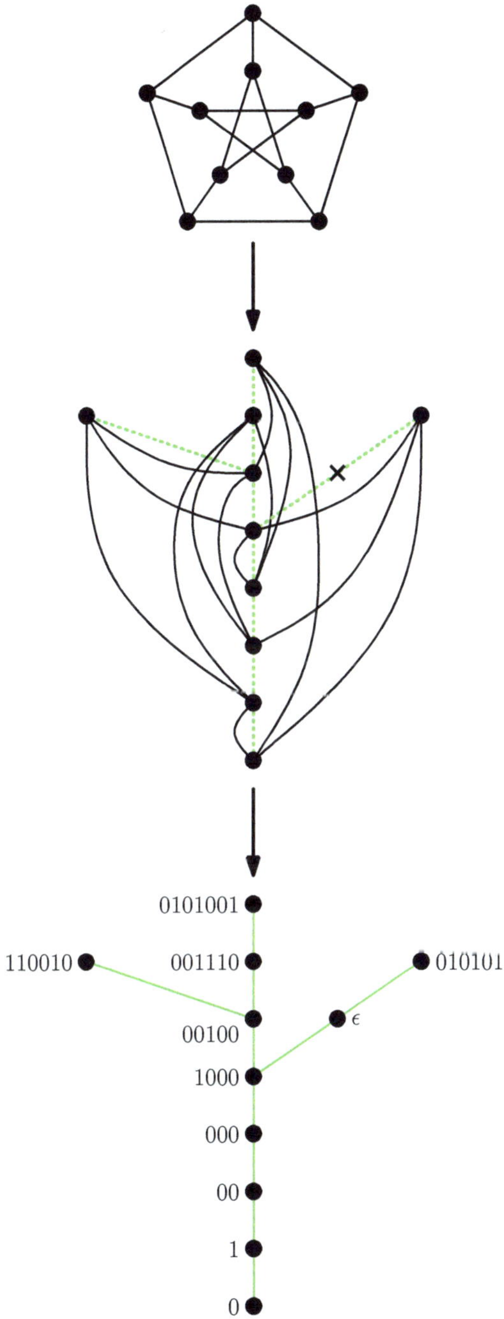

Fig. 12.4 A td-representation of the Petersen graph

$$\min |\mathcal{F}_i| \geq |G|^{(3|F|t)^{-\frac{t(t+1)}{2}}\epsilon^t} \tag{12.17}$$

$$k \geq \left\lceil \frac{\log(\#F \subseteq G) - p^2|F|\log(p^2|F|)}{\log|G|} - \epsilon \right\rceil. \tag{12.18}$$

Proof. As $\mathrm{td}(G) \leq t$, there exists a td-representation (Y, ν^Y, γ^Y) of G such that $\mathrm{height}(Y) \leq t$. According to Lemma 12.2, there exists a td-representation (T, ν^T, γ^T) such that $\sigma_\zeta(T, Y) \geq (\#F \subseteq G)/\Xi(T, t) \geq (\#F \subseteq G)/(p^2|F|)^{p^2|F|}$ (as $|T| \leq p|F|$). The result then follows from Theorem 12.3. □

Corollary 12.2. *Let* F *be a graph and let* $0 < \epsilon < 1$*. Then there exist a positive integer* N *and a positive real* τ *(depending on both* F *and* ϵ*) such that every graph* G *of order at least* N *and tree-depth at most* $|F|$ *contains a* (k, F)*-sunflower* $(C, \mathcal{F}_1, \dots, \mathcal{F}_k)$ *where*

$$\min |\mathcal{F}_i| \geq |G|^\tau \quad \text{and} \quad k \geq \left\lceil \frac{\log(\#F \subseteq G)}{\log|G|} - \epsilon \right\rceil.$$

12.5 Counting Subgraphs in Graphs

Recall that for integer p and a graph G, a p-*tree-depth coloring* of a graph G by N colors is a coloration of the vertices of G by N colors, such that each $p' \leq p$ color classes induce a subgraph of tree-depth at most p'. Accordingly, a sequence of chromatic numbers $\chi_1(G) \leq \chi_2(G) \leq \cdots \leq \mathrm{td}(G)$ is associated to a graph G, where $\chi_1(G)$ is the usual chromatic number $\chi(G)$ of G and where for every integer p the value $\chi_p(G)$ is the minimum number of colors of a p-tree-depth coloring of G (see Chap. 7).

As a consequence of previous sections we shall now prove that every large graph with a small χ_p contains a large (k, F)-sunflower:

Theorem 12.5 (Clearing and Stepping up). *Let* F *be a graph of order* p *and let* $0 < \epsilon < 1$*. Then there exist positive reals* c *and* τ *(depending on both* F *and* ϵ*) such that every graph* G *which contains more than* $|G|^{k+\epsilon}$ *copies of* F *and for which* $\chi_p(G) < c|G|^{\epsilon/p}$ *contains a* $(k+1, F)$*-sunflower* $(C, \mathcal{F}_1, \dots, \mathcal{F}_{k+1})$ *where*

$$\min |\mathcal{F}_i| \geq \left(\frac{|G|}{\chi_p(G)^{p/\epsilon}}\right)^\tau.$$

Proof. If $(\#F \subseteq G) < |G|^\epsilon$ then the statement is straightforward, so we can assume $(\#F \subseteq G) \geq |G|^\epsilon$. Consider a p-tree-depth coloring of G using $\chi_p(G)$ colors. Obviously any copy of F in G belongs to some subgraph of G induced by p colors (maybe even several such ones). Hence there exists a subset of p colors inducing a subgraph G' such that

$$(\#F \subseteq G') \geq (\#F \subseteq G)/\binom{\chi_p(G)}{p}$$

and

$$\begin{aligned}|G| \geq |G'| &\geq \left((\#F \subset G)/\binom{\chi_p(G)}{p}\right)^{1/p}\\ &\geq (\#F \subseteq G)^{(1/p)}/\chi_p(G)\\ &\geq |G|^{\epsilon/p}/\chi_p(G).\end{aligned}$$

By adapting the constants of Corollary 12.2 we conclude the proof. □

12.6 Counting Subgraphs in Graphs in a Class

We now apply the characterization of nowhere dense classes based on χ_p of Theorem 7.9.

Theorem 12.6. *For every hereditary nowhere dense class of graphs $\mathcal{C}$ and every fixed graph* F*:*

$$\limsup_{G\in\mathcal{C}} \frac{\log(\#F \subseteq G)}{\log|G|} \in \{-\infty, 0, 1, \dots, \alpha(F)\}. \tag{12.19}$$

Proof. Let

$$\alpha = \limsup_{G\in\mathcal{C}} \frac{\log(\#F \subseteq G)}{\log|G|}.$$

If $\alpha = -\infty$, we are done. Otherwise, consider an infinite sequence $(G_n)_{n\in\mathbb{N}}$ of graphs with strictly increasing orders such that

$$\lim_{n\to\infty} \frac{\log(\#F \subseteq G_n)}{\log|G_n|} = \alpha.$$

For every $\epsilon > 0$, let c and τ be the positive reals appearing in Theorem 12.5. According to Theorem 7.9, as $\mathcal{C}$ is nowhere dense, we have $\log\chi_{|F|}(G) = o(\log|G|)$. Hence there exists $N(\epsilon)$ such that every graph G_n with $n > N(\epsilon)$ is such that $\chi_p(G_n) < \max(c,1)|G|^{\epsilon/2p}$ and G_n contains at least $|G_n|^{\alpha-\epsilon}$

copies of F. Then, according to Theorem 12.5, G_n contains a (k, F)-sunflower $(C, \mathcal{F}_1, \dots, \mathcal{F}_k)$ where

$$\min |\mathcal{F}_i| \geq \left(\frac{|G_n|}{\chi_p(G_n)^{p/\epsilon}}\right)^{\tau} \geq |G_n|^{\tau/2}$$
$$\text{and} \quad k \geq \left\lceil \frac{\log(\#F \subseteq G)}{\log |G|} - \epsilon \right\rceil \geq \lceil \alpha - 2\epsilon \rceil.$$

Let H_n be the subgraph of G_n induced by this sunflower. As $\lim_{n\to\infty} |H_n| = \infty$, we have

$$\limsup_{n\to\infty} \frac{\log(\#F \subseteq H_n)}{\log |H_n|} \geq k.$$

Due to the maximality of α we have $\alpha \geq \lceil \alpha - 2\epsilon \rceil$ hence α is an integer (as the inequality holds for every $\epsilon > 0$). Moreover, $k \leq \alpha(F)$ (as noticed in Fact 1) hence $\alpha \in \{0, 1, \dots \alpha(F)\}$. □

Finally, we get a new characterization of nowhere dense classes by counting. This is stated as Theorem 12.1 and (for the benefit of the reader) again in the following form:

Theorem 12.7. *For every infinite class of graphs $\mathcal{C}$ and every fixed graph F with at least one edge:*

$$\mathrm{free}(F, \mathcal{C}) = \sup_r \limsup_{G \in \mathcal{C} \,\widetilde{\nabla}\, r} \frac{\log(\#F \subseteq G)}{\log |G|} \in \{-\infty, 0, 1, \dots, \alpha(F), |F|\}. \quad (12.20)$$

Moreover, $\mathrm{free}(F, \mathcal{C}) = |F|$ *if and only if* $\mathcal{C}$ *is somewhere dense.*

Proof. If the class $\mathcal{C}$ is nowhere dense then, according to Theorem 12.6, we have for every integer r:

$$\limsup_{G \in \mathcal{C} \,\widetilde{\nabla}\, r} \frac{\log(\#F \subseteq G)}{\log |G|} \in \{-\infty, 0, 1, \dots, \alpha(F)\}.$$

Hence the same holds for the supremum.

If $\mathcal{C}$ is somewhere dense then there exists a time $\widetilde{\tau}(\mathcal{C})$ such that $\mathcal{C} \,\nabla\, t$ contains every complete graph, hence for every $r \geq \widetilde{\tau}(\mathcal{C})$ we have

$$\limsup_{G \in \mathcal{C} \,\widetilde{\nabla}\, r} \frac{\log(\#F \subseteq G)}{\log |G|} = |F|.$$

□

And we have a similar result when we consider the resolution $\mathcal{C}^{\nabla}$ instead of the topological resolution $\mathcal{C}^{\widetilde{\nabla}}$:

Theorem 12.8. *For every infinite class of graphs* $\mathcal{C}$ *and every fixed graph* F *with at least one edge:*

$$\sup_{r\to\infty} \limsup_{G\in\mathcal{C}\,\triangledown\, r} \frac{\log(\#F \subseteq G)}{\log|G|} \in \{-\infty, 0, 1, \dots, \alpha(F), |F|\}. \qquad (12.21)$$

Moreover, $\lim_{r\to\infty} \limsup_{G\in\mathcal{C}\,\triangledown\, r} \frac{\log(\#F\subseteq G)}{\log|G|} = |F|$ *if and only if* $\mathcal{C}$ *is somewhere dense.*

Proof. This is essentially the same proof as Theorem 12.7. □

Exercises

12.1. Let $\mathcal{C}$ be a class of k-degenerate graphs and let F be a connected graph. Then for every $\epsilon > 0$ there exists $C = C(F, k, \epsilon)$ such that every graph $G \in \mathcal{C}$ has a subset of vertices A with

$$|A| \leq \epsilon |G|$$
$$(\#F \subseteq G - A) \leq C|G|$$

12.2. The *homomorphism domination exponent* HDE(F, G) of graphs F and H is defined as the supremum of all real numbers c such that for every graph T it holds

$$\hom(F,T) \geq \hom(G,T)^c.$$

This parameter has been introduced and studied in [282]. (The exercise follows this paper.)

Prove the following dual expression for HDE(F, G):

$$\mathrm{HDE}(F,G) = \inf_{T:\ \hom(G,T)\geq 2} \frac{\log \hom(F,T)}{\log \hom(G,T)}.$$

Prove that the supremum is always attained, but the infimum is not always attained.
Prove that the relation $F \succeq G$ defined by $\mathrm{HDE}(F,G) \geq 1$ is a partial order that extends the order induced by existence of a surjective homomorphism.

12.3. Prove that Theorem 12.1 holds when the number of copies of F is replaced by the number of homomorphisms from F, that is:

Let $\mathcal{C}$ be an infinite class of graphs and let F be a fixed graph with at least one edge and stability number $\alpha(F)$. Then the limit

$$\sup_r \limsup_{G \in \mathcal{C} \,\widetilde{\nabla}\, r} \frac{\log \hom(F,G)}{\log |G|}$$

is $|F|$ if and only if $\mathcal{C}$ is somewhere dense and otherwise (i.e. if $\mathcal{C}$ is nowhere dense) the limit belongs to $\{-\infty, 0, 1, \dots, \alpha(F)\}$.

12.4. The aim of this Exercise is to extend Theorem 12.5 to k-rooted graphs.

Let F be a graph and let $\mathbf{f} = (f_1, \dots, f_p)$ be a p-tuple of distinct vertices of F. We say that a p-tuple $\mathbf{v} = (v_1, \dots, v_p)$ of vertices G extends to F if there exists a subset A of vertices of G which contains $v_1, \dots, v_p$ and an isomorphism $g : F \to G[A]$ with $g(f_i) = v_i$. We denote by $(\#(F, \mathbf{f}) \subseteq G)$ the number of p-tuples of vertices of G that extend to F.

Following the proofs of this chapter, prove that the following extension of Theorem 12.5 holds for k-rooted graphs.

For every $0 < \epsilon < 1$, there exist positive reals c and τ (depending on p, F and ϵ) such that every graph G which contains more than $|G|^{k+\epsilon}$ p-tuple that extend to F and which is such that $\chi_{|F|}(G) < c|G|^{\epsilon/|F|}$ there exists a partition $A_1, \dots, A_{k+1}$ of $\{f_1, \dots, f_p\}$, a partition $F_0, \dots, F_{k+1}$ of the vertex set of F (here F_0 is allowed to be empty), and a k+2 families $\mathcal{F}_0, \mathcal{F}_1, \dots, \mathcal{F}_{k+1}$ of subset of vertices of G such that:

- For every $1 \leq i \leq k+1$, $A_i \subseteq F_i$,
- There is no edge between F_i and F_j if $0 < i < j \leq k+1$,
- $\mathcal{F}_0$ contains a single (possibly empty) set V_0,
- The sets in $\bigcup \mathcal{F}_i$ are pairwise disjoint,
- For every choice of $k+1$ sets $V_i \in \mathcal{F}_i$, there exists an isomorphism $g : F \to G[V_0 \cup \dots \cup V_{k+1}]$ such that $g(F_i) = V_i$,
- For $i > 0$, the family $\mathcal{F}_i$ contains at least $\left(\frac{|G|}{\chi_{|F|}(G)^{|F|/\epsilon}}\right)^\tau$ sets.

12.5. Deduce from Exercise 12.4 that the following extension of Theorem 12.6 holds:

For every hereditary nowhere dense class of graphs $\mathcal{C}$ and every fixed existential formula $\phi(x_1, \dots, x_p)$:

$$\limsup_{G \in \mathcal{C}} \frac{\log |\{(v_1, \dots, v_p) \in V(G)^p : \ G \models \phi(v_1, \dots, v_p)\}|}{\log |G|} \in \{-\infty, 0, 1, \dots, p\}.$$

Chapter 13
Back to Classes

When the class is sparse, the teacher is far away...

In this chapter we summarize the results on sparsity of classes with all their characterizations. The multiplicity of the equivalent characterizations that can be given for the nowhere dense–somewhere dense dichotomy is mainly a consequence of several related aspects:

- The relationships between the different type of resolutions, namely of minor resolution, topological resolution, and immersion resolution;
- The relationships between shallow minors, shallow topological minors, lexicographic products and shallow immersions;
- The polynomial dependence (and weak polynomial dependence) of key graph invariants, like ∇_r, $\widetilde{\nabla}_r$, χ_p, col_p, wcol_p, etc.
- The characterization of uniformly quasi-wide classes.

This will be elaborated in detail in this chapter.

13.1 Resolutions

Our main classification, the nowhere dense–somewhere dense dichotomy, is based on resolutions. We defined several types of resolutions: the minor resolution

$$\mathcal{C}^{\nabla} = (\mathcal{C} \mathbin{\nabla} 0, \mathcal{C} \mathbin{\nabla} 1/2, \mathcal{C} \mathbin{\nabla} 1, \dots),$$

the topological resolution

$$\mathcal{C}^{\widetilde{\nabla}} = (\mathcal{C} \mathbin{\widetilde{\nabla}} 0, \mathcal{C} \mathbin{\widetilde{\nabla}} 1/2, \mathcal{C} \mathbin{\widetilde{\nabla}} 1, \dots),$$

J. Nešetřil and P.O. de Mendez, *Sparsity*, Algorithms and Combinatorics 28, DOI 10.1007/978-3-642-27875-4_13, © Springer-Verlag Berlin Heidelberg 2012

and the immersion resolution

$$\mathcal{C}^{\widetilde{\nabla}} = (\mathcal{C} \mathbin{\widetilde{\nabla}} (1,0), \mathcal{C} \mathbin{\widetilde{\nabla}} (2,1/2), \mathcal{C} \mathbin{\widetilde{\nabla}} (3,1), \ldots).$$

As expected, these resolutions may behave differently. For instance, consider the minor resolution $\mathcal{C}^{\nabla}$ of a class $\mathcal{C}$ and its limit $\mathcal{C} \nabla \infty$ (i.e. its minor closure):

The first possibility is that the class $\mathcal{C} \nabla \infty$ is strictly included in $\mathcal{G}raph$:

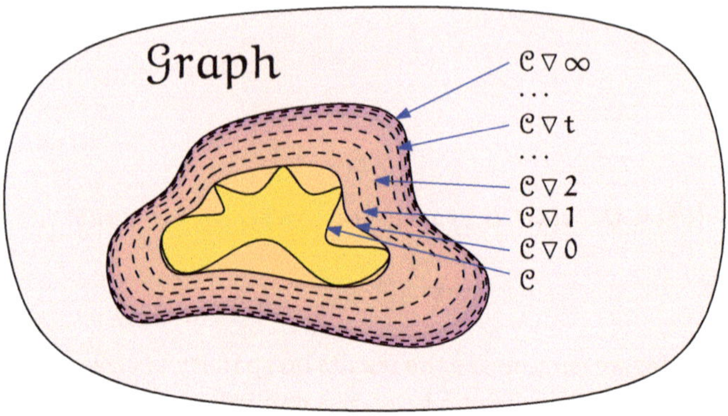

This is the case when $\mathcal{C}$ is included in a proper minor closed class. The minor resolution may then be used to get a finer information about subclasses with smaller density. For instance, if $\mathcal{P}lanar$ is the class of all planar graphs and $\mathcal{C}_1$ is the subclass of $\mathcal{P}lanar$ with graphs of girth at least g, then $\mathcal{C}_1 \nabla \frac{1}{2}\lceil \log_2(g/3)\rceil = \mathcal{C}_1 \nabla \infty = \mathcal{P}lanar$. However, if $\mathcal{C}_2$ is the subclass of $\mathcal{P}lanar$ with graphs of maximum degree 3, then $\mathcal{C}_2 \nabla t$ is the class of planar graphs with maximum degree 3.2^t and thus the class $\mathcal{P}lanar$ is only reached at the limit.

The second possibility is that the class $\mathcal{C} \nabla \infty$ is equal to $\mathcal{G}raph$, although $\mathcal{C} \nabla t$ is strictly included in $\mathcal{G}raph$ for each t:

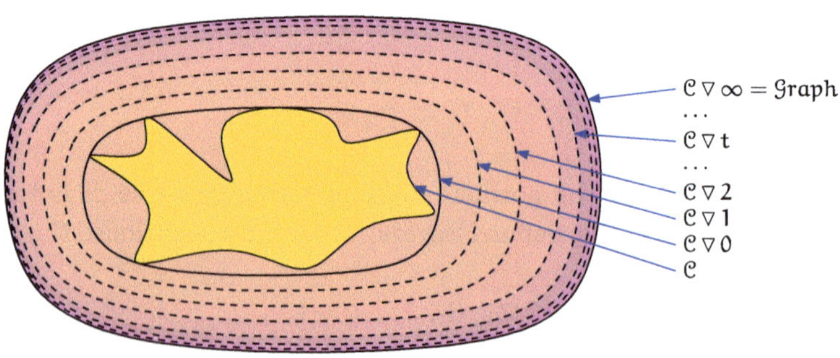

This is the case when $\mathcal{C}$ is nowhere dense but is not included in a proper minor closed class. Such a situation allows us to parametrize graphs by the minor resolution of $\mathcal{C}$, by associating to each graph G the minimum integer t (interpreted as "time") such that $G \in \mathcal{C} \triangledown t$. The interest of this scaling stands in its universality (it applies to every graph) and its extension (its set of values is not finite). In this book, we have seen that many hard problems become tractable in this parametrization.

The third possibility is that there exists a time t_c such that $\mathcal{C} \triangledown t_c = \mathcal{C} \triangledown \infty = \mathcal{G}raph$:

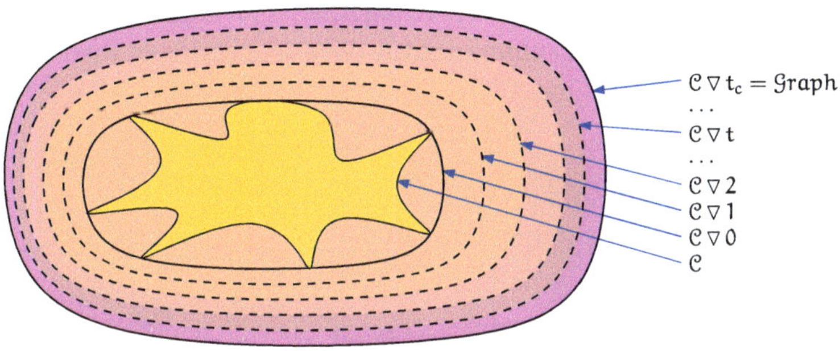

This is the case when $\mathcal{C}$ is somewhere dense. An extremal case is, of course, when $\mathcal{C}$ is the class $\mathcal{G}raph$ itself. Such a situation suggests that properties of the class $\mathcal{C}$ could be obtained by transporting general properties of dense graphs backward. We have seen in Chap. 5 that the classes for which the third case applies, that is somewhere dense classes, are those which have the property that there exists a critical value t_c at which the resolution stabilizes to $\mathcal{G}raph$, whatever resolution we consider (the minor resolution, the topological resolution, or the immersion resolution). However, the value t_c depends on the considered resolution. Not only that, but also for nowhere dense classes the asymptotic behavior of our resolutions varies. For instance:

- The class $\mathcal{T}_3$ of graphs with tree-depth at most 3 is such that (see Chap. 4, Fig. 4.11)

$$\mathcal{T}_3 \overset{\propto}{\triangledown} \infty = \mathcal{G}raph, \text{ but } \mathcal{T}_3 \triangledown \infty = \mathcal{T}_3 \widetilde{\triangledown} \infty = \mathcal{T}_3;$$

- The class $\mathcal{D}_3$ of graphs with maximum degree at most 3 is such that

$$\mathcal{D}_3 \triangledown \infty = \mathcal{G}raph \text{ but both } \mathcal{D}_3 \overset{\propto}{\triangledown} \infty = \mathcal{D}_3 \text{ and } \mathcal{D}_3 \widetilde{\triangledown} \infty = \mathcal{D}_3.$$

Looking at these examples (and similar other ones) it is perhaps surprising how stable the nowhere dense–somewhere dense dichotomy is.

13.2 Parameters

The equivalence between minor resolution and topological resolution in the definition of the nowhere dense–somewhere dense dichotomy is based on the relations linking the clique numbers in both resolution. These relations have been stated in Proposition 5.2. The equivalence between the shallow minor approach and the topological minor approach (from the point of view of our classification of classes) is confirmed by the polynomial equivalence of grads and top-grads (see Sect. 4.5). Altogether, the dependencies of graph parameters shown Fig. 13.1 summarize the deep connection of shallow minors and shallow topological minors.

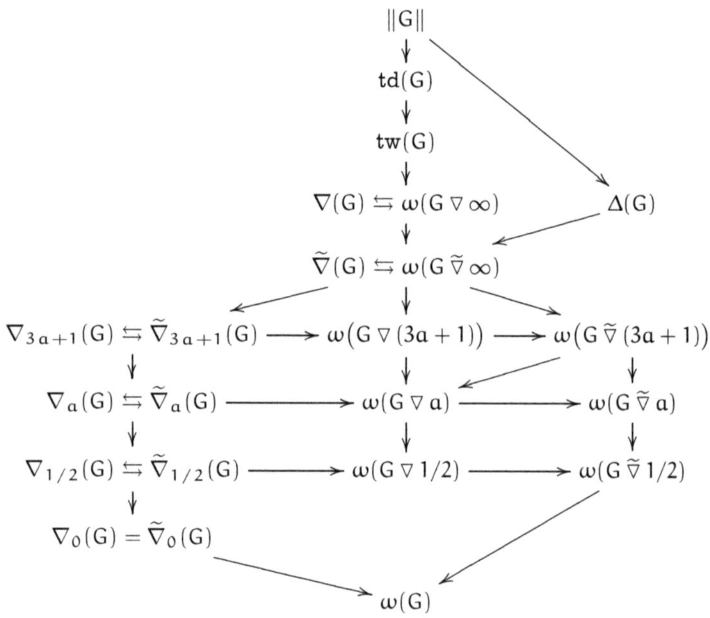

Fig. 13.1 Dominance of some graph parameters (up to polynomial parameters)

In Fig. 13.1, arrows mean polynomial dependencies. For instance, $\mathrm{tw}(G) \to \nabla(G)$ means that there exists a polynomial P such that $\nabla(G) \leq P(\mathrm{tw}(G))$. In this particular case, $P(X) = X$ as $\|H\|/|H| \leq \mathrm{tw}(H) \leq \mathrm{tw}(G)$ holds for every minor H of G.

We summarize in Table 13.1 the relationship of the graph parameters.

Table 13.1 Dependencies between several parameters (p is any positive integer)

$\widetilde{\nabla}_r(G) \le \widetilde{\nabla}_r(G \bullet K_p)$	(obvious)	(13.1)
$\widetilde{\nabla}_r(G) \le \nabla_r(G)$	(Corollary 4.1)	(13.2)
$\widetilde{\nabla}_r(G) \le \overset{\propto}{\nabla}_{p,r}(G)$	(obvious)	(13.3)
$\widetilde{\nabla}_r(G \bullet K_p) \le p(p+2r)\widetilde{\nabla}_r(G)$	(by Proposition 4.6)	(13.4)
$\widetilde{\nabla}_r(G) = O(\chi(G \,\widetilde{\nabla}\, (2r+\frac{1}{2}))^4)$	(Proposition 4.4)	(13.5)
$\nabla_r(G) \le 4(4\widetilde{\nabla}_r(G))^{(r+1)^2}$	(Corollary 4.1)	(13.6)
$\nabla_r(G) \le \mathrm{wcol}_{2r+1}(G) - 1$	(Lemma 7.11)	(13.7)
$\nabla_r(G) \le (2r+1)\binom{\chi_{2r+2}(G)}{2r+2}$	(Proposition 7.1)	(13.8)
$\overset{\propto}{\nabla}_{p,r}(G) \le (2r(p-1)+1)\widetilde{\nabla}_r(G)$	Eq. 4.20	(13.9)
$\chi(G \,\widetilde{\nabla}\, r) \le \chi(G \,\nabla\, r)$	(obvious)	(13.10)
$\chi(G \,\widetilde{\nabla}\, r) \le \chi(G \,\overset{\propto}{\nabla}\, (p,r))$	(obvious)	(13.11)
$\chi(G \,\overset{\propto}{\nabla}\, (p,r)) \le \chi(G \,\widetilde{\nabla}\, r)^{2r(p-1)+1}$	Eq. 4.21	(13.12)
$\chi(G \,\widetilde{\nabla}\, r) \le 2\widetilde{\nabla}_r(G) + 1$	(Proposition 4.4)	(13.13)
$\chi(G \,\nabla\, r) \le 2\nabla_r(G) + 1$	(Proposition 4.5)	(13.14)
$\mathrm{col}_r(G) \le \mathrm{wcol}_r(G)$	(Proposition 4.8)	(13.15)
$\mathrm{col}_r(G) \le A_r(\nabla_{\frac{r-1}{2}}(G))$	(Theorem 7.11)	(13.16)
$\mathrm{wcol}_r(G) \le \mathrm{col}_r(G)^r$	(Proposition 4.8)	(13.17)
$\chi_r(G) \le \mathrm{wcol}_{2^{r-1}}(G)$	(Theorem 7.10)	(13.18)
$\chi_r(G) \le B_r(\widetilde{\nabla}_{2^{r-2}+1/2}(G))$	(Theorem 7.8)	(13.19)
$\omega(G \,\widetilde{\nabla}\, r) \le \omega(G \,\nabla\, r)$	(obvious)	(13.20)
$\omega(G \,\widetilde{\nabla}\, r) \le \omega(G \,\overset{\propto}{\nabla}\, (p,r))$	(obvious)	(13.21)
$\omega(G \,\nabla\, r) \le 2\omega(G \,\widetilde{\nabla}\, (3r+1))^{\lfloor r \rfloor + 1}$	(Proposition 5.2)	(13.22)
$\omega(G \,\overset{\propto}{\nabla}\, (p,r)) < R(\overbrace{\omega(G \,\widetilde{\nabla}\, r)+1, \ldots}^{2r(p-1)+1})$	Eq. 4.22	(13.23)

To facilitate a uniform treatment of these bounds and several other dependencies between graph parameters (which will allow us to give a multiplicity of characterizations for nowhere dense classes or for classes with bounded expansion) we find it convenient to introduce the family $\mathcal{F}$ as the following graph parameters (parametrized by integral parameter r):

- $G \mapsto \nabla_r(G)$,
- $G \mapsto \widetilde{\nabla}_r(G)$,
- $G \mapsto \chi_r(G)$,
- $G \mapsto \chi(G \,\widetilde{\triangledown}\, r)$,
- $G \mapsto \chi(G \triangledown r)$,
- $G \mapsto \mathrm{col}_r(G)$,
- $G \mapsto \mathrm{wcol}_r(G)$.

We also add to the family $\mathcal{F}$ all the parametrized graph parameters defined from an arbitrary integral polynomial $P(X)$ taking only strictly positive values for $X \geq 0$ by:

- $G \mapsto \nabla_r(G \bullet K_{P(r)})$,
- $G \mapsto \widetilde{\nabla}_r(G \bullet K_{P(r)})$,
- $G \mapsto \overset{\propto}{\nabla}_{P(r),r}(G)$.

In a sense $\mathcal{F}$ is the family of all the nice parameters we have considered in this book.

13.3 Nowhere Dense Classes

Figure 13.2 schematically depicts the variety of important hereditary classes covered by our approach. It summarizes examples of nowhere dense classes scattered through the earlier chapters. Similarly, we shall now gather almost all of our characterizations of nowhere dense classes obtained in Chaps. 5–12 in a single theorem:

Theorem 13.1. *Let $\mathcal{C}$ be an infinite class of graphs.*

Let F be a graph with at least one edge, let f_r be a parametrized graph parameter in $\mathcal{F}$, and let $\mathfrak{C}$ be either $\mathcal{C}^{\triangledown}, \mathcal{C}^{\widetilde{\triangledown}}$ or $\mathcal{C}^{\overset{\propto}{\widetilde{\triangledown}}}$. Then the following conditions are equivalent:

1. *$\mathcal{C}$ is a class of nowhere dense graphs,*
2. *No class in the resolution $\mathfrak{C}$ has unbounded clique number,*
3. *There exists a weakly topological monotone graph parameter ϱ bounding the clique number parameter ω such that $\varrho(\mathcal{C}) < \infty$,*
4. *$\overline{\ell\mathrm{dens}}(\mathfrak{C}) \leq 1$,*
5. *For every integer r, $\displaystyle\limsup_{G \in \mathfrak{C}_r} \frac{\log(\#F \subseteq G)}{\log|G|} < |F|$,*
6. *For every integer r, $\displaystyle\limsup_{G \in \mathcal{C}} \frac{\log f_r(G)}{\log|G|} = 0$,*
7. *For every integer r, $\displaystyle\limsup_{G \in \mathfrak{C}_r} \frac{\log \chi(G)}{\log|G|} = 0$,*

8. For every integer r, $\liminf_{G \in \mathfrak{C}_r} \frac{\log \alpha(G)}{\log |G|} = 1$,

9. $\mathcal{C}$ *is a uniformly quasi-wide class,*

10. $H(\mathcal{C})$ *is a quasi-wide class.*

Proof. The equivalence (1) $\iff$ (2) follows from the definition for $\mathfrak{C} = \mathcal{C}^{\nabla}$; it follows from Proposition 5.2 for $\mathfrak{C} = \mathcal{C}^{\widetilde{\nabla}}$. The equivalence of (2) for $\mathcal{C}^{\overset{\propto}{\nabla}}$ and $\mathcal{C}^{\widetilde{\nabla}}$ follows from (4.22).

The equivalence (2) $\iff$ (3) follows from Proposition 5.1.

The equivalence (1) $\iff$ (4) follows from Theorem 5.4 and Proposition 5.4.

The equivalence (1) $\iff$ (5) follows from Theorem 12.7 and Theorem 12.8.

Let us prove the equivalence (4) $\iff$ (6): First notice that in (6) the choice of $f_r \in \mathcal{F}$ and $\mathfrak{C}$ among $\mathcal{C}^{\nabla}, \mathcal{C}^{\widetilde{\nabla}}$ and $\mathcal{C}^{\overset{\propto}{\nabla}}$ does not matter according to the dependencies recalled on Table 13.1. Thus assume that for some integer r_0 we have

$$\limsup_{G \in \mathcal{C}} \frac{\log \widetilde{\nabla}_{r_0}(G)}{\log |G|} = \epsilon > 0.$$

Then for every integer N there exists $G \in \mathcal{C} \,\widetilde{\nabla}\, r_0$ such that $|G| > N$ and $\frac{\|G\|}{|G|} > |G|^{\epsilon/2}$ hence $\frac{\log \|G\|}{\log |G|} > 1 + \epsilon/2$, a contradiction. The opposite direction follows from the easy inequality $\limsup_{G \in \mathcal{C}} \frac{\log \widetilde{\nabla}_r(G)}{\log |G|} \leq \limsup_{G \in \mathcal{C} \,\nabla\, r} \frac{\log \|G\|}{\log |G|} - 1$.

Let us consider the equivalence (6) $\iff$ (7): For $\mathfrak{C} = \mathcal{C}^{\widetilde{\nabla}}$, it follows from Proposition 4.4 while for $\mathfrak{C} = \mathcal{C}^{\nabla}$, it follows from Proposition 4.5. The case where $\mathfrak{C} = \mathcal{C}^{\overset{\propto}{\nabla}}$ reduces to the case where $\mathfrak{C} = \mathcal{C}^{\widetilde{\nabla}}$ thanks to (4.18).

Let us consider the equivalence (7) $\iff$ (8): If $\limsup_{G \in \mathfrak{C}_r} \frac{\log \chi(G)}{\log |G|} = 0$ then, as $\alpha(G)\chi(G) \geq |G|$ we have $\liminf_{G \in \mathfrak{C}_r} \frac{\log \alpha(G)}{\log |G|} = 1$. Now if $\limsup_{G \in \mathfrak{C}_r} \frac{\log \chi(G)}{\log |G|} > 0$ then $\mathcal{C}$ is somewhere dense thus there exists r_0 such that every complete graph belongs to $\mathfrak{C}_{r_0}$. It follows that $\liminf_{G \in \mathfrak{C}_{r_0}} \frac{\log \alpha(G)}{\log |G|} = 0$.

Finally, the equivalence (1) $\iff$ (9) $\iff$ (10) follows from Theorem 8.2.

□

13.4 Bounded Expansion Classes

The parameter equivalences proved in the first part of this book imply the following characterizations of classes with bounded expansions, much in the same style as Theorem 13.1.

Theorem 13.2. *Let $\mathcal{C}$ be a class of graphs. Let f_r be a parametrized graph parameter in $\mathcal{F}$, and let $\mathfrak{C}$ be either $\mathcal{C}^{\triangledown}, \mathcal{C}^{\widetilde{\triangledown}}$ or $\mathcal{C}^{\overset{\propto}{\triangledown}}$. Then the following conditions are equivalent:*

1. *$\mathcal{C}$ has bounded expansion,*
2. *No class in $\mathfrak{C}$ has unbounded average degree,*
3. *No class in $\mathfrak{C}$ has unbounded chromatic number,*
4. *There exists a weakly topological monotone parameter ϱ bounding the average degree parameter, such that $\varrho(\mathcal{C}) < \infty$,*
5. *There exists a weakly topological monotone parameter ϱ bounding the chromatic number, such that $\varrho(\mathcal{C}) < \infty$,*
6. *For every integer r, $\sup_{G \in \mathcal{C}} f_r(G) < \infty$,*
7. *For every integer r, $\limsup_{G \in \mathcal{C}} f_r(G) < \infty$,*
8. *$\mathcal{C}$ has low tree-width colorings,*

The proof is similar to the one of Theorem 13.1, using results related to bounded expansion classes (see Table 13.1).

13.5 Bounded Tree-Depth Classes

It is a bit surprising that the classes with bounded tree depth have a characterization much in the same style as bounded expansion classes (Theorem 13.2) and nowhere dense classes (Theorem 13.1). This we believe shows a coherence of our theory.

Theorem 13.3. *Let $\mathcal{C}$ be a class of graphs. The following conditions are equivalent:*

1. *$\mathcal{C}$ has bounded tree-depth,*
2. *There exists an integer $l(\mathcal{C})$ such that no graph $G \in \mathcal{C}$ includes a path of length greater than $l(\mathcal{C})$,*
3. *$\mathcal{C}$ is degenerate (i.e. $\nabla_0(\mathcal{C}) < \infty$) and there exists an integer $L(\mathcal{C})$ such that no graph $G \in \mathcal{C}$ includes an induced path of length greater than $L(\mathcal{C})$,*
4. *$\mathcal{C}$ is nowhere dense and there exists an integer $L(\mathcal{C})$ such that no graph $G \in \mathcal{C}$ includes an induced path of length greater than $L(\mathcal{C})$,*
5. *$\lim_{p \to \infty} \chi_p(\mathcal{C}) < \infty$ (which means that the $\chi_p(G)$ for $G \in \mathcal{C}$ are uniformly bounded by a constant independent of p).*
6. *$\lim_{p \to \infty} \mathrm{col}_p(\mathcal{C}) < \infty$,*
7. *$\lim_{p \to \infty} \mathrm{wcol}_p(\mathcal{C}) < \infty$.*

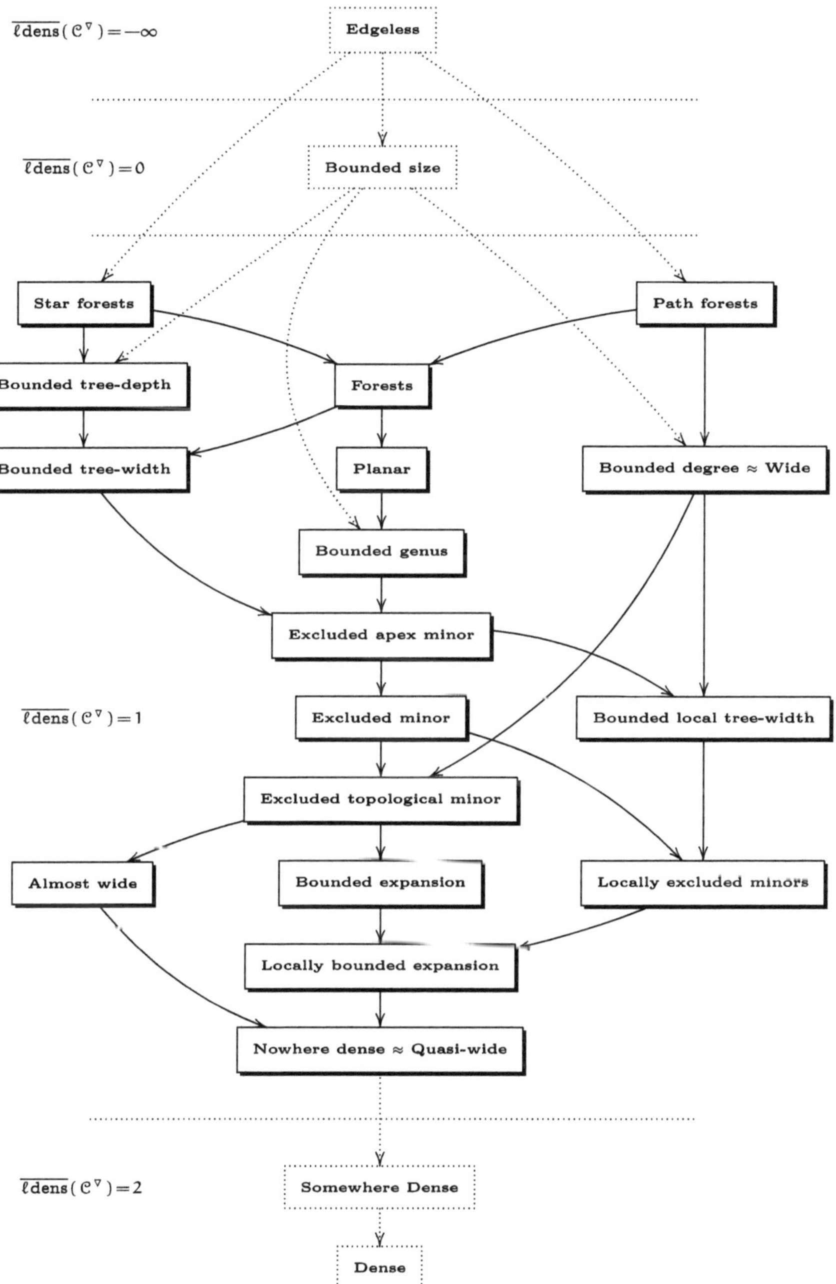

Fig. 13.2 Inclusion map of some hereditary classes

At this stage of the book, the proof of this theorem can be left as an exercise.

13.6 Remarks on Structures

As mentioned in Sects. 3.8 and 5.8, there are several possibilities to define a classification of classes of structures by means of our classification of classes of graphs: Gaifman graphs of the structures or Incidence graphs of the structures.

If we define classes of structures that are nowhere dense, bounded exansion classes, quasiwide, etc. by the corresponding properties of their Gaifman graphs, then , of course, the characterization theorems 13.1, 13.2, and 13.3 remain valid. We do not have to state this explicitely. However this should be regarded as a first (and often good) approximation of the properties of an infinite class of structures. However this is perhaps the beginning only and one should aim for more fitting notions. Consider for example the notion of quasi-wideness: a class of structures $\mathcal{C}$ is said to be *quasi-wide* if there exist functions f and g such that for every integers d, m and every structure $\mathbf{A} \in \mathcal{C}$ of order at least $f(d, m)$ there exists in the ground set of $\mathbf{A}$ a subset S of at most $g(d)$ elements and a subset I of at least m elements such that in $\mathbf{A} - S$, any two elements of I are at distance at least m. Also, a class of structures is *almost wide* if it is quasi-wide and one can require that the function g is constant. Notice that deleting an element in a relational structure implies to deletion of all the relations to which it belongs hence $\mathrm{Gaifman}(\mathbf{A} - S)$ is a subgraph of $\mathrm{Gaifman}(\mathbf{A}) - S$. Thus, as the distance in $\mathbf{A}$ and $\mathrm{Gaifman}(\mathbf{A})$ are the same, we have:

Proposition 13.1. *Let $\mathcal{C}$ be a class of relational structures.*

- *If* $\mathrm{Gaifman}(\mathcal{C})$ *is almost wide then* $\mathcal{C}$ *is almost wide;*
- *If* $\mathrm{Gaifman}(\mathcal{C})$ *is quasi wide then* $\mathcal{C}$ *is quasi wide.*

However the converse implications are not true in general, see Fig. 13.3.

It follows that G-nowhere dense structures are quasi-wide. This property can be used to prove that every G-nowhere dense class of relational structures admit a relativized version of the homomorphism preservation theorem.

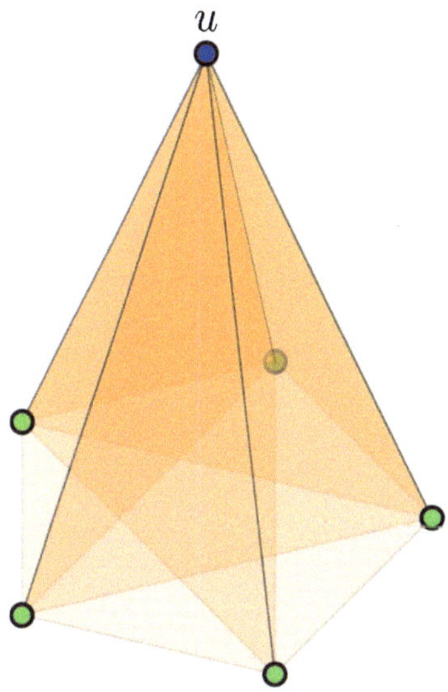

Fig. 13.3 The class of triple systems $\mathbf{A}_n$ with triples $\{v_i, v_j, u\}$ $(1 \leq i < j < n, n \geq 3)$ is almost wide (deleting vertex u fully disconnects the structure) although the class of the Gaifman graphs of the $\mathbf{A}_n$ is the (somewhere dense) class of all complete graphs of order at least 3

Applications

Chapter 14
Classes with Bounded Expansion – Examples

Bound to be free...

Bounded expansion classes are the focus of this chapter and one of the leitmotivs of the whole book. In this chapter, we shall give many examples of classes with bounded expansion. The examples which we cover are schematically depicted on Fig. 14.1. These classes cover most classes considered in structural graph theory and the relevant parts of logic and discrete geometry. This will be explained for several of these classes in a greater detail in this chapter.

In Sect. 14.1, we show that the notion of bounded expansion is compatible with Erdős-Rényi model of random graphs with constant average degree (that is, for random graphs of order n with edge probability d/n). Then, we provide a number of examples of classes with bounded expansion that appear naturally in the context of graph drawing or graph coloring. In particular, we prove that each of the following classes have bounded expansion, even though they are not contained in a (proper) topologically-closed class:

- Graphs that can be drawn with a bounded number of crossings per edge (Sect. 14.2),
- Graphs with bounded queue-number (Sect. 14.4),
- Graphs with bounded stack-number (Sect. 14.5),
- Graphs with bounded non-repetitive chromatic number (Sect. 14.6).

We also prove that graphs with "linear" crossing number are contained in a topologically-closed class, and graphs with bounded crossing number are contained in a minor-closed class (Sect. 14.2). Many of these results were obtained in collaboration with David Wood, see [359].

J. Nešetřil and P.O. de Mendez, *Sparsity*, Algorithms and Combinatorics 28,
DOI 10.1007/978-3-642-27875-4_14,

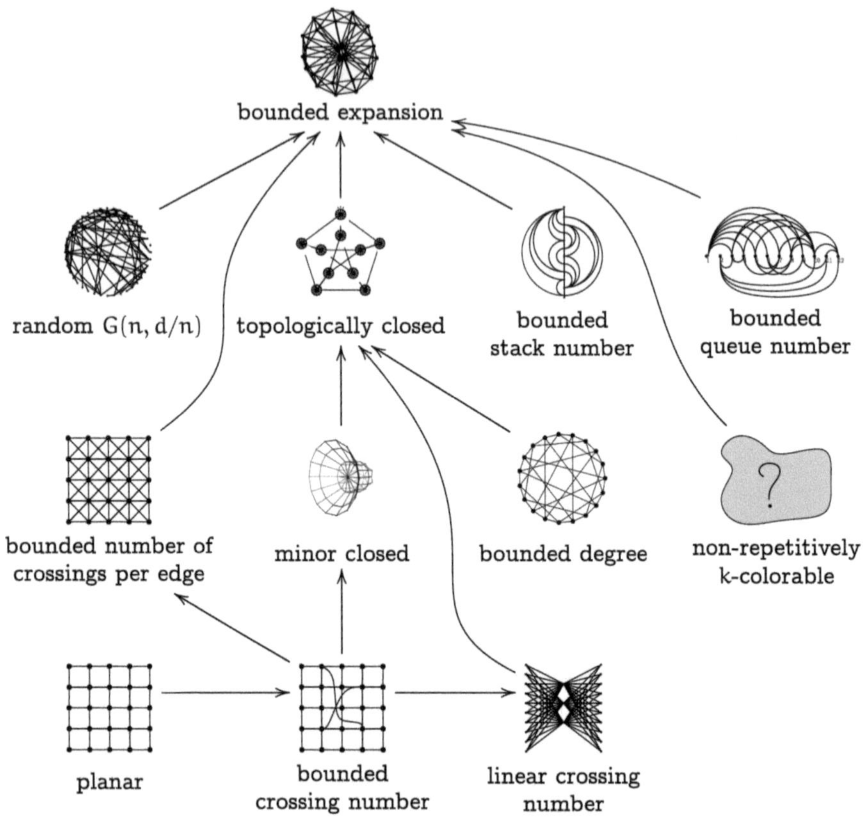

Fig. 14.1 Classes with bounded expansion. The results about classes with bounded crossings, bounded queue-number, bounded stack-number, and bounded non-repetitive chromatic number are proved in this chapter. Arrows represent class inclusion

14.1 Random Graphs (Erdős-Rényi Model)

The $G(n, p)$ model of random graphs was introduced by Gilbert [217] and Erdős and Rényi [170]. It is the most common random graph model, see e.g. [76]. In this model, a graph with n vertices is built, with each edge appearing independently with probability p. It is frequently considered that p may be a function of n, hence the notation $G(n, p(n))$ (see Fig. 14.2).

Let us review some basic facts about $G(n, d/n)$ and $G(n, p(n))$. The order of the largest complete (topological) minor in $G(n, p/n)$ was studied intensively. It is known since the work of [318] that random graphs $G(n, p(n))$ with $p(n) - 1/n \ll n^{-4/3}$ are asymptotically almost surely planar, whereas those with $p(n) - 1/n \gg n^{-4/3}$ asymptotically almost surely contain unbounded clique minors. Recall that a property of random graphs holds *asymptotically almost surely* (*a.a.s.*) if, over a sequence of draws, its probability converges to 1. Fountoulakis et al. [182] proved that for every $c > 1$ there exists a constant $\delta(c)$ such that asymptotically almost surely the maximum order $h(G(n, c/n))$ of a complete minor of a graph in $G(n, c/n)$ satisfies the inequality $\delta(c)\sqrt{n} \leq h(G(n, c/n)) \leq 2\sqrt{cn}$. Also, Ajtai et al. [6] proved that, as long as the expected degree $(n-1)p$ is at least $1+\epsilon$ and is $o(\sqrt{n})$, then asymptotically almost surely the order of the largest complete topological minor of $G(n, p)$ is almost as large as the maximum degree, which is $\Theta(\log n / \log \log n)$.

On the other hand, it is known that the number of short cycles of $G(n, d/n)$ is bounded. More precisely, the expected number of cycles of length t in $G(n, d/n)$ is at most $(e^2 d/2)^t$. It follows that the expected value $E(\omega(G \,\widetilde{\triangledown}\, r))$ of the clique size of a shallow topological minor of G at depth r is bounded by approximately $(Cd)^{2r}$.

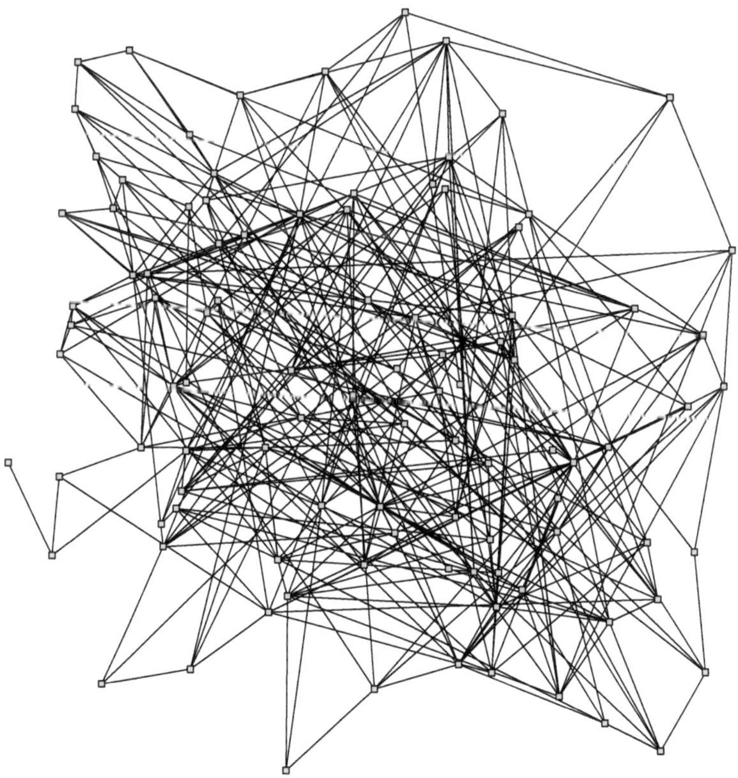

Fig. 14.2 A random graph with edge probability $6/n$ (here $n = 100$)

Fox and Sudakov [183] proved that $G(n, d/n)$ is asymptotically almost surely $(16d, 16d)$-degenerate. Here this undefined notion has the following meaning: a graph H is said to be (d, Δ)-*degenerate* if there exists an ordering $v_1, \dots, v_n$ of its vertices such that for each v_i, there are at most d vertices v_j adjacent to v_i with $j < i$, and there are at most Δ subsets of the form $N(v_j) \cap \{v_1, \dots, v_i\}$ for some neighbor v_j of v_i with $j > i$ (recall that the *neighborhood* $N(v_j)$ is the set of vertices that are adjacent to v_j). We refine [183] in order to prove that for each integer d there exists a bounded expansion class $\mathcal{R}_d$ such that $G(n, d/n)$ almost surely belongs to $\mathcal{R}_d$ (Theorem 14.1).

We shall proceed as follows: Using the characterization of bounded expansion given in Proposition 5.6 we first prove that graphs in $G(n, d/n)$ asymptotically almost surely have the property that only a small proportion of vertices have sufficiently large degree. We then prove that asymptotically almost surely subgraphs having sufficiently dense sparse topological minors must span some positive fraction of the vertex set of the whole graph. Thanks to Lemma 4.2, this last property will follow from the following two facts:

- As the random graph with edge probability d/n has a bounded number of short cycles, it follows that if one of its subgraphs is a $\leq r$-subdivision of a sufficiently dense graph it should asymptotically almost surely span at least some positive fraction $F_{\mathrm{prop}}(r)$ of the vertices (Lemma 14.2);
- For every $\epsilon > 0$, the proportion of the vertices of the random graph with edge probability d/n which have sufficiently large degree ($> F_{\mathrm{deg}}(\epsilon)$) is asymptotically almost surely less than ϵ (Lemma 14.3).

Let us give details.

Lemma 14.1. *Let* $\epsilon > 0$. *Asymptotically almost surely every subgraph* G' *of* $G(n, d/n)$ *with at most* $(4d)^{-(1+1/\epsilon)}n$ *vertices satisfies* $\widetilde{\nabla}_0(G') \leq 1+\epsilon$.

Proof. It is sufficient to prove that almost surely every subgraph G' of $G(n, d/n)$ with at most $4^{-(1+1/\epsilon)}n$ vertices satisfies $\|G'\|/|G'| \leq 1 + \epsilon$. Let G' be an induced subgraph of G of order t with $t \leq 4^{-(1+1/\epsilon)}n$. The probability that G' has size at least $m = (1+\epsilon)t$ is at most $\binom{\binom{t}{2}}{m}(d/n)^m$. Therefore, by the union bound, the probability that G has an induced subgraph of order t with size at least $m = (1 + \epsilon)t$ is

$$\begin{aligned}\binom{n}{t}\binom{\binom{t}{2}}{m}(d/n)^m &\le \left(\frac{en}{t}\right)^t\left(\frac{et^2}{2m}\right)^m\left(\frac{d}{n}\right)^m \\ &= e^t\left(\frac{e}{2(1+\epsilon)}\right)^{(1+\epsilon)t}\left(\frac{n}{t}\right)^t\left(\frac{dt}{n}\right)^{(1+\epsilon)t} \\ &= \left(\frac{e^{2+\epsilon}}{(2+2\epsilon)^{1+\epsilon}}\right)^t\left(\frac{d^{1+1/\epsilon}\ t}{n}\right)^{\epsilon t} \\ &< 4^t\left(\frac{d^{1+1/\epsilon}\ t}{n}\right)^{\epsilon t}.\end{aligned}$$

Summing over all $t \le (4d)^{-(1+1/\epsilon)}n$, one easily checks that the probability that G has an induced subgraph G' of order at most $(4d)^{-(1+1/\epsilon)}n$ such that $\|G\|/|G'| \ge 1+\epsilon$ is $o(1)$, completing the proof. □

Lemmas 4.2 and 14.1 imply:

Lemma 14.2. *Let* $r \in \mathbb{N}$. *Asymptotically almost surely every subgraph* G' *of* $G(n, d/n)$ *with at most* $(4d)^{-(1+1/(4r+1))}n$ *vertices satisfies* $\widetilde{\nabla}_r(G') \le 2$. *That is, for every positive integer* r *every subgraph* H *of* $G(n, d/n)$ *asymptotically almost surely satisfies:*

$$\widetilde{\nabla}_r(H) > 2 \Longrightarrow |H| > (4d)^{-(1+\frac{1}{4r+1})}|G| \ .$$

Lemma 14.3. *Let* $\alpha > 1$ *and let* $c_\alpha = 4e\alpha^{-4\alpha d}$. *Asymptotically almost surely there are at most* $c_\alpha n$ *vertices of* $G(n, d/n)$ *with degree greater than* $8\alpha d$.

Proof. Put $s = c_\alpha n$ (rounded to an even integer) and let A be the subset of the s vertices of largest degree in $G = G(n, d/n)$, and let D be the minimum degree of vertices in A. Thus there are at least $sD/2$ edges that have at least one endpoint in A. Consider a random subset A' of A with size $|A|/2$. Every edge that has an endpoint in A has probability at least $\frac{1}{2}$ of having exactly one endpoint in A'. So there is a subset $A' \subset A$ of size $|A|/2$ such that the number m of edges between A' and $V(G)\setminus A'$ satisfies $m \ge sD/4 = |A'|D/2$.

We now give an upper bound on the probability that $D \ge 8\alpha d$. Each set A' of $\frac{s}{2}$ vertices in $G = G(n, d/n)$ has probability at most

$$\binom{\frac{s}{2}(n-\frac{s}{2})}{m}(d/n)^m \le \left(\frac{esn}{2m}\right)^m(d/n)^m \le \left(\frac{2sd}{m}\right)^m \le \left(\frac{8d}{D}\right)^m \le \alpha^{-2\alpha ds}$$

of having at least $m \ge (s/2)(8\alpha d)/2 = 2\alpha ds$ edges between A' and $V(G)\setminus A'$. Therefore the probability that there is a set A' of $s/2$ vertices in G that has at least $2\alpha sd$ edges between A' and $V(G) \setminus A'$ is at most

$$\binom{n}{s/2}\alpha^{-2\alpha ds} < \left(\frac{2en}{s}\right)^{s/2}\alpha^{-2\alpha ds} \leq \left(\frac{(2e\alpha^{-4\alpha d})n}{s}\right)^{s/2} = o(1),$$

completing the proof. □

The following is then the main result of this section.

Theorem 14.1. *For each integer* d *there exists a bounded expansion class* $\mathcal{R}_d$ *such that* $G(n, d/n)$ *almost surely belongs to* $\mathcal{R}_d$.

Proof. We first prove that for each integer r $G = G(n, d/n)$ asymptotically almost surely satisfy $\widetilde{\nabla}_r(G) < f_d(r)$ for some suitable function f_d. For a fixed positive integer d define the functions $F_{\mathrm{ord}}, F_{\mathrm{deg}}, F_{\nabla}, F_{\mathrm{prop}} : \mathbb{R}^+ \to \mathbb{R}$ by

$$\begin{aligned} F_{\mathrm{ord}}(x) &= 0, \\ F_{\mathrm{deg}}(x) &= 8dg(x), \\ F_{\nabla}(x) &= 2, \\ \text{and } F_{\mathrm{prop}}(x) &= (4d)^{-1+\frac{1}{4x+1}}, \end{aligned}$$

where $g(x)$ is implicitly defined by

$$4eg(x)^{-4dg(x)} = x,$$

or in the other way:

$$g(x)\log g(x) = \frac{\log 4e - \log x}{4d}.$$

Then, according to Lemma 14.2, Lemma 14.3 and Proposition 5.6 we have that for every r the graphs in $G(n, d/n)$ asymptotically almost surely satisfy

$$\widetilde{\nabla}_r(G) < f_d(r)$$

where

$$f(r) = 2\max\left(8dg\left(\frac{(4d)^{-1+\frac{1}{4r+1}}}{4r+2}\right), 4r+2\right).$$

It follows that for each integer r there exists $N_d(r)$ such that graphs in $G(n, d/n)$ with $n > N_d(r)$ almost surely satisfy $\widetilde{\nabla}_r(G) \leq 2f(r)$. Therefore we can define our class $\mathcal{R}_d$ as follows:

$$\mathcal{R}_d = \{G \in \mathcal{G}raph : \forall r \in \mathbb{N},\ \widetilde{\nabla}_r(G) \leq \max(N_d(r), 2f(r))\}.$$

□

14.2 Crossing Number

For a graph G, let $\mathrm{cr}(G)$ denote the *crossing number* of G, defined as the minimum number of crossings in a drawing of G in the plane; this is one of the frequently studied parameters in geometric graph theory, see the surveys [378, 450]. Note the beautiful applications of this parameter to additive combinatorics and discrete geometry [82, 152, 449]. It is easily seen that $\mathrm{cr}(H) = \mathrm{cr}(G)$ for every subdivision H of G. Thus crossing number is weakly topological. The following "crossing lemma", independently due to [301] and [4], implies that crossing number bounds the average degree parameter (i.e. is *degree-bound*). We include it together with its book proof [3].

Lemma 14.4. *If* $\|G\| \geq 4|G|$ *then* $\mathrm{cr}(G) \geq \frac{\|G\|^3}{64|G|^2}$.

Proof. Consider a minimal drawing of G and a positive real $0 < p \leq 1$. Let G_p be a random induced subgraph of G obtained by selected each vertex of G with probability p.

Let n_p, m_p, X_p be the random variables counting the number of vertices, edges and crossings in G_p. By Euler formula, the inequality $\mathrm{cr}(G) - \|G\| + 3|G| \geq 0$ holds for any graph G hence

$$E(X_p - m_p + 3n_p) \geq 0.$$

Clearly, $E(n_p) = p|G|$ and $E(m_p) = p^2\|G\|$, since an edge appears in G_p if and only if both end-vertices do. And finally, $E(X_p) = p^4\mathrm{cr}(G)$, since a crossing is present in G_p if and only if all four involved vertices are there.

By linearity of expectation, we get

$$0 \leq E(X_p) - E(m_p) + 3E(n_p) = p^4\mathrm{cr}(G) - p^2\|G\| + 3p|G|,$$

that is:

$$\mathrm{cr}(G) \geq \frac{p^2\|G\| - 3p|G|}{p^4} = \frac{\|G\|}{p^2} - \frac{3|G|}{p^3}.$$

Let $p = \frac{4|G|}{\|G\|}$ (hence $p \leq 1$). We get

$$\mathrm{cr}(G) \geq \frac{1}{64}\left(\frac{4\|G\|}{(|G|/\|G\|)^2} - \frac{3|G|}{(|G|/\|G\|)^3}\right) = \frac{1}{64}\frac{\|G\|^3}{|G|^2}.$$

□

As the crossing number is a weakly topological parameter, Theorem 13.2 (4) and Lemma 14.4 imply that a class of graphs with bounded crossing number has bounded expansion. In fact, the following theorem implies that any class with bounded crossing number is included in a minor-closed class:

Theorem 14.2. *For every graph* G,

$$\nabla(G) \in \mathcal{O}(\sqrt{\mathrm{cr}(G)\ \log \mathrm{cr}(G)}) \ .$$

Proof. If G is planar then $\nabla(G) < 3$. Thus we may assume that $\mathrm{cr}(G) \geq 1$. Let K_n be a minor of G. Bokal et al. [75] proved that if H is a minor of G then

$$\mathrm{cr}(H) \leq \left\lfloor \tfrac{1}{2}\Delta(H) \right\rfloor^2 \mathrm{cr}(G) \ .$$

Now apply this result with $H = K_n$. We claim that $n(n-1) \leq 128 \cdot \mathrm{cr}(G)$. This is immediate if $n \leq 8$. Now assume that $n \geq 9$. Thus Lemma 14.4 implies that $\mathrm{cr}(K_n) \geq \frac{1}{512}n(n-1)^3$. (In fact, [276] proved that $\mathrm{cr}(K_n) \geq \frac{1}{80}n(n-1)(n-2)(n-3)$ for sufficiently large n.) Thus

$$\tfrac{1}{512}n(n-1)^3 \leq \mathrm{cr}(K_n) \leq \left\lfloor \tfrac{1}{2}(n-1) \right\rfloor^2 \mathrm{cr}(G) \leq \tfrac{1}{4}(n-1)^2 \mathrm{cr}(G) \ .$$

Thus $n(n-1) \leq 128 \cdot \mathrm{cr}(G)$. Hence $\nabla(G) \leq \mathcal{O}(\sqrt{\mathrm{cr}(G) \log \mathrm{cr}(G)})$ by (4.11). □

The following result establishes that graphs with linear crossing number (in a sense made precise below) are contained in a topologically-closed class, and thus also have bounded expansion. Let $G_{\geq 3}$ denote the subgraph of G induced by the vertices of G that have degree at least 3.

Theorem 14.3. *For a constant* $c \geq 1$, *let* $\mathcal{C}_c$ *be the class of graphs* G *such that* $\mathrm{cr}(H) \leq c|H_{\geq 3}|$ *for every subgraph* H *of* G. *Then* $\mathcal{C}_c$ *is contained in a topologically-closed class of graphs* $\mathcal{C}'_c$ *with* $\widetilde{\nabla}(\mathcal{C}'_c) \leq 4c^{1/3}$.

Proof. Let $G \in \mathcal{C}_c$ and let H be a topological minor of G such that $\|H\|/|H| = \widetilde{\nabla}(G)$. Let $S \subseteq G$ be a witness subdivision of H in G. We prove that $\|H\| \leq 4c^{1/3}|H|$ by contradiction. Were it false, then $\|H\| > 4c^{1/3}|H|$ and by Lemma 14.4,

$$\frac{\|H\|^3}{64|H|^2} \leq \mathrm{cr}(H) = \mathrm{cr}(S) \leq \mathrm{cr}(S_{\geq 3}) = c|H| \ .$$

Thus $\|H\|^3 < 64c|H|^3$, a contradiction. Hence $\widetilde{\nabla}(G) \leq 4c^{1/3}$ for every $G \in \mathcal{C}_c$. □

The assumption involving $\mathcal{C}_c$ is necessary. To see this, consider the class of graphs that admit drawings with at most one crossing per edge. Obviously this includes large subdivisions of arbitrarily large complete graphs. Thus this class is not contained in a proper topologically-closed class. Note however that this class has bounded expansion. This holds generally: bounded number of crossings per edges is sufficient for bounded expansion.

Theorem 14.4. *Let* $c \geq 1$ *be a constant. The class of graphs* G *that admit a drawing with at most* c *crossings per edge has bounded expansion. Precisely, for every integer* d *we have*

$$\widetilde{\nabla}_d(G) \in \mathcal{O}(\sqrt{cd}).$$

Proof. Assume G admits a drawing with at most c crossings per edge. Consider a subgraph H of G that is a $(\leq 2d)$-subdivision of a graph X. So X has a drawing with at most $c(2d+1)$ crossings per edge. Pach and Tóth [377] proved that if an n-vertex graph has a drawing with at most k crossings per edge, then it has at most $4.108\sqrt{kn}$ edges. Thus $\|X\| \leq 4.108\sqrt{c(2d+1)}\,|X|$ hence $\widetilde{\nabla}_d(G) \leq 4.108\sqrt{c(2d+1)}$. □

Thus all these classes are also covered by our theory.

14.3 Queue and Stack Layouts

A graph G is *ordered* if $V(G) = \{1, 2, \dots, |G|\}$. Let G be an ordered graph. Let $\ell(e)$ and $r(e)$ denote the endpoints of each edge $e \in E(G)$ such that $\ell(e) < r(e)$. Two edges e and f are *nested* and f is *nested inside* e if $\ell(e) < \ell(f)$ and $r(f) < r(e)$. Two edges e and f *cross* if $\ell(e) < \ell(f) < r(e) < r(f)$.

An ordered graph is a *queue* if no two edges are nested. An ordered graph is a *stack* if no two edges cross. Observe that the left and right endpoints of the edges in a queue are in first-in-first-out order, and are in last-in-first-out order in a stack—hence the names 'queue' and 'stack'.

An ordered graph G is a k-*queue* if there is a partition $\{E_1, E_2, \dots, E_k\}$ of $E(G)$ such that each $G[E_i]$ is a queue. An ordered graph G is a k-*stack* if there is a partition $\{E_1, E_2, \dots, E_k\}$ of $E(G)$ such that each $G[E_i]$ is a stack.

Let G be an (unordered) graph. A k-*queue layout* of G is a k-queue that is isomorphic to G. A k-*stack layout* of G is a k-stack that is isomorphic to G. A k-stack layout is often called a k-*page book embedding*. The *queue-number* $\mathrm{qn}(G)$ of G is the minimum integer k such that G has a k-queue layout. The *stack-number* $\mathrm{sn}(G)$ of G is the minimum integer k such that G has a k-queue layout.

Stack layouts are more commonly called *book embeddings*, and stack-number has been called *book-thickness*, *fixed outer-thickness*, and *page-number*. See [131] for references and applications of queue and stack layouts. In theoretical computer science, these are frequently studied notions.

It is known (see [61]) that a graph has stack number 1 if and only if it is outerplanar, and it has stack number at most 2 if and only if it is a subgraph of a Hamiltonian planar graph (see Fig. 14.3). Thus every 4-connected planar graph has stack number at most 2. Yannakakis [473, 474] proved that every planar graph has stack number at most 4. In fact, every proper minor-closed class has bounded stack-number [64]. On the other hand, even though stack and queue layouts appear to be somewhat "dual", it is unknown whether planar graphs have bounded queue-number [246, 248], and more generally, it is unkown whether queue-number is bounded by stack-number [133]. It is known [133] that planar graphs have bounded queue-number if and only if 2-stack graphs have bounded queue-number, and that queue-number is bounded by stack-number if and only if 3-stack graphs have bounded queue-number. The largest class of graphs for which queue-number is known to be bounded is the class of graphs with bounded tree-width [129]. We greatly generalize these results: In the following two sections, we prove that graphs of bounded queue-number or bounded stack-number have bounded expansion. The closest previous result in this direction is that graphs of bounded queue-number or bounded stack-number have bounded acyclic chromatic number. In particular, [130] proved that every k-queue graph has acyclic chromatic number at most $4k \cdot 4^{k(2k-1)(4k-1)}$, and every k-stack graph has acyclic chromatic number at most $80^{k(2k-1)}$.

14.4 Queue Number

Every 1-queue graph is planar [130, 248]. However, the class of 2-queue graphs is not contained in a proper topologically-closed class since every graph has a 2-queue subdivision, as proved by [133]. The same authors proved the following connection between subdivisions and queue layouts:

Theorem 14.5. *(a) For all $k \geq 2$, every graph G has a k-queue subdivision with at most $c \log_k \mathrm{qn}(G)$ division vertices per edge, for some absolute constant c.*
(b) If some $(\leq t)$-subdivision of a graph G has a k-queue layout, then $\mathrm{qn}(G) \leq \frac{1}{2}(2k+2)^{2t} - 1$, and if $t = 1$ then $\mathrm{qn}(G) \leq 2k(k+1)$.

Also, queue-number bounds the average degree parameter [131, 248, 380]:

Lemma 14.5. *Every k-queue graph has average degree less than $4k$.*

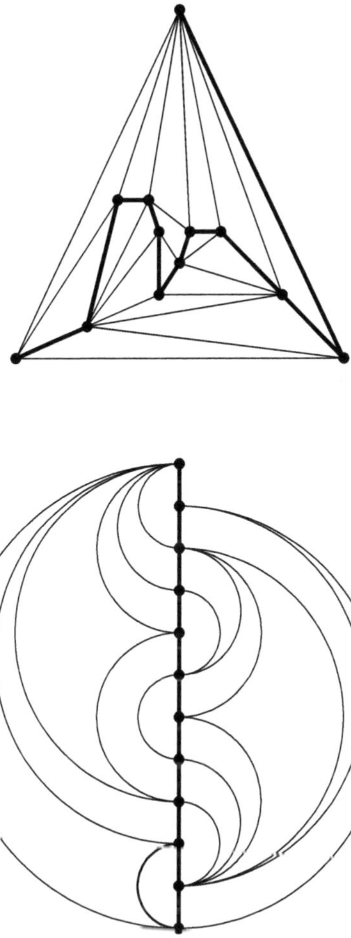

Fig. 14.3 Every 4-connected planar graph has stack number at most 2 (since it is Hamiltonian)

According to Theorem 13.2–4, it now follows that:

Theorem 14.6. *Graphs of bounded queue-number have bounded expansion. In particular*

$$\widetilde{\nabla}_d(G) < (2k+2)^{4d}$$

for every k*-queue graph* G.

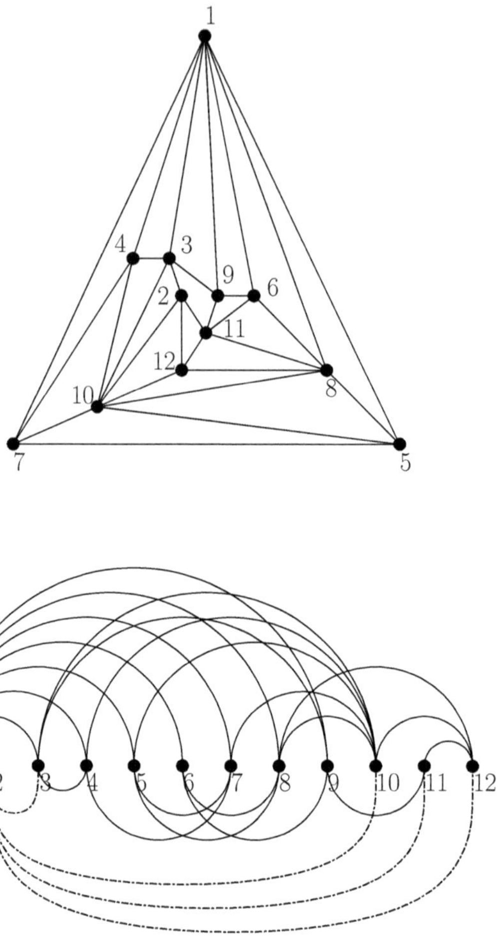

Fig. 14.4 A 3-queue layout of a given planar graph

Proof. Consider a subgraph H of G that is a $(\leq 2d)$-subdivision of a graph X. Thus $\mathsf{qn}(\mathsf{H}) \leq k$, and $\mathsf{qn}(\mathsf{X}) < \frac{1}{2}(2k+2)^{4d}$ by Theorem 14.5(b). Thus the average degree of X is less than $\delta := 2(2k+2)^{4d}$ by Lemma 14.5. Thus $\widetilde{\nabla}_d(\mathsf{G}) \leq (2k+2)^{4d}$. □

From Theorem 14.6 not only follows that graphs with bounded queue-number form a class with bounded expansion, but also that graphs G with queue-number of order $|\mathsf{G}|^{o(1)}$ form a nowhere dense class. For an application to posets, see Exercise 14.2. Note that an alternate (in a sense local) proof is given below in Theorem 14.7.

Note that there is also an exponential lower bound on $\widetilde{\nabla}_d$ for graphs of bounded queue-number. Fix integers $k \geq 2$ and $d \geq 1$. Let G be the graph obtained from K_n by subdividing each edge $2d$ times, where $n = k^d$. Dujmović and Wood [133] constructed a k-queue layout of G. Observe that $\widetilde{\nabla}_d(G) \sim n = k^d$. We now set out to give a direct proof of an exponential bound for the grad $\widetilde{\nabla}_d(G)$ (instead as for the top-grad $\widetilde{\nabla}_r(G)$) for graphs with bounded queue-number.

Consider a k-queue layout of a graph G. For each edge vw of G, let $q(vw) \in \{1, 2, \ldots, k\}$ be the queue containing vw. For each ordered pair (v, w) of adjacent vertices in G, let

$$Q(v,w) := \begin{cases} q(vw) & \text{if } v < w, \\ -q(wv) & \text{if } w < v. \end{cases}$$

Note that $Q(v, w)$ has at most $2k$ possible values.

Lemma 14.6. *Let G be a graph with a k-queue layout.*

(a) Let vw and xy be disjoint edges of G such that $Q(v, w) = Q(x, y)$. Then $v < x$ if and only if $w < y$.
(b) Let $(v_1, v_2, \ldots, v_r)$ and $(w_1, w_2, \ldots, w_r)$ be disjoint paths in G, such that $Q(v_i, v_{i+1}) = Q(w_i, w_{i+1})$ for each $i \in [1, r-1]$. Then $v_1 < w_1$ if and only if $v_r < w_r$.

Proof. (a) Without loss of generality, $v < w$ and $x < y$ since $|Q(v, w)| = |Q(x, y)|$.
Say $v < x$. If $y < w$ then $v < x < y < w$. Thus xy is nested inside vw, which is a contradiction since $q(vw) = q(xy)$. Hence $w < y$.
Say $w < y$. If $x < v$ then $x < v < w < y$. Thus vw is nested inside xy, which is a contradiction since $q(xy) = q(vw)$. Hence $v < x$.
(b) (b) is proved by induction using (a).

□

The following result provides an alternative proof of the fact (established by Theorem 14.6) that graphs with bounded queue-number form a bounded expansion class:

Theorem 14.7. *Let G be a graph with a k-queue layout. Let F be a subgraph of G such that each component of F has radius at most r. Let H be obtained from G by contracting each component of F. Then H has a $f_r(k)$-queue layout, where*

$$f_r(k) := 2k\left(\frac{(2k)^{r+1} - 1}{2k - 1}\right)^2 .$$

Proof. We can assume that F is spanning by allowing 1-vertex components in F. For each component X of F fix a *centre* vertex v of X at distance at most r from every vertex in X. Call X the v-*component*.

Consider a vertex v' of G in the v-component of F. Fix a shortest path $P(v') = (v = v_0, v_1, \ldots, v_s = v')$ between v and v' in F. Thus $s \in [0, r]$. Let

$$Q(v') := \big(Q(v_0, v_1), Q(v_1, v_2), \ldots, Q(v_{s-1}, v_s)\big) \ .$$

Consider an edge $v'w'$ of G, where v' is in the v-component of F, w' is in the w-component of F, and $v \neq w$. Such an edge survives in H. Say $v < w$. Color $v'w'$ by the triple

$$\big(Q(v'), Q(v', w'), Q(w')\big) \ .$$

Observe that the number of colors is at most

$$2k\left(\sum_{s=0}^{r}(2k)^s\right)^2 = 2k\left(\frac{(2k)^{r+1}-1}{2k-1}\right)^2 \ .$$

From the linear order of G, contract each component of F into its centre. That is, the linear order of H is determined by the linear order of the centre vertices in G. After contracting there might be parallel edges with different edge colors. Replace parallel edges by a single edge and keep one of the colors.

Consider disjoint monochromatic edges vw and xy of H, where $v < w$ and $x < y$. By construction, there are edges $v'w'$ and $x'y'$ of G such that v' is in the v-component, w' is in the w-component, x' is in the x-component, y' is in the y-component, and

$$\big(Q(v'), Q(v', w'), Q(w')\big) = \big(Q(x'), Q(x', y'), Q(y')\big).$$

Thus $|P(v')| = |P(x')|$ and $|P(w')| = |P(y')|$. Consider the paths

$$\begin{aligned}&(v = v_0, v_1, \ldots, v_s = v', w' = w_t, w_{t-1}, \ldots, w_0 = w) \text{ and}\\ &(x = x_0, x_1, \ldots, x_s = x', y' = y_t, y_{t-1}, \ldots, y_0 = y),\end{aligned}$$

Since $Q(v') = Q(x')$, we have $Q(v_i, v_{i+1}) = Q(x_i, x_{i+1})$ for each $i \in [0, s-1]$. Similarly, since $Q(w') = Q(y')$, we have $Q(w_i, w_{i+1}) = Q(y_i, y_{i+1})$ for each $i \in [0, t-1]$. Since $Q(v', w') = Q(x', y')$, Lemma 14.6(b) is applicable to these two paths. Thus $v < x$ if and only if $x < y$. Hence vw and xy are not nested. Thus the edge coloring of H defines a queue layout. □

Theorem 14.7 implies Theorem 14.6 (with a better bound on the expansion function) since by Lemma 14.5, the graph H in the statement of Theorem 14.7 has bounded density. In particular, if G has a k-queue layout then

$$\nabla_d(G) \le 8k\left(\frac{(2k)^{d+1}-1}{2k-1}\right)^2 .$$

Theorem 14.7 basically says that minors and queue layouts are compatible, in the same way that queue layouts are compatible with subdivisions; see Theorem 14.5(b). This motivated us to give two proofs.

14.5 Stack Number

The class of 3-stack graph (see Sect. 14.3 for the definiion) is not contained in a proper topologically-closed class since every graph has a 3-stack subdivision [37, 65, 155, 334, 335] (the first proof was by [37], similar ideas were present in the earlier work of [261, 262] on knot projections). Many authors studied bounds on the number of divisions vertices per edge in 3-stack subdivisions, especially of K_n. The most general bounds on the number of division vertices are by the following [133]:

Theorem 14.8. *For all* $s \ge 3$*, every graph* G *has an* s*-stack subdivision with at most* $c\log_{s-1}\min\{\mathrm{sn}(G), \mathrm{qn}(G)\}$ *division vertices per edge, for some absolute constant* c.

Blankenship and Oporowski [65] conjectured that a result like Theorem 14.5(b) holds for stack layouts:

Conjecture 14.1. There is a function f such that $\mathrm{sn}(G) \le f(\mathrm{sn}(H))$ for every graph G and (≤ 1)-subdivision H of G.

The validity of this conjecture implies that stack-number is topological. This conjecture holds for $G = K_n$ as proved by [65], [155], and [160]. The proofs by [65] and [160] use essentially the same Ramsey-theoretic argument.

Enomoto et al. [156] proved the following bound for the density of graphs having a $\le t$-subdivision with a k-stack layout:

Theorem 14.9. *Let* G *be a graph such that some* $(\le t)$*-subdivision of* G *has a* k*-stack layout for some* $k \ge 3$*. Then*

$$\|G\| \le \frac{4k(5k-5)^{t+1}}{5k-6}|G| .$$

It follows that graphs with bounded stack number form a class with bounded expansion:

Theorem 14.10. *Graphs of bounded stack number have bounded expansion. In particular:*

$$\widetilde{\nabla}_r(G) \leq \frac{4k(5k-5)^{2r+1}}{5k-6}$$

for every k*-stack graph* G.

Proof. (≤ 2)-stack graphs have bounded expansion since they are planar. Let G be a graph with stack-number $\mathrm{sn}(G) \leq k$ for some $k \geq 3$. Consider a subgraph H of G that is a ($\leq 2r$)-subdivision of a graph X. Thus $\mathrm{sn}(H) \leq k$, and by Theorem 14.9,

$$\|X\| \leq \frac{4k(5k-5)^{2r+1}}{5k-6}|X| \ .$$

It follows that $\widetilde{\nabla}_r(G) = \frac{\|H\|}{|H|} \leq \frac{4k(5k-5)^{2r+1}}{5k-6}$. □

It is not known whether the exponential bound for $\widetilde{\nabla}_r$ is necessary. This is related to the following problem, which is equivalent to some problems in computational complexity [207, 208, 270].

Problem 14.1. Do 3-stack n-vertex graphs have $o(n)$ separators?

(See Chap. 16 for results relating expansion and separators).

14.6 Non-repetitive Colorings

Let f be a coloring of a graph G. Then f is called *repetitive* on a path $(v_1,\ldots,v_{2s})$ in G if $f(v_i) = f(v_{i+s})$ for each $i \in [1,s]$. If f is not repetitive on every path in G, then f is called *non-repetitive*. Let $\pi(G)$ be the minimum number of colors in a non-repetitive coloring of G. Nonrepetitive colorings are recently intensively studied [16, 17, 49, 50, 83, 84, 106, 234–236, 294, 323, 327]. This study started with the classical result of Thue [461] which, in the above terminology, states that $\pi(P_n) \leq 3$; see [105] for a survey of related results. Note that a non-repetitive coloring is a proper coloring (consider $s=1$). Moreover, a non-repetitive coloring contains no bichromatic P_4 ($s=2$), and thus it is a star coloring. Hence $\pi(G) \geq \chi_{st}(G) \geq \chi(G)$.

The main result in this section (formulated as Theorems 14.11 and 14.12 at the end of this section) is that the parameter π is weakly topological, and that every class of graphs with bounded π has bounded expansion. We shall establish these results in a sequence of several lemmas. The closest previous result is by [470] who proved that if G' is the 1-subdivision of a graph G then $\chi_{st}(G') \geq \sqrt{\chi(G)}$, and thus $\pi(G') \geq \sqrt{\chi(G)}$.

Lemma 14.7. *(1) For every* (≤ 1)*-subdivision* H *of a graph* G*,*

$$\pi(H) \leq \pi(G) + 1.$$

(2) For every (≤ 3)*-subdivision* H *of a graph* G*,*

$$\pi(H) \leq \pi(G) + 2.$$

(3) For every subdivision H *of a graph* G*,*

$$\pi(H) \leq \pi(G) + 3.$$

Proof. First we prove (a). Given a non-repetitive k-coloring of G, introduce a new color for each division vertex of H. Since this color does not appear elsewhere, a repetitively colored path in H defines a repetitively colored path in G. Thus H contains no repetitvely colored path. Part (b) follows by applying (a) twice.

Now we prove (c). Let n be the maximum number of division vertices on some edge of G. Thue [461] proved that P_n has a non-repetitive 3-coloring $(c_1, c_2, \ldots, c_n)$. Arbitrarily orient the edges of G. Given a non-repetitive k-coloring of G, choose each c_i to be one of three new colors for each arc vw of G that is subdivided d times, color the division vertices from v to w by $(c_1, c_2, \ldots, c_d)$. Suppose H has a repetitively colored path P. Since $H - V(G)$ is a collection of disjoint paths, each of which is non-repetitively colored, P includes some principal vertices of G. Let P' be the path in G obtained from P as follows. If P includes the entire subdivision of some edge vw of G then replace that subpath by vw in P'. If P includes a subpath of the subdivision of some edge vw of G, then without loss of generality, it includes v, in which case replace that subpath by v in P'. Since the colors assigned to division vertices are distinct from the colors assigned to principal vertices, a t-vertex path of division vertices in the first half of P corresponds to a t-vertex path of division vertices in the second half of P. Hence P' is a repetitively colored path in G. This contradiction proves that H is non-repetitively colored. Hence $\pi(H) \leq k + 3$. □

Note that Lemma 14.7(a) is best possible in the weak sense that $\pi(C_5) = 4$ and $\pi(C_4) = 3$; see [105].

Loosely speaking, Lemma 14.7 says that non-repetitive colorings of subdivisions are not much harder than non-repetitive colorings of the original graph. This intuition is made more precise if we subdivide each edge many times. Then non-repetitive colorings of subdivisions are much easier than non-repetitive coloringings of the original graph. In particular, [234] proved that every graph has a non-repetitively 5-colorable subdivision. This bound was improved to 4 by [51, 327], and very recently to 3 by [382]; see [83, 105] for related results. This implies that the class of nonrepetively 3-colorable graphs is not contained in a proper topologically-closed class.

We now set out to prove a converse of Lemma 14.7; that is, $\pi(G)$ is bounded by a function of $\pi(H)$. The following easy tool [363] will be useful.

Lemma 14.8. *For every* k*-coloring of the arcs of an oriented forest* T*, there is a* $(2k+1)$*-coloring of the vertices of* T*, such that between each pair of (vertex) color classes, all arcs go in the same direction and have the same color.*

Proof. The colors assigned to the vertices will be $0, \ldots, 2k$ with the property that, at a vertex colored i, outgoing edges colored j reach vertices colored $i + j \bmod 2k+1$ and incoming edges colored j come from vertices colored $i - j \bmod 2k+1$.

Without loss of generality, we may assume that T is a tree. Existence of a coloring of the vertices of T with the prescribed properties is proved by induction. If T has a single vertex, we color it any color. Assume that the coloring exists for every oriented tree T of order $n < n_0$ and let T be an oriented tree of order n. Let x be a leaf of T. By induction, there exists a prescribed coloring for $T - x$. If x has an incoming (resp. outgoing) arc of color j incident to a vertex of color i we color x $i + j \bmod 2k+1$ (resp. $i - j \bmod 2k+1$). The obtained coloring of T matches our requirements. □

A *rooting* of a forest F is obtained by selecting one vertex in each component tree of F as a *root* vertex.

Lemma 14.9. *Let* T' *be the* 1*-subdivision of a forest* T*, such that* $\pi(T') \leq k$*. Then*

$$\pi(T) \leq k(k+1)(2k+1).$$

Moreover, for every non-repetitive k*-coloring* c *of* T'*, and for every rooting of* T*, there is a non-repetitive* $k(k+1)(2k+1)$*-coloring* q *of* T*, such that:*

(1) For all edges vw *and* xy *of* T *with* $q(v) = q(x)$ *and* $q(w) = q(y)$*, the division vertices corresponding to* vw *and* xy *have the same color in* c*.*

(2) For all non-root vertices v *and* x *with* $q(v) = q(x)$*, the division vertices corresponding to the parent edges of* v *and* x *have the same color in* c*.*

(3) For every root vertex r *and every non-root vertex* v*, we have* $q(r) \neq q(v)$*.*

(4) For all vertices v *and* w *of* T*, if* $q(v) = q(w)$ *then* $c(v) = c(w)$*.*

Proof. Let c be a non-repetitive k-coloring of T', with colors $[1, k]$. Color each edge of T by the color assigned by c to the corresponding division vertex. Orient each edge of T towards the root vertex in its component. By Lemma 14.8, there is a $(2k + 1)$-coloring f of the vertices of T, such that between each pair of (vertex) color classes in f, all arcs go in the same direction and have the same color in c. Consider a vertex v of T. If v is a root, let $g(r) := 0$; otherwise let $g(v) := c(vw)$ where w is the parent of v. Let $q(v) := (c(v), f(v), g(v))$. The number of colors in q is at most $k(k+1)(2k+1)$. Observe that claims (c) and (d) hold by definition.

We claim that q is non-repetitive. Suppose on the contrary that there is a path $P = (v_1, \ldots, v_{2s})$ in T that is repetitively colored by q. That is, $q(v_i) = q(v_{i+s})$ for each $i \in [1, k]$. Thus $c(v_i) = c(v_{i+s})$ and $f(v_i) = f(v_{i+s})$ and $g(v_i) = g(v_{i+s})$. Since no two root vertices are in a common path, (c) implies that every vertex in P is a non-root vertex.

Consider the edge $v_i v_{i+1}$ of P for some $i \in [1, s-1]$. We have $f(v_i) = f(v_{i+s})$ and $f(v_{i+1}) = f(v_{i+s+1})$. Between these two color classes in f, all arcs go in the same direction and have the same color. Thus the edge $v_i v_{i+1}$ is oriented from v_i to v_{i+1} if and only if the edge $v_{i+s} v_{i+s+1}$ is oriented from v_{i+s} to v_{i+s+1}. And $c(v_i v_{i+1}) = c(v_{i+s} v_{i+s+1})$.

If at least two vertices v_i and v_j in P have indegree 2 in P, then some vertex between v_i and v_j in P has outdegree 2 in P, which is a contradiction. Thus at most one vertex has indegree 2 in P. Suppose that v_i has indegree 2 in P. Then each edge $v_j v_{j+1}$ in P is oriented from v_j to v_{j+1} if $j \leq i - 1$, and from v_{j+1} to v_j if $j \geq i$ (otherwise two vertices have indegree 2 in P). In particular, $v_1 v_2$ is oriented from v_1 to v_2 and $v_{s+1} v_{s+2}$ is oriented from v_{s+2} to v_{s+1}. This is a contradiction since the edge $v_1 v_2$ is oriented from v_1 to v_2 if and only if the edge $v_{s+1} v_{s+2}$ is oriented from v_{s+1} to v_{s+2}. Hence no vertex in P has indegree 2. Thus P is a directed path.

Without loss of generality, P is oriented from v_1 to v_{2s}. Let x be the parent of v_{2s}. Now $g(v_{2s}) = c(v_s x)$ and $g(v_s) = c(v_s v_{s+1})$ and $g(v_s) = g(v_{2s})$. Thus $c(v_s v_{s+1}) = c(v_{2s} x)$.

Summarizing, the path

$$\big(\underbrace{v_1, v_1 v_2, v_2, \ldots, v_s, v_s v_{s+1}}, \underbrace{v_{s+1}, v_{s+1} v_{s+2}, v_{s+2}, \ldots, v_{2s}, v_{2s} x}\big)$$

in T' is repetitively colored by c. (Here division vertices in T' are described by the corresponding edge.) Since c is non-repetitive in T', we have the desired contradiction. Hence q is a non-repetitive coloring of T.

It remains to prove claims (a) and (b). Consider two edges vw and xy of T, such that $q(v) = q(x)$ and $q(w) = q(y)$. Thus $f(v) = f(x)$ and $f(w) = f(y)$. Thus vw and xy have the same color in c. Thus the division vertices corresponding to vw and xy have the same color in c. This proves claim (a). Finally consider non-root vertices v and x with $q(v) = q(x)$. Thus $g(v) = g(x)$. Say w and y are the respective parents of v and x. By construction, $c(vw) = c(xy)$. Thus the division vertices of vw and xy have the same color in c. This proves claim (b). □

We now extend Lemma 14.9 to apply to graphs with bounded acyclic chromatic number; see [21, 363] for similar methods.

Lemma 14.10. *Let* G' *be the* 1*-subdivision of a graph* G*, such that* $\pi(G') \le k$ *and* $\chi_a(G) \le \ell$*. Then*

$$\pi(G) \le \ell\big(k(k+1)(2k+1)\big)^{\ell-1}.$$

Proof. Let p be an acyclic ℓ-coloring of G, with colors $[1,\ell]$. Let c be a non-repetitive k-coloring of G'. For distinct $i, j \in [1,\ell]$, let $G_{i,j}$ be the subgraph of G induced by the vertices colored i or j by p. Thus each $G_{i,j}$ is a forest, and c restricted to $G'_{i,j}$ is non-repetitive.

Apply Lemma 14.9 to each $G_{i,j}$. Thus $\pi(G_{i,j}) \le k(k+1)(2k+1)$, and there is a non-repetitive $k(k+1)(2k+1)$-coloring $q_{i,j}$ of $G_{i,j}$ satisfying Lemma 14.9(a)–(d).

Consider a vertex v of G. For each color $j \in [1,\ell]$ with $j \ne p(v)$, let $q_j(v) := q_{p(v),j}(v)$. Define

$$q(v) := \Big(p(v), \big\{(j, q_j(v)) : j \in [1,\ell], j \ne p(v)\big\}\Big).$$

Note that the number of colors in q is at most $\ell\big(k(k+1)(2k+1)\big)^{\ell-1}$. We claim that q is a non-repetitive coloring of G.

Suppose on the contrary that some path $P = (v_1, \dots, v_{2s})$ in G is repetitively colored by q. That is, $q(v_a) = q(v_{a+s})$ for each $a \in [1,s]$. Thus $p(v_a) = p(v_{a+s})$ and for each $a \in [1,s]$. Let $i := p(v_a)$. Choose any $j \in [1,\ell]$ with $j \ne i$. Thus $(j, q_j(v_a)) = (j, q_j(v_{a+s}))$ and $q_j(v_a) = q_j(v_{a+s})$. Hence $c(v_a) = c(v_{a+s})$ by Lemma 14.9(d).

Consider an edge $v_a v_{a+1}$ for some $i \in [1, s-1]$. Let $i := p(v_a)$ and $j := p(v_{a+1})$. Now $q(v_a) = q(v_{a+s})$ and $q(v_{a+1}) = q(v_{a+s+1})$. Thus $p(v_{a+s}) = i$ and $p(v_{a+s+1}) = j$. Moreover, $(j, q_j(v_a)) = (j, q_j(v_{a+s}))$ and

$(i, q_i(v_{a+1})) = (i, q_i(v_{a+s+1}))$. That is, $q_{i,j}(v_a) = q_{i,j}(v_{a+s})$ and $q_{i,j}(v_{a+1}) = q_{i,j}(v_{a+s+1})$. Thus $c(v_a v_{a+1}) = c(v_{a+s} v_{a+s+1})$ by Lemma 14.9(a).

Consider the edge $v_s v_{s+1}$. Let $i := p(v_s)$ and $j := p(v_{s+1})$. Without loss of generality, v_{s+1} is the parent of v_s in the forest $G_{i,j}$. In particular, v_s is not a root of $G_{i,j}$. Since $q_{i,j}(v_s) = q_{i,j}(v_{2s})$ and by Lemma 14.9(c), v_{2s} also is not a root of $G_{i,j}$. Let y be the parent of v_{2s} in $G_{i,j}$. By Lemma 14.9(b) applied to v_s and v_{2s}, we have $c(v_s v_{s+1}) = c(v_{2s} y)$.

Summarizing, the path

$$\big(\underbrace{v_1, v_1v_2, v_2, \ldots, v_s, v_sv_{s+1}}, \underbrace{v_{s+1}, v_{s+1}v_{s+2}, v_{s+2}, \ldots, v_{2s}, v_{2s}y}\big)$$

is repetitively colored in G'. This contradiction proves that G is repetitively colored by q. □

Lemmas 14.10 and 14.7(a) imply:

Lemma 14.11. *Let* H *be a* (≤ 1)*-subdivision of a graph* G*, such that* $\pi(H) \leq k$ *and* $\chi_a(G) \leq \ell$*. Then*

$$\pi(G) \leq \ell\big((k+1)(k+2)(2k+3)\big)^{\ell-1}.$$

We get the following interesting dependence of acyclic coloring and non-repetitive coloring.

Proposition 14.1. *Assume that the* 1*-subdivision* G' *of a graph* G *has a non-repetitive* k*-coloring. Then*

$$\chi_a(G) \leq k \cdot 2^{2k^2}.$$

Proof. Let c be a non-repetitive k-coloring of the 1-subdivision G' of a graph G. Orient the edges of G arbitrarily. Let $A(G)$ be the set of oriented arcs of G. So c induces a k-coloring of $V(G)$ and of $A(G)$. For each vertex v of G, let

$$q(v) = \{c(v)\} \cup \{(+, c(vw), c(w)) : vw \in A(G)\}$$
$$\cup \{(-, c(wv), c(w)) : wv \in A(G)\}.$$

The number of possible values for $q(v)$ is at most $k \cdot 2^{2k^2}$. We claim that q is an acyclic coloring of G.

Suppose on the contrary that $q(v) = q(w)$ for some arc vw of G. Thus $c(v) = c(w)$ and $(+, c(vw), c(w)) \in q(v)$, implying $(+, c(vw), c(w)) \in q(w)$. That is, for some arc wx, we have $c(wx) = c(vw)$ and $c(x) = c(w)$. Thus the path (v, vw, w, wx) in G' is repetitively colored. This contradiction shows that q properly colors G.

It remains to prove that G contains no bichromatic cycle (with respect to q). First consider a bichromatic path $P = (u, v, w)$ in G with $q(u) = q(w)$. Thus $c(u) = c(w)$.

Suppose on the contrary that P is oriented (u, v, w), as illustrated in Fig. 14.5(a). By construction, $(+, c(uv), c(v)) \in q(u)$, which implies $(+, c(uv), c(v)) \in q(w)$. That is, $c(uv) = c(wx)$ and $c(v) = c(x)$ for some arc wx (and thus $x \neq v$). Similarly, $(-, c(vw), c(v)) \in q(w)$, implying $(-, c(vw), c(v)) \in q(u)$. Thus $c(vw) = c(tu)$ and $c(v) = c(t)$ for some arc tu (and thus $t \neq v$). Hence the 8-vertex path $(tu, u, uv, v, vw, w, wx, x)$ in G' is repetitively colored by c, as illustrated in Fig. 14.5(b). This contradiction shows that both edges in P are oriented toward v or both are oriented away from v.

Consider the case in which both edges in P are oriented toward v. Suppose on the contrary that $c(uv) \neq c(wv)$. By construction, $(+, c(uv), c(v)) \in q(u)$, implying $(+, c(uv), c(v)) \in q(w)$. That is, $c(uv) = c(wx)$ and $c(v) = c(x)$ for some arc wx (implying $x \neq v$ since $c(uv) \neq c(wv)$). Similarly, $(+, c(wv), c(v)) \in q(w)$, implying $(+, c(wv), c(v)) \in q(u)$. That is, $c(wv) = c(ut)$ and $c(t) = c(v)$ for some arc ut (implying $t \neq v$ since $c(ut) = c(wv) \neq c(uv)$). Hence the path $(ut, u, uv, v, wv, w, wx, x)$ in G' is repetitively colored in c, as illustrated in Fig. 14.5(c). This contradiction shows that $c(uv) = c(wv)$. By symmetry, $c(uv) = c(wv)$ when both edges in P are oriented away from v.

Hence in each component of G', all the division vertices have the same color in c. Every bichromatic cycle contains a 4-cycle or a 5-path. If G contains a bichromatic 5-path (u, v, w, x, y), then all the division vertices in (u, v, w, x, y) have the same color in c, and $(u, uv, v, vw, w, wx, x, xy)$ is a repetitively colored path in G', as illustrated in Fig. 14.5(d). Similarly, if G contains a bichromatic 4-cycle (u, v, w, x), then all the division vertices in (u, v, w, x) have the same color in c, and $(u, uv, v, vw, w, wx, x, xu)$ is a repetitively colored path in G', as illustrated in Fig. 14.5(e).

Thus G contains no bichromatic cycle, and q is an acyclic coloring of G. □

Note that the above proof establishes the following stronger statement: If the 1-subdivision of a graph G has a k-coloring that is non-repetitive on paths with at most 8 vertices, then G has an acyclic $k \cdot 2^{2k^2}$-coloring in which each component of each 2-colored subgraph is a star or a 4-path.

Lemma 14.12. *(a) If some* (≤ 1)*-subdivision of a graph* G *has a non-repetitive* k*-coloring, then* $\chi_a(G) \leq (k+1) \cdot 2^{2(k+1)^2}$.

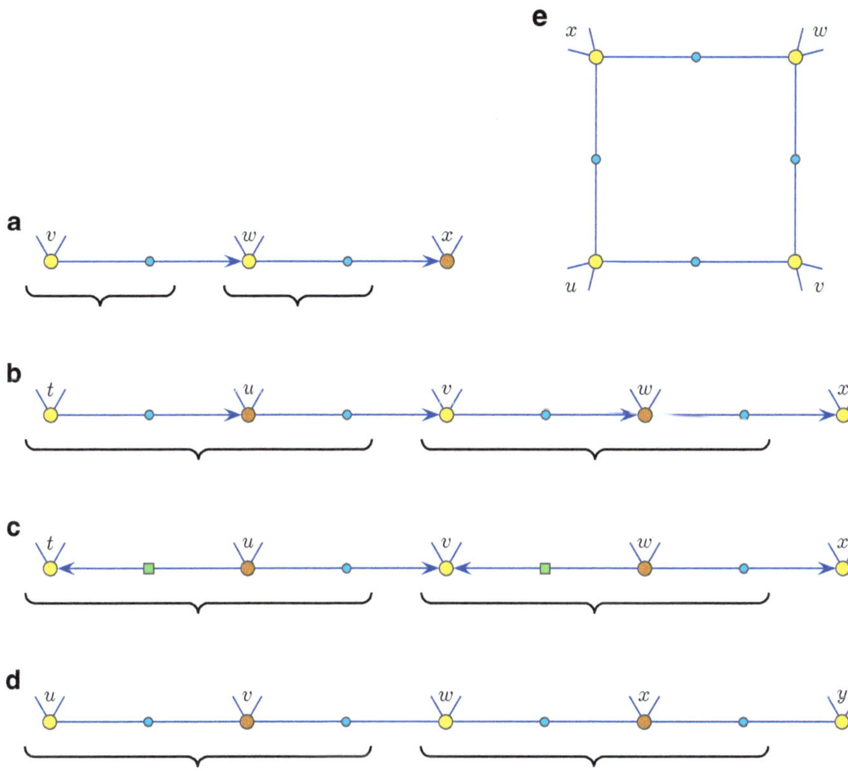

Fig. 14.5 Illustration for Proposition 14.1

(b) If $\pi(\mathsf{H}) \leq \mathsf{k}$ *for some* (≤ 1)*-subdivision of a graph* G*, then*

$$\pi(\mathsf{G}) \leq (\mathsf{k}+1) \cdot 2^{2(\mathsf{k}+1)^2} \left((\mathsf{k}+1)(\mathsf{k}+2)(2\mathsf{k}+3)\right)^{(\mathsf{k}+1)\cdot 2^{2(\mathsf{k}+1)^2}-1}.$$

(c) There is a function f *such that* $\pi(\mathsf{G}) \leq \mathsf{f}(\pi(\mathsf{H}), \mathsf{d})$ *for every* $(\leq \mathsf{d})$*-subdivision* H *of a graph* G.

Proof. (a) follows from Proposition 14.1 and Lemma 14.7(a). By (a) we have $\chi_a(\mathsf{G}) \leq (\mathsf{k}+1) \cdot 2^{2(\mathsf{k}+1)^2}$, thus (b) follows from Lemma 14.11 with $\ell = (\mathsf{k}+1) \cdot 2^{2(\mathsf{k}+1)^2}$. (c) is then a direct corollary. □

One of the most interesting open problems regarding non-repetitive colorings is whether planar graphs have bounded π (as mentioned in most papers regarding non-repetitive colorings). Lemma 14.12(c) implies that to prove that planar graphs have bounded π it suffices to show that every planar graph has a subdivision with bounded π and a bounded number of division vertices per edge. This shows that Conjectures 4.1 and 5.2 in [234] are equivalent.

We now get to the main results of this section. Lemmas 14.12(b) and 14.7(a) imply:

Theorem 14.11. *The parameter π is weakly topological.*

π is degree-bound since every graph G has a vertex of degree at most $2\pi(G)-2$ (see [51], Proposition 5.1). Since π is hereditary, Theorems 13.2 and 14.11 imply:

Theorem 14.12. *For every* k, *the class of all graphs* G *with* $\pi(G) \leq k$ *has bounded expansion.*

Actually, we obtained this result as a corollary of the fact that π is weakly topological, which is an interesting result by itself. It is possible to give a direct proof that the class of all graphs G with $\pi(G) \leq k$ has bounded expansion and to give an explicit bound for $\widetilde{\nabla}_r(G)$ in terms of $\pi(G)$ (see Exercise 14.3).

Note that perhaps the most important in this area asks whether the class of planar graphs has bounded non-repetitive coloring number. Theorems 14.11 and 14.12 may be viewed as contributions to this problem.

Exercises

14.1. Prove that for every integer D and every positive real $d > 0$ there exists a class $\mathcal{R}_{d,D}$ with bounded expansion such that if G is a graph of order n whose edge set is the union of the edge set of a D-regular graph and the one of a random graph in $G(n, d/n)$ then $G \in \mathcal{R}_{d,a}$ asymptotically almost surely.

Such graphs are called *liquid graphs.*

14.2. This exercise relates our study of classes with bounded queue number with classes of posets with bounded jump-number, showing that the Hasse diagrams of these posets form a class with bounded expansion [359].

Let P be a poset. The *Hasse diagram* $H(P)$ of P is the graph whose vertices are the elements of P and whose edges correspond to the *cover relation* of P. Here x *covers* y in P if $x >_P y$ and there is no element z of P such that $x >_P z >_P y$.

A *linear extension* of P is a total order $\prec$ of P such that $x <_P y$ implies $x \prec y$ for every $x, y \in P$. The *jump number* $\mathrm{jn}(P)$ of P is the minimum number of consecutive elements of a linear extension of P that are not comparable in P, where the minimum is taken over all possible linear extensions of P.

Assume that P has jump number k and let $\prec$ by an optimal total order of P. Then the linear order can be split into $k+1$ consecutive sequences $S_0, \dots, S_k$ without jumps. Consider a linear embedding of P corresponding to $\prec$.
Prove that if (u_1, v_1) and (u_2, v_2) are arcs of $H(P)$, $u_1, u_2 \in S_i$, $v_1, v_2 \in S_j$, and $i < j$ then (u_1, v_1) and (u_2, v_2) cross. (Hint: otherwise, one of the two arcs is a transitivity edge)
Deduce that if arcs between S_i and S_j are assigned to queue $|i - j|$ then no two arcs in a same queue nest, hence for every poset P it holds [247].

$$\mathrm{qn}(H(P)) \leq \mathrm{jn}(P) + 1.$$

Deduce from Theorem 14.6 that if $\mathcal{P}$ is a class of posets with bounded jump number then the class $H(\mathcal{P})$ of the Hasse diagrams of the posets in $\mathcal{P}$ has bounded expansion.

14.3. Let G be a graph. Assume that for some integer r it holds $\widetilde{\nabla}_r(G) \geq \pi(G)^{2r+2}$ and let $H \in G \,\widetilde{\triangledown}\, r$ be such that $\|H\|/|H| = \widetilde{\nabla}_r(G)$ and let G' be the $\leq 2r$-subdivision of H in G. Consider a non-repetitive coloring c of G by $\pi(G)$ colors.

Fix an orientation $\vec{H}$ of H. As every non-repetitive coloring is proper, show that the arcs of $\vec{H}$ can be colored by $\sum_{i=0}^{2r} \pi(G)(\pi(G) - 1)^{i+1} \leq \pi(G)^{2r+2}$ colors in such a way that two arcs (u, v) and (x, y) of $\vec{H}$ get the same color if

the color of the vertices of the branches (in G') from u to v and from x to y are the same;
Deduce that (in this edge coloring) $\vec{H}$ contains a monochromatic subgraph $\vec{H}_0$ which underlying undirected graph is a cycle;
Prove that $\vec{H}_0$ cannot contain a directed path of length 2, and that the underlying undirected graph H_0 of $\vec{H}_0$ both cannot include a path of length 4 and cannot be a cycle of length 4. (Hint: only here we really use the assumption that the coloration of G is non-repetitive.)
Deduce that for every graph G and every integer r it holds

$$\widetilde{\nabla}_r(G) < \pi(G)^{2r+2}.$$

Hence any class of graphs with bounded Thue number π has bounded expansion. This provides a more direct proof of Theorem 14.12.

Chapter 15
Some Applications

And now for something completely different.

15.1 Finding Matching and Paths

15.1.1 Introduction

A *matching* of a graph G is a set of pairwise non-intersecting edges. An *induced matching* of a graph G is a matching of G which is an induced subgraph of G, that is a matching with the property that no endpoint of an edge in the matching is adjacent to an endpoint of another edge in the matching.

The problem of finding a *maximum matching* (that is: a matching with maximum cardinality) is known as the "marriage problem". Its variant, which consists in finding a *maximum induced matching* has been introduced by Stockmeyer and Vazirani [444] as the "risk-free marriage problem" and it was studied extensively [128, 165, 173, 223, 441]. As a particular nice result let us mention [395] which provides an asymptotic solution of an extremal problem (due to Erdős and Nešetřil) on graphs without induced matchings of size 2.

A vertex v of a graph G is a *clone* if G has a vertex $u \neq v$ with the same neighborhood as v. In that say we say that v is a *clone* of u. The size of a maximum induced matching (resp. of a maximum induced matching) will be denoted by $\beta(G)$ (resp. $\beta^*(G)$)

There are two simple conditions which may prevent the existence of a large induced matching in a graph:

J. Nešetřil and P.O. de Mendez, *Sparsity*, Algorithms and Combinatorics 28, DOI 10.1007/978-3-642-27875-4_15, © Springer-Verlag Berlin Heidelberg 2012

- If a graph G contains "many" clones, adding a new clone vertex will increase neither $\beta(G)$ nor $\beta^*(G)$; for instance, a star graph has a maximum (induced) matching of size 1, whatever its order is.
- If a graph is sufficiently dense, it may have no large induced matchings even if it is *rigid* (that is: if it has no non-trivial automorphisms);

We shall actually prove that every sufficiently sparse clone-free graph G has a linear size matching (of size $\Omega(|G|)$) (Theorem 15.1) and consequently an induced matching of size $\Omega(|G|)$ as well (Theorem 15.2). As a consequence, for every fixed surface S, every graph G with minimum degree at least 3 embedded on S has a linear size induced matching (Theorem 15.1). Also, we will show that the stronger the assumption on the sparsity and on the forbidden automorphisms will be, the bigger will be the integer k such that G will necessarily contain a subset of $\Omega(|G|)$ vertices inducing a forest formed by paths of length k (Theorem 15.4).

It is this second application which justifies our somewhat cumbersome proof. For matchings, we have a min-max theorem (Tutte [462], Berge [58]). This can be used for a simpler proof for the existence of linear size matchings in clone-free graphs with bounded $\nabla_{\frac{1}{2}}$. However we claim a much more general result and thus have to take a more complicated way. Our approach is based on the analysis of graphs with bounded tree-depth and low tree-depth decompositions. It is also related to analysis of rigidity and symmetries (see also Corollary 8.1, Sect. 8.6).

15.1.2 Finding a Big Subgraph with Low Degrees

We start with the following.

Lemma 15.1. *Let* G *be a clone-free graph of order* n *and let* $0 < \epsilon < 1$.

Let d_0 *be the average degree of* G *(hence* $d_0 \leq 2\nabla_0(G)$*), let* $d_1 = 2\widetilde{\nabla}_{1/2}(G)$ *and let* $d = d_0(1 + d_1/2 + 2^{d_1})/\epsilon$.

Then the sum of the orders of the non-trivial (i.e. those having order at least 2) connected components of the graph $G_{<d}$ *is at least* $(1-\epsilon)n$.

Proof. Let X be the subset of the vertices of G having degree at least d. Then (see (3.3) in Sect. 3.2) $|X| \leq (d_0/d)n$. Let Y be the set of the vertices of $G_{<d}$ having at least a neighbor in $G_<$.

Assume for contradiction that $|Y| < (1-\epsilon)n$. Let $Z = V(G) - X - Y$. Notice that the vertices in Z have all their neighbors in X. Let Z' be a maximal subset of Z such that there exists a mapping $\phi : Z' \to \binom{X}{2}$ with the following properties:

$$\forall z_1 \neq z_2 \in Z' : \quad \phi(z_1) \neq \phi(z_2),$$
$$\forall z \in Z' : \quad \phi(z) \subseteq N_G(z),$$

where $N_G(z)$ denotes the set of the neighbors of z in the graph G.

Then, consider the graph H with vertex set $\bigcup_{z \in Z'} \phi(z)$ and edges $\{\phi(z) : z \in Z'\}$. By construction, H is a simple graph and G contains a 1-subdivision of H. Moreover, the vertex set of H is included in X. Hence, the size of H is at most $(d_1/2)|X|$. As ϕ is a bijection from Z' to the edges set of H, we conclude that $|Z'| \leq (d_1/2)|X|$.

Let $z \in Z - Z'$. Then $N_G(z)$ induces a clique in H. By definition of d_1, we have $\nabla_0(H) \leq (d_1/2)$. Hence, according to Lemma 3.1, H includes at most $2^{d_1}|X|$ cliques. As G is clone-free, no two vertices of $Z - Z'$ have the same neighborhood. Hence $|Z - Z'| \leq 2^{d_1}|X|$.

Altogether, we get:

$$\begin{aligned} n &= |X| + |Y| + |Z - Z'| + |Z'| \\ &< |X| + (1-\epsilon)n + 2^{d_1}|X| + (d_1/2)|X| \end{aligned}$$

Thus:

$$\epsilon n < (1 + d_1/2 + 2^{d_1})(d_0/d)n = \epsilon n,$$

which is a contradiction. □

We can continue in this line and prove a stronger version (for paths on length 3):

Lemma 15.2. *Let G be a graph of order n with no automorphisms exchanging exactly two cliques of order at most 2 and let $0 < \epsilon < 1$.*

Let d_0 be the average degree of G, let $d_2 = 2\widetilde{\nabla}_1(G))$, and let $d = d_0(1 + d_2/2 + 2^{d_2} + 6 \cdot 4^{d_2})/\epsilon$.

Then the sum of the orders of those connected components of $G_{<d}$ which include a path of length at least 3 is at least $(1-\epsilon)n$.

Proof. We proceed similarly as for the proof of Lemma 15.1. Let X be the subset of the vertices of G having degree at least d ($|X| \leq (d_0/d)n$) and let Y be the set of the vertices of $G_{<d}$ belonging to a path of length at least 3 of $G_{<d}$.

Assume for contradiction that $|Y| < (1-\epsilon)n$. Let $Z = V(G) - X - Y$. Notice that the vertices in Z have all their neighbors in $X \cup Z$ and that all the connected components of $G[Z]$ have order at most 2. Let $\mathcal{Z}$ be the family of the connected components of $G[Z]$ and let $\mathcal{Z}'$ be a maximal subset of $\mathcal{Z}$ such that there exists a mapping $\phi : \mathcal{Z}' \to \binom{X}{2}$ with the following properties:

$$\forall B_1 \neq B_2 \in \mathcal{Z}' : \quad \phi(B_1) \neq \phi(B_2),$$
$$\forall B \in \mathcal{Z}' : \quad \phi(B) \subseteq N_G(B),$$

where $N_G(B)$ denotes the set of the neighbors of the vertices of B in the graph G (i.e. $N_G(B) = \bigcup_{z \in B} N_G(z) \setminus B$).

Then, consider the graph H with vertex set $\bigcup_{B \in \mathcal{Z}'} \phi(B)$ and edges $\{\phi(B) : B \in \mathcal{Z}'\}$. By construction, H is a simple graph and G contains a ≤ 2-subdivision of H. Moreover, the vertex set of H is included in X. Hence, the size of H is at most $(d_2/2)|X|$. As ϕ is a bijection from $\mathcal{Z}'$ to the edges set of H, we conclude that $|\mathcal{Z}'| \leq (d_2/2)|X|$.

Let $B \in \mathcal{Z} - \mathcal{Z}'$. Then $N_G(B)$ induces a clique in H. By definition of d_2, we have $\nabla_0(H) \leq (d_2/2)$. Hence, according to Lemma 3.1, H includes at most $\binom{d_2}{t-1}|X|$ cliques of size t for $1 \leq t \leq d_2 + 1$.

As G has no automorphism exchanging exactly two cliques of order at most 2, at most $3^t + 1$ connected components in $\mathcal{Z} - \mathcal{Z}'$ may be adjacent to a same clique of size t: 1 single vertex and 3^t K_2's being linked differently to the t vertices of the clique. Thus $\sum_{B \in \mathcal{Z} - \mathcal{Z}'} |V(B)| \leq (2^{d_2} + 6 \cdot 4^{d_2})|X|$.

Altogether, we get:

$$\begin{aligned} n &= |X| + |Y| + \sum_{B \in \mathcal{Z} - \mathcal{Z}'} |V(B)| + \sum_{B \in \mathcal{Z}'} |V(B)| \\ &< |X| + (1 - \epsilon)n + (2^{d_2} + 6 \cdot 4^{d_2})|X| + (d_2/2)|X|. \end{aligned}$$

Thus we get:

$$\epsilon n < (1 + d_2/2 + 2^{d_2} + 6 \cdot 4^{d_2})(d_0/d)n = \epsilon n.$$

This is a contradiction. □

Now, we are ready for the general case (paths of length l). The proof is similar but the bounds we get are less explicit:

Lemma 15.3. *Let $\mathcal{F}(x, y)$ be the function introduced in Theorem 6.5, which evaluates the maximum order a graph G may have if the vertices of G are colored by x colors, the tree-depth of G is y and G has no non-trivial color preserving automorphism.*

Let ℓ be a positive integer, let G be a graph of order n and average degree d_0 and let $0 < \epsilon < 1$. Let

$$C = 1 + \widetilde{\nabla}_{\frac{\ell}{2}}(G) + \sum_{t=0}^{2\widetilde{\nabla}_{\frac{\ell}{2}}(G)} \binom{2\widetilde{\nabla}_{\frac{\ell}{2}}(G)}{t} \mathcal{F}(2^t, \ell)$$

Let $d = Cd_0/\epsilon$. *Then, either* G *has a non-trivial automorphism, or the sum of the orders of the connected components of the subgraph of* $G_{<d}$ *induced by the vertices included (in* $G_{<d}$*) in at least one path of length at least* ℓ *is at least* $(1-\epsilon)n$.

Proof. Let X be the subset of the vertices of G having degree at least d ($|X| \leq (d_0/d)n$) and let Y be the set of the vertices of $G_{<d}$ which belong to some path of length ℓ of $G_{<d}$.

Assume for contradiction that $|Y| < (1-\epsilon)n$. Let $Z = V(G) - X - Y$. Notice that the vertices in Z have all their neighbors in $X \cup Z$ and that no connected components of $G[Z]$ includes a path of length ℓ. Let $\mathcal{Z}$ be the family of the connected components of $G[Z]$ and let $\mathcal{Z}'$ be a maximal subset of $\mathcal{Z}$ such that there exists a mapping $\phi : \mathcal{Z}' \to \binom{X}{2}$ with the following properties:

$$\begin{aligned} \forall B_1 \neq B_2 \in \mathcal{Z}' : &\quad \phi(B_1) \neq \phi(B_2), \\ \forall B \in \mathcal{Z}' : &\quad \phi(B) \subseteq N_G(B), \end{aligned}$$

where $N_G(B)$ denotes the set of the neighbors of the vertices of B in the graph G (i.e. $N_G(B) = \bigcup_{z \in B} N_G(z) \setminus B$).

Then, consider the graph H with vertex set $\bigcup_{B \in \mathcal{Z}'} \phi(B)$ and edges $\{\phi(B) : B \in \mathcal{Z}'\}$. By construction, H is a simple graph and G contains a $\leq \ell$-subdivision of H. Moreover, the vertex set of H is included in X. Hence, the size of H is at most $\widetilde{\nabla}_{\frac{\ell}{2}}(G)|X|$. As ϕ is a bijection from $\mathcal{Z}'$ to the edges set of H, we conclude that $|\mathcal{Z}'| \leq \widetilde{\nabla}_{\frac{\ell}{2}}(G)|X|$.

Let $B \in \mathcal{Z} - \mathcal{Z}'$. Then $N_G(B)$ induces a clique in H. By definition of $\widetilde{\nabla}_{\frac{\ell}{2}}(G)$, we have $\nabla_0(H) \leq \widetilde{\nabla}_{\frac{\ell}{2}}(G)$. Hence, according to Lemma 3.1, H includes at most $\binom{2\widetilde{\nabla}_{\frac{\ell}{2}}(G)}{t-1}|X|$ cliques of size t for $1 \leq t \leq 2\widetilde{\nabla}_{\frac{\ell}{2}}(G) + 1$. Let K be a set of t vertices of X inducing a clique in H and let Q be the disjoint union of all the graphs in $\mathcal{Z} - \mathcal{Z}'$ having K as their neighbor set. Color the vertices of Q by 2^t colors, according to the subset of K to which they are adjacent. According to Theorem 6.5, every 2^t-colored graph with no P_ℓ (hence tree-depth at most ℓ, according to Lemma 6.1) and order greater than $\mathcal{F}(2^t, \ell)$ has a non-trivial color-preserving automorphism. Hence Q has order at most $\mathcal{F}(2^t, \ell)$. It follows that the sum of the orders of the graphs in $\mathcal{Z} - \mathcal{Z}'$ is at most $\sum_{t=0}^{2\widetilde{\nabla}_{\frac{\ell}{2}}(G)} \binom{2\widetilde{\nabla}_{\frac{\ell}{2}}(G)}{t} \mathcal{F}(2^t, \ell) \leq 2^{2\widetilde{\nabla}_{\frac{\ell}{2}}(G)} \mathcal{F}(2^{2\widetilde{\nabla}_{\frac{\ell}{2}}(G)}, \ell)$.

Altogether, we get:

$$n = |X| + |Y| + \sum_{B \in \mathcal{Z} - \mathcal{Z}'} |V(B)| + \sum_{B \in \mathcal{Z}'} |V(B)|$$

$$< |X| + (1-\epsilon)n + (\sum_{t=0}^{2\widetilde{\nabla}_{\frac{\ell}{2}}(G)} \binom{2\widetilde{\nabla}_{\frac{\ell}{2}}(G)}{t} F(2^t, \ell))|X| + \widetilde{\nabla}_{\frac{\ell}{2}}(G)|X|$$

Thus:

$$\epsilon n < (1 + \widetilde{\nabla}_{\frac{\ell}{2}}(G) + \sum_{t=0}^{2\widetilde{\nabla}_{\frac{\ell}{2}}(G)} \binom{2\widetilde{\nabla}_{\frac{\ell}{2}}(G)}{t} F(2^t, \ell))(d_0/d)n = \epsilon n,$$

which is a contradiction. □

15.1.3 Finding Matchings

Theorem 15.1. *Let* G *be a clone free-graph of order* n. *Then* G *has a matching of size*

$$\beta(G) \geq \frac{n}{2(d+1)}$$

where

$$d = 2\nabla_0(G)\left(1 + \widetilde{\nabla}_{\frac{1}{2}}(G) + 2^{2\widetilde{\nabla}_{\frac{1}{2}}(G)}\right).$$

Proof. Let $\epsilon = 1/2$. According to Lemma 15.1, G has a subset S of size at least $n/2$ such that every vertex in S has degree at most d and is adjacent to at least one vertex of degree at most d. □

From this bound we will deduce a linear lower bound on the induced matching number of G, tanks to the following (probably folkloric) simple lemma.

Lemma 15.4. *Let* G *be a graph. Then the maximum size* $\beta^*(G)$ *of an induced matching of* G *and the maximum size* $\beta(G)$ *of a matching of* G *are related by*

$$\frac{\beta(G)}{\lfloor 4\nabla_0(G) \rfloor - 1} \leq \beta^*(G) \leq \beta(G)$$

In other words this means that β *and* β^* *are related by a simple function of degeneracy.*

Proof. Let M be a matching of G of size $\beta(G)$ and let H be the graph obtained from G/M by simplifying it (i.e. removing parallel edges). Let A be a subset of vertices of H and let B be the corresponding subset of vertices of G: each vertex of A arising from the contraction of an edge in M is replaced in B by the two endpoints of the edge and each other vertex of A is kept in B. Then $\|H[A]\| \leq \|G[B]\| - (|B| - |A|)$ (as one edge is contracted each time a vertex of A corresponds to two vertices in B). Thus

$$\begin{aligned}\frac{\|H[A]\|}{|A|} &\leq \frac{\|G[B]\| - (|B| - |A|)}{|A|}\\ &= \frac{\|G[B]\|}{|B|} + \frac{|B| - |A|}{|A|}\left(\frac{\|G[B]\|}{|B|} - 1\right)\\ &\leq 2\nabla_0(G) - 1\end{aligned}$$

Hence $\nabla_0(H) \leq 2\nabla_0(G) - 1$. Moreover, as H is also obviously $(\lfloor 2\nabla_0(H)\rfloor + 1)$-colorable and as any subset of M corresponding to a monochromatic subset of vertices of H will form an induced matching of G, we conclude. □

Theorem 15.2. *Let G be a clone free-graph of order n. Then G has an induced matching of size*

$$\beta^*(G) \geq \frac{n}{2(\lfloor 4\nabla_0(G)\rfloor - 1)(d+1)}$$

where

$$d = 2\nabla_0(G)\left(1 + \widetilde{\nabla}_{\frac{1}{2}}(G) + 2^{2\widetilde{\nabla}_{\frac{1}{2}}(G)}\right).$$

Proof. This is a direct consequence of Theorem 15.1 and Lemma 15.4. □

The induced matching number provides a lower bound for the *product dimension* of a graph (see [311]); this dimension is mentioned in Sect. 3.7. It follows that clone-free graphs with bounded $\nabla_{\frac{1}{2}}$ have a large dimension ($\geq \log \beta^*(G)$).

Denote by $\sim$ be the equivalence relation defined by $x \sim y$ if x and y have the same neighbors (i.e. are clones). Let $G/\!\sim$ be the graph obtained by keeping exactly one vertex per equivalence class of $\sim$; the vertex kept in a class is identified with the class it belongs to, so that for a vertex x of G belonging to a class represented by a vertex $\hat{y}$ of $G/\!\sim$ we shortly write $x \in \hat{y}$. We take time out for another simple lemma which we apply to topological graphs.

Lemma 15.5. *Let G be a graph. Then $G/\!\sim$ has no clones and the sizes of maximum induced matchings in G and $G/\!\sim$ are related by*

$$\beta^*(G) = \beta^*(G/\sim).$$

Proof. The fact that $\sim$ is an equivalence relation is obvious. Moreover, it is straightforward that $G/\sim$ has no clones: Assume $\hat{x}$ and $\hat{y}$ are vertices of $G/\sim$ corresponding to the class of the vertices x and y of G. The neighbors of $\hat{x}$ (resp. $\hat{y}$) are the classes of the neighbors of x (resp. y in G). If $\hat{x}$ and $\hat{y}$ have the same neighbors in G then x and y have the same neighbors in G hence belong to a same equivalence class of $\sim$. Thus the inequality $\beta^*(G) \geq \beta^*(G/\sim)$ is obvious as $G/\sim$ is isomorphic to an induced subgraph of G.

For the second inequality, consider any maximum induced matching M of G. Let F be the set of edges of $G/\sim$ defined by $\hat{x}\hat{y} \in F$ if there exists $x \in \hat{x}$ and $y \in \hat{y}$ such that $xy \in M$. Assume two edges of F are adjacent, namely $\hat{x}\hat{y}_1$ and $\hat{x}\hat{y}_2$. Then there exists $x_1, x_2 \in \hat{x}, y_1 \in \hat{y}_1, y_2 \in \hat{y}_2$ such that x_1y_1 and x_2y_2 belong to M. But y_1 is also a neighbor of x_2 (as it is a neighbor of x_1) hence M is not an induced matching, contradiction. It follows that F is an induced matching of $G/\sim$ and $\beta^*(G) \leq \beta^*(G/\sim)$. □

Corollary 15.1. *Define*

$$d(g) = \begin{cases} 3, & \text{if } g = 0, \\ \frac{5+\sqrt{48g+1}}{4}, & \text{otherwise.} \end{cases}$$

Then every graph G *with minimum degree* 3 *and genus* g *has a matching of order*

$$\beta(G) \geq \frac{n}{4(g+1)(2d(G)2^{2d(G)} + 2d(G)^2 + 2d(G) + 1)}$$

and an induced matching of order

$$\beta^*(G) \geq \frac{n}{4(g+1)\lceil 4d(g)+1\rceil(2d(G)2^{2d(G)} + 2d(G)^2 + 2d(G) + 1)} = n2^{-\sqrt{12g}(1+o(1))}$$

(where $o(1)$ *refers to an error term which only depends on* g *and goes to* 0 *when* g *tends to infinity).*

Proof. Every vertex of G has at most $4g+1$ clones, as the minimum degree of G is at least 3 and the genus of $K_{3,s}$ is $\lceil \frac{s-2}{4} \rceil$. Hence $|G/\sim| \geq |G|/(4g+2)$. Then every connected graph G with genus g has average degree at most $2d(g)$: According to Euler formula, if G has order n and size m then $m \leq 3n-6+6g$. If $g = 0$ then the average degree of G is obviously at most 6. Otherwise, assume for contradiction that $m > d(g)n$. Then $m - 3n - 6g + 6 > (d(g) - 3)(2d(g) + 1) - 6g + 6 > 0$ (as G has a vertex of degree at least $2d(g)$ hence

$n \geq 2d(g)+1$) which contradicts Euler formula. As every topological minor of a graph with genus g has genus at most g we deduce that $\widetilde{\nabla}_r(G) \leq d(g)$ for every integer r (and in particular $\widetilde{\nabla}_0(G) = \nabla_0(G) \leq d(g)$). The result then follows from Theorem 15.2 and Lemma 15.5. □

15.1.4 Finding Paths

For the general case of packing length l paths we have to consider higher order grads, yet in all instances we get a linear number of disjoint l-paths.

Theorem 15.3. *There exists a function* $f_{\text{path}} : \mathbb{N} \times \mathbb{R}^+ \to]0;1]$ *such that every rigid graph* G *of order* n *includes at least*

$$f_{\text{path}}(L, \widetilde{\nabla}_{L/2}(G))|G|$$

vertex disjoints paths of length L.

Proof. Let $\alpha > 0$. According to Lemma 15.3 (with $\epsilon = L/(L+1)$), there exists $\phi(L,\alpha) > 0$ such that every graph G of order n with $\widetilde{\nabla}_{L/2}(G) \leq \alpha$ is such that for $d = (1+1/L)\phi(K,\alpha)$, the sum of the orders of the connected components of the subgraph H of $G_{<d}$ induced by vertices belonging (in $G_{<d}$) to a path of length L is at least $n/(L+1)$. As the maximum degree of H is at most $d-1$, each connected component of H_i contains at least $\lceil |H_i|/d^L \rceil$ vertices pairwise at distance at least L in H_i. As each of these vertices belong to some path of length L of H we deduce that H contains at least $f_{\text{path}}(L,\alpha)n$ vertex disjoint paths of length L, where $f_{\text{path}}(L,\alpha) = \frac{L^L}{(L+1)^{(L+1)}}/\phi(L,\alpha)$. □

Theorem 15.4. *There exists a function* $f_{\text{induced}} : \mathbb{N} \times \mathbb{R}^+ \to]0;1]$ *such that every rigid graph* G *of order* n *includes at least*

$$f_{\text{induced}}(\ell, \widetilde{\nabla}_{\ell/2}(G))\,|G|$$

vertex disjoints induced paths of length ℓ.

Proof. Let $d = \widetilde{\nabla}_{\ell/2}(G) + 1 \geq \nabla_0(G) + 1$. By a minor variation of the proof of Theorem 15.3, G includes at least $f'(d^{d^\ell}, \widetilde{\nabla}_{\ell/2}(G))|G|$ vertex disjoint paths of length $d^{d^{\ell}}$ such that no vertex in one path is adjacent to a vertex of a different path. According to Lemma 6.4, the graphs induced by the vertex set of each such path includes the vertex set of an induced path of length ℓ. □

15.1.5 A Particular Application: Strong Star Chromatic Number

In this book we contributed to the explosion of coloring variations defined in recent years. It is a strong sign of the vitality of graph theory that most of these definitions are based on problems steming from outside of graph theory. Actually, many of these were motivated by applied problem such as the *Chanel Assignment Problem* and problems arising from biology (to name just two such cases).

Dujmović and Wood introduced in [132] the following concept: a vertex coloring is a *strong star coloring* if between every pair of color classes, all edges (if any) are incident to a single vertex. That is, each bichromatic subgraph consists of a star and possibly some isolated vertices. The *strong star chromatic number* of a graph G, denoted by $\chi_{sst}(G)$, is the minimum number of colors in a strong star coloring of G. The following result is proved in [132]:

Lemma 15.6. *Every graph G with m edges and maximum degree $\Delta \geq 1$ has strong chromatic number $\chi_{sst}(G) \leq 14\sqrt{\Delta m}$.*

If the maximum degree is not bounded, the following upper bound is also proved in [132]:

Lemma 15.7. *Every graph G with m edges has strong star chromatic number $\chi_{sst}(G) \leq 15m^{2/3}$.*

An asymptotically stronger result was known for graphs with bounded maximum degree in another context: a *harmonious coloring* of a simple graph G is a proper vertex coloring such that each pair of colors appears together on at most one edge. The *harmonious chromatic number* $\mathrm{harm}(G)$ is the least number of colors in such a coloring. Obviously, $\chi_{sst}(G) \leq \mathrm{harm}(G)$. The following bounds are proved in [143]:

Theorem 15.5. *Let Δ be a fixed integer, and $\epsilon > 0$. There is a natural number M such that if G is any graph with $m \geq M$ edges and maximum degree at most Δ, then the harmonious chromatic number $\mathrm{harm}(G)$ satisfies*

$$\sqrt{2m} \leq \mathrm{harm}(G) \leq (1+\epsilon)\sqrt{2m}$$

The upper bound also applies to $\chi_{sst}(G)$, but the derived lower bound for $\chi_{sst}(G)$ is $\sqrt{\frac{2m}{\Delta}}$. Notice that in the most general case, we cannot expect any good lower bound for $\chi_{sst}(G)$ as the strong chromatic number of a star graph is 2. However, if we forbid clones we are able to prove a $\Theta(\sqrt{n})$ lower bound for graphs with bounded $\widetilde{\nabla}_{\frac{1}{2}}$:

Theorem 15.6. *Let* G *be a connected clone-free graph of order* n *and let*

$$C(G) = 4\nabla_0(G)\left(1 + \widetilde{\nabla}_{\frac{1}{2}}(G) + 2^{2\widetilde{\nabla}_{\frac{1}{2}}(G)}\right) + 2.$$

Then the strong star chromatic number of G *is bounded by*

$$\chi_{sst}(G) \geq \sqrt{\frac{n}{C(G)}}.$$

Proof. According to Theorem 15.1, G has a matching of size $\frac{n}{C(G)}$. No two edges of the matching may have its endpoints colored by the same pair of colors. The result follows. □

Remark 15.1. The condition that two color classes induce at most a star may be weakened to the condition that any two color classes induce a $2K_2$-free bipartite graph. The $2K_2$-free graphs got alternative names and definitions:

- *Bipartite chain graphs* [472]: A graph is a bipartite chain graph if and only if it is bipartite and for each color class the neighborhoods of the nodes in that color class can be ordered linearly with respect to inclusion (subset or equal);
- *Difference graphs* [242]: A graph is a difference graph if every vertex v_i can be assigned a real number a_i and there exists a positive real number T such that (a) $|a_i| < T$ for all i and (b) $(v_i, v_j) \in E \iff |a_i - a_j| \geq T$;
- *Non-separable bipartite graph* [122]: A bipartite graph is non-separable if each pair of edges either share an end vertex or are connected by an edge.

Also, these graphs may be defined by the property that they are bipartite with at most one non-trivial connected component which is P_5-free.

15.2 Burr–Erdős Conjecture

Ramsey theory is a domain of (very) large numbers, see e.g. [227], [340]. However there are exceptions from this seemingly universal rule. One such exception concerns game versions of Ramsey problems, see [54], and the another is detailed analysis of (generalized) Ramsey numbers. Generalized Ramsey number is defined for an arbitrary graph G as the least integer $r(G)$, the *Ramsey number* of G, so that for every graph H of order at least $r(G)$ either H or its complement contains G as a subgraph (of course, not necessarily induced). When the graph G is sparse then we can expect small Ramsey numbers (and in many cases exact results). Such results often belong more to graph theory than to Ramsey theory. But this is not the case with the linear Ramsey numbers where the analysis involves techniques from the very heart of Ramsey theory. A family of graphs $\mathcal{F}$ is a *Ramsey linear family* if there exists a constant $c = c(\mathcal{F})$ such that $r(G) \leq cn$ for every $G \in \mathcal{F}$ of order n. In 1973, Burr and Erdős formulated the following conjecture.

Conjecture 15.1 (Burr–Erdős conjecture, [87]). For each positive integer p, there exists a constant c_p so that if G is a p-degenerate graph on n vertices then $r(G) < c_p n$.

Conjecture 15.1 may be restated as:

Conjecture 15.2. *(Alternate form of Burr–Erdős conjecture).*
There exists a function $f : \mathbb{R} \to \mathbb{R}$ such that for any graph G of order n:

$$\frac{r(G)}{n} < f(\nabla_0(G))$$

In 1983 Chvátal et al. [97] proved that the conjecture holds for graphs with bounded maximum degree (improved in [225], tight bounds for bipartite case in [226]). This result has been extended to p-arrangeable graphs by Chen and Schelp [93].

Recall that a graph G is p-*arrangeable* (concept introduced in [93] if its vertices can be ordered as $v_1, v_2, \dots, v_n$ in such a way that $|N_{L_i}(N_{R_i}(v_i))| \leq q$ for each $1 \leq i \leq b_1$, where $L_i = \{v_1, v_2, \dots, v_i\}$, $R_i = \{v_{i+1}, v_{i+2}, \dots, v_n\}$, and $N_A(B)$ denotes the neighbors of B which lie in A.

Thus, according to the definition of generalized coloring numbers (see Sect. 4.9), for every graph G we have:

$$\text{G is p-arrangeable} \quad \Longleftrightarrow \quad \mathrm{col}_2(G) \leq p + 1. \tag{15.1}$$

Up to now, the better bounds are given by the following result of Graham et al. [225] (extending earlier results of Eaton [140]):

Theorem 15.7. *For some positive constant* c *and all integers* $p \geq 2$ *and all* $n \geq p+1$, *if* H *is a* p*-arrangeable graph with* n *vertices then*

$$\log_2 \frac{r(G)}{n} \leq cp(\log_2 p)^2 \tag{15.2}$$

It is proved in [93] that planar graphs are p-arrangeable for some p. In [424], Rödl and Thomas prove that graphs included no subdivision of K_q are p-arrangeable for some p depending on q. The Burr – Erdős conjecture is further known to hold for subdivided graphs [12] (improved in [302]). Moreover, some further progress toward the conjecture may also be found in [286–288]. A general survey of what is known on Ramsey numbers may be found in [392].

The arrangeability of a graph is polynomially related to the grad of rank $\frac{1}{2}$ as follows from (15.1) and the results of Sect. 7.5 on generalized coloring numbers. Precisely, it follows from Lemma 7.11 and Proposition 4.8 that if G is p-arrangeable then $\widetilde{\nabla}_{\frac{1}{2}}(G) \leq (p+1)^2 - 1$ and, conversely, it follows from Theorem 7.11 that every graph G is $8\nabla_{\frac{1}{2}}(G)^3$-arrangeable.

Combining with [225] we obtain:

Theorem 15.8. *For the positive constant* c *of Theorem 15.7, all integers* $p \geq 2$, *all* $n \geq p+1$, *and all graphs* G *of order* n:

$$\log_2\left(\frac{r(G)}{n}\right) \leq 72c\nabla_{\frac{1}{2}}(G)^3(\log_2 \nabla_{\frac{1}{2}}(G) + 3)^2)$$

Also, combining our results with those of [436, 437] we get:

Corollary 15.2. *There exists a function* $g : \mathbb{R}^2 \to \mathbb{R}$ *such that for any graphs* G_1, G_2:

$$\frac{r(G_1, G_2)}{\max(|V(G_1)|, |V(G_2)|)} \leq g(\nabla_0(G_1), \nabla_{\frac{1}{2}}(G_2))$$

Presently this is as close as we may get to the Conjecture 15.1.

15.3 The Game Chromatic Number

Let G be a graph and let Γ be a set of colors. Alice and Bob play the following coloring game on G. Alice and Bob alternately color the vertices of G. At each turn, the player choose an uncolored vertex x and color it with a color from Γ which is not used by any of the colored neighbors of x. Alice has the first move. Alice wins if all the vertices of G are eventually colored. Bob wins if at some point there is an uncolored vertex such that each of the k colors appears at least once in its neighborhood, i.e. if the current partial coloring cannot be extended to a complete coloring of G. The *game chromatic number* $\chi_g(G)$ is the least integer k for which Alice has a winning strategy.

In this context, Kierstead and Trotter [273] introduced the concept of *admissibility*, which is closely related to the one of arrangeability we considered in the previous section:

Definition 15.1. Let G be a graph, let $M \subseteq V(G)$, and let $v \in M$. A set $A \subseteq V(G)$ is called an M-*blade* with center v if either

1. $A = \{a\}$ and $a \in M$ is adjacent to v, or
2. $A = \{a, b\}$, $a \in M - \{v\}$, $b \in V(G) - M$, and b is adjacent to both v and a.

An M-*fan* with center v is a set of pairwise disjoint M-blades with center v. Let k be an integer. A graph G is k-*admissible* if the vertices of G can be numbered $v_1, v_2, \dots, v_n$ in such a way that for every $i = 1, 2, \dots, n$, G has no $\{v_1, v_2, \dots, v_i\}$-fan with center v_i of size $k + 1$.

One of the interest of the notion of admissibility is its relation with game chromatic number, as shown by Kierstead and Trotter [273]:

Theorem 15.9. *Let* k *and* t *be positive integers. If a* k-*admissible graph has chromatic number* t, *then its game chromatic number is at most* $kt + 1$.

The concepts of arrangeability (introduced in the previous section) and of admissibility are almost equivalent, as shown in the same paper [273]:

Lemma 15.8. *Let* k *be an integer. Any* k-*arrangeable graph is* $2k$-*admissible; any* k-*admissible graph is* $(k^2 - k + 1)$-*arrangeable.*

This property is central to the proof that graphs with no K_p subdivisions is $(\frac{1}{2}p^2(p^2 + 1))$-admissible [424]. We give here some new characterization of admissibility in terms of transitive fraternal augmentations (for details on transitive fraternal augmentations see Sect. 7.3).

Lemma 15.9. *Let* p *be an integer, let* G *be a graph, let* $\vec{G}$ *be an orientation of* G *and let* $\vec{H}$ *be a* 1*-transitive fraternal augmentation of* G.

Then G *is* $(\Delta^-(\vec{G}) + 2\nabla_0(H))$*-admissible.*

Proof. Let $p = \Delta^-(\vec{G}) + 2\nabla_0(H)$. Let $i \in \{0, 1, \dots, n\}$ be the least integer such that there exists (distinct) vertices $v_{i+1}, v_{i+2}, \dots, v_n$ with the following property: for all $j = i, i+1, \dots, n$, G has no $(V(G) - \{v_{j+1}, v_{j+2}, \dots, v_n\}$-fan with center v_j of size $p+1$. If $i \neq 0$, let $M = V(G) - \{v_{i+1}, v_{i+2}, \dots, v_n\}$. This set is non-empty and such that for every $v \in M$ there is an M-fan in G with center v of size $p+1$.

For $v \in M$, let $F(v)$ be an M-fan with center v with cardinality at least $p+1$. Associate a type with M-blades B in $F(v)$ as follows:

- Type $= 1$ if $B = \{a\}$ is a singleton and $(a, v) \in E(\vec{G})$,
- Type $= 2$ if $B = \{a\}$ is a singleton and $(v, a) \in E(\vec{G})$,
- Type $= 3$ if $B = \{a, b\}, b \notin M, (b, v) \in E(\vec{G}), (b, a) \in E(\vec{G})$,
- Type $= 4$ if $B = \{a, b\}, b \notin M, (b, v) \in E(\vec{G}), (a, b) \in E(\vec{G})$,
- Type $= 5$ if $B = \{a, b\}, b \notin M, (v, b) \in E(\vec{G}), (b, a) \in E(\vec{G})$,
- Type $= 6$ if $B = \{a, b\}, b \notin M, (v, b) \in E(\vec{G}), (a, b) \in E(\vec{G})$.

Then if B has type $1, 2, 4, 5, 6$ the vertices a and v are adjacent in H. Thus $\sum_{v \in M} |\{B \in F(v), \text{type}(v) \neq 3\}| \leq \sum_{v \in M} d_H(v) \leq 2\nabla_0(H)|M|$. Now remark that two distinct M-blades of type 3 with center v use two different arcs entering v. As the maximum indegree of $\vec{G}$ is $\Delta^-(G)$, we get $\sum_{v \in M} |\{B \in F(v), \text{type}(v) = 3\}| \leq \sum_{v \in M} d^-(v) \leq \Delta^-(G)|M|$. Altogether, as $|F(v)| \geq p$ for any $v \in M$:

$$p + 1 \leq 2\nabla_0(H) + \Delta^-(G) \tag{15.3}$$

what contradicts the definition of p.

Hence $i = 0$, and $v_1, v_2, \dots, v_n$ is an enumeration of the vertices of G showing that G is p-admissible. □

As we are only interested in a single 1-transitive fraternal augmentation, it is possible to improve the general bound of Lemma 7.2.

Lemma 15.10. *Let* $\vec{G}$ *be an acyclically oriented simple directed graph.*

Then there exists an edge coloring Υ *using at most* $2\Delta^-(\vec{G})$ *colors such that any color induce a star forest oriented outward.*

Proof. Let v be a sink. Color $G - v$ by induction with colors in $\{1, \dots, 2\Delta^-(\vec{G})\}$. For each arc (x, v) entering v, at least $\Delta^-(G)$ colors among $\{1, \dots, 2\Delta^-(\vec{G})\}$ are not present in an arc entering x. As there are at most $\Delta^-(\vec{G})$ arcs entering v, one can choose a suitable color for each arc entering v such that all these arcs get a different color. □

Lemma 15.11. *Let $\vec{G}$ be an acyclically directed graph with maximum indegree $\Delta^-(\vec{G})$. Then $\vec{G}$ has a 1-fraternal augmentation $\vec{H}$ such that $\Delta^-(\vec{H}) \leq \Delta^-(G)(1+2\nabla_{\frac{1}{2}}(G))$.*

Proof. Let Υ be the edge coloring defined in Lemma 15.10.

For any color α in $1,\dots,\Upsilon(E(\vec{G}))$, let G_α be the graph whose vertices are those vertices of G which have at least one outgoing edge colored α, and such that two vertices x, y are adjacent in $\vec{G}_\alpha$ if there exists a path of length at most two linking x and y which contains an arc colored α going out of x or y. Notice that $G_\alpha \in G \,\triangledown\, \frac{1}{2}$.

Let $(x,z),(y,z)$ be arcs of $\vec{G}$ (x,y,z being distinct vertices) so that $\Upsilon((x,z)) = \alpha$. Then x and y are distinct and adjacent in G_α. As G_α may be oriented with indegree at most $\nabla_0(G_\alpha) \leq \nabla_{\frac{1}{2}}(G)$ we get that $\vec{G}$ has a 1-fraternal augmentation $\vec{H}$ with indegree bounded by $\Delta^-(\vec{H}) \leq \Delta^-(G) + |\Upsilon(E(\vec{G}))|\nabla_{\frac{1}{2}}(G)$. □

Theorem 15.10. *Every graph G is p-admissible for*

$$p = 4\nabla_0(G)(\nabla_{\frac{1}{2}}(G) + \nabla_0(G) + 1).$$

Proof. This is a direct consequence of Lemma 15.11, Lemma 15.9 and the fact that G has an acyclic orientation with indegree at most $2\nabla_0(G)$ (Proposition 3.2). □

Corollary 15.3. *Every graph G with $\nabla_0(G) \geq 1$ has game chromatic number bounded by:*

$$\begin{aligned}\chi_g(G) &\leq 4\nabla_0(G)(2\nabla_0(G)+1)(\nabla_{\frac{1}{2}}(G) + \nabla_0(G) + 1) + 1\\ &= O(\nabla_0(G)^2\nabla_{\frac{1}{2}}(G)).\end{aligned}$$

Notice that Dinsky and Zhu proved [124] that the game chromatic number of a graph G is bounded by $\chi_a(G)(\chi_a(G)+1)$ (where $\chi_a(G)$ is the *acyclic chromatic number* of G) , which is also bounded by a polynomial in $\nabla_{\frac{1}{2}}(G)$). They conjectured that, conversely, the acyclic chromatic number may be bounded by a function of the game chromatic number of a graph.

15.4 Fiedler Value of Classes with Sublinear Separators

The *Laplacian* $L(G)$ of a graph G of order n is the $n \times n$ matrix with degrees on the diagonal and -1 for adjacent pairs of vertices (i.e. $L(G) = D(G) - A(G)$). This matrix is real and symmetric hence has all of its n eigenvalues real. As Laplacian matrices are positive semi-definite, all the eigenavalues are non-negative. The all-one vector is clearly an eigenvector of this matrix, with associated eigenvalue 0. The second smallest eigenvalue λ_2 of $L(G)$ is called the *algebraic connectivity* of G [177], or the *Fiedler value* of G [439, 440] (see also Sect. 3.6).

Let $\mathcal{C}$ be a class of graphs. The *Fiedler value* of the class is

$$\lambda_{2\max}(\mathcal{C}, n) = \max_{G \in \mathcal{C}} \lambda_2(G).$$

The Fiedler value of a graph G is related to embeddings of G is Euclidean space by the following lemma of Spielman and Teng [439, 440].

Lemma 15.12.

$$\lambda_2(G) = \min \frac{\sum_{ij \in E(G)} \|\vec{v_i} - \vec{v_j}\|^2}{\sum_{i \in V(G)} \|\vec{v_i}\|^2},$$

where the minimum is taken over all possible choices of $\vec{v_1}, \ldots, \vec{v_n} \in \mathbb{R}^m$ *such that* $\sum_i \vec{v_i} = \vec{0}$.

In particular, if $\phi : V(G) \to \mathbb{C}$ and $\sum_{u \in V(G)} \phi(u) = 0$ we have

$$\lambda_2(G) \leq \frac{\sum_{uv \in E(G)} |\phi(u) - \phi(v)|^2}{\sum_{u \in V(G)} |\phi(u)|^2}.$$

Barrière et al. [52] obtained the following bound for K_h minor free graphs:

$$\lambda_{2\max}(K_h \text{ minor free}, n) \leq \begin{cases} h - 2 + O(\frac{1}{\sqrt{n}}) & \text{if } 4 \leq h \leq 9 \\ \gamma h \sqrt{\log h} + O(\frac{1}{\sqrt{n}}) & \text{otherwise} \end{cases}$$

We extend here their results in the context of ω-expansion. Following the proof of Barrière et al., we state two lemmas allowing to bound $\lambda_2(G)$ by the density of edges incident to a small subset of vertices of G.

Lemma 15.13. *Let* $n_{1,1}, n_{1,2}, n_{2,1}, n_{2,2}$ *be positive integers such that*

$$\begin{aligned} n_{1,1} &\leq n_{1,2} \leq 2n_{1,1} \\ n_{2,1} &\leq n_{2,2} \leq 2n_{2,1} \\ n_{1,1} + n_{1,2} &\leq n_{2,1} + n_{2,2} \leq 2(n_{1,1} + n_{1,2}) \end{aligned}$$

Then there exist $z_{1,1}, z_{1,2}, z_{2,1}, z_{2,2} \in \mathbb{S}^2$ *(where* $\mathbb{S}^2 = \{z \in \mathbb{C} : |z| = 1\}$*) such that*

$$n_{1,1}z_{1,1} + n_{1,2}z_{1,2} + n_{2,1}z_{2,1} + n_{2,2}z_{2,2} = 0$$

Proof. Let $n_1 = n_{1,1} + n_{1,2}, n_2 = n_{2,1} + n_{2,2}$ and $n = n_1 + n_2$. Define the real numbers $x_1 = 2/3$ and $x_2 = -\frac{n_2}{n_1}z_1$, so that $n_1x_1 + n_2x_2 = 0$ and $-2/3 \leq x_2 \leq -1/3$.

For $0 < x < 1$ define the function $f_x : \mathbb{S}^2 \to \mathbb{S}^2$ such that $f_x(z)$ is the intersection of the unit circle and of the line through z and x. Let $g :]0,1[\times\mathbb{S}^2 \to \mathbb{R}$ be defined by $g(x,z) = \frac{f_x(z)-x}{x-z}$. Notice that g is continuous. As $g(x_1, 1) = 5$ and $g(x_1, -1) = 1/5$ there exists $z_{1,1}$ such that $g(x_1, z_{1,1}) = n_{1,2}/n_{1,1}$. Also, as $g(x_2, 1) = 1/2$ and $g(x_2, -1) = 2$ there exists $z_{2,1}$ such that $g(x_2, z_{2,1}) = n_{2,2}/n_{2,1}$. Let $z_{1,2} = f_{x_1}(z_{1,1})$ and $z_{2,2} = f_{x_2}(z_{2,1})$. Then $x_1 = \frac{n_{1,1}z_{1,1}+n_{1,2}z_{1,2}}{n_1}$ and $x_2 = \frac{n_{2,1}z_{2,1}+n_{2,2}z_{2,2}}{n_2}$. Thus $n_{1,1}z_{1,1} + n_{1,2}z_{1,2} + n_{2,1}z_{2,1} + n_{2,2}z_{2,2} = 0$. □

Lemma 15.14. *Let* $\mathcal{C}$ *be a monotone class of graphs and let* $s(n)$ *denote the maximum size of a vertex separator of a graph* $G \in \mathcal{C}$ *with order at most* n.

Then, for every graph $G \in \mathcal{C}$ *with order* n *there exists a subset* $S \subset V(G)$ *of cardinality at most* $s(n) + 2s(2n/3)$ *such that:*

$$\lambda_2(G) \leq \frac{e(S, V - S)}{n - |S|}.$$

Proof. Let S_0 be a vertex separator of G of size at most $s(n)$ and let (Z_1, Z_2) be a partition of $V - S_0$ such that $|Z_1| \leq |Z_2| \leq 2|Z_1|$ and no edge exists between Z_1 and Z_2. Let S_1 (resp. S_2) be separators of size at most $s(2n/3)$ of $G[Z_1]$ (resp. $G[S_2]$), let $(Z_{1,1}, Z_{1,2})$ (resp. $(Z_{2,1}, Z_{2,2})$) be a partition of Z_1 (resp. Z_2) such that $|Z_{i,1}| \leq |Z_{i,2}| \leq 2|Z_{i,1}|$ and no edge exists between $Z_{i,1}$ and $Z_{i,2}$ in $G[Z_i] - S_i$. According to Lemma 15.13, there exists four complex numbers $z_{1,1}, z_{1,2}, z_{2,1}$ and $z_{2,2}$ with $|z_{i,j}| = 1$ and $\sum_{i=1}^{2}\sum_{j=1}^{2} |Z_{i,j}|\, z_{i,j} = 0$. Define $\phi : V(G) \to \mathbb{C}$ as follows

$$\phi(v) = \begin{cases} z_{i,j} & \text{if } v \in Z_{i,j} \\ 0 & \text{otherwise} \end{cases}$$

Then we have

$$\lambda_2(G) \leq \frac{\sum_{uv\in E(G)} |\phi(u) - \phi(v)|^2}{\sum_{u\in V(G)} |\phi(u)|^2} = \frac{e(S, V - S)}{n - |S|},$$

where $S = S_0 \cup S_1 \cup S_2$ has cardinality at most $s(n) + 2s(2n/3)$. □

Remark 15.2. If G has maximum average degree d then $\lambda_2(G) \le d(d+1)n$. Indeed, G has a proper coloration with $d+1$ colors. If S is the union of the d smallest color classes, then $V(G) - S$ is disconnected hence may be easily split into four parts having approximately the same size.

We now give a general bound for the size of a bipartite subgraph of a graph G in terms of the maximum average degree and maximum clique size of shallow topological minors of G.

Lemma 15.15. *Let* A, B *be disjoint vertices of a graph* G, *with* $|A| \ge |B|$ *and* $n = |A| + |B|$. *Then*

$$e(A,B) \le (\omega(G \mathbin{\widetilde{\triangledown}} 1/2) - 1)\,n + (\nabla_0(G) - \omega(G \mathbin{\widetilde{\triangledown}} 1/2) + 1)(\widetilde{\nabla}_{1/2}(G) + 1)\,|B|.$$

In particular, if $G \in \mathcal{C}$ *and* $\mathcal{C}$ *is a minor closed class with maximum average degree* $d = 2\nabla_0(\mathcal{C})$ *and clique number* $\omega = \omega(\mathcal{C})$, *we get:*

$$e(A,B) \le (\omega - 1)\,n + (d/2+1)(d/2+1-\omega)\,|B|.$$

Proof. Let $\omega = \omega(G \mathbin{\widetilde{\triangledown}} \frac{1}{2})$. Partition A into A_1 and A_2 such that A_1 contains the vertices with degree at most $\omega - 1$ and A_2 contains the vertices with degree at least ω.

Consider any linear ordering $x_1, \dots, x_p$ of A_2. We construct $H \in G \mathbin{\widetilde{\triangledown}} \frac{1}{2}$ as follows. At the beginning, H is the empty graph with vertex set B. For each vertex x_i in A_2, if x_i has two neighbours u, v in B that are not adjacent in H we (choose one such pair of vertices and) make them adjacent in H and we continue with the next vertex of A_2. If we cannot continue, this means that all the neighbours of x_i are adjacent in H. Then by construction we have $H \oplus K_1 \in G \mathbin{\widetilde{\triangledown}} \frac{1}{2}$ although $\omega(H \oplus K_1) > \omega$, a contradiction. Hence we can continue until A_2 is exhausted. Then we obtain $H \in G \mathbin{\widetilde{\triangledown}} \frac{1}{2}$ such that $\|H\| = |A_2|$ and $|H| = |B|$. Hence we have $|A_2| \le \widetilde{\nabla}_{1/2}(G)|B|$ and

$$e(A_2, B) \le \|G[A_2 \cup B]\| \le \nabla_0(G)(|A_2| + |B|).$$

As the maximum degree of vertices in A_1 is $\omega - 1$ we have $e(A_1, B) \le (\omega - 1)(n - |A_2| - |B|)$. Altogether, we get

$$e(A,B) = e(A_1,B) + e(A_2,B) \le (\omega-1)n + (\nabla_0(G) - \omega + 1)(\widetilde{\nabla}_{1/2}(G)+1)|B|.$$

□

Theorem 15.11. *Let $\mathcal{C}$ be a monotone class with sub-linear separators and bounded $\widetilde{\nabla}_{1/2}$. Let $s(n)$ denote the maximum size of a vertex separator of graphs $G \in \mathcal{C}$ of order at most n. Then*

$$\lambda_{2\,\max}(\mathcal{C}, n) \leq \omega\left(\mathcal{C}\,\widetilde{\nabla}\,\frac{1}{2}\right) - 1 + O\left(\frac{s(n)}{n}\right).$$

Proof. According to Lemma 15.14 there exists, for every graph $G \in \mathcal{C}$ with order n, a subset $S \subset V(G)$ of cardinality at most $s(n) + 2s(2n/3)$ such that $\lambda_2(G) \leq \frac{e(S,V-S)}{n-|S|}$. According to Lemma 15.15, we have

$$e(V-S,S) \leq (\omega(G\,\widetilde{\nabla}\,1/2)-1)\,n+(\nabla_0(G)-\omega(G\,\widetilde{\nabla}\,1/2)+1)(\widetilde{\nabla}_{1/2}(G)+1)\,|S|.$$

As $s(n) = o(n)$, it follows that

$$\lambda_{2\,\max}(\mathcal{C}, n) \leq \omega\left(\mathcal{C}\,\widetilde{\nabla}\,\frac{1}{2}\right) - 1 + O\left(\frac{s(n)}{n}\right).$$

□

This result is nearly optimal as we shall show now. We take time out for a lemma.

Lemma 15.16. *Let H_1, H_2 be graphs and let $H_1 \oplus H_2$ denote the complete join of H_1 and H_2. Then*

$$\lambda_2(H_1 \oplus H_2) = \min(\lambda_2(H_1) + |H_2|, \lambda_2(H_2) + |H_1|).$$

Proof. Let $G = H_1 \oplus H_2$ be the complete join of H_1 and H_2. Then

$$L(G) = \begin{pmatrix} L(H_1) + |H_2|I & -J \\ -J & L(H_2) + |H_1|I \end{pmatrix}$$

Hence if x_1 is an eigenvector of $L(H_1)$ with eigenvalue α_1 and if x_2 is an eigenvector of $L(H_2)$ with eigenvalue α_2, both being orthogonal to the all-one vectors, we have:

$$L(G)\begin{pmatrix} x_1 \\ 0 \end{pmatrix} = \begin{pmatrix} L(H_1) + |H_2|I & -J \\ -J & L(H_2) + |H_1|I \end{pmatrix}\begin{pmatrix} x_1 \\ 0 \end{pmatrix} = (\alpha_1 + |H_2|)\begin{pmatrix} x_1 \\ 0 \end{pmatrix}$$

and

$$L(G)\begin{pmatrix}0\\x_2\end{pmatrix} = \begin{pmatrix}L(H_1)+|H_2|I & -J\\ -J & L(H_2)+|H_1|I\end{pmatrix}\begin{pmatrix}0\\x_2\end{pmatrix} = (\alpha_2+|H_1|)\begin{pmatrix}0\\x_2\end{pmatrix}$$

Moreover, if x is the vector with $|H_1|$ first entries equal to $|H_2|$ and the remaining $|H_2|$ entries equal to $-|H_1|$ we have

$$L(G)x = nx$$

With the all-one vector, which is an eigenvector of $L(G)$ with associated eigenvalue 0, we have determined the full spectrum of G. It follows that the second smallest eigenvalue of G is

$$\lambda_2(G) = \min(\lambda_2(H_1)+|H_2|, \lambda_2(H_2)+|H_1|).$$

□

Hence we have, for $n > h$ (as $G \oplus K_1$ is K_{h+1}-minor free if G is K_h-minor free):

$$\lambda_{2\,\max}(K_{h+1}\text{ minor free}, n+1) \geq \lambda_{2\,\max}(K_h\text{ minor free}, n)+1$$

In particular, we get the following corollary of Theorem 15.11.

Corollary 15.4. *For every integer* $h \geq 2$ *we have*

$$h-2 \leq \lambda_{2\,\max}(K_h\ \textit{minor free}, n) \leq h-2+O\left(\frac{1}{\sqrt{n}}\right)$$

Proof. The upper bound comes from Theorem 15.11. According to Lemma 15.16 we have, for $n \geq h$:

$$\lambda_{2\,\max}(K_h\text{ minor free}, n) \geq \lambda_2(K_{h-2}\oplus(n-h+2)K_1) = h-2.$$

□

For graphs on surfaces, one can improve the bounds using the following:

Lemma 15.17. *Let* A, B *be disjoint vertices of a graph* G*, with* $|A| \geq |B|$ *and* $n = |A|+|B|$*. Let* $p \in \mathbb{N}$ *be such that* $K_{3,p}$ *is not a subgraph of* G*. Then*

$$e(A,B) \leq 2n + (\nabla_0(G)-2)((p-1)\widetilde{\nabla}_{1/2}(G)^2+\widetilde{\nabla}_{1/2}(G)+1)|B|.$$

Proof. Partition A into A_1 and A_2 such that A_1 contains the vertices with degree at most 2 and A_2 contains the vertices with degree at least 3.

Consider any linear ordering $x_1, \dots, x_p$ of A_2. We construct a partition $Z_0, Z_1, \dots, Z_{p-1}$ of A_2, $p-1$ sets $T_1, \dots, T_{p-1}$ of triples of vertices in B and a graph $H \in G \,\widetilde{\triangledown}\, \frac{1}{2}$ as follows. At the beginning, H is the empty graph with vertex set B. For each vertex x_i in A_2, if x_i has two neighbours u, v in B that are not adjacent in H we (choose one such pair of vertices and) make them adjacent in H, put x_i in Z_0 and continue with the next vertex of A_2. If x_i has three neighbours u, v, w such that $\{u, v, w\}$ is not in Z_1, we put $\{u, v, w\}$ in Z_i and continue with the next vertex in A_2. Otherwise, we try to find a triples of neighbours of x_i not in $Z_2, Z_3, \dots, Z_{p-1}$. With this construction, all the vertices of A_2 are exhausted for otherwise we would exhibit a $K_{3,p}$ subgraph of G.

Then we obtain $H \in G \,\widetilde{\triangledown}\, \frac{1}{2}$ such that $\|H\| = |Z_0|$ and $|H| = |B|$. Hence we have $|Z_0| \leq \widetilde{\nabla}_{1/2}(G)|B|$ and the number of triangles in H is at most $2\nabla_0(H)^2\,|H| \leq 2\widetilde{\nabla}_{1/2}(G)^2\,|B|$. Each of the sets $Z_1, \dots, Z_{p-1}$ contains only triples corresponding to triangles of H. Hence for $1 \leq i \leq p-1$ we have $|Z_i| \leq 2\widetilde{\nabla}_{1/2}(G)^2\,|B|$. Altogether, we get

$$|A_2| \leq \widetilde{\nabla}_{1/2}(G)(1 + (p-1)\widetilde{\nabla}_{1/2}(G))|B|.$$

Moreover

$$e(A_2, B) \leq \|G[A_2 \cup B]\| \leq \nabla_0(G)(|A_2| + |B|).$$

As the maximum degree of vertices in A_1 is 2 we have $e(A_1, B) \leq 2(n - |A_2| - |B|)$. As $e(A, B) = e(A_1, B) + e(A_2, B)$, we get

$$e(A, B) \leq 2n + (\nabla_0(G) - 2)((p-1)\widetilde{\nabla}_{1/2}(G)^2 + \widetilde{\nabla}_{1/2}(G) + 1)|B|.$$

□

We deduce the following extension of the inequalities obtained by Barrière et al. [52] for planar graphs.

Corollary 15.5. *Let* $g \in \mathbb{N}$. *Then*

$$2 + \Omega\left(\frac{1}{n^2}\right) \leq \lambda_{2\,\max}(\text{genus } g, n) \leq 2 + O\left(\frac{1}{\sqrt{n}}\right)$$

Proof. According to Lemma 15.14 there exists, for every graph G of genus g with order n, a subset $S \subset V(G)$ of cardinality at most $s(n) + 2s$

$(2n/3) = O(\sqrt{n})$ such that $\lambda_2(G) \leq \frac{e(S,V-S)}{n-|S|}$. As G has genus g, it does not contain $K_{3,4g+3}$ as a subgraph [396]. Hence, according to Lemma 15.17, we have $e(V-S,S) \leq 2n + O(\sqrt{n})$. It follows that

$$\lambda_{2\max}(\text{genus } g, n) \leq 2 + O\left(\frac{1}{\sqrt{n}}\right).$$

For the lower bound, consider the planar graph $K_2 \oplus P_{n-2}$, for which $\lambda_2 = 4 - 2\cos(\frac{\pi}{n-1}) = 2 + \Theta(\frac{1}{n^2})$. □

Remark 15.3. The same kind of argument could be applied to prove that graphs that do not contain $K_{p,q}$ for some $p \leq q$ but have bounded $\widetilde{\nabla}_{1/2}$ and sub-linear separators actually have λ_2 bounded by $p - 1 + o(1)$ (as $n \to \infty$).

This chapter contains applications of our theory from the first part of this book, particularly exercising its generality and effectivity. Thus no exercises are given.

Chapter 16
Property Testing, Hyperfiniteness and Separators

More finite than finite,
but asymptotically...

16.1 Property Testing

Since the introduction by Alhazen and Avicenna of the experimental method and of the combination of observations, experiments and rational arguments in the early eleventh century, the scientific method gained in significance and became, after the works of Bacon, Descartes, Boyle and Newton, the standard methodology to come close to the Truth.

The fundamental idea of the experimental approach is that a limited number of experiments should be sufficient to determine whether some property is "close to be true" or not with a "good likelihood". These aspects took a particular importance with statistical likelihood-ratio tests (like χ^2-test), for testing whether there is evidence of the need for a move from a simple model to a more complicated one. In the nineteenth century, Pierce introduced a core principle of modern statistical theory, the randomization.

In Computer Science, such an approach naturally leads to the notion of *property testing*. A property testing algorithm for a decision problem is an algorithm whose query complexity to its input is much smaller than the instance size of the problem. Typically property testing algorithms are used to decide if some mathematical object (such as a graph) has a "global" property, or is "far" from having this property, using only a small number of "local" queries to the object. This is indeed a privileged tool to study large networks.

J. Nešetřil and P.O. de Mendez, *Sparsity*, Algorithms and Combinatorics 28, DOI 10.1007/978-3-642-27875-4_16,

However, different models are usually considered when dealing with dense and sparse structures.

Property testing has been introduced (surprisingly only recently) by Blum et al. [66] and Rubinfeld and Sudan [427] (in relation to program testing), and by Arora et al. [35] and Arora and Safra [36] (in relation to probabilistically checkable proofs). Testing graph properties was first investigated by Goldreich et al. [221]. Also, property testing was studied in the context of computational geometry by Czumaj et al. [107] and in the context of language theory by Alon et al. [20].

From a "mathematical" point of view, the main ingredients of property testing are:

- A random sampling of a large structure,
- A suitable notion of distance between objects.

Let $\mathcal{P}$ be a class of graphs (called *graph property* in this context). A graph G is said to have property $\mathcal{P}$ if $G \in \mathcal{P}$. The graph G is said to be ϵ*-far* from satisfying $\mathcal{P}$ if no graph at distance at most ϵ from G satisfies $\mathcal{P}$. A testing algorithm (or *tester*) for graph property $\mathcal{P}$ and accuracy ϵ is an algorithm that distinguishes with probability at least 2/3 between graphs satisfying $\mathcal{P}$ from graphs that are ϵ-far from satisfying it. More precisely, the property testing algorithm

- Should accept with probability at least 2/3 every input graph that belongs to $\mathcal{P}$,
- Should reject with probability at least 2/3 every input graph that has distance more than ϵ from any graph in $\mathcal{P}$, i.e. if its ϵ-far from satisfying $\mathcal{P}$.

Here, the probabilities are taken over the coin tosses of the tester.

Of course, from the "computational" point of view, the notion of property testing is a bit trickier: We first have to define an encoding of the objects, and then define the distance of two objects as the ratio of the edit distance between the encodings by the "length" of the encoding. Also, one has to precise which "local" queries are allowed to the tester. But these technicalities are not considered here.

A graph property $\mathcal{P}$ is *testable* if for any $\epsilon > 0$, there is a constant time randomized algorithm that can distinguish with high probability between graphs satisfying $\mathcal{P}$ from those that are ϵ-far from satisfying it.

One should notice that the introduction of the parameter ϵ will make some properties impossible to distinguish. Precisely, two properties $\mathcal{P}$ and $\mathcal{Q}$ are *indistinguishable* if for every $\epsilon > 0$ there exists $N = N(\epsilon)$ such that:

- For every graph $G \in \mathcal{P}$ with order at least N there exists $H \in \mathcal{Q}$ with the same order such that $\mathrm{dist}(G, H) < \epsilon$,
- For every graph $H \in \mathcal{Q}$ with order at least N there exists $G \in \mathcal{P}$ with the same order such that $\mathrm{dist}(G, H) < \epsilon$.

As proved in [15] (in the context of dense graphs, but easily extended to the general case), if two properties are indistinguishable then either they are both testable or none of them is testable.

16.1.1 The Dense Model

For comparison, we briefly mention the dense case. In the context of dense graphs, the standard encoding is given by the adjacency matrix of the graph. The local queries correspond to checking the adjacency of two sample vertices. The encoding of a graph of order n is then of length $\binom{n}{2}$, and two graphs G and G' of order n are ϵ-far if one has to change at least $\epsilon\binom{n}{2}$ adjacencies in G to get a graph isomorphic to G'.

Despite the apparent symmetry of the definition, it should be noticed that the fact that a property P is testable is not related to the fact that $\neg P$ is testable. For instance, Alon et al. [15] proved that any first-order property of the form "$\exists\forall$" is testable while there exist some first-order properties of the form "$\forall\exists$" which are not testable.

The key observation here is that any first-order properties of the form

$$\exists x_1 \dots \exists x_s \forall y_1 \dots \forall y_t \, A(x_1, \dots, x_s, y_1, \dots, y_t)$$

(where A is a first-order quantifier-free formula) is indistinguishable from properties of the form

$$\exists \gamma : V(G) \to [c] \ \forall (F, \gamma_F) \in \mathcal{F} \ \forall A \subseteq V(G) \ (G[A], \gamma|A) \not\simeq (F, \gamma_F)$$

Notice that, as a special case, this includes the properties defined by homomorphism $F \to G, F \nrightarrow G$ and $G \to F$ (for fixed F).

The study of property testing in the context of dense graphs is usually based on Szemerédi's regularity lemma [451]. Using this structural lemma, Alon and Shapira proved that every monotone property is testable [27] and then extended this result to hereditary properties [26, 28]. The generalization of Szemerédi's regularity lemma and of the removal lemma to hypergraphs [422, 423] allowed Rödl and Schacht and Avart to prove that every monotone 3-graph property is testable [44]. Rodl and Schacht then proved that every hereditary hypergraph property is testable [421]. The testability of hereditary properties has been further extended to partite hypergraph properties by Ishigami [265] and to multiple directed polychromatic graphs and hypergraphs by Austin and Tao [43]. So (thanks to this spectacular activity in one of the star area of contemporary combinatorics) the dense case seems to be well understood!

16.1.2 The Bounded Degree Model

The *bounded degree model* for graph property testing has been introduced by Goldreich and Ron [222]. In this model, we fix a degree bound d and represent graphs using adjacency lists.

Let $\mathcal{P}$ be a graph property. A graph G with maximum degree at most d is said to be ϵ*-far* for satisfying $\mathcal{P}$ if one needs to modify at least ϵdn adjacencies to make it satisfy $\mathcal{P}$.

In such a model, the tester is given the order of the graph as input and is provided with access to the adjacency tables, and the query complexity $q_{\mathcal{T}}(n)$ of the tester $\mathcal{T}$ is defined as the maximal number of access to the adjacency tables that the tester can execute on any graph G with n vertices.

In [57], Benjamini, Schramm and Shapira showed that every minor-closed graph property can be tested with a constant number of queries in the bounded degree model (see also [107]). For instance, planarity is testable in the bounded degree model.

Actually, they prove a much stronger theorem. To formulate this we first introduce the concept of hyperfiniteness.

A class $\mathcal{C}$ of (finite) graphs is *hyperfinite* if for every positive real $\epsilon > 0$ there exists a positive integer $K(\epsilon)$ such that every graph $G \in \mathcal{C}$ has a subset of at most $\epsilon\,|G|$ edges whose deletion leaves no connected component of order greater than $K(\epsilon)$. Although this notion appeared implicitly in the literature (i.e., [305]), Elek [148–150] was the first to give it a name and propose its systematic study. This notion may be parameterized as follows: $\mathcal{C}$ is (ϵ, k)-*hyperfinite* if every graph $G \in \mathcal{C}$ has a subset F of at most $\epsilon|G|$ edges such that no connected component of $G - F$ has order greater than k.

Example 16.1. Assume classes $\mathcal{A}$ and $\mathcal{B}$ are respectively (ϵ_1, k_1)-hyperfinite and (ϵ_2, k_2)-hyperfinite. Then $\mathcal{C} = \{G \times H : G \in \mathcal{A}, H \in \mathcal{B}\}$ is $(\epsilon_1 + \epsilon_2, k_1 k_2)$-hyperfinite.

Proof. Let $G \in \mathcal{A}$ and $H \in \mathcal{B}$. Then there exist a subset F_1 of at most $\epsilon_1|G|$ edges of G such that no connected component of $G - F_1$ has order greater than k_1, and a subset F_2 of at most $\epsilon_2|H|$ edges of H such that no connected component of $H - F_2$ has order greater than k_2. Let F be the subset of the edges $\{(x, y), (x', y')\}$ of $G \times H$ such that $\{x, x'\} \in F_1$ or $\{y, y'\} \in F_2$. Then $|F| \leq |F_1|\,|H| + |F_2|\,|G|$ hence $|F| \leq (\epsilon_1 + \epsilon_2)|G \times H|$. It is easily checked that if two vertices (a, b) and (a', b') of $G \times H$ belong to the same connected component of $G \times H - F$ then a and a' belong to the same connected component of $G - F_1$ and b and b' belong to the same connected component of $H - F_2$. Thus the connected components of $G \times H - F$ have order at most $k_1 k_2$. □

If a class $\mathcal{C}$ has bounded degree d, then $\mathcal{C}$ is hyperfinite if, and only if, for every positive real $\epsilon > 0$, there exists a finite class $\mathcal{F}_\epsilon$ (of finite graphs) such that every graph in $\mathcal{C}$ is ϵ-close to a disjoint union of some graphs belonging to $\mathcal{F}_\epsilon$. Loosely speaking, while graphs in $\mathcal{C}$ are not finite, they are not far from being so. An important example of hyperfinite classes are bounded degree proper minor closed classes of graphs (as a combination of a theorem of Lipton and Tarjan [305] with a result of Alon et al. [24], regarding separators in minor-free graphs). In this setting, Benjamini et al. [57] proved the following:

Theorem 16.1. *Every monotone hyperfinite graph property is testable.*

Proof (Rough sketch of the proof). Fix an integer d and consider graphs with maximum degree at most d. For graphs G, G' (with maximum degree at most d) and for integer r define the pseudometric

$$\rho_r(G, G') = \sum_H \left| \frac{|\{v,\ B_G(v, r) \cong H\}|}{|G|} - \frac{|\{v',\ B_{G'}(v', r) \cong H\}|}{|G'|} \right|.$$

The first main step is to prove that this pseudometric allows to distinguish hyperfinite properties from those which are not hyperfinite. More formally, $\forall k$ and $\epsilon > 0$. Define

- A: set of (ϵ, k)-hyperfinite graphs G with maximum degree at most d;
- B: set of non $(4\epsilon \log(4d/\epsilon), k)$-hyperfinite graphs with maximum degree at most d.

Then there exists $R = R(d, k, \epsilon)$ such that $\rho_R(A, B) > 0$.

The second main step of the proof is to show that if $\mathcal{P}$ and ϵ-far$(\mathcal{P})$ are distinguishable for every $\epsilon > 0$ then $\mathcal{P}$ is testable, where two graph properties $\mathcal{P}, \mathcal{Q}$ are *distinguishable* if there is an integer r such that

$$\inf_{G \in \mathcal{P}, G' \in \mathcal{Q}} \rho_r(G, G') > 0.$$

□

Using a detailed analysis of bounded expansion classes with an sub-exponential growth we can extend the range of applications of this result. Towards this end we define in the next section the notion of weakly hyperfinite classes.

16.2 Weakly Hyperfinite Classes

A class $\mathcal{C}$ of graphs is *weakly hyperfinite* if for any $\epsilon > 0$ there exists $K(\epsilon)$ such that every $G \in \mathcal{C}$ has a subset of at most $\epsilon|G|$ vertices whose deletion leaves no connected component of order greater than K.

Although it is obvious that in order to be hyperfinite a monotone class of graphs needs to bound the maximum degrees of its elements. However, weakly hyperfinite classes may contain graphs with unbounded maximum degrees. Moreover, it is clear that any hyperfinite class is also weakly hyperfinite and that these two notions coincide for classes of graphs with bounded maximum degrees.

The relation between the two notions is made precise by the following easy result:

Theorem 16.2. *For a positive integer D, denote by Δ_D the class of the graphs having maximum degree at most D. Let $\mathcal{C}$ be a monotone class of graphs with bounded average degree.*

The class $\mathcal{C}$ is weakly hyperfinite if and only if for every integer D the class $\mathcal{C} \cap \Delta_D$ is hyperfinite.

Proof. It is straightforward that if $\mathcal{C}$ is weakly hyperfinite then for every integer D the class $\mathcal{C} \cap \Delta_D$ is hyperfinite (the deletion of a vertex corresponds to the deletion of at most D edges). Conversely, assume that for each integer D the class $\mathcal{C} \cap \Delta_D$ is hyperfinite and let $\epsilon > 0$ be a positive integer. Let C be the supremum of the average degrees of the graphs in $\mathcal{C}$ and let $D = 2C/\epsilon$. As $\mathcal{C} \cap \Delta_D$ is hyperfinite there exists an integer K such that every $H \in \mathcal{C} \cap \Delta_D$ has a subset of at most $(\epsilon/2)|H|$ vertices whose deletion leaves no connected component of order greater than K. Consider a graph $G \in \mathcal{C}$. As the average degree of graphs in $\mathcal{C}$ is bounded by C we get:

$$C \geq \frac{\sum_{i \geq 1} i \, |\{v \in G : d(v) = i\}|}{|G|} = \frac{\sum_{i \geq 1} |\{v \in G : d(v) \geq i\}|}{|G|}$$
$$\geq D \frac{|\{v \in G : d(v) \geq D\}|}{|G|} .$$

Hence $|\{v \in G : d(v) \geq D\}| \leq (\epsilon/2)|G|$. Let $G_{<D}$ be the subgraph of G induced by the vertices of degree smaller than D of G. As $\mathcal{C}$ is monotone, $G_{<D} \in \mathcal{C} \cap \Delta_D$ hence there exists a subset X of at most $(\epsilon/2)|G_{<D}| \leq (\epsilon/2)|G|$ vertices of $G_{\leq D}$ whose deletion leaves no connected component of order greater than K. It follows that the set $S = \{v \in V(G) : d(v) \geq D\} \cup X$ has cardinality at most $\epsilon|G|$ and that its deletion leaves no connected component of order greater than K. □

16.3 Vertex Separators

A key advantage of the notion of weak hyperfinite class is its connection with the existence of sublinear vertex separators. Let G be a graph of order n. Recall that an *α-vertex separator* of G is a subset S of vertices such that every connected component of $G - S$ contains at most αn vertices (we assume, of course, $0 < \alpha < 1$). In Sect. 6.4, we introduced, for each graph G of order n, the function $s_G : \{1, \dots, n\} \to \mathbb{N}$ by

$$s_G(i) = \max_{\substack{|A| \leq i,\\ A \subseteq V(G)}} \min\{|S| : S \text{ is a } \tfrac{1}{2}\text{-vertex separator of } G[A]\}.$$

We established there some connections between s_G, the tree-width and the tree-depth of G. Here, we are interested in the relative sizes of vertex separators for graphs in a class $\mathcal{C}$. This may be studied using the function $\sigma : \mathbb{N} \to \mathbb{N}$ defined by:

$$\sigma(n) = \sup_{G \in \mathcal{C}, |G| \leq n} \min\{|S| : S \text{ is a } \tfrac{1}{2}\text{-vertex separator of } G\}.$$

Hence, if $\mathcal{C}$ is hereditary we have

$$\sigma(n) = \sup_{G \in \mathcal{C}} s_G(n).$$

Instead of σ, it will be convenient to consider a real valued concave sublinear approximation of σ. This will be achieved by the following standard construction from convex analysis (see e.g. [85]): The *convex conjugate* (or *Legendre–Fenchel transform*) of a lower semi-continuous function $\phi : X \to \mathbb{R} \cup \{\infty\}$ is the function $\phi^\star : X^\star \to \mathbb{R} \cup \{\infty\}$ (where X is a real normed vector space and X^* is its dual space) defined by

$$\phi^\star(x^\star) = \sup\{\langle x^\star, x\rangle - \phi(x) : x \in X\}.$$

The *convex biconjugate* ϕ^{**} of ϕ (i.e. the convex conjugate of the convex conjugate of ϕ) is also the *closed convex hull* of ϕ, i.e. the largest lower semi-continuous convex function smaller than ϕ. Hence if ϕ is convex and lower semi-continuous then $\phi^{**} = \phi$.

For a non-decreasing function $f : \mathbb{N} \to \mathbb{R}^+$ we define $\widehat{f}(x) = -g^{**}(-x)$, where

$$g(x) = f(\lfloor x \rfloor) + (x - \lfloor x \rfloor)(f(\lceil x \rceil) - f(\lfloor x \rfloor)).$$

The function $\widehat{f}$ is then the smallest upper continuous concave function greater or equal to f. It is non-decreasing if f is non-decreasing. Hence, according to Lemma 6.6, we have for a hereditary class $\mathcal{C}$:

$$\begin{aligned}\sigma(n) = \sup_{G\in\mathcal{C}} s_G(n) &\leq \sup_{G\in\mathcal{C},|G|\leq n} \mathrm{td}(G)\\ &\leq \sup_{G\in\mathcal{C}} \sum_{i=0}^{\log_2 n} s_G\left(\frac{n}{2^i}\right) \leq \sum_{i=0}^{\log_2 n} \sigma\left(\frac{n}{2^i}\right) \leq \sum_{i=0}^{\log_2 n} \widehat{\sigma}\left(\frac{n}{2^i}\right)\\ &\leq 2\widehat{\sigma}(n).\end{aligned}$$

Moreover, as σ and $\widehat{\sigma}$ are bounded by the same linear functions, $\widehat{\sigma}$ is sublinear if and only if σ is sublinear. This is the basis of the following:

Theorem 16.3. *Let $\mathcal{C}$ be a monotone class of graphs. The following properties are equivalent:*

1. *The graphs in $\mathcal{C}$ have sublinear vertex separators:*
$$\limsup_{G\in\mathcal{C}} \frac{\min\{|S| : S \text{ is a } \frac{1}{2}\text{-vertex separator of } G\}}{|G|} = 0;$$
2. *The graphs in $\mathcal{C}$ have sublinear s_G:* $\lim_{n\to\infty} \sup_{G\in\mathcal{C}} \frac{s_G(n)}{n} = 0$;
3. *The function σ defined by $\mathcal{C}$ is sublinear:* $\limsup_{n\to\infty} \frac{\sigma(n)}{n} = 0$;
4. *The graphs in $\mathcal{C}$ have sublinear tree-width:* $\limsup_{G\in\mathcal{C}} \frac{\mathrm{tw}(G)}{|G|} = 0$;
5. *The graphs in $\mathcal{C}$ have sublinear tree-depth:* $\limsup_{G\in\mathcal{C}} \frac{\mathrm{td}(G)}{|G|} = 0$.

Proof. Items (1) and (2) are clearly equivalent thanks to the monotony of $\mathcal{C}$. Items (1), (3), and (5) are equivalent as

$$\sigma(n) = \sup_{G\in\mathcal{C}} s_G(n) \leq \sup_{G\in\mathcal{C},|G|\leq n} \mathrm{td}(G) \leq 2\widehat{\sigma}(n)$$

and as $\widehat{\sigma}$ is sublinear if and only if σ is sublinear. Item (4) implies item (1) as every graph G has a $\frac{1}{2}$-vertex separator of size $\mathrm{tw}(G)+1$. Item (5) implies item (4) as $\mathrm{td}(G) \geq \mathrm{tw}(G)$ for every graph G. □

Let $\mathcal{C}$ be a monotone class of graphs, such that each $G \in \mathcal{C}$ has a $\frac{1}{2}$-vertex separator of size at most $\sigma(|G|) = o(|G|)$. We shall state a consequence of the existence of sublinear vertex-separators. Our result will follow from an extension of an optimization result of Lipton and Tarjan [304], which is stated as Theorem 16.4 and which is based on the property of $\widehat{\sigma}$ expressed by the following lemma.

Lemma 16.1. *Let $G \in \mathcal{C}$ have order n and let $\mu : V(G) \to [0,1]$ be a probability measure. Then $V(G)$ can be split into parts A, B, C such that $|C| \leq \widehat{\sigma}(2n)$, $\max(|A|,|B|) \leq 2n/3$ and $\max(\mu(A),\mu(B)) \leq 2/3\,\mu(G)$ and no vertex in A has a neighbor in B.*

Proof. Find a $\frac{1}{2}$-vertex separator C_0 and let A_0, B_0 be a partition of $V(G) - C_0$ including each at most $n/2$ vertices. If $\max(\mu(A_0),\mu(B_0) \leq 2/3\,\mu(G)$ let $A = A_0, B = B_0$ and $C = C_0$. Otherwise, we recurse on the part having the largest measure until the biggest part. The total number of vertices which will be in the separator will be $\widehat{\sigma}(n) + \widehat{\sigma}(n/2) + \cdots \leq \widehat{\sigma}(2n)$ (by concavity). □

Theorem 16.4. *Let $G \in \mathcal{C}$ have order n, let $\mu : V(G) \to [0,1]$ be a probability measure. and let $0 < \epsilon < 1$ be a positive real.*

Then there exists a set C of cardinality at most $3\widehat{\sigma}(2\epsilon n/3)/\epsilon$ such that no connected component of $G - C$ has a measure greater than ϵ.

Proof. We apply the following algorithm to G:

- Initialization: Let $C = \emptyset$.
- General Step: Find some connected component K in $G - C$ with $\mu(K) > \epsilon$. Apply Lemma 16.1 to K, producing a partition A_1, B_1, C_1 of its vertices. Let $C \leftarrow C \cup C_1$. If one of A_1 and B_1 (say A_1) has measure exceeding two-thirds the measure of K, apply Lemma 16.1 to $G[A_1]$, producing a partition A_2, B_2, C_2 of A_1 . Let $C \leftarrow C \cup C_2$.
- We repeat the general step until $G - C$ has no component with measure exceeding ϵ.

The effect of one execution of the general step is to divide the component K into smaller components, each with no more than two-thirds the measure of K and each with no more than two-thirds as many vertices as K. We consider all components which arise during the course of the algorithm. We assign a level to each component as follows: If the component exists when the algorithm halts, the component has level zero. Otherwise the level of the component is one greater than the maximum level of the components formed when it is split by the general step. With this definition, any two components on the same level are vertex-disjoint. Each level one component has cost greater than ϵ, since it is eventually split by the general step. It follows that, for $i > 1$, each level i component has measure at least $(3/2)^{i-1}\epsilon$ and contains at least $(3/2)^i$ vertices. Since the total measure of G is at most one, the total number of components of level i is at most $(2/3)^{i-1}/\epsilon$. Since a component of level i contains at least $(3/2)^i$ vertices, the maximum level k must satisfy $(3/2)^k < n$, or $k < \log_{3/2} n$. The total size of the set C produced by the algorithm is then bounded by:

$$
\begin{aligned}
|C| &\leq \sum\{\widehat{\sigma}(2|K|) : K \text{ is a component split by the general step}\} \\
&\leq \sum_{i=1}^{\lfloor \log_{3/2} n \rfloor} \max\left\{ \sum_{j=1}^{\lfloor (2/3)^{i-1}/\epsilon \rfloor} \widehat{\sigma}(2n_j) : \sum_{j=1}^{\lfloor (2/3)^{i-1}/\epsilon \rfloor} n_j \leq n \text{ and } n_j \geq 0 \right\} \\
&\leq \sum_{i=1}^{\lfloor \log_{3/2} n \rfloor} \frac{(2/3)^{i-1}}{\epsilon} f\left(\frac{2n\epsilon}{(2/3)^{i-1}}\right) \\
&\leq \frac{3}{\epsilon} f\left(\frac{2\epsilon n}{3}\right)
\end{aligned}
$$

□

Theorem 16.5. *Every monotone class $\mathcal{C}$ of graphs with sublinear vertex separators is weakly hyperfinite.*

Explicitly, for every positive real $\epsilon > 0$ there exists an integer K such that every graph $G \in \mathcal{C}$ has a subset of vertices S of cardinality at most $\epsilon|G|$ whose deletion leaves no connected component of order greater than K.

Proof. For $G \in \mathcal{C}$, let $\mu : V(G) \to [0, 1]$ be defined by $\mu(v) = 1/|G|$ for every $v \in V(G)$. As $\lim_{x\to\infty} \widehat{\sigma}(x)/x = 0$ there exists an integer K such that $\widehat{\sigma}(2K/3)/(2K/3) < \epsilon/2$. As any graph G has a $\frac{1}{2}$-vertex separator of cardinality $\sigma(n) \leq \widehat{\sigma}(n)$, we deduce from Theorem 16.4 that $V(G)$ has a subset S of cardinality at most $(3n/K)\widehat{\sigma}(2K/3) \leq \epsilon n$ such that no connected component of $G - S$ has a measure greater than K/n, that is such that no connected component of $G - S$ has an order greater than K. □

As a direct consequence of Theorems 16.2 and 16.5 we get

Corollary 16.1. *Let $\mathcal{C}$ be a monotone class of graphs with sublinear vertex-separators and bounded average degree and let* D *be a positive integer.*

Then the subclass of $\mathcal{C}$ including those graphs in $\mathcal{C}$ which have maximum degree at most D *is hyperfinite.*

16.4 Sub-exponential ω-Expansion

Another celebrated theorem of Lipton and Tarjan [304] states that any planar graph has a separator of size $O(\sqrt{n})$. Alon et al. [24] showed that graphs excluding K_h as a minor have a separator of size at most $O(h^{3/2}\sqrt{n})$. Gilbert et al. [218] further proved that graphs with genus g have a separator of size $O(\sqrt{gn})$ (this result is optimal). Eventually, Kawarabayashi and Reed [271] proved that excluding K_h as a minor ensures the existence of a separator of size at most $O(h\sqrt{n})$. This is again best possible as 3-regular expander graphs with n vertices have no K_t-minor for $t = cn^{1/2}$ and no separator of size dn (for appropriately chosen positive constants c, d). Plotkin et al. [386] introduced the concept of *limited-depth minor* exclusion and have shown that exclusion of small limited-depth minors implies the existence of a small separator: they prove that any graph excluding K_h as a depth l minor has a separator of size $O(lh^2 \log n + n/l)$ hence proving that excluding a K_h minor ensures the existence of a separator of size $O(h\sqrt{n \log n})$. More precisely, Plotkin et al. [386] proved the following:

Theorem 16.6. *Given a graph with* m *edges and* n *nodes, and integers* l *and* h*, there is an* $O(mn/l)$ *time algorithm that either produces a* K_h*-minor of depth at most* $l \log n$ *or finds a separator of size at most* $O(n/l + 4lh^2 \log n)$.

We sketch the proof of Theorem 16.6. Following [386], we begin with the following lemma. We use $N(S)$ to denote the set of vertices that are adjacent to the vertices in S. Roughly speaking, the lemma states that either a shallow tree containing at least one representative from each of several given subsets can be found, or that there exists a subset R of vertices that can be separated from the rest by removing a relatively small (with respect to $|R|$) number of vertices.

Lemma 16.2. *Let* G *be a graph with* n *vertices, let* $A_1, \dots, A_k$ *be* k *subsets of the vertex set* V(G) *and let* $l \geq 1$ *be an integer. Then either*

1. *there is a rooted tree* T *in* G *such that the depth of* T *is at most* $4l \log n$ *and such that* $V(T) \cap A_i \neq \emptyset$ *for* $i = 1, \dots, k$,
2. *or there exists a* $S \subseteq V(G)$ *where*

   a. $|N(S) \cap V - S| \leq \min(|S|, |V - S|)/l$, *and*
   b. $|N(V - S) \cap S| < \min(|S|, |V - S|)/l$.

Proof (Sketch of the proof).
Initialize subset R to be some vertex $v \in V$. Each step we will augment R by including vertices that are within distance 2 of the vertices in R. We repeat this augmentation step as long as it either causes $|R|$ to grow by a factor of $(1 + 1/l)$, or causes $|V - R|$ to shrink by a factor $(1 + 1/l)$.

Consider the set R after the execution of the last step. If R contains at least one vertex from each of the sets A_i, then we output a tree consisting of shortest paths from these to v. Otherwise, we output $R \cap N(R)$.

It is easy to see that after $2l \log n$ iterations, either S would contain more than n vertices, or $V-S$ would be empty, and hence the augmentation process has to stop after at most $2l \log n$ iterations. Thus, the length of a path from v to each of the representative vertices of the sets $\{A_i\}$ is bounded by $4l \log n$ as required by condition 1.

The termination condition implies that if a tree is not produced, then the size of the sets N(R) and N(N(R)) is bounded by $\min(|R|, |V - R|)/l$, which implies conditions 2a and 2b. □

Proof (Theorem 16.6). We are given a graph G, a weight function w on its vertices and edges, and integers l, h. The algorithm maintains a three-way partition $(V_r, M, V - V_r - M)$. We will prove by induction that this partition conforms to the following conditions:

1. The set M can be partitioned into $k < h$ subsets $\mathcal{A}_1, \mathcal{A}_2, \dots, \mathcal{A}_k$, where the size of each subset is bounded by $4lh \log n$. The radius of each one of the subsets is bounded by $4l \log n$. For every two of these subsets there is an edge that connects a vertex in one to a vertex in the other; In other words, M has K_k as a minor at depth $4l \log n$.
2. $|N(V_r) \cap (V - V_r - M)| \leq |V_r|/l$, i.e. V_r can be separated from the rest of the graph by removing at most $|V_r|/l$ vertices.
3. $w(G_r) \leq 2w(G)/3$, where G_r is the graph induced by V_r in G.

The algorithm proceeds in phases, where at each phase it either augments M so that it contains a larger minor, or adds vertices from M and/or vertices from $V - V_r - M$ to V_r. The algorithm terminates when the graph induced by $V - V_r - M$ contains less than 2/3-fraction of the total weight. The separator returned by the algorithm consists of $M \cup (N(V_r) - V_r)$. Observe that if the above invariants are satisfied, the size of the returned separator is bounded by $|M| + |V_r|/l \leq O(lh^2 \log n + n/l)$, as required by Theorem 16.6.

Clearly, the invariants are satisfied initially. Assume that at the beginning of some phase, M contains a K_k minor whose vertices come from the contraction of the subgraphs induced by $\mathcal{A}1, \ldots, \mathcal{A}_k$ (which we further call *supernodes*. For all i, define A_i to be the neighbors of $\mathcal{A}_i$ that lie in $V - V_r - M$, i.e., $A_i = N(\mathcal{A}_i) - V_r - M$. If some A_i is empty then we remove the corresponding supernode from M and add it to V_r. Observe that this operation maintains the invariants.

If all of A_i's are nonempty, we apply Lemma 16.2 on the graph induced by $V - V_r - M$ in G to produce either the minimal tree T that contains a node in each A_i or a subset of nodes S.

Case 1. *Tree* T *produced:* By Lemma 16.2, the depth of the tree is limited by $4l \log n$. Notice that a minimal tree contains at most $4kl \log n$ vertices since a minimal tree consists of at most k paths of length $l \log n$. By construction, every set A_i has a vertex adjacent to some vertex in the tree. Hence, if we define $\mathcal{A}_{k+1}$ as the vertices of the tree, the fact that Invariant 1 was satisfied at the start of the phase implies that it will be satisfied in the end of the phase. The only possible problem can occur if $k + 1 = h$. In this case the algorithm terminates and outputs the sets $\mathcal{A}_1, \ldots, \mathcal{A}_h$ as a proof hat G has a small-depth K_h minor.

The set V_r is not changed and hence Invariant 3 is maintained.

Case 2. *Set* S *returned:* The algorithm adds the set S or the set $V - V_r - M - S$ to V_r, depending on which set contains the least weight. Let S' denote the set that was added to V_r. Part 2 of Lemma 16.2 implies that at most $|S'|/l$ vertices are added to $N(V_r) \cap V - V_r - M$. Since at least $|S'|$ vertices are added to V_r, Invariant 2 is maintained. Invariant 3 is maintained since, at the end of the phase, the weight of V_r is bounded by $w(V_r) + (w(V) - w(V_r))/2$, which is at most $2w(V)/3$ as long as $w(V_r)$ was previously less than $w(V)/3$. (If it was not, the algorithm would have terminated in the previous iteration.) Invariant 1 is trivially maintained since the set M remains unchanged in this case.

□

The ω-*expansion* of a class $\mathcal{C}$ is the mapping

$$i \mapsto \sup_{G \in \mathcal{C} \triangledown i} \omega(G),$$

where $\omega(G)$ stands for the *clique number* of G, i.e. the order of the largest complete subgraph of G. Notice that a class has bounded ω-expansion if and only if it is nowhere dense.

A class $\mathcal{C}$ has sub-exponential ω-expansion if

$$\limsup_{i\to\infty} \sup_{G\in\mathcal{C}\,\triangledown\, i} \frac{\log \omega(G)}{i} = 0.$$

The following results extend the result of [355]:

Lemma 16.3. *There exists a constant* C *such that any graph* G *of order* n *has a separator of size at most* $C\frac{n\log n}{z}$ *whenever* z *is a multiple of* $\log n$ *such that*

$$2z(\omega(G\,\triangledown\, z)+1) \le \sqrt{n\log n}. \tag{16.1}$$

Proof. Let $l = z/\log n$ and let $h = \lfloor \omega(G\,\triangledown\, z)+1\rfloor$. As l is an integer and $\omega(G\,\triangledown\, z) \le f(z) < h$, G has no K_h minor of depth at most $z = l\log n$. According to Theorem 16.6, G has a separator of size at most $(C/2)(n/l + 4lh^2\log n)$ for some fixed constant C, i.e. a separator of size at most $(C/2)(\frac{n\log n}{z} + 4z(\omega(G\,\triangledown\, z)+1)^2) \le C\frac{n\log n}{z}$. □

We shall also need a construction similar to the one introduced in Sect. 16.3:

Lemma 16.4. *Let* $f:\mathbb{N}\to\mathbb{R}^+$ *be a non-decreasing function such that* $f(n) = o(n)$. *Then there exists a continuous concave function* $\hat{f}:\mathbb{R}^+\to\mathbb{R}^+$ *such that* $\hat{f}(n) \ge f(n)$ *holds for every* $n\in\mathbb{N}$ *and* $\hat{f}(x) = o(x)$.

Proof. If f is bounded, put $\hat{f}(x) = \sup_n f(n)$.

Otherwise, let $a_0 = 0$. We define recursively real numbers $a_1 \le a_2 \le \dots \le a_i \le \dots$ and the non-negative real numbers $b_1, b_2, \dots, b_i, \dots$ in such a way that

$$\forall n,\quad f(n) \le f(a_{i-1}) + b_i + \frac{n-a_{i-1}}{i}.$$

We put $b_1 = \max_{n\in\mathbb{N}} f(n) - f(0) - n$ and define a_1 as the maximum integer such that $f(a_1) = a_1 + b_1 + f(0)$ (both a_1 and b_1 are finite as $f(n) = o(n)$). If a_i and b_i have been defined, we put

$$b_{i+1} = \max_{n\ge a_i}(f(n) - f(a_i)) - (n-a_i)/i$$

and define a_{i+1} as the maximum integer such that

$$f(a_{i+1}) = f(a_i) + (a_{i+1} - a_i)/i$$

(again, both a_{i+1} and b_{i+1} are finite as $f(n) = o(n)$; moreover, $a_{i+1} \ge a_i$ as $(f(n) - f(a_i)) - (n-a_i)/i) < 0$ if $n < a_i$).

As f is not bounded, we have $\lim_{i\to\infty} a_i = \infty$. We define $\hat{f}$ as the lower envelope of the affine functions

$$L_i(x) = f(a_i) + (x - a_i)/i.$$

This function is clearly continuous and concave.

Theorem 16.7. *Let $\mathcal{C}$ be a class of graphs with sub-exponential ω-expansion.*

Then the graphs of order n in $\mathcal{C}$ have separators of size $s(n) = o(n)$ which may be computed in time $O(ns(n)) = o(n^2)$.

Proof. Let $f(x) = \omega(\mathcal{C} \mathbin{\triangledown} x)$ and let $\phi(x) = \log f(x)$. By assumption, we have $\phi(x) = o(x)$. According to Lemma 16.4 there exists a continuous concave function $\phi' \geq \phi$ such that $\phi'(x) = o(x)$.

Without loss of generality, we can hence assume that the function f bounding the ω-expansion of the class $\mathcal{C}$ is continuous, positive, defined on $\mathbb{R}$, such that $\log f(x)$ is concave and such that $\log f(x) = o(x)$. It follows that for every integer n it holds $\frac{\log f(n) + \log f(n+2)}{2} \leq \log f(n+1)$ thus

$$\begin{aligned}\frac{\log f(n+2)}{\log f(n+1)} &= 1 + \frac{\log f(n+2) - \log f(n+1)}{\log f(n+1)}\\ &\leq 1 + \frac{\log f(n+1) - \log f(n)}{\log f(n)}\\ &= \frac{\log f(n+1)}{\log f(n)}\end{aligned}$$

Hence

$$\sup_{n\in\mathbb{N}} \frac{\log f(n+1)}{\log f(n)} = \frac{\log f(1)}{\log f(0)}.$$

Let $A = \frac{\log f(1)}{\log f(0)}$ and let $g(x) = \frac{\log f(x)}{x}$. By assumption, $g(x) = o(1)$. Define $\zeta(n)$ as the greatest integer such that

$$\log f(\zeta(n)) < \frac{\log n}{3}$$

As $\log f(\zeta(n) + 1) \geq \frac{\log n}{3}$ and $\frac{\log f(\zeta(n)+1)}{\log f(\zeta(n))} \leq A$ we deduce

$$\log f(\zeta(n)) \geq \frac{\log n}{3A}.$$

Notice that ζ is increasing and $\lim_{n\to\infty}\zeta(n) = \infty$. From the definition of $g(x)$, we deduce

$$\zeta(n) = \frac{\log f(\zeta(n))}{g(\zeta(n))} \geq \frac{\log n}{3Ag(\zeta(n))} \gg \log n.$$

Let $l(n) = \lfloor \frac{\zeta(n)}{\log n} \rfloor$. We have $l(n)\log n = \zeta(n)(1+o(1))$ and

$$\log(2l(n)\log n(f(l(n)\log n)+2)) < \frac{\log n}{3}(1+o(1)).$$

It follows that if n is sufficiently large (say $n > N$), we have

$$\log(2l(n)\log n(f(l(n)\log n)+2)) < \frac{\log n + \log\log n}{2},$$

that is: $2l(n)\log n(f(l(n)\log n)+2) < \sqrt{n\log n}$. Thus if $n > N$, a graph $G \in \mathcal{C}$ with order n has a separator of size at most

$$C\frac{n\log n}{l(n)\log n} = 3g(\zeta(n)(1+o(1)))n = o(n).$$

□

As random cubic graphs almost surely have bisection width at least $0.101n$ (see e.g. [285]), they have almost surely no separator of size smaller than $n/20$. It follows that if $\log f(x) = (\log 2)x$, the graphs have no sublinear separators any more. This shows the optimality of Theorem 16.7.

As a consequence, we obtain the main result of this section:

Theorem 16.8. *Let $\mathcal{P}$ be a monotone class of graphs with subexponential ω-expansion.*

Then the property $G \in \mathcal{P}$ is testable in the bounded degree model.

Proof. This is a direct consequence of Theorem 16.1, Theorem 16.7 and Corollary 16.1. □

This generalizes all known instances of property testing for sparse graphs.

Exercises

16.1. A class of graphs is small if it contains at most $n!\alpha^n$ different labeled graphs on n vertices, for some constant α. Answering a question of Welsh et al. [372] showed that every proper minor-closed classes of graphs is small. This result has been extended by Dvořak and Norine [139]. They proved that every class $\mathcal{C}$ closed under taking induced subgraphs such that every graph $G \in \mathcal{C}$ with $n \geq 3$ vertices has a separator of size at most

$$s(n) = O\left(\frac{n}{(\log n \, \log\log n)^2}\right)$$

is small.

They deduce from both this result and a theorem of [355] bounding $s(n)$ in terms of grads that every class $\mathcal{C}$ such that for some $k > 0$ and every integer r

$$\nabla_r(\mathcal{C}) \leq f(r) = \frac{k}{r} e^{\frac{1}{2}\sqrt[3]{\frac{9r}{\log^2(r+e)}}} - 2$$

is small, which is the case if $\nabla_r(\mathcal{C}) \leq e^{r^{1/3-\epsilon}}$ for some $\epsilon > 0$.

Deduce from Lemma 16.3 that a stronger result holds for classes with ω-expansion bounded by $f(r) + 1$.

16.2. On the opposite side from graphs with sublinear vertex separators are expanders. It is important in both combinatorics, combinatorial number theory, and computer science that expanders with bounded degree exist.

Let Δ_d be the class of all graphs with maximum degree at most d. Determine the ω-expansion of Δ_d. This class, for every $d \geq 3$, contains expanders (see e.g. [259]).

It is an open problem whether (under some mild conditions) the existence of sublinear vertex separators is equivalent to a sub-exponential ω-expansion.

Chapter 17
Core Algorithms

Fast, robust, strong... youthful modern times.

An essential part of this book deals with estimates of complexity of algorithms. The aim of this chapter is to describe core algorithms, like the computation of a p-tree-depth decomposition. We shall describe this particular algorithm in a sufficiently precise way to allow an actual implementation of the described algorithms. In order to base our complexity results, we specify our computational model in this chapter.

17.1 Data Structures and Algorithmic Aspects

Our algorithms are often fast (linear and nearly linear) and thus we have to specify the way we encode and handle (often sparse) graphs. This will be done in this section. A reader interested in structural results only may decide to skip this chapter.

In computer science, several data structures have been used to represent the concept of graph. For instance, a graph may be represented by means of:

- An *adjacency list*, usually implemented as an array or a linked list of pairs of vertices,
- An *incidence list*, allowing loops and multiple edges,
- An *adjacency matrix* more suitable for dense graphs, usually implemented by means of uninitialized arrays,
- An *incidence matrix* particularly adapted to dense multigraphs and to algebraic computations.

J. Nešetřil and P.O. de Mendez, *Sparsity*, Algorithms and Combinatorics 28, 381
DOI 10.1007/978-3-642-27875-4_17, © Springer-Verlag Berlin Heidelberg 2012

Whatever implementation is chosen, three kind of operations will be of importance:

- Traversal operations, on which will be based standard algorithms like the *breadth-first search* and the *depth-first search*;
- Proximity tests, which allow to test whether two vertices are adjacent or not, whether two edges intersect, whether the distance between two vertices is smaller than some constant, etc.;
- Modifications of the graph, like the addition or the deletion of vertices and edges, the extraction of a subgraph or the copy of a graph.

Depending on the context, the importance of the time complexity of these operations and of the space required by the data structure will lead to different choices. In this book, we are concerned with the structural analysis of sparse graphs. This suggests the use of a data structure which could benefit from the low density of the considered graphs and could optimize the traversal operations and the proximity tests. Also, as we will consider orientations of graphs, operations of *reorientation* (i.e. the operation consisting of reversing the orientation of an arc) should be fast. The modification operations which are of interest in our setting will usually be local and rare. For all these reasons, we singled out a data structure used in topological graph theory, namely the representation of graphs by means of combinatorial maps. This choice, which is surprisingly unusual, is the one of the graph library PIGALE [194] (developed by a team including the second author of this book) and we believe it is particularly effective.

Combinatorial maps will be used as a "working data structure". The input graphs are given in a very simple way: An *input graph* G will be a pair (n, L) where n is the order of G and L is a linked list of pairs of integers (i, j) corresponding to an edge $\{i, j\}$ (or an arc (i, j) if the input graph is assumed to be directed). The main point here is that the vertex set is implicitly defined as $\{1, \dots, n\}$ and the edge set is also implicitly ordered by the list L, see Fig. 17.1.

A *combinatorial map* (or *rotation scheme*) is intuitively a graph with a circular order of the edges around the vertices [142, 251]. This concept has been introduced in topological graph theory to study graphs embedded on surfaces, and more particularly planar maps [100, 101]. Formally, a *combinatorial map* is a triple (D, σ, α) where D is an even finite set of *darts* (which intuitively correspond to "half-edges"), σ is a permutation on D encoding the circular order of the darts around the vertices and α is a fixed-point free involution exchanging the two darts composing an edge. In order to simplify the computations, the set D for a graph of size m will be $\{-m, \dots, -1, 1, \dots, m\}$ and we will fix the involution α as the involution exchanging i and $-i$. Hence the only information we will have to encode is the permutation σ.

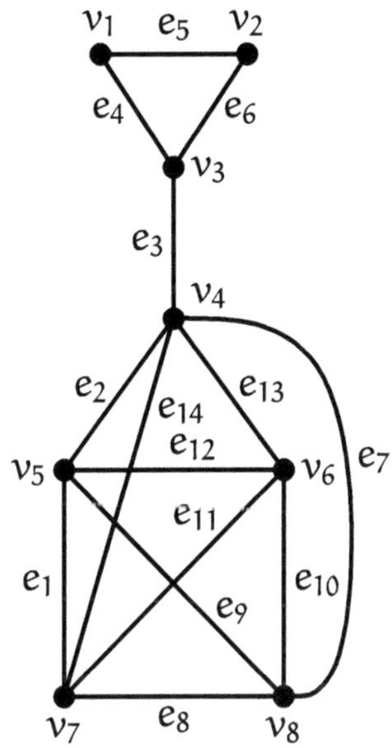

$$L = \big((5,7),(5,4),(4,3),(1,3),(1,2),(2,3),(4,8),(8,7),(8,5),(8,6),(6,7),(6,5),(6,4),(4,7)\big)$$

Fig. 17.1 An input graph encoded as a pair (n, L) where n is the order of the graph (here 8) and L is the list of the graph edges

To get an efficient encoding of the map, we will actually encode σ as two arrays indexed by $\{-m, \ldots, m\}$ which we call Cir (for circular order) and Acir (for anti-circular order) such that $\text{Cir}[i] = \sigma(i)$ and $\text{Acir}[i] = \sigma^{-1}(i)$. The vertices of a combinatorial map are implicitly defined as the orbits of σ. So, we introduce two other arrays: FirstDart, indexed by $\{1, \ldots, n\}$ associate to each vertex number the number of a first dart incident to it, and Vin, indexed by $\{-m, \ldots, m\}$ which associates to each dart number the number of the vertex to which it is incident. See Fig. 17.1.

The construction of a combinatorial map from an input graph and, conversely, the enumeration of the edges of a combinatorial map, present no particular difficulty and may be achieved in $O(m + n)$-time, where n and m are respectively the order and the size of the graph. This we formalize as Procedure 1:

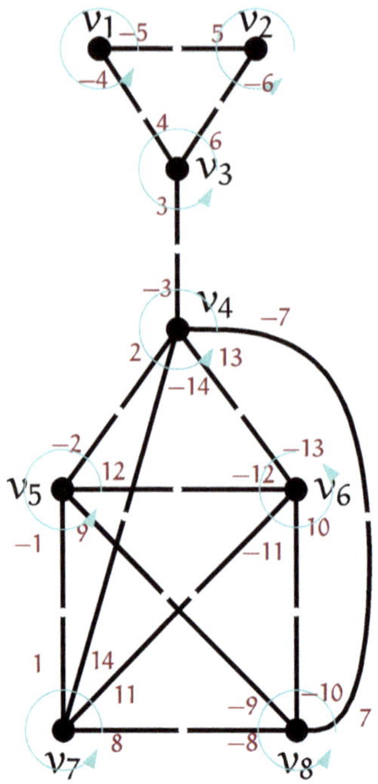

dart #	Cir	Acir	Vin
−14	13	2	4
−13	−12	10	6
−12	−11	−13	6
−11	10	−12	6
−10	−9	7	8
−9	−8	−10	8
−8	7	−9	8
−7	−3	13	4
−6	5	5	2
−5	−4	−4	1
−4	−5	−5	1
−3	2	−7	4
−2	−1	12	5
−1	9	−2	5
1	8	14	7
2	−14	−3	4
3	6	4	3
4	3	6	3
5	−6	−6	2
6	4	3	3
7	−10	−8	8
8	11	1	7
9	12	−1	5
10	−13	−11	6
11	14	8	7
12	−2	9	5
13	−7	−14	4
14	1	11	7

Fig. 17.2 A combinatorial map and the corresponding arrays Cir, Acir and Vin

Procedure 1 Computation of a combinatorial map from an input graph

Require: input graph G is given by a pair (n, L).
Ensure: the created arrays encode a combinatorial map of G.

```
  Create the arrays for the combinatorial map.
  Initialize FirstDart to 0.
  Let d = 0.
  for all e = (i, j) in L do
     Let d ← d + 1. {number the dart}
     {Insertion of dart d at vertex i}
     Let f = FirstDart[i].
     if f = 0 then
        Put FirstDart[i] = Cir[d] = Acir[d] = d.
     else
        Put Acir[d] = Acir[f] and Cir[Acir[f]] = d.
        Put Cir[d] = f and ACir[f] = d.
     end if
     Put Vin[d] = i.
     {Insertion of dart −d at vertex j}
     Let f = FirstDart[j].
     if f = 0 then
        Put FirstDart[j] = Cir[−d] = Acir[−d] = −d.
     else
        Put Acir[−d] = Acir[f] and Cir[Acir[f]] = −d.
        Put Cir[−d] = f and ACir[f] = −d.
     end if
     Put Vin[−d] = j.
  end for
```

In this book we will be interested in ordering the vertices of a k-degenerate graph G so that every vertex has back degree at most k. Procedure 2 achieves such a task.

Procedure 2 Computation of a vertex-ordering of a k-degenerate graph G so that each vertex has back degree at most k.

Require: The input graph G is given as a pair (n, L).
Ensure: Rank is the rank associated to the linear order and Order is the ordered array of the vertices.

```
Construct a combinatorial map (Procedure 1).
Create an array Degree initialized with 0
for all d ∈ {−m, . . . , −1, 1, m} do
  Increase Degree[Vin[i]] by one.
end for
{Initial sort}
Create SCir, SAcir and SFirst (initialized with 0) implementing a partition of
{1, . . . , n} into n doubly-linked circular lists (SFirst[i] gives first element in list
#i, SCir and SAcir link each element to the previous and the next in the same
list).
for all 1 ≤ i ≤ n do
  Add i to the list #Degree[i].
end for
{Construction of the elimination order}
a = 0.
for k = 1 to n do
  while SFirst[a] = 0 do
    a ← a + 1
  end while
  Pop i from list #a, let Rank[i] = k and Order[k] = i.
  for all dart d incident to vertex i do
    Let j = Vin[−d] and x = Degree[j]
    if x > a then
      Move j from the list #x to list #(x − 1).
      Put Degree[j] = x − 1.
    end if
    Delete vertex i (and its incident edges).
  end for
end for
```

We shall see that Procedure 2 follows almost the same lines of a classical topological sort procedure for directed graphs. Its running time is obviously $O(n + m)$ (where n and m denote respectively the order and the size of the input graph). Notice that the procedure does not get as an input the degeneracy of the graph, but actually computes it. Indeed, the variable a maintains the maximum of the minimum degrees met in the iterative deletion of vertices of low degrees. At the end of the procedure, the value of a is hence at least k. As the procedure computes an ordering of the vertices with back-degree bounded by a, we also get $a \leq k$ hence $a = k$.

Such an algorithm can be used to color a k-degenerate graph using at most $k+1$ colors. The corresponding Procedure 3 obviously runs in time $O(m+n)$. In the pseudo code describing this procedure we don't go into unnecessary details, as the algorithm presents no real difficulty.

Procedure 3 Computation of a vertex coloring

Require: The input graph G is given by a pair (n, L).
Ensure: the computed coloring c using at most $k+1$ colors if G is k-degenerated.

```
Order the vertices of the graph (n, L) using Procedure 2.
Initialize an array Dep of lists with ( ).
for all edge (i, j) ∈ L do
  if Rank[i] < Rank[j] then
    Add i to Dep[j].
  else
    Add j to Dep[i].
  end if
end for
for r = 1 to n do
  Let i = Order[r].
  Let c[i] be the smallest integer different not in {c[i], j ∈ Dep[i]}.
end for
```

17.2 p-Tree-Depth Coloring

It is established in Theorem 7.8 that for every integer p there exists a polynomial P_p such that every graph G has a p-tree-depth coloring requiring at most $P_p(\widetilde{\nabla}_{2^{p-2}+1/2}(G))$ colors.Recall that a *p-tree-depth coloring* is a coloring such that every $p' \leq p$ color classes induce a subgraph of tree-depth at most p' (see Chap. 7). (For properties of tree-depth, see Chap. 6.) The strong benefit of the proof of Theorem 7.8 is that it actually leads to a simple linear time algorithm, and that it not only produces a p-tree-depth coloring, but actually a $(p+1)$-centered coloring, that is a vertex coloring such that, for any (induced) connected subgraph H, either some color appears exactly once in H, or H gets at least $p+1$ colors (see Sect. 7.2). Also, the proof of Theorem 7.8 will actually imply that for every positive integer p there exists a polynomial F_p such that we can compute for every graph G a p tree-depth coloring of G using $N_p(G) \leq F_p(\nabla_{2^{p-2}+\frac{1}{2}}(G))$ colors in time $O(N_p(G)|G|)$. This result, which should be seen as an algorithmic version of Theorem 7.8, will be stated as Theorem 17.1.

Input graphs will be given as a pair (n, L), where n will be the order of the graph and L will be the list of the edges, each edge being a pair (i, j) with $1 \le i \neq j \le n$ (the vertex set of the graph is implicitly $\{1, \dots, n\}$). This type of encoding allows to represent a graph G using $2\|G\| + 1$ integers. However, it assumes that the vertices of the graph have been numbered. For details on the chosen data structure and basic procedures, see above Sect. 17.1.

17.2.1 Fraternal Augmentation

We follow the lines of Sect. 7.4 to design an efficient algorithm to compute p tree-depth colorings of graphs. See Sect. 7.4 for here undefined notions and notations. In the detail description of our algorithms this part is distinguished from the rest of the book.

For an integer k, a fraternity function w on a $\{1, \dots, n\}$ will be encoded by means of a function $\widetilde{w}$ such that $\widetilde{w}(j)$ is the list of the pairs $(i, w(i, j))$ for $1 \le i \le n$ such that $w(i, j) \le k$.

We shall define three procedures, Procedures 4–6. They form together an algorithm for p-tree-depth coloring of graphs.

Our first step is to construct a function $\widetilde{w_1}$ corresponding to a 1-fraternity function w such that $G_1^w = G$ and $\Delta_1^-(w_1) \le 2\nabla_0(G)$. This is done by Procedure 4 in linear time.

Procedure 4 Computation of $\widetilde{w_1}$

Require: The input graph G is given by a pair (n, L).
Ensure: the function w_1 associated to $\widetilde{w_1}$ is such that $G_1^w = G$ and $\Delta_1^-(w_1) \le 2\nabla_0(G)$.

Compute the rank Rank of a vertex ordering of G so that the maximum back-degree will be at most $2\nabla_0(G)$ (Procedure 2, page 385).
for all (i, j) in L **do**
  if Rank[i] < Rank[j] **then**
    Add $(i, 1)$ to $\widetilde{w_1}(j)$.
  else
    Add $(j, 1)$ to $\widetilde{w_1}(i)$.
  end if
end for

Lemma 17.1. *Let k be an integer and let w_k be a k-fraternity function encoded by $\widetilde{w_k}$. Then an encoding $\widetilde{w_{k+1}}$ of a $(k+1)$-fraternity function w_{k+1} can be computed in time $O((\Delta_k^-(w_k)^2 + \Delta_{k+1}^-(w_{k+1}))n)$, which is such that:*

$$\forall(x,y) \in V^2, \quad w_k(x,y) \le k \implies w_{k+1}(x,y) = w_k(x,y)$$
$$\Delta^-_{k+1}(w_{k+1}) \le 2\widetilde{\nabla}_{k/2}(G \bullet \overline{K}_{1+N_{w_{k+1}}(k+1)})$$

Proof. Consider Procedure 5 below. This algorithm first computes all pairs (i,j) such that $\min(w_{k+1}(i,j), w_{k+1}(j,i)) = k+1$ (in time $O(\Delta^-_k(w_k)^2 n)$) and then orients the fraternity edges (in time $O(\Delta^-_k(w_{k+1})n)$). □

Procedure 5 Augmentation of a k-fraternity function into a (k+1)-fraternity function

Require: $\widetilde{w_k}$ is a k-fraternity function on $\{1,\dots,n\}$.
Ensure: $\widetilde{w_{k+1}}$ is the augmented $(k+1)$-fraternity function.

```
{Computation of the list of edges of G_{k+1}^{w_{k+1}}.}
Initialize L = ().
for all i ∈ {1,...,n} do
  for all (a, l_a) ∈ w̃_k(i) do
    for all (b, l_b) ∈ w̃_k(i) after (a, l_a) do
      if l_a + l_b = k + 1 then
        add (a, b) to L
      end if
    end for
  end for
end for
{Computation of the orientation of G⃗_{k+1}^{w_{k+1}}.}
Compute the rank Rank of a vertex ordering of the graph with edge list L mini-
mizing the maximum back degree (Procedure 2, page 385).
Initialize w̃_{k+1} with w̃_k.
for all (i, j) ∈ L do
  if Rank[i] < Rank[j] then
    add (i, k + 1) to w̃_{k+1}(j)
  else
    add (j, k + 1) to w̃_{k+1}(i)
  end if
end for
```

Notice that our augmentation is such that $\Delta^-_{k+1}(w_{k+1})$ is at most the double of the bound computed in Lemma 7.6. This factor of 2 comes from the choice of a simple orientation algorithm based on the recursive elimination of the vertices of minimal degrees.

Procedure 6 Computation of a $(p+1)$-centered coloring

Require: The input graph G is given by a pair (n, L); the integer p is part of the input.

Ensure: the computed coloring c is $(p+1)$-centered and uses $N_p(G) \leq P_p(\widetilde{\nabla}_{2^{p-2}+\frac{1}{2}}(G))$ colors.

Compute $\widetilde{w_1}$ such that associated 1-fraternity function w_1 is such that $G_1^w = G$ and $\Delta_1^-(w_1) \leq 2\nabla_0(G)$ (by Procedure 4).
Let $q = 2^{p-1} + 1$.
for $k = 1$ **to** q **do**
  Compute the function $\widetilde{w_{k+1}}$ from the function $\widetilde{w_k}$ using the augmentation Procedure 5.
end for
{Computation of depth p transitivity}
for all vertex $i \in \{1, \ldots, n\}$ **do**
  Compute the list S[i] of the vertices j such that there exists a sequence $x_0 = j, \ldots, x_l = i$ such that $l \leq p$ and $w(x_{a-1}, x_a) < \infty$ for every $1 \leq a \leq p$.
end for
{Computation and coloring of the conflict graph}
Compute the list L' of the (i, j) such that $1 \leq i \leq n$ and $j \in S[i]$.
Compute a coloring c of (n, L') using Procedure 3, page 386.

Next we can apply Procedure 6 which yields straightforwardly

Theorem 17.1. *For every integer* p *there exists a polynomial* P_p *(of degree about* 2^{2^p}*) such that for every graph* G *Procedure 6 computes a* $(p+1)$*-centered coloring of* G *with* $N_p(G) \leq P_p(\widetilde{\nabla}_{2^{p-2}+\frac{1}{2}}(G))$ *colors in time* $O(N_p(G)n)$*-time.*

17.2.2 Computing the Forest

We now construct a rooted forest of height p including G in its closure using Procedure 7.

Procedure 7 runs in $O(pm)$ time. If G is connected, it returns a rooted tree Y of height at most p such that $G \subseteq \mathrm{clos}(Y)$ (thus proving that tree depth of G is at most p).

Procedure 7 Computation of a tree of height p including in its closure a given graph with a centered coloring using p colors

Require: c is a centered-coloring of the graph G using colors $1, \dots, p$.
Ensure: F is a rooted forest of height p such that $G \subset \mathrm{clos}(F)$.

```
Set F = ∅.
Let Big be an array of size p.
for all Connected component G_i of G do
  Initialize Big to false.
  Set root_color ← 0.
  for all v ∈ V(G_i) do
    if Big[c[v]] = false then
      if c[v] = root_color then
        root_color ← 0, Big[c[v]] ← true.
      else
        root ← v; root_color ← c[v].
      end if
    end if
  end for
  Recurse on G − root thus getting some rooted forest F' = {Y'_1, ..., Y'_j}.
  Add to F the tree with root root and subtrees Y_1, ..., Y_j, where the sons of root
  are the roots of Y_1, ..., Y_j.
end for
```

17.3 Computing and Approximating Tree-Depth

There is an (easy) polynomial algorithm to decide whether $\mathrm{td}(G) \leq k$ for any fixed k. However, assuming P≠NP, there is no polynomial time approximation algorithm for the tree-depth can guarantee an error bounded by n^{ϵ}, where ϵ is a constant with $0 < \epsilon < 1$ and n is the order of the graph [73].

Nevertheless, there is a simple linear time algorithm which allows to approximate tree-depth, up to an exponentiation:

Lemma 17.2. *The Depth-First Search algorithm computes in linear time, for each input graph* G*, a rooted forest* Y *such that:*

$$Y \subseteq G \subseteq \mathrm{clos}(Y),$$
$$\mathrm{td}(G) \leq \mathrm{height}(Y) \leq 2^{\mathrm{td}(G)}.$$

Proof. Perform a depth-first search on G and let h be the height of the obtained DFS rooted forest F. As $G \subseteq \mathrm{clos}(F)$ we have $h \geq \mathrm{td}(G)$. As G includes a path of length h, we have $\mathrm{td}(G) \geq \log_2 h$. □

As noticed in Sect. 6.10, the property to have tree-depth at most t is expressible in first-order logic, as a consequence of Lemma 6.13. Actually, a sentence τ_t of reasonable size such that for every graph G holds $G \models \tau_t$ if and only if $td(G) \leq t$ can be constructed, as shown in Exercise 6.6. This property allows to check efficiently if a graph has tree-depth at most t, for some fixed integer t:

Theorem 17.2. *For every fixed* t*, there exists a linear time algorithm that test whether a graph* G *has tree-depth at most* t.

Proof. Compute a depth-first search tree Y on G. If the height of Y is greater than 2^t then $td(G) > t$.

Otherwise, we construct a tree-decomposition (T, λ) of G having width at most $(2^t - 1)$: Set $T = Y$ and define $\lambda(x) = \{v \leq_Y x\}$. Then for any v, $\{x \in V(T) : v \in \lambda(x)\} = \{x \geq_Y v\}$ induces the subtree of Y rooted at v (hence a subtree of T). Moreover, as $G \subseteq \text{clos}(Y)$, any edge $\{x, y\}$ with $x <_Y y$ is a subset of $\lambda(y)$. Hence (T, λ) is a tree-decomposition of G. As $\max_{v \in V(G)} |\lambda(v)| = \text{height}(Y) \leq p$, this tree-decomposition has width at most $(p-1)$. This tree-decomposition may be obviously constructed in linear time.

Then, we can apply Courcelle's theorem (Theorem 3.9) and test in linear time whether a graph has tree-depth at most t or not. □

Theorem 17.3. *For a fixed integer* t*, there is a linear time algorithm which computes, for an input graph* G *with tree-depth at most* t*, a rooted forest of height* t *whose closure includes* G.

Proof. Without loss of generality, we can assume that G is connected. For fixed t, there exists a first-order formula $\phi_t(x)$ such that $G \models \Phi_t(v)$ if $td(G - v) < t$. Making use of an extension of Courcelle's Theorem 3.9 (see for instance [178]), we can compute in linear time a vertex v with this property, which will be the root of our tree. Then we recursively construct a rooted tree for each connected component of $G - v$ (in global linear time). □

The preceding two theorems make use of Courcelle's theorem or its derivatives. It is a natural problem to find a simpler methods.

Problem 17.1. Is there a simple linear time algorithm to check $td(G) \leq t$ for fixed t?
Is there a simple linear time algorithm to compute a rooted forest Y of height t such that $G \subseteq Clos(Y)$ (provided that such a rooted forest exist)?

The decision problem "$tw(G) \leq k$" (for fixed k) belongs to the class L. This problem is actually complete for L [146]. Recall that the complexity class L contains decision problems which can be solved by a deterministic Turing machine using a logarithmic amount of memory space (see Sect. 3.11).

In terms of space complexity, it appears that deciding $td(G) \leq k$ is more easy than deciding $tw(G) \leq k$, as shown in [147]

Theorem 17.4. *For every $k \geq 1$, the decision problem "$td(G) \leq k$" belongs to the class AC^0.*

In [147] is also proved an analog of Theorem 3.9 for graphs with bounded tree-depth:

Theorem 17.5. *For every MSO-sentence ϕ over some signature σ and every $d \in \mathbb{N}$, there is a DLOGTIME-uniform AC^0-circuit family that, on input of an arbitrary σ-structure $\mathbf{S}$, outputs 1 if, and only if, the tree depth of $\mathbf{S}$ is at most d and $\mathbf{S} \models \phi$ holds.*

Note that in contrast, it is known that the same problem for graphs of bounded tree width is L-complete [108, 146].

17.4 Counting Homomorphisms to Graphs with Bounded Tree-Depth

We first introduce a list variant of the graph homomorphism counting problem:

Let F, G be graphs and let $L : V(G) \to 2^{V(F)}$ be a *list assignment* which associates to each vertex $x \in G$ a list $L(x)$ of admissible pre-images of x. The corresponding *to-list homomorphism counting problem* is to determine the number of homomorphisms $f : F \to G$ such that $u \in L(f(u))$ for every $u \in F$. Notice that lists restricts pre-images and not images, like in the standard list coloring problem. However, if $L(v) = V(F)$ for every $v \in V(H)$, we have the problem of counting homomorphisms.

Lemma 17.3. *Let* $L : V(G) \to 2^{V(F)}$ *be a list assignment. The number of homomorphisms* $f : F \to G$ *such that* $u \in L(f(u))$ *for every* $u \in F$ *can be computed in time* $O(|F|\mathrm{td}(G)2^{|F|\mathrm{td}(G)}|G|)$.

Proof. We prove the proposition by induction over the tree-depth of G. If $\mathrm{td}(G) = 1$ then we check whether F is edgeless. Then the number of homomorphisms $f : F \to G$ such that $u \in L(f(u))$ for every $u \in F$ is simply the product of $|\{v \in G, u \in L(v)\}|$ for $u \in F$. This product may be computed in $O(|G| + |F|)$ time.

Now assume that the proposition has been proved for graphs G of tree-depth at most $t \geq 1$ and consider graphs G with tree depth $t + 1$.

- If both F and G are connected and if r is a vertex of G such that $\mathrm{td}(G-r) \leq t$, we consider all the possible stable sets S of $L(r)$ (there are obviously at most $2^{|F|-1}$ such sets). Then we modify the list assignment of G as follows: for every $x \in F$ adjacent to some $y \in S$, remove x from all the lists of vertices of G which are not adjacent to r. This can be done in time $O(|F||G|)$;
- If F is connected but G is not and if $G_1, \dots, G_p$ are the connected components of G, we compute the sum of the number of homomorphisms $f : F \to G_i$ such that $u \in L(f(u))$ for every $u \in F$. Thus we reduce the computation to the case where G is connected;
- If F is not connected and $F_1, \dots, F_p$ are the connected components of F, we compute the product of the number of homomorphisms $f : F_i \to G$ such that $u \in L_i(f(u))$ for every $u \in F_i$, where $L_i(x) = L(x) \cap V(F_i)$. The computation of the lists L_i is easily performed in $O(|F||G|)$-time, and we reduce the computation to the case where F is connected.

□

17.5 First-Order Cores of Graphs with Bounded Tree-Depth

We shall now prove a result which is specific to graphs with bounded tree-depth: every "large" graph with bounded tree-depth may be approximated by a "small" induced subgraph. Here, the notion of approximation will be based on the p-back-and-forth equivalence $\equiv^p$ (see Sect. 3.8.4).

Theorem 17.6. *For every integers* p, t *there is a linear time algorithm which, given a graph* G*:*

- *Either outputs a subgraph of* G *which is a path of order* 2^t *(hence a certificate that* $\mathrm{td}(G) > t$*),*
- *Or outputs a subset of vertices* A *of cardinality at most* $C(p,t)$ *such that* $G[A] \equiv^p G$ *and* $G \longrightarrow G[A]$.

Proof. As noticed in Lemma 17.2, a depth-first search on G computes in linear time a rooted forest F such that $G \subseteq \mathrm{clos}(F)$. If the height of F is at least 2^t then outputs a tree path of order 2^t. Otherwise, Theorem 17.6 will follow from the following stronger statement we formulate now.

Claim. For every integers p, c, h there is a linear time algorithm which, given a triple (G, F, γ) formed by a graph G, a rooted forest F of height h such that $G \subseteq \mathrm{clos}(F)$ and a coloring $\gamma : G \to [c]$ of the vertices of G, computes a subset A of vertices of G such that the graphs G and $G[A]$ are hom-equivalent in a way which preserves both the coloring γ and the ancestor relation, and also such that $G \equiv^p G[A]$.

Proof of the claim: To each vertex x of G assign a $0-1$ vector $\gamma_0(x)$ of length $\mathrm{height}_F(x) - 1$ encoding the adjacency of x with its ancestors in F. Inductively define $\gamma_1(x)$ by $\gamma_1(x) = (\gamma(x), \gamma_0(x))$ for every x such that $\mathrm{height}_F(x) = h$ and $\gamma_1(x) = (\gamma(x), \gamma_0(x), S(x))$ for the other vertices, where $S(x)$ is the multiset of the values $\gamma_1(y)$ for y son of x, where each value is kept only at most p times. Then $\gamma_2(x)$ is defined as the couple $(\gamma_1(r), \gamma_1(x))$ where r is the root of the connected component of F which contains x.

The set A is built as follows: as long as there exists a connected component Y of $F[A]$ with root r such that the value $\gamma_2(r)$ is shared by at least p other vertices in A, delete r. Then, inductively consider higher and higher level: as long as there exists a subtree Y of $F[A]$ rooted at a soon y of a vertex $x \in A$ whose γ-value is shared by at least p others sons of x, delete y.

Now, let us describe the strategy of Duplicator in a p-back and forth game between G and $G[A]$. Assume that the strategy successfully found a local isomorphism π_i from $G[A_i]$ to $G[B_i]$ where $A_i \subseteq A$ and B_i are closed under the ancestor relation and assume $i < p$.

- Assume Spoiler chooses a vertex $a_{i+1} \in A \setminus A_i$. If a_{i+1} is a root of $F[A]$ then let b_{i+1} be a root of F with the same γ_2-value as a_{i+1} which does not yet belong to B and define $A_{i+1} = A_i \cup \{a_{i+1}\}, B_{i+1} = B_i \cup \{b_{i+1}\}$ and π_{i+1} is the extension of π_i on A_{i+1} such that $\pi_{i+1}(a_{i+1}) = b_{i+1}$;

otherwise, let a_j be the highest ancestor of a_{i+1} in A_i and let $b_j = \pi_i(a_j)$. Let $a^1_{i+1}, \dots, a^k_{i+1} = a_{i+1}$ be the chain of Y[A] from a son of a_j to a_{i+1}. According to the construction of γ_2, there exists a chain of Y $b^1_{j+1}, \dots, b^k_{j+1}$ starting from a son of b_j such that $b^1_{j+1} \notin B_j$ and $\gamma_2(b^l_{i+1}) = \gamma_2(a^l_{i+1})$ for every $1 \leq l \leq k$. Define $A_{i+1} = A_i \cup \{a^1_{j+1}, \dots, a^k_{j+1}\}$, $B_{i+1} = B_i \cup \{b^1_{j+1}, \dots, b^k_{j+1}\}$ and let π_{i+1} be the extension of π_i to A_{i+1} such that $\pi_{i+1}(a^l_{i+1}) = b^l_{i+1}$ for every $1 \leq l \leq k$.

- The case where Spoiler chooses a vertex b_{i+1} is handled in a similar way, the construction of the set A ensuring that the vertices $a^1_{i+1}, \dots, a^k_{i+1}$ can be chosen in $A \setminus A_i$.

That $G \longrightarrow G[A]$ is clear from the construction of the coloring γ_2.

□

Note that the property stated in Theorem 17.6 for graphs with bounded tree-depth is not true for graphs with tree-width at most 2. As a direct consequence of Theorem 17.6 we have:

Corollary 17.1. *There is a computable function* F *such that for every graph* G *and every First-Order sentence* ϕ*, it is possible to check whether* G *satisfies* ϕ *or not in time*

$$F(\mathrm{td}(G), \mathrm{qrank}(\phi))\ |G|.$$

(See Sect. 3.8 for a definition of the quantifier rank qrank of a formula.) As a corollary we obtain an alternative proof of Theorem 17.2:

Corollary 17.2. *For every fixed* t *there is a linear time algorithm to decide whether* $\mathrm{td}(G) \leq t$.

Proof. We can use Corollary 17.1 to decide $\mathrm{td}(G) \leq t$ in linear time, as this property is First-Order definable. □

Exercises

17.1. It has been proved by Bodlaender [68, 69] that graph isomorphism can be decided in time $O(n^{t+4.5})$ for graph with tree-width at most t.

Prove that there exists an algorithm A running in time $O(n^t \log n)$ that computes, for an input graph G of tree depth at most t and order n, a colored plane lexicographic monotone forest $\mathrm{Canon}(G)$ of height t such that G' is isomorphic to G if and only if $\mathrm{Canon}(G') = \mathrm{Canon}(G)$.

Deduce that there exists an $O(n^t \log n)$-time isomorphism testing algorithm for the class of graphs with tree-depth at most t.

17.2.
- Prove that for each integer t there exists an algorithm deciding whether two input graphs G and H with tree-depth at most t are homomorphically equivalent or not, which runs in time $O(|G| + |H|)$;
- Show that the multiplicative constant hidden in the big O reflects the size of the minimal graphs with tree-depth t hence is not an elementary function of t.

Note that examples combinatorial problems with even more paradoxical almost linear algorithms were given in [307].

Chapter 18
Algorithmic Applications

Check the model without a serial number.

18.1 Introduction

In 1928, Hilbert posed the following challenge, known as the *Entscheidungsproblem*:

> Does there exist an algorithm which takes as input a description of a formal language and a mathematical statement in the language and outputs either "true" or "false" according to whether the statement is true or false?

As a particular interpretation, this problem contains the *decision problem for first-order logic* (that is, the problem of algorithmically deciding whether a first-order formula is universally valid). A negative answer to the Entscheidungsproblem was given by Church and Turing, who proved that it is impossible to decide algorithmically whether statements in arithmetic are true or false.

We consider here the following related *model-checking problem*.

J. Nešetřil and P.O. de Mendez, *Sparsity*, Algorithms and Combinatorics 28, DOI 10.1007/978-3-642-27875-4_18, © Springer-Verlag Berlin Heidelberg 2012

Model-checking problem

Given a sentence ϕ of some logic L and a structure $\mathbf{A}$, decide whether ϕ holds in $\mathbf{A}$, that is:

$$\text{Is it true that } \mathbf{A} \models \phi?$$

This is a typical problem studied by model theory and theoretical computer science. Of particular interest for computer science is the case where the input structure is restricted to be finite. The techniques from finite model theory are especially applied in databases, artificial intelligence, formal languages, and complexity theory. The study of the relations between the expressiveness of a logic and the complexity of the decision problems that can be expressed with this logic is the aim of *descriptive complexity*. The first main result of descriptive complexity was Fagin's theorem [171], which established that the complexity class NP coincides with the class of the decision problems expressible in existential second-order logic.

It is known that model-checking for first-order logic, that is checking whether $\mathbf{A} \models \Phi$ where ϕ is a first-order formula and both $\mathbf{A}$ and ϕ are parts of the input is a PSPACE-complete problem (this is even the case if $\mathbf{A}$ is fixed and equal to a graph with two vertices) and thus most likely not solvable in polynomial time, see [443, 465] for a vast literature related to this subject.

If Φ is a fixed first-order formula of quantifier rank k it is clear that deciding whether $\mathbf{A} \models \Phi$, where $\mathbf{A}$ has size n, may be done in time $O(n^k)$. However, it is not known whether there exists a universal constant c such that for every first-order formula, deciding whether $\mathbf{A} \models \Phi$ may be done in time $O(n^c)$ [446]. Such a statement would hold, for instance, if PSPACE=P, but is strongly believed to be false as it would imply a collapse of the full parametrized complexity hierarchy. Therefore, a particular attention has been paid to consider the typical case where one has to check whether a relatively "small" sentence holds in a "large" structure under the light of the recently very active parametrized complexity theory. Recall that a problem with parameter k is called *fixed-parameter tractable* if it can be solved in time $f(k)\,P(n)$ for an arbitrary computable function f and some polynomial function P (here n denotes the size of the input structure). The class of fixed-parameter tractable decision problems is denoted by FPT. Actually, there is a hierarchy of intractable classes

$$\mathrm{FPT} \subseteq \mathrm{W}[1] \subseteq \mathrm{W}[2] \subseteq \cdots \subseteq \mathrm{W}[*] \subseteq \mathrm{AW}[*].$$

For instance, deciding if a graph G contains an independent set of size k is W[1]-complete and deciding if G has a dominating set of size k is W[2]-complete. Also, for input formed by a graph G and a first-order formula Φ, deciding $G \models \Phi$, parametrized by the length of Φ, is AW[*]-complete.

One way to reduce the complexity of model-checking is to restrict the input graph (or structure) to a fixed class $\mathcal{C}$. An example is provided by Theorem 3.9. However, even for first-order logic, there are some limits to this approach: Under the complexity theory assumption FPT $\neq$ AW[*], Dawar and Kreutzer proved [113] that if a monotone class $\mathcal{C}$ is somewhere dense and satisfies some effectivity conditions then first-order model checking is not fixed-parameter tractable. It follows that the best we shall expect is fixed-parameter tractability for nowhere dense classes. Also, despite restricting model-checking to a class of very sparse graphs, the dependence to the parameter in the running time may be greater than a single exponential. For instance, Frick and Grohe proved [203] that there is no model-checking algorithm for first-order logic on the class of binary trees (still under the assumption FPT $\neq$ AW[*]) whose running time is bounded by $2^{2^{2^{o(k)}}} P(n)$ where P is a polynomial, k denotes the size of the input sentence and n the size of the input structure.

Much work has gone into establishing so-called meta-theorems for variants of monadic second-order logic [104] and for first-order logic. For first-order logic, Seese [433] proved that first-order model-checking is fixed-parameter tractable on graph classes with bounded degree. This has been generalized by Frick and Grohe [202] to graph classes of bounded local tree width, by Flum and Grohe [179] to graph classes excluding a fixed minor, and by Dawar et al. [111] to graph classes locally excluding a fixed minor. (See for instance [442] for examples of applications of such generic results for proving fixed-parameter tractability to restricted classes of problems.) This chapter will be concerned with generalizations of all these results to bounded expansion and nowhere dense classes. Actually, these classes seem to provide a natural setting for meta-theorems.

18.2 Truncated Distances

The fixed parameter linear time algorithm for p-tree depth coloring has a number of algorithmic consequences. Let us start our tour of algorithmic applications of our theory with the following result. It is just a weighted extension of the basic observation that bounded orientations allow $O(1)$-time checking of adjacency [96] and is a basic example of First-Order Boolean query.

Theorem 18.1. *For any class $\mathcal{C}$ with bounded expansion and for any integer k, there exists a linear time preprocessing algorithm so that for any preprocessed $G \in \mathcal{C}$ and any pair $\{x, y\}$ of vertices of G the value $\min(k, \text{dist}(x, y))$ may be computed in $O(1)$-time.*

Proof. The proof goes by a variation of our augmentation algorithm so that each arc e gets a weight $w(e)$ and each added arc gets weight $\min(w(e_1) + w(e_2))$ over all the pairs (e_1, e_2) of arcs which may imply the addition of e and simplification should keep the minimum weighted arc.

Then, after k augmentation steps, two vertices at distance at most k have distance at most 2 in the augmented graph. The value $\min(k, \text{dist}(x, y))$ then equals

$$\min(k, w((x, y)), w((y, x)), \min_{(z,x),(z,y) \in \vec{G}} (w(z, x) + w(z, y))).$$

$\square$

This problem is an easy example of an algorithm checking the existence of a given subgraph in G with some of its vertices prescribed: checking whether $\text{dist}(x, y) \leq k$ amounts in testing whether G includes as a subgraph a path of length at most k with prescribed end-vertices x and y). Theorem 18.1 naturally extends to nowhere dense classes, see Exercise 18.1.

18.3 The Subgraph Isomorphism Problem and Boolean Queries

The technique of p-tree-depth coloring is locally sensitive and it facilitates the detection of most local graph properties. Particularly it can be used to check existential first-order properties. A standard example is provided by the *subgraph isomorphism problem*. For a fixed pattern H, the problem is to check whether an input graph G has an induced subgraph isomorphic to H. This problem is known to have complexity at most $O(n^{\omega l/3})$ where l is the order of H and where ω is the exponent of square matrix fast multiplication algorithm [361] (hence $O(n^{0.792\, l})$ using the fast matrix algorithm of [99]). The particular case of subgraph isomorphism in planar graphs have been studied by Plehn and Voigt [385], Alon [29] with super-linear bounds and then by Eppstein [157, 158] who gave the first linear time algorithm for fixed pattern H and G planar. This was extended to graphs with bounded genus in [159].

Table 18.1 Subgraph isomorphism problem: complexity for a fixed pattern H and for an input graph restricted to some class of graphs

Subgraph isomorphism problem		
Context	*Complexity*	*Reference(s)*
General	$O(n^{0.792\,\lvert V(H)\rvert})$	[361] using [99]
Bounded tree-width	$O(n)$	[158] (also [102, 103])
Planar	$O(n)$	[157, 158]
Bounded genus	$O(n)$	[159]
Bounded expansion (includes the three previous classes)	$O(n)$	[355]
Nowhere dense	$n^{1+o(1)}$	

The model-checking problem of an existential first-order sentence is easily solved by combining Theorem 17.1 with Courcelle's theorem (Theorem 3.9). This way, we get:

Theorem 18.2. *Let* p *be a fixed integer. Let* ϕ *be a* $FOL(\tau_2)$ *sentence. Then there exists a polynomial* $P_p(X)$ *and an algorithm which checks* $\exists X : (|X| < p) \wedge (G[X] \models \phi)$ *and runs in time* $O(P_p(\widetilde{\nabla}_{2^{p-2}+\frac{1}{2}}(G)\ |G|))$.

In particular, the algorithm runs in $O(n)$ *time if* G *is bound to a class with bounded* $\widetilde{\nabla}_{2^{p-2}+\frac{1}{2}}(G)$ *(such as a bounded expansion class), and in* $n^{1+o(1)}$ *time if* G *is bound to a class where* $\widetilde{\nabla}_{2^{p-2}+\frac{1}{2}}(G) = |G|^{o(1)}$ *(such as a nowhere dense class).*

For instance:

Corollary 18.1. *Let* $\mathcal{K}$ *be a class with bounded expansion and let* H *be a fixed graph. Then, for each of the next properties there exists a linear time algorithm to decide whether a graph* $G \in \mathcal{K}$ *satisfies them:*

- H *has a homomorphism to* G,
- H *is a subgraph of* G,
- H *is an induced subgraph of* G.

18.4 The Distance-d Dominating Set Problem

We shall now consider a problem which is not expressible by an existential first-order sentence, but by a more complicated first-order sentences (such as $\exists\forall$ first-order sentences). The DOMINATING SET problem (DSP) is defined as follows.

Input A graph $G = (V, E)$ and an integer parameter k.
Question Does there exist a dominating set of size k or less for G, i.e., a set $V' \subseteq V$ with $|V'| \leq k$ and such that for all $u \in V - V'$ there is a v in V' for which $uv \in E$?

This is a classic NP-complete problem [212] which is also apparently not fixed parameter tractable (with respect to the parameter k) because it is known to be W[2]-complete in the W-hierarchy of fixed parameter complexity theory [127].

DSP is fixed parameter tractable with respect to, for example, tree-width [33] and tree decompositions are computable in linear time, for fixed tree-width [70]. DSP is similar in definition to the vertex cover problem (VCP), but they seem to differ considerably in their fixed-parameter tractability properties. The Robertson-Seymour theory of graph minors [406] can be used to show that VCP is a fixed parameter tractable problem because vertex cover is closed with respect to taking minors, and fixed-parameter tractable algorithms have been described [127] for VCP. But DSP is not closed with respect to taking minors.

DSP remains NP-complete when restricted to planar graphs [212]. Fellows and Downey [126, 127] gave a search tree algorithm for this problem which has time complexity $O(11^k n)$, when the input is restricted to planar graphs, improved to $O(8^k n)$ in [8], to $O(4^{\sqrt[6]{34k}} n)$ in [7], and to $O(2^{27\sqrt{k}} n)$ in [269].

In [153] it is shown, using the search tree approach, that the dominating set problem is fixed parameter tractable for graphs of bounded genus, with time complexity of $O((4g+40)^k n^2)$ for graphs of genus $g \geq 1$. Our results imply that a time complexity of $O(f(g,k)n)$ can be achieved, for some computable function f.

More recently, Alon and Gutner [18] gave a linear time parametrized algorithm for dominating sets on d-degenerate graphs running in time $k^{O(dk)} n$. Although the result proved in Theorem 18.3 is weaker than this last result, it is a nice illustration on how reduction to subgraphs of bounded tree-depth may be used.

Let $G = (V, E)$ be a graph. A subset $X \subseteq V$ of G is a *dominating set* of G if every vertex of G not in X is adjacent to some vertex in X. We note $\mathfrak{D}(G)$ the family of all dominating sets of G and by $\mathfrak{D}_k(G)$ the family of the dominating sets of G having cardinality at most k.

For subsets $X, W \subseteq V$, we say that X *dominates* W if every vertex in $W \setminus X$ has a neighbor in X. We denote $\mathfrak{D}_k(G, W)$ the family of the subsets dominating W and having cardinality at most k.

Lemma 18.1. *For every integers $k, l \geq 1$ for every graph $G = (V, E)$ with tree-depth at most l and for every subset $W \subseteq V$ of vertices, there exists a blocker $A = A(G, W) \subseteq V$ of at most kl vertices meeting every $X \in \mathfrak{D}_k(G, W)$. Moreover, if a rooted forest Y of height l is given such that $G \subseteq \mathrm{clos}(Y)$ then we can find the blocker set A in $O(kl)$-time.*

From this Lemma, using a p tree-depth coloring, we deduce:

Lemma 18.2. *Let $\mathcal{C}$ be a class with bounded expansion. Then there exists a function $f : \mathbb{N} \to \mathbb{N}$ such that for every integer k, for every $G = (V, E) \in \mathcal{C}$ and for every $W \subseteq V$ a set $A(G, W)$ of cardinality at most $f(k)$ may be computed in $O(n)$-time (where n is the order of G) which meets every set in $\mathfrak{D}_k(G, W)$.*

Hence, by an easy induction on k:

Theorem 18.3. *Let $\mathcal{C}$ be a class with bounded expansion. Then there exists a function $g : \mathbb{N} \to \mathbb{N}$ such that for every integer k, every $G = (V, E) \in \mathcal{C}$ and every $W \subseteq V$ one may compute in time $O(g(k)n)$ a set X which is either minimal set cardinality at most k dominating W or the empty set if G has no dominating set of cardinality at most k.*

Actually, we also deduce that any graph G has at most $F(k, \nabla_{k^k}(G))$ dominating sets of size at most k and that they may be all enumerated in time $O(\phi(k, \nabla_{k^k}(G))n)$. Notice that the result does not extend to the problem of finding a set X of cardinality at most k such that every vertex not in X is at distance at most 2 from X (consider k disjoint stars of order n/k, giving $(n/k)^k$ possible solutions to the problem.

The *distance-d dominating set problem* (or (k, d)*-center problem*) is a generalization of the dominating-set problem where we are given a graph G and integer parameters d and k and where we have to determine whether G contains a subset X of at most k vertices such that every vertex of G has distance at most d to a vertex of X (see [48] and references in [115]). Dawar and Kreutzer [112] proved that this problem is fixed-parameter tractable on any *effectively nowhere dense class*, that is on any nowhere dense class $\mathcal{C}$ if there is a computable function $f : \mathbb{N} \to \mathbb{N}$ such that $f(r) > \omega(\mathcal{C} \,\nabla\, r)$.

Theorem 18.4. *The following is fixed-parameter tractable for any effectively nowhere dense class of graphs.*

DISTANCE-d-DOMINATING SET

Input: *A graph* $G \in \mathcal{C}, W \subseteq V(G), k, d \geq 0$
Parameter: $k + d$
Problem: *Determine whether there is a set* $X \subseteq V(G)$ *of* k *vertices which* d*-dominates* G.

Notice that this result does not use p tree-depth coloring but instead relies on the equivalence of nowhere dense classes with uniformly quasi-wide classes proved in Chap. 8.

To the opposite, Dvořák [136] (again!) proved the following intriguing connection between generalized weak coloring numbers (see Sect. 4.9 and 7.5), generalized independence numbers (see Chap. 8), and the distance d-domination numbers:

Theorem 18.5. *Let* $1 \leq m \leq 2k + 1$, *and let* G *be a graph of order* n *such that* $\mathrm{wcol}_m(G) \leq c$. *Then* $\mathrm{dom}_k(G) \leq c^2\, \alpha_m(G)$. *Furthermore, if an ordering of* $V(G)$ *witnessing* $\mathrm{wcol}_k(G) \leq c$, *then a distance-*$k$ *dominating set* D *and an* m*-independent set* A *such that* $|D| \leq c^2|A|$ *can be found in* $O(c^2 \max(k, m)\, n)$ *time.*

By making use of the general algorithm for first-order model checking described in the next section, the author deduce [136]:

Theorem 18.6. *Let* $\mathcal{C}$ *be a class with bounded expansion, and let* $d \geq 1$ *be a constant. There exists a linear time algorithm which computes, for each input graph* $G \in \mathcal{C}$, *a distance-*d *dominating set* D *and a* $(2d + 1)$*-independent set* A *such that* $|D| = O(|A|)$.

18.5 General First-Order Model Checking

We now consider the general problem of First-Order model-checking, examples of which are the subgraph isomorphism problem (with fixed pattern) and the distance-d dominating set problem we considered above.

Recently, Dvořák et al. [138] have given a linear-time algorithm for deciding first-order properties in classes with bounded expansion, as well as an almost linear time algorithm for deciding first-order properties in classes with locally bounded expansion. Here, a class $\mathcal{C}$ of σ-structures $\mathrm{Rel}(\sigma)$ has locally bounded expansion if Gaifman($\mathcal{C}$) has locally bounded expansion.

Precisely, the following is proved in [138]:

Theorem 18.7. *Let $\mathcal{C} \subseteq \mathrm{Rel}(\sigma)$ be a class of σ-structures with bounded expansion (resp. locally bounded expansion), and ϕ be a First-Order sentence (on the natural language of $\mathrm{Rel}(\sigma)$). There exists a linear time (resp. an almost linear time) algorithm that decides whether a structure $\mathbf{A} \in \mathcal{C}$ satisfies ϕ.*

More generally, they design a dynamic data structure for finitely colored graphs belonging to a fixed class of graphs of bounded expansion with the following properties (for a fixed first-order formula $\phi(x)$ with one free variable):

- The data structure is initialized in linear time;
- The color of a vertex or an edge can be changed in constant time;
- We can find in constant time a vertex $v \in V(G)$ such that $\phi(v)$ holds

All this extends to relational structures. This is based on the existence of low tree-depth colorings and on a procedure of quantifier elimination [138]. Recently, Grohe and Kreutzer [229], instead of eliminating all the quantifiers by means of the introduction of new functional symbols and the augmentation of the original structure in a complicated way, proposed a procedure allowing to eliminate universal quantifications by means of the additions of new relations while preserving the Gaifman graph of the structure.

Lemma 18.3. *Let $\mathcal{C}$ be a class of σ-structures with bounded expansion, and let $\Phi(\vec{x}) = \forall \vec{y}\ \Psi(\vec{x}, \vec{y})$ be a universal formula (where Ψ is quantifier-free).*

Then there exists an existential formula $\Phi'(\vec{x})$ such that for every $\mathbf{A} \in \mathcal{C}$ there exists a σ'-structure $\mathbf{A}'$ with the same universe such that:

(1) $\mathbf{A}'$ may be constructed in linear time from $\mathbf{A}$,
(2) $\mathrm{Gaifman}(\mathbf{A}) = \mathrm{Gaifman}(\mathbf{A}')$*;*
(3) for every $\vec{x}$, we have

$$\mathbf{A} \models \Phi(\vec{x}) \quad \Longleftrightarrow \quad \mathbf{A}' \models \Phi'(\vec{x}).$$

We give a high-level sketch of this lemma, based on [229]:

Proof (Very rough sketch of the proof). Let $\mathbf{A} \in \mathcal{C}$ be a structure. Two q-tuples $\vec{a}$ and $\vec{b}$ of elements of $\mathbf{A}$ have the same *full type* if they satisfy the same formulas of quantifier rank at most q. The interest of full types is that it allows to check formulas on the full types representatives instead of checking it on all tuples, leading to a constant time evaluation algorithm. Computation of the full types proceeds in stages.

(1) Compute a tree-depth decomposition;
(2) For each k-tuple C of colors is defined the *local type* of quantifier-rank q of a tuple $\vec{a}$ of elements in the substructure $\mathbf{A}_C$ induced by the colors in C, corresponding to equivalence classes in $\mathbf{A}_C$ of tuples satisfying the same formulas of quantifier rank at most q.
(3) The *global type* of a tuple $\vec{a}$ is defined as the collection of all local types of $\vec{a}$ in the individual substructures $\mathbf{A}_C$, for all C of length at most k. It is shown that global types can be defined by existential first-order formulas.
(4) The global types then serve as the basis for the definition of full types. It is shown that each full type can be described by an existential first-order formula. The existential formulas describing full types in a structure $\mathbf{A}$ is not be over the structure $\mathbf{A}$ itself, but over an expansion of $\mathbf{A}$ by the edges of tree-depth decompositions.

□

From this lemma and Theorem 18.2 we deduce

Theorem 18.8. *Let $\mathcal{C} \subseteq \mathrm{Rel}(\sigma)$ be a class of σ-structures with bounded expansion, and $\phi(\vec{x})$ be a First-Order formula. There exists a linear time algorithm that counts, for an input σ-structure $\mathbf{A} \in \mathcal{C}$ the number of vectors $\vec{x}$ such that $\mathbf{A}$ satisfies $\phi(\vec{x})$.*

This may not be the end of the story and we are naturally led to the following question:

Conjecture 18.1. Is it true that first-order model-checking is fixed-parameter tractable on nowhere dense classes of graphs (or structures)?

18.6 Counting Versions of Model Checking

18.6.1 Enumerating Isomorphs

It appears that one of the main lemmas of [158] (Lemma 2) actually induces a linear time algorithm to solve the problem of counting all the isomorphs of H in a graph G as soon as we have a linear time algorithm allowing to compute a low-tree width partition of G:

Lemma 18.4. *Assume we are given graph* G *with* n *vertices along with a tree decomposition* T *of* G *with width* w. *Let* S *be a subset of vertices of* G, *and let* H *be a fixed graph with at most* w *vertices. Then in time* $2^{O(w \log w)}n$ *we can count all isomorphs of* H *in* G *that include some vertex in* S. *We can list all such isomorphs in time* $2^{O(w \log w)}n + O(kw)$, *where* k *denotes the number of isomorphs and the term* kw *represents the total output size.*

We deduce from this lemma and Theorem 17.1 an extension of Eppstein's result of [158, 159] to all classes with bounded expansion:

Corollary 18.2. *Let* $\mathcal{C}$ *be a class with bounded expansion and let* H *be a fixed graph. Then there exists a linear time algorithm which computes, from a pair* (G, S) *formed by a graph* $G \in \mathcal{C}$ *and a subset* S *of vertices of* G, *the number of isomorphs of* H *in* G *that include some vertex in* S. *There also exists an algorithm running in time* $O(n) + O(k)$ *listing all such isomorphism where* k *denotes the number of isomorphs (thus represents the output size).*

18.6.2 Counting Versions

Actually, such an improvement holds for a quite general class of counting problems, namely to the problems having the following form: Let ϕ be a quantifier free first-order sentence (i.e. a Boolean query) with free variables $x_1, \ldots, x_p$ built using two binary relations Adj ("is adjacent to") and = (equality).

> How many assignments of the free variables $x_1, \ldots, x_p$ of ϕ to vertices of G are such that $G \models \phi(v)$?

We deal with this in Sect. 18.6.3. This includes for instance, for a fixed graph H and for input graphs G the problems

- "How many homomorphisms does H have to G?",
- "How many subgraphs of G are isomorphic to H?",
- "How many induced subgraphs of G are isomorphic to H?",

The counting version of Courcelle's theorem by Arnborg et al. [32] states that for each monadic second order property $\Phi(X_1,\dots,X_l)$, and for each class of graphs $\mathcal{C}$ with tree-width at most k, computing

$$|\{(A_1,\dots,A_l):\ G\models\Phi[A_1,\dots,A_l]\}|$$

may be done in linear time (if a tree-decomposition of width at most k of G is given). Of course, this theorem allows to prove, using low-tree-depth decompositions and using inclusion/exclusion, that counting the satisfying assignments of a Boolean query for a graph in a class with bounded expansion (resp. a nowhere dense class) may be done in linear time (resp. in time $n^{1+o(1)}$). However, this can be proved directly without the help of the above theorem by means of an efficient linear time counting algorithm for classes of graphs with bounded tree-depth.

18.6.3 Counting the Number of Solutions of a Boolean Query

A *Boolean query* is a quantifier free first-order formula. Given a graph G and a Boolean query ϕ with free variables $x_1,\dots,x_p$, we would like to count the number of vectors $\mathbf{v}$ of p vertices of G such that $G\models\phi(\mathbf{v})$. Notice that this number may be as large as $|G|^p$. Of course, it follows from Corollary 17.1 that if ϕ is fixed and input graphs G have bounded tree-depth, then it may be decided in linear time whether the number of vectors $\mathbf{v}$ such that $G\models\phi(\mathbf{v})$ is zero or not. However counting may also be performed in linear time:

Theorem 18.9. *Let p be a fixed integer. Let ϕ be a Boolean query with free variables $x_1,\dots,x_p$. Then there exists a polynomial $Q_p(X)$ and an algorithm which count the number of vectors $\mathbf{v}$ of p vertices of G such that $G\models\phi(\mathbf{v})$ and runs in time $O(Q_p(\widetilde{\nabla}_{2^{p-2}+\frac{1}{2}}(G)\ |G|))$.*

In particular, the algorithm runs in $O(n)$ time if G is bound to a class with bounded $\widetilde{\nabla}_{2^{p-2}+\frac{1}{2}}(G)$ (such as a bounded expansion class), and in $n^{1+o(1)}$ time if G is bound to a class where $\widetilde{\nabla}_{2^{p-2}+\frac{1}{2}}(G)=|G|^{o(1)}$ (such as a nowhere dense class).

Proof. According to Theorem 17.1, a p tree-depth coloring of a graph G of order n with $N_p(G)\le P_p(\widetilde{\nabla}_{2^{p-2}+\frac{1}{2}}(G)\ |G|)$ colors may be computed in

$O(N_p(G)\,n)$ time. For each subset I of $p' \leq p$ colors, counting the number of isomorphisms from a graph F of order at most p to the subgraph G_I of G induced by vertices with color in I may be computed in time $O(p^2 2^p n)$, according to Lemma 17.3. By inclusion/exclusion we easily derive the number of vectors $\mathbf{v}$ of p vertices of G such that $G \models \phi(\mathbf{v})$. □

Notice that for graphs bound in a bounded expansion class, the running time of the algorithm counting the number of vectors $\mathbf{v}$ of p vertices of G such that $G \models \phi(\mathbf{v})$ is linear although this number may be as large as $\Theta(n^p)$. The point here is that the algorithm count solution vectors but do not enumerate them. Note that algorithms of this part can be generalized straightforwardly to relational structures.

Exercises

18.1. Prove that for every nowhere dense class $\mathcal{C}$ and for any integer k, there exists an $O(n^{1+o(1)})$ time preprocessing algorithm so that for any preprocessed $G \in \mathcal{C}$ and any pair $\{x, y\}$ of vertices of G the value $\min(k, \mathrm{dist}(x, y))$ may be computed in $n^{o(1)}$-time.

18.2. Deduce from Exercise 12.4 and Lemma 18.3 that for every class of graphs $\mathcal{C}$ with bounded expansion, every first-order formula $\phi(x_1, \ldots, x_p)$, and every $0 < \epsilon < 1$ there exist positive reals C, N_0, τ such that the following holds for every $G \in \mathcal{C}$ of order at least N_0:

If $|\{\mathbf{x} \in V(G)^p : G \models \phi(\mathbf{x})\}| > |G|^{k+\epsilon}$ then (up to relabelling of the free variables of ϕ) there exist a p_0-tuple $\mathbf{t}_0$ of vertices of G (with $p_0 \geq 0$) and, for each i ($1 \leq i \leq k+1$) a family $\mathcal{F}_i$ of p_i-tuples of vertices of G (with $p_i \geq 1$) such that $\sum_{i=0}^{k+1} p_i = p$ and such that:

- The tuples in $\{\mathbf{t}_0\} \cup \bigcup_{i=1}^{k+1} \mathcal{F}_i$ are pairwise disjoint;
- For each $1 \leq i \leq k+1$ it holds $|\mathcal{F}_i| \geq (|G|/C)^\tau$;
- For every choice $\mathbf{t}_i \in \mathcal{F}_i$, it holds

$$G \models \phi(\mathbf{t}_0, \mathbf{t}_1, \ldots, \mathbf{t}_{k+1}).$$

Chapter 19
Further Directions

"Frankly speaking, my friends,
Things could not be better!"
(John Gordon Gimbel)

Like every active area of mathematics our book leaves some intriguing and challenging problems open. They relate for example to better computations and improvements of provided bounds. In a way most of our results on logarithmic density are just the first order approximations which certainly can be improved on many places.

In this chapter we collect some particular problems from areas, which seem promising and challenging to us.

Low tree-depth (vertex) decomposition is one of our main tools. It is natural to ask whether such decompositions generalize to edge partitions. This question has a very different flavour and a *matroidal* character. Some results in this direction were recently obtained in [360].

As stated above in several areas our results may be seen as a first order approximation results. In some cases, improvements may lead to interesting structural theorems. For example, the extremal result for subgraph counting in nowhere dense classes gives the existence of a sunflower of size n^{ϵ} in graphs containing at least n^k copies of a fixed subgraph. Can this be improved? Is it possible, by refining the sunflower structure to bigger depth branching regular structures, to capture almost all the copies of the fixed subgraph (up to, say, a polylog factor)?

Another challenging problem is a refinement of random analysis. Is it possible to generate general bounded expansion classes exactly (not just as the liquid graphs introduced in Exercise 14.1). Perhaps this is too ambitious

J. Nešetřil and P.O. de Mendez, *Sparsity*, Algorithms and Combinatorics 28, 411
DOI 10.1007/978-3-642-27875-4_19, © Springer-Verlag Berlin Heidelberg 2012

and one should first consider the key "building blocks", that is graphs with tree-depth at most t. The probabilistic properties of random star forest is already involved [384], and what we need are general bounded height finitely colored rooted forests. In a way, the probabilistic analysis of random colored star forests is a key step for a general analysis. Generally, the random aspects of sparse graphs present a very challenging problem (as it is also known in other contexts [76, 267, 471]).

Model theory and mathematical logic is an important aspect of many parts of this book. In many cases, extension of graph results to structures are routine (and expected) but we listed several areas (such as homomorphism preservation theorems or dualities) where the connection seems to be more profound. And there are problems too. One of them is to find a proper setting for a notion of nowhere dense class of relational structures in the context of first-order logic (see for instance Problem 5.1 in Sect. 5.7).

Let us mention three specific problems in a greater detail.

We believe that the results of this book give to an interested reader an impression that Nowhere Dense classes are somewhat well understood and they multiple characterizations lead to interesting results and applications. But the somewhere dense classes we did not analyze in a greater detail and mostly we considered them like one large bag.

Problem 19.1. Is there a good parametrization of somewhere dense classes?

For example, one can consider the smallest time $\tau(\mathcal{C})$ such that $\mathcal{C} \nabla \tau(\mathcal{C})$ is the class of all graphs, or the smallest time $\widetilde{\tau}(\mathcal{C})$ such that $\mathcal{C} \widetilde{\nabla} \widetilde{\tau}(\mathcal{C})$ is the class of all graphs. We could regard them as (minor and topological minor) *phase transition times* of the class $\mathcal{C}$. Are phase transition times good parametrizations?

The difference between the two versions of the phase transition time may be arbitrarily large. An example is depicted on Fig. 19.1. However, according to Proposition 5.2, the minor and topological minor phase transition times are related by:

$$\frac{\widetilde{\tau}(\mathcal{C})-1}{3} \leq \tau(\mathcal{C}) \leq \widetilde{\tau}(\mathcal{C}).$$

The topological phase transition time $\widetilde{\tau}$ naturally defines a sub-classification of somewhere dense graphs. It follows from a standard Ramsey argument that, for a somewhere dense class of graphs $\mathcal{C}$, the value $\widetilde{\tau}(\mathcal{C})$ is the minimum half-integer such that the class $\mathcal{K}_\alpha$ of the exact $(2\alpha+1)$-subdivisions of the complete graphs is included in $\mathcal{C} \widetilde{\nabla} 0$ (an exact $(2\alpha+1)$-subdivision

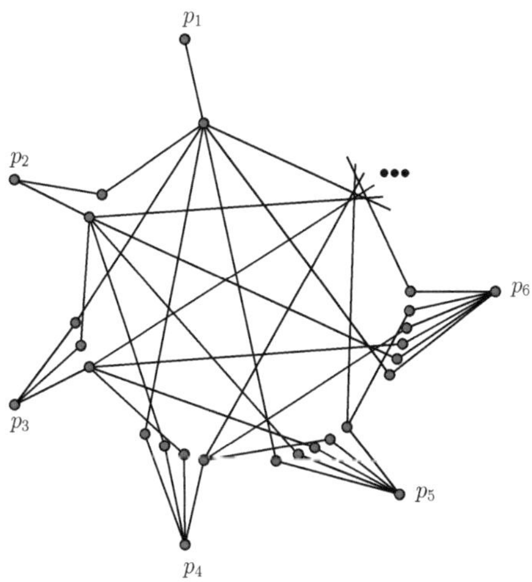

Fig. 19.1 By subdividing the graphs whose construction is shown above we can build a sequence of classes $\mathcal{C}_i$ such that $\widetilde{\tau}(\mathcal{C}_i) \sim \frac{4}{3}\tau(\mathcal{C}_i)$ and $\tau(\mathcal{C}_i) \to \infty$ as $i \to \infty$ thus proving that $\widetilde{\tau}(\mathcal{C}) - \tau(\mathcal{C})$ may be arbitrarily large

subdivides every edge by exactly $(2\alpha+1)$ points). As a consequence, we have the following stability result:

> The collection of the nowhere dense classes with topological phase transition time $\widetilde{\tau}_0$ is closed under any intersections and under finite unions.

Let us approach this yet from another side: Assume a class $\mathcal{C}$ has a phase transition at time $\widetilde{\tau}(\mathcal{C})$. Then all the classes $\mathcal{C} \mathbin{\widetilde{\nabla}} t$ for $t < \widetilde{\tau}(\mathcal{C})$ are different from $\mathcal{G}raph$ (thus all of them have a bounded clique number). According to Theorems 5.3 and 7.8, for every positive integer $p \ll \widetilde{\tau}(\mathcal{C})$ and $G \in \mathcal{C}$, we have

$$\chi_p(G) = O\big(n^{\frac{10p2^{2^p}}{\widetilde{\tau}(\mathcal{C})}}\big).$$

Thus we may think of $\mathcal{C}$ as a *truncated nowhere dense class* in the following sense:

Recall the local versions of Theorem 17.1 (Lemma 17.3 and Theorem 12.5). We see that many of the properties of nowhere dense classes are preserved far from the transition time. Consider the following example: Let $\epsilon > 0$ be a small positive real. Assume

$$|F| \leq \log_2 \log_2(\epsilon\widetilde{\tau}(\mathcal{C})).$$

Then:

- There exists an algorithm of time complexity $O(n^{1+\epsilon(|F|+1)})$ to count the number of copies of F in $G \in \mathcal{C}$;
- The upper logarithmic density of F in a monotone class $\mathcal{C}$

$$\ell = \limsup_{G \in \mathcal{C}} \frac{\log(\#F \subseteq G)}{\log |G|}$$

is either $-\infty$ or belongs to one of the intervals

$$[0, \epsilon],\ [1, 1+\epsilon], \ldots, [\alpha(F), \alpha(F)+\epsilon].$$

(see Fig. 19.2).

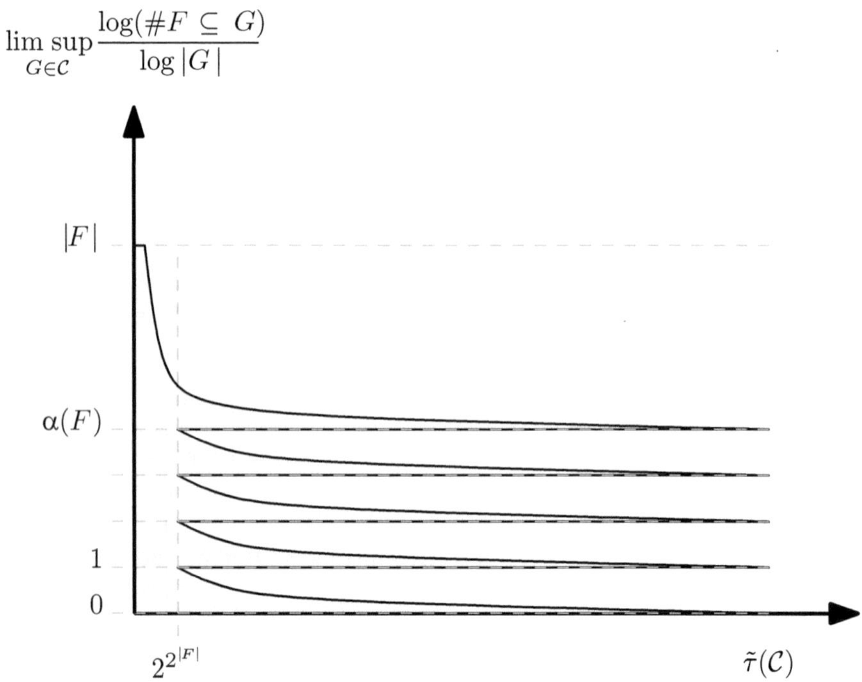

Fig. 19.2 Possible values of the logarithmic densities of a small graph F in a monotone class $\mathcal{C}$ depending on the transition time $\tilde{\tau}(\mathcal{C})$ (if not $-\infty$)

This (and other similar examples) shows that the somewhere dense classes share some properties with nowhere dense classes at the beginning of their resolution. In fact this is one of our motivations for formulating and studying local versions. These properties are then lost if we move closer to the phase transition (and of course beyond it). Our world seems to be then suddenly governed by statistics of dense graphs. Can one formulate the statistics in terms of the original class $\mathcal{C}$? Can we then develop a version of statistics for dense graphs by means of frequencies of paths or powers of incidence matrices or yet something else? Perhaps more concretely: is there a version of Szemerédi Regularity Lemma for somewhere dense classes?

We know that nowhere dense classes are very broad and cover many interesting instances.

But is there a nowhere dense class $\mathcal{C}$ which is *algebraically universal*?

More concretely: a mapping Φ which assigns to every graph G a graph $\Phi(G)$ belonging to $\mathcal{C}$ and to every homomorphism $f : G \longrightarrow H$ a homomorphism $\Phi(f) : \Phi(G) \longrightarrow \Phi(H)$ is called an *embedding* if the following holds:

1. $\Phi(\mathrm{id}_G) = \mathrm{id}_{\Phi(G)}$ (id denotes the identity);
2. $\Phi(f) \circ \Phi(g) = \Phi(f \circ g)$ whenever the right hand side makes sense;
3. Φ is injective;
4. For every $g : \Phi(G) \longrightarrow \Phi(H)$ there exists $f : G \longrightarrow H$ such that $\Phi(f) = g$.

Our problem then takes the following form:

Problem 19.2. Is there an embedding of the category of all finite graphs into a nowhere dense class $\mathcal{C}$?

This question is interesting already at the beginning even for just one graph (where it amounts to representing groups and monoids by graphs in a bounded expansion class). In these classes papers [45, 46] provide a negative answer for every proper minor closed classes (for groups) and every proper topological minor closed classes (for monoids). Nevertheless the answer is positive for a nowhere dense class in both cases (and even for any finite set of graphs; i.e. any finite category). It is also positive for infinite partial orders by means of Theorem 3.14. But perhaps in general one should expect that the answer is negative.

Let us finish this book on a more speculative note.

Problem 19.3. What are the building blocks of nowhere dense classes?
What are the building blocks of bounded expansion classes?

What we are looking for here are structure theorems for these class, in the line of the following known results of structural graph theory:

Robertson and Seymour proved a structure theorem [398] for the class of H-minor free graphs: every H-minor free graph can be decomposed in a way such that each part is "almost embeddable" into a fixed surface;
Recently, Grohe et al. [228] gave an extension of this structure theorem to classes excluding a topological minor. The latest version of this structure theorem by Grohe and Marx [231] states that graphs excluding a fixed graph H as a topological minor have a tree decomposition where each part is either "almost embeddable" to a fixed surface or has bounded degree with the exception of a bounded number of vertices.

Can we go further? Can all (typical) bounded expansion classes be generated by a finite list of such building blocks?

What are the nowhere dense classes which fail to have bounded expansion? By Theorems 13.1 and 13.2 we know that these are exactly classes for which at some time of a resolution χ (or $\overline{d}$) becomes infinite while ω stays bounded. Hence these classes should contain shallow subdivisions of graphs with arbitrarily large girth and chromatic number (or minimum degree), assuming that Erdős-Hajnal Conjecture 11.3 or Thomassen Conjecture 11.5 holds.

At this fine landscape we decided to end.

Chapter 20
Solutions and Hints for some of the Exercises

Thy Holly Thread of Ariadne
Shalt thou follow as far as the Minotaur.

Exercises of Chapter 3

3.1 For $A \subseteq V(G)$ let $f(A)$ be the number of edges between A and $V(G) \setminus A$. Let A be such that $f(A)$ is maximal. No vertex $v \in A$ has more neighbors in A than in $V(G) \setminus A$ as we would have $f(A - v) > A$. Also, no vertex $v \notin A$ has more neighbors in $V(G) \setminus A$ than in A as we would have $f(A + v) > A$. Hence the minimum degree of the bipartite subgraph of G obtained by deleting the edges within A and $V(G) \setminus A$ is at least $\lceil d/2 \rceil$.

3.2 The argument is similar to the one of the proof of Lemma 3.1. Consider an acyclic orientation $\vec{G}$ of G with indegree at most k. The orientation of G naturally defines an acyclic orientation of each copy of H in G. Let $\vec{H}$ be some acyclic orientation of H and let $h_1, \ldots, h_t$ be the sinks of $\vec{H}$. Notice that no two sinks of $\vec{H}$ are adjacent hence $t \leq \alpha(H)$. Let $x_1, \ldots, x_t$ be t vertices of G. If we require that h_i is mapped to x_i when looking for a copy of $\vec{H}$ in $\vec{G}$, then there are at most k possible choices for each in-neighbors of these vertices and, then, at most k possible choices for each of the in-neighbors of the in-neighbors, and so on.

J. Nešetřil and P.O. de Mendez, *Sparsity*, Algorithms and Combinatorics 28, 417
DOI 10.1007/978-3-642-27875-4_20, © Springer-Verlag Berlin Heidelberg 2012

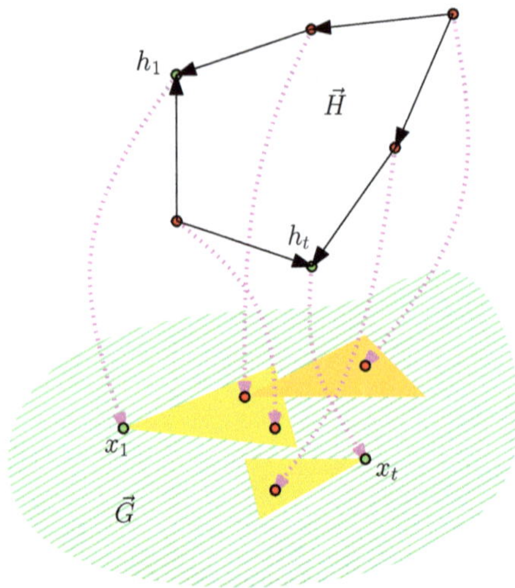

It follows that the number of copies of $\vec{H}$ where h_i is mapped to x_i is at most $k^{|H|-t}$. As there are at most $|G|^t$ choices for $x_1, \ldots, x_t$ we obtain, by summing up over all possible acyclic orientations of H, that the number of copies of H in G is at most

$$\sum_{t=1}^{\alpha(H)} \mathrm{Acyc}_t(H) k^{|H|-t} |G|^t$$

3.8

- Let H be a graph of order n and let $M(H)$ be its Mycielskian. It is clear that $|M(H)| = 2|H| + 1$ and that $M(H)$ is triangle-free if H is. Assume $M(H)$ is k-chromatic. Consider a proper k-coloring of $M(H)$ with k-colors, in which that vertex z is colored k. If a vertex x (in the copy of H in $M(H)$) has color k, then we can safely recolor it with the color of x' (which has color different from k) as these vertices have the same neighbors. It follows that H has chromatic number at most $k-1$. Conversely, any proper $k-1$-coloring of H obviously defines a proper k-coloring of $M(H)$.
- Consider a linear order on the vertices of a graph G and the corresponding natural orientation of G.
  1. If there exists a vertex v such that $\chi(G[N^-(v)]) > h(\omega(G)-1, c)$, then by induction $G[N^-(v)]$ (hence G) has a triangle-free subgraph with chromatic number at least c;
  2. If every vertex v is such that $\chi(G[N^-(v)]) \leq h(\omega(G)-1, c)$, color all the edges with maximum endpoint v by the colors of their lower endpoints in a proper coloring of $G[N^-(v)]$ with at most $h(\omega(G)-1, c)$ colors.

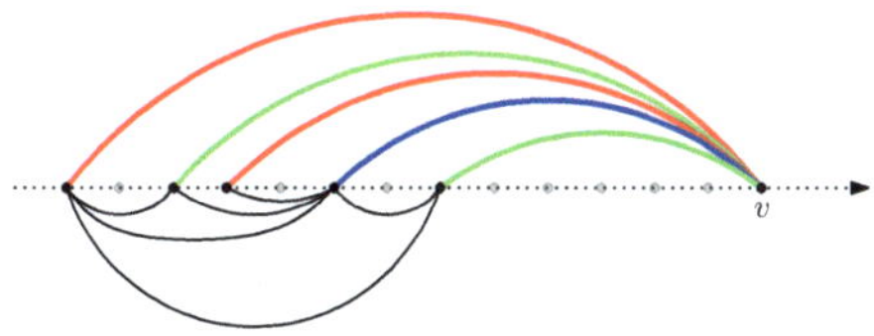

Then one of the monochromatic subgraphs G_i $(1 \leq i \leq h(\omega(G)-1, c))$ of G has chromatic number

$$\chi(G_i) \geq \chi(G)^{\frac{1}{h(\omega(G)-1,c)}} \geq c.$$

Moreover, this monochromatic subgraph is triangle-free by construction.

3.7 The proof of the statement follows along the same lines as the proof of Proposition 3.4 (and a similar statement holds for every graph G satisfying $\chi(G) \leq k$, for any fixed k).

3.9 Let $R(x, y)$ denote the adjacency relation. Define

- $U[v_1, v_2]$ as $(v_1 \simeq v_2) \vee R(v_1, v_2)$;
- $E[u_1, u_2, v_1, v_2]$ as $(u_1 \simeq u_2) \wedge (v_1 \simeq v_2)$;
- $F_R[u_1, u_2, v_1, v_2]$ as

$$(u_1 \simeq u_2) \wedge (u_1 \simeq v_1) \vee (u_1 \simeq v_2) \wedge (u_2 \simeq v_1) \vee (v_1 \simeq v_2) \wedge (v_1 \simeq u_1).$$

Exercises of Chapter 4

4.1 If G is planar, then every minor of G is also planar (thus has average degree less than 6) hence

$$\widetilde{\nabla}_r(G) \leq \nabla_r(G) < 3.$$

If G has maximum degree D, so has every topological minor of G hence $\widetilde{\nabla}_r(G) \leq D$. Shallow minors of G at depth r have maximum degree at most $D(D-1)^{r-1}$ hence $\nabla_r(G) \leq D(D-1)^{r-1}$.

If G may be drawn in the plane in such a way that every edge is crossed by at most one other edge, then G is a ≤ 1-subdivision of $H \bullet K_2$, for some planar graph H. Thus, by monotony and by Proposition 4.2 we have

$$\widetilde{\nabla}_r(G) \leq \widetilde{\nabla}_{2r+1/2}(H \bullet K_2).$$

According to Proposition 4.6:

$$\widetilde{\nabla}_{2r+1/2}(H \bullet K_2) \leq \max(4r+1,4)\widetilde{\nabla}_{2r+1/2}(H) + 1 \leq \max(12r+4,13).$$

Thus:

$$\widetilde{\nabla}_r(G) \leq \max(12r+4,13).$$

(A better bound for $\widetilde{\nabla}_r(G)$ is given by Theorem 14.4 in Chap. 14.) According to Corollary 4.1 we have

$$\nabla_r(G) \leq 4(4\widetilde{\nabla}_r(G))^{(r+1)^2} \leq 4(\max(48r+16,52)^{(r+1)^2}.$$

4.2

Let $Z = \{z_1, \dots, z_p\}$ be a maximal subset of vertices of G pairwise at distance at least $2d+1$. Define the mapping $\pi : V(G) \to Z$ as follows: for $x \in V(G)$, $\pi(x)$ is, among the vertices $z_i \in Z$ such that $\mathrm{dist}(x,z)$ is minimum, the one with minimum index i. Let T_i be the subgraph induced by $\pi^{-1}(z_i)$. Notice that $\mathrm{dist}(x,\pi(x)) \leq 2d$. The graph T_i is connected and, as the girth of G is at least $8d+3$, each T_i is a tree. Moreover, for $1 < i < j \leq p$ there is at most one edge linking a vertex of T_i and a vertex of T_j (as the girth of G is at least $8d+3$). As $\delta(G) \geq 3$, every leaf of T_i is adjacent to at least two vertices out of T_i. Moreover, the number of leaves of T_i is at least equal to the number of vertices of T_i at distance d from z_i. As these vertices are closer to z_i than to any other z_j (as $\mathrm{dist}(z_i,z_j) > 2d$) and as $\delta(G) \geq 3$, T_i contains at least 3.2^{d-1} such vertices. By contracting each T_i (which has radius at most 2d) we therefore get a graph with minimum degree 3.2^d. Hence $\nabla_{2d}(G) \geq 3.2^{d-1} > 2^d$.

Exercises of Chapter 5

5.2 Let $f^+ : \mathbb{N} \to \mathbb{R}^+$ be defined as $f^+(n) = \sup_{i \geq n} f(i) + 1/\log_2 n$. Then we have $f^+(n) \geq f(n)$, f^+ is decreasing, and $\lim_{n\to\infty} f^+(n) = 0$. Let $h : \mathbb{N} \to \mathbb{N}$ be a non decreasing function such that for every integer g it holds $f^+(h(g)) < 1/g$. Then define the class $\mathcal{C}_f$ as the class of the graphs G such that $\Delta(G) \leq h(\mathrm{girth}(G))$. This class is nowhere dense: assume $K_t \in \mathcal{C}_f \,\widetilde{\nabla}\, p$. Because the corresponding shallow subdivision of K_t is a subgraph of a graph $G \in \mathcal{C}_f$ with $\Delta(G) \geq t-1$ and $\mathrm{girth}(G) \leq 6p+3$ we have $t \leq h(6p+3)+1$ thus $\omega(\mathcal{C} \,\widetilde{\nabla}\, p) < \infty$. Let $n \in \mathbb{N}$. Fix $d = n^{f^+(n)}$. There exists a graph G_n of order n, girth $g \geq \frac{\log n}{\log d} = \frac{1}{f^+(n)}$ and minimum degree between d. Thus:

$$\Delta(G_n) \leq n \leq h(g) \leq h(\mathrm{girth}(G_n)).$$

Moreover, we have $|G_n| \geq n$ and

$$\|G_n\| \geq d|G_n|/2 = |G_n|^{1+f^+(n)-\frac{1}{\log_2 n}} \geq |G_n|^{1+f(|G_n|)}.$$

Exercises of Chapter 6

6.1 The inequality $td(T_n) \geq n$ follows from an easy induction:

- $td(T_1) = 1$,
- Let r be a vertex of T_n such that $td(T_n - r) = td(T_n) - 1$. One of the connected components of $T_n - r$ contains a copy of $td(T_{n-1})$. As td is monotone we deduce (using the induction hypothesis)

$$td(T_n) = td(T_n - r) + 1 \geq td(T_{n-1}) + 1 = n.$$

6.2 We prove the result by induction on t. Either G already contains at least m connected components (and we are done) or one of the connected components G_0 has order at least $\frac{m^t-1}{m-1} \geq m^{t-1} + 1$. Let r_0 be a vertex of G_0 such that $td(G_0 - r_0) < td(G_0)$. By induction, there exists a subset S_0 of at most $t-2$ vertices of $G_0 - r_0$ such that $G_0 - r_0 - S_0$ has at least m connected components. Put $S = S_0 \cup \{r_0\}$.

6.3 We consider three cases:

- G and H are both bipartite. Then prove a stronger result by replacing the product by a "semi-product" where only the white–white and black–black vertices are kept in the product;
- The case where $H \approx K_2$. Use a centered coloring in the product, find a minimum vertex (r, i) and then conclude as

$$td(G \times K_2) \geq td((G - r) \times K_2) + 1 \geq td(G - r) + 1 \geq td(G).$$

- The case where $td(H) > 2$. Use a centered coloring in the product, find a minimum vertex (u, v) and then conclude as

$$td(G \times H) \geq td(G \times (H - v)) + 1 \geq td(G) + td(H - v) - 2 + 1 \geq td(G) + td(H) - 2.$$

6.4

1. Follow an elimination order of G on every orientation $\vec{G}$. At depth $td(G) - 1$, only oriented stars remain (hence no strongly connected components). Thus $cr^+(G) \leq td(G) - 1$.
2. Fix an integer $n \geq 2$. Let $G = (A, B)$ be the complete bipartite graph with $|A| = |B| = n$, let M be a perfect matching of G and let $\vec{G}$ be the orientation of G such that every edge $e = \{a, b\}$ with $a \in A$ and $b \in B$ is oriented from a to b if $e \notin M$ and from b to a if $e \in M$.

- $\mathrm{td}(G) = n+1$: If the removal of a subset S of vertices disconnects the graph then S fully contains either A or B.
- $\mathrm{cr}^+(G) = n$: Let S be a minimal set of vertices such that $\vec{G} - S$ is not strongly connected. Let X and Y be a partition of $V(G) - S$ into two non-empty subsets such that no arc is oriented from Y to X. If none of X and Y has cardinality at least 2, we are done. Otherwise, at least one element of X is not matched with all elements of Y thus $X \cap A \neq \emptyset$ and $Y \cap B \neq \emptyset$. It follows that $X \subseteq A$ and $Y \subseteq B$ and that no edge in M is incident to a vertex in X and a vertex in Y. Thus $|X \cup Y| \leq n$ and $|S| \geq n$.
- Let H be a minor of G. Let $\vec{H}$ be an orientation of H such that $\mathrm{cr}(\vec{H}) = \mathrm{cr}^+(H)$. There exists an orientation $\vec{G}$ of G such that $\vec{H}$ is a minor of $\vec{G}$. According to Theorem 6.2, we have $\mathrm{cr}(\vec{H}) \leq \mathrm{cr}(\vec{G})$. Hence

$$\mathrm{cr}^+(H) = \mathrm{cr}(\vec{H}) \leq \mathrm{cr}(\vec{G}) \leq \mathrm{cr}^+(G).$$

6.6

- Define inductively $\vartheta_{0,k}(r_1, \ldots, r_k, x, y)$ as $((x = y) \vee (x \sim y)) \wedge \bigwedge_i \neg(x = r_i) \wedge \neq (y = r_i)$, $\ldots$, and $\vartheta_{t+1,k}(r_1, \ldots, r_k, x, y)$ as $\exists z\ \vartheta_{t,k}(r_1, \ldots, r_k, x, z) \wedge \theta_{t,k}(r_1, \ldots, r_k, z, y)$;
- Cases where $t = 1$ or $t = 2$ are handled directly. For $t \geq 3$ define $\delta_{t,k}(r_1, \ldots, r_k)$ as $\forall x \forall y\ \vartheta_{t+1,k}(r_1, \ldots, r_k, x, y) \to \vartheta_{t,k}(r_1, \ldots, r_k, x, y)$.

6.8 Define inductively

$$\Lambda_1(w_1, \ldots, w_l, x, y) \stackrel{\text{def}}{=} (x \sim y) \vee (x = y)$$
$$\Lambda_d(w_1, \ldots, w_l, x, y) \stackrel{\text{def}}{=} \Lambda_{d-1}(w_1, \ldots, w_l, x, y) \vee$$
$$\exists z \left(\bigwedge_{i=1}^{l} \neg(z = w_i) \wedge \Lambda_{\lfloor d/2 \rfloor}(w_1, \ldots, w_l, x, z) \wedge \Lambda_{\lceil d/2 \rceil}(w_1, \ldots, w_l, x, z)) \right)$$

The second item is proved by induction on $t = \mathrm{td}(G[C_{G-\{a_1,\ldots,a_l\}}(r)] - r)$. Let $C = C_{G-\{a_1,\ldots,a_l\}}(r)$. Assume that $t = 0$. Then $C = \{r\}$. Define

$$\Psi(z_1, \ldots, z_l, w) \stackrel{\text{def}}{=} \bigwedge_{i=1}^{l} \neg(w = z_i) \wedge \left(\forall x \left((x \sim w) \leftrightarrow \bigvee_{i: a_i \sim r} (x = z_i) \right) \right).$$

Then $\mathrm{qrank}(\Psi) = 1$ and conditions (1) and (2) are clearly equivalent.

Now assume that the property has been proved for all the situations where $0 \leq \mathrm{td}(G[C_{G-\{x_1,\ldots,x_k\}}(y)]) < t$ and assume $\mathrm{td}(G[C_{G-\{a_1,\ldots,a_l\}}(r)] - r) = t \geq 1$. Let $C = C_{G-\{a_1,\ldots,a_l\}}(r)$ and let $C_1, \ldots, C_m$ be the vertex sets of the connected components of $G - \{a_1, \ldots, a_l, r\}$. Say that C_i is *equivalent* to C_j

if there exists an isomorphism $g : G[\{a_1, \dots, a_l, r\} \cup C_i] \to [\{a_1, \dots, a_l, r\} \cup C_j]$ that fixes $a_1, \dots, a_l$ and r. Let n be the number of non-equivalent sets among $C_1, \dots, C_m$. Without loss of generality, we can assume that these non-equivalent sets are $C_1, \dots, C_n$. We denote by N_i (for $1 \le i \le n$) the number of sets among $C_1, \dots, C_m$ which are equivalent to C_i. In each C_i we choose a vertex r_i such that $td(G[C_i - r_i]) < td(G[C_i])$. Finally, we denote by σ_i the number of vertices x in C_i such that there exists an automorphism of $G[\{a_1, \dots, a_l, r\} \cup C_i]$ which fixes $a_1, \dots, a_l, r$ and sends x to r_i. By induction there exists, for each $1 \le p \le n$ a formula $\Psi_p(z_1, \dots, z_{l+1}, w)$ such that

- $\text{qrank}(\Psi_p) = td(G[C_p] - r_p) + 1 \le td(G[C] - r) = t$;
- For every graph H and for every $b_1, \dots, b_l, s, s_p$ in $V(H)$ the following conditions are equivalent:
  1. The mapping $g_0 : a_i \mapsto b_i, r \mapsto s$ is an isomorphism from $G[a_1, \dots, a_l, r]$ to $H[b_1, \dots b_l, s]$ and $H \models \Psi_p(b_1, \dots, b_l, s, s_p)$;
  2. There exists an isomorphism
  $$g_p : G[C_p \cup \{a_1, \dots, a_l, r\}] \to H[C_{H-\{b_1, \dots, b_l, s\}}(s_p) \cup \{b_1, \dots, b_l, s\}]$$
  such that $g_p(r_p) = s_p$, $g_p(r) = s$ and $g_p(a_i) = b_i$ (for $1 \le i \le l$).

We define $\Psi(z_1, \dots, z_l, w)$ as the conjunction of the $\Phi_i(z_1, \dots, z_l, w)$ (for $1 \le i \le 4 + n$), where

$$\begin{aligned}
\Phi_1(z_1, \dots, z_l, w) &\overset{\text{def}}{=} \bigwedge_{i=1}^{l} \neg(w - z_i) \\
\Phi_2(z_1, \dots, z_l, w) &\overset{\text{def}}{=} \bigwedge_{i: a_i \sim r} (w \sim z_i) \wedge \bigwedge_{i: \neg(a_i \sim r)} \neg(w \sim z_i) \\
\Phi_3(z_1, \dots, z_l, w) &\overset{\text{def}}{=} \forall v\ (\Lambda(2^t, z_1, \dots, z_l, w, v) \\
&\qquad \to \Lambda(2^t - 1, z_1, \dots, z_l, w, v)) \\
\Phi_4(z_1, \dots, z_l, w) &\overset{\text{def}}{=} \exists^{=|C|} v\ \Lambda(2^t - 1, z_1, \dots, z_l, w, v) \\
\Phi_{4+p}(z_1, \dots, z_l, w) &\overset{\text{def}}{=} \exists^{N_p \sigma_p} y\ \Psi_p(z_1, \dots, z_l, w, y)
\end{aligned}$$

First notice that $\text{qrank}(\Psi) = t + 1$. Let H be a graph and let $b_1, \dots, b_l, s$ be vertices of H. Then:

- $H \models \Phi_1(b_1, \dots, b_l, s)$ if s is different from $b_1, \dots, b_l$;
- If $f_0 : a_i \mapsto b_i$ is an isomorphism from $G[a_1, \dots, a_l]$ to $H[b_1, \dots, b_l]$ and $s \notin \{b_1, \dots, b_l\}$ then $H \models \Phi_2(b_1, \dots, b_l, s)$ if $g_0 : a_i \mapsto b_i, r \mapsto s$ is an isomorphism from $G[a_1, \dots, a_l, r]$ to $H[b_1, \dots b_l, s]$;
- $H \models \Phi_3(b_1, \dots, b_l, s)$ if no vertex in the connected component of $H - \{b_1, \dots, b_l\}$ which contains s is at distance greater than $2^t - 1$ from s

(in this component), what is the case in particular if $\mathrm{td}(H[C_{H-\{b_1,\dots,b_l\}}(s)] - s) \le t$;

- Assuming $H \models \Phi_3(b_1,\dots,b_l,s)$, then we have $H \models \Phi_4(b_1,\dots,b_l,s)$ if $|C_{H-\{b_1,\dots,b_l\}}(s)| = |C|$.
- Assuming $H \models \Phi_1(b_1,\dots,b_l,s)$ and $H \models \Phi_2(b_1,\dots,b_l,s)$, then $H \models \Phi_{4+p}(b_1,\dots,b_l,s)$ implies (according to induction hypothesis) that there exists $N_p\sigma_p$ vertices s_p such that there exists an isomorphism from $G[C_p \cup \{a_1,\dots,a_l,r\}]$ to $H[C_{H-\{b_1,\dots,b_l,s\}}(s_p) \cup \{b_1,\dots,b_l,s\}]$ which send a_i to b_i, r to s and r_p to s_p.

The last item follows by defining

$$\hat{\Psi} \stackrel{\text{def}}{=} (\exists^{=|G|}x) \wedge \bigwedge_p \exists^{N_p\sigma_p} v\, \Psi_p(v).$$

Exercises of Chapter 7

7.1 Consider the countable directed graph $\vec{G}$ with vertex set $\{v_1,\dots,v_n,\dots\}$ constructed as follows: $(v_1,v_2),(v_1,v_3)$ and (v_2,v_3) are arcs of $\vec{G}$; for each $i \ge 3$, if the two in-neighbors of v_i are v_a and v_b (with $a < b$) then v_{2i-2} has in-neighbors v_a and v_i and v_{2i-1} has in-neighbors v_b and v_i. Consider any 2-coloring of the vertices of G. For $n \in \mathbb{N}$ define $l(n)$ as the maximum for $2^{n-1}+2 \le i \le 2^n+1$ of $|\vec{P}_1| + |\vec{P}_2|$ where $\vec{P}_1$ (resp. $\vec{P}_2$) is a monochromatic directed path with color 1 (resp. with color 2) ending at an in-neighbor of v_i. By easy induction, $l(n) \ge n-1$. Hence, denoting by G_n the subgraph of G induced by the vertices $v_1,\dots,v_{2^n+1}$, no 2-coloring of the vertices of G_n can avoid the existence of a monochromatic path of length $n/2-1$.

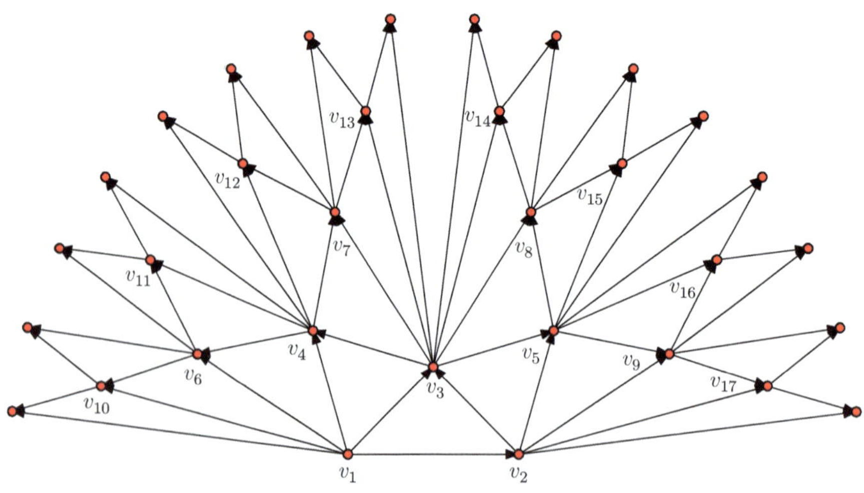

7.2

1. For each pair $\{i, j\}$ of colors, the subgraph of G induced by colors i and j is (by definition of a star coloring) a star forest. Orient each star of the forest from its center (or arbitrarily when stars are reduced to single edges). It is easily checked that if H is a tight 1-transitive fraternal augmentation of G (with respect to the just defined orientation) then $\chi(H) \leq N = \chi_s(G)$.
2. We deduce a construction of planar subcubic graphs of arbitrary high girth and star chromatic number 4:

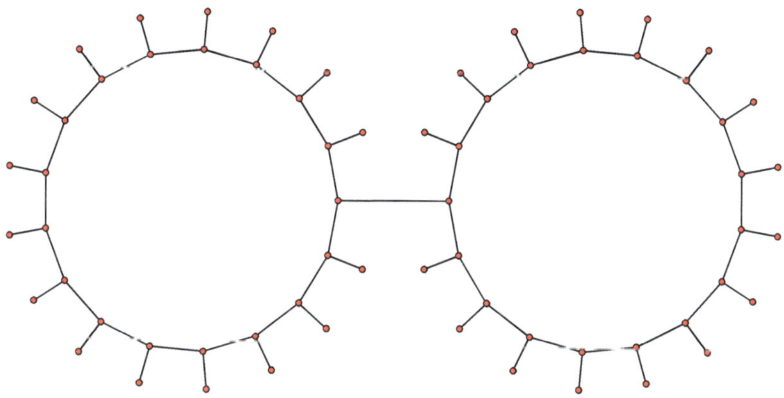

3. Color the vertices of G in such a way that two vertices get different colors (in G) if they are adjacent in the augmented graph H. It is easily checked that whatever orientation a path of length 3 of G may have, its vertex set will induce in H a subgraph with at least one triangle.

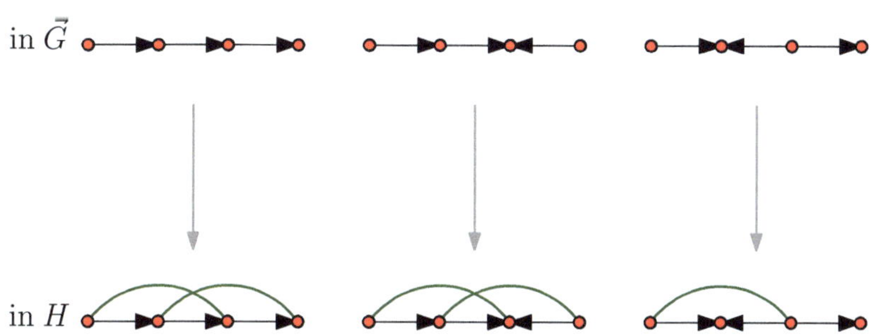

It follows that the vertices of every path of length 4 of G get at least three colors, hence the coloration is a star coloring of G.

7.4 Consider the following example:

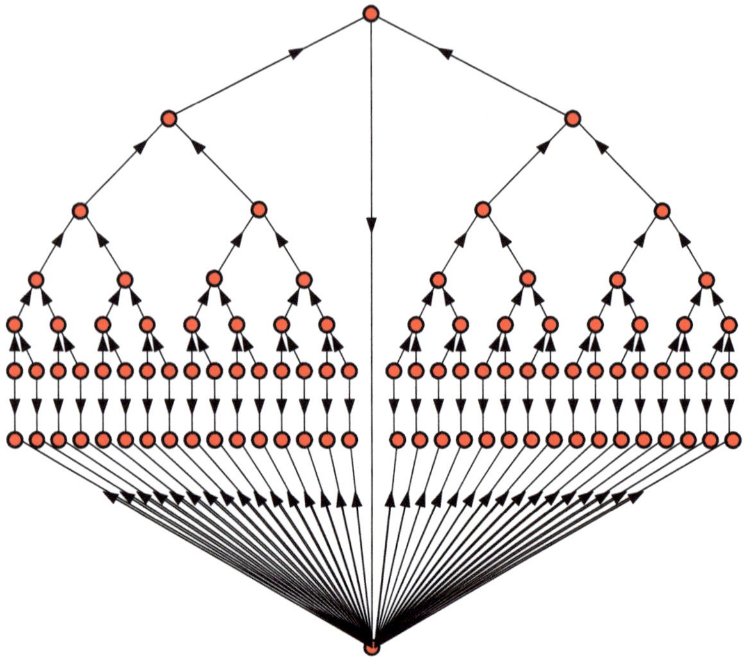

Exercises of Chapter 8

8.1 Recall that the *girth* of a graph G is the minimal length of its cycles, and that it is denoted by girth(G). Consider the class $\mathcal{C}$ of all 2-connected graphs G satisfying $\Delta(G) \leq \text{girth}(G)$ (see some examples of this elusive class on the figure below).

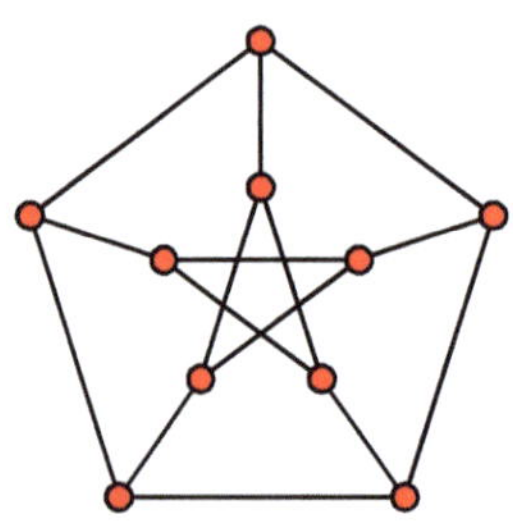

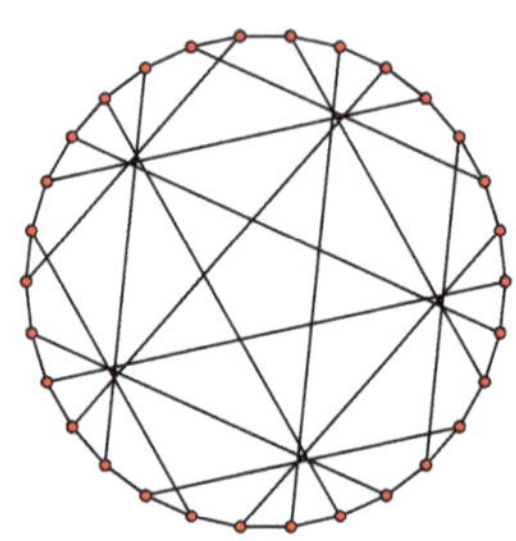

Then the class $\mathcal{C}$ does not have a bounded average degree. However, it is easy to see that $\mathcal{C}$ is a wide class:

Assume that a graph $G \in \mathcal{C}$ has diameter at most D. As G is 2-connected, it includes a cycle of length $\text{girth}(G) \leq 2D + 1$. It follows that $\Delta(G)$ is at most $2D+1$ thus G has at most about $(2D+1)^D$ vertices (because diameter is at most D). Hence for every integer d and m, every graph in the class with at least (about) $(2dm+1)^{dm}$ vertices has a d-independent set of size m.

8.2

1. The class of all the 2-degenerate graphs (which include 1-subdivisions of arbitrarily large cliques) is degenerate but not uniformly quasi wide;
2. The class of all graphs whose girth is larger than their maximum degree is uniformly quasi-wide but not degenerate (graphs in this class do not even have a bounded average degree).

Exercises of Chapter 9

9.1

- $\vec{\gamma}$ circuit of $\vec{G} \Longrightarrow$ loop at $V(\vec{\gamma})$ in $U(\vec{G})$
- $(A_1, \ldots, A_k)$ circuit of $U(\vec{G}) \Longrightarrow \vec{G}$ contains a circuit whose length is a multiple of k.

More generally, if $U(\vec{G})$ contains a cycle γ_U then $\vec{G}$ contains a cycle γ such that for some $k \in \mathbb{N}$ it holds:

$$|\gamma^+| = k\,|\gamma_U^+| \qquad \text{and} \qquad |\gamma^-| = k\,|\gamma_U^-|.$$

9.3

- Assume that $[\mathbf{A}] = [\mathbf{N}_1 \times \mathbf{N}_2 \times \cdots \times \mathbf{N}_q]$ where the $\mathbf{N}_j$'s are multiplicative and pairwise non comparable. For each $1 \le a \le p$ we have

$$\prod_{j=1}^{q} \mathbf{N}_j \rightleftarrows \prod_{i=1}^{p} \mathbf{M}_i \to \mathbf{M}_a$$

  As $\mathbf{M}_a$ is multiplicative we deduce that there exists $1 \le f(a) \le q$ such that $\mathbf{N}_{f(a)} \to \mathbf{M}_a$. Similarly, for each $1 \le b \le q$ there exists $1 \le g(b) \le p$ such that $\mathbf{M}_{g(b)} \to \mathbf{N}_b$. Thus, for every $1 \le a \le p$ we have $\mathbf{M}_{g \circ f(a)} \to \mathbf{N}_{f(a)} \to \mathbf{M}_a$ hence $g \circ f(a) = a$ (as the $\mathbf{M}_i$'s are pairwise non-comparable). Similarly, for every $1 \le b \le q$ we have $f \circ g(b) = b$ hence $p = q$, f is a bijection, $g = f^{-1}$ and $\mathbf{N}_{f(i)} \rightleftarrows \mathbf{M}_i$ for every $1 \le i \le p$. That is: the decomposition of $[\mathbf{A}]$ is unique.

9.5 To prove that the set of non-connected elements of $[\mathrm{Rel}(\sigma)]$ is dense in $\overline{[\mathrm{Rel}(\sigma)]}_L$, we prove that for every connected $\mathbf{G}$ and every $\epsilon > 0$ there exists non-connected $\mathbf{H}$ such that $\mathbf{G} \to \mathbf{H}$ and $\mathrm{dist}_L([\mathbf{G}], [\mathbf{H}]) < \epsilon$.

According to the Lemmas 9.9 and 9.3-(d), there exist $\mathbf{A}$ and $\mathbf{B}$ such that $\mathbf{A} \not\rightleftarrows \mathbf{B}$, $\mathbf{G} \to \mathbf{A} \times \mathbf{B}$ and $\mathrm{dist}_L([\mathbf{G}], [\mathbf{A} \times \mathbf{B}]) < \epsilon$. Notice that $\mathbf{A} \times \mathbf{B} \rightleftarrows\!\!\!\!\!/\;\, \mathbf{A} + \mathbf{B}$.

According to ambivalence lemma there exists a graph $\mathbf{H}$ such that $\mathbf{A}\times\mathbf{B} \mathrel{\substack{\longrightarrow \\ \not\longleftarrow}} \mathbf{H} \mathrel{\substack{\longrightarrow \\ \not\longleftarrow}} \mathbf{A}+\mathbf{B}$ such that $\mathrm{dist}_L([\mathbf{A}]\times[\mathbf{B}],[\mathbf{H}])<\epsilon$ and $\mathrm{dist}_R([\mathbf{H}],[\mathbf{A}+\mathbf{B}])<2^{-|\mathbf{A}|+|\mathbf{B}|}$. We deduce that $\mathrm{dist}_L([\mathbf{G}],[\mathbf{H}])<\epsilon$ (as dist_L is an ultrametric), $\mathbf{G}\to\mathbf{H}$ and $\mathbf{H}\to\mathbf{A}+\mathbf{B}$ although $\mathbf{H}\nrightarrow\mathbf{A}$ and $\mathbf{H}\nrightarrow\mathbf{B}$ (which means that $\mathbf{H}$ is not connected).

Exercises of Chapter 11

11.2 The property is a direct consequence of the existence of the following homomorphisms:

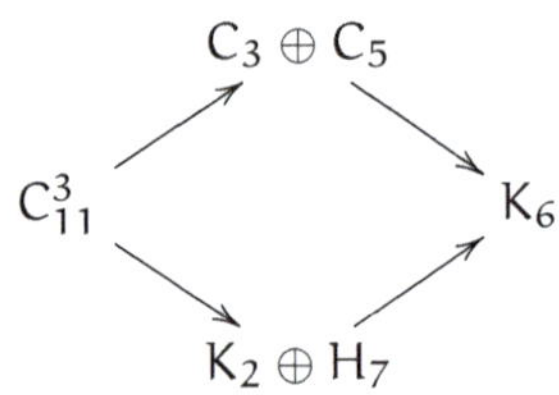

11.5 Assume that G is planar and that $G^{[\sharp 2p+1]}$ contains a K_5. As K_5 is not planar, two minimum distance paths linking disjoint pairs $\{a,b\}$ and $\{a',b'\}$ of main vertices of the K_5 intersect. However, this implies that two vertices among a,b,a',b' are linked by a path of length strictly smaller that $2p+1$.

An evidence that the clique number of the odd-distance graphs of outer-planar graphs is not bounded follows from the following construction:

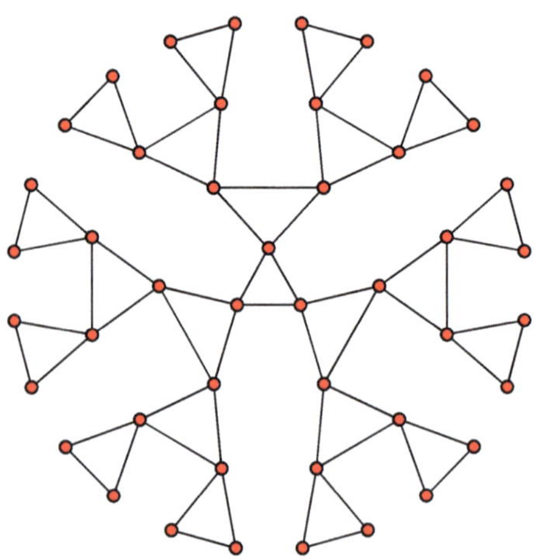

Exercises of Chapter 12

12.1 Let A be the set of the vertices of G with degree greater than $\frac{2k}{1-\epsilon}$. As G is k-degenerate, its average degree is at most 2k. Moreover, according to (3.3) (Sect. 3.2), we have $|A| \le \epsilon\,|G|$.

As $G - A$ has maximum degree at most $\frac{2k}{1-\epsilon}$, it contains at most $(\frac{2k}{1-\epsilon})^{|F|-1}\,|G|$ copies of F.

Exercises of Chapter 16

16.1 Consider a graph G of order n. Apply Lemma 16.3 with

$$z = \left\lfloor \frac{1}{2k}\log^2 n \log^2 \log n \right\rfloor \log n \approx \frac{1}{2k}\log^3 n \log^2 \log n.$$

The assumptions are satisfied, as

$$\begin{aligned}
2z(\omega(G \triangledown z) + 1) &\le 2z(f(z) + 2)\\
&\le 2z(f(\log^3 n \log^2 \log n) + 2)\\
&\le e^{\frac{1}{2}\log n}\\
&\le \sqrt{n \log n}.
\end{aligned}$$

Hence G has a separator of order $C\frac{n \log n}{z}$. Apply the lemma of Dvořak and Norine [139] and conclude that $\mathcal{C}$ is small.

References

1. F.N. Abu-Khzam, M.A. Langston, Graph coloring and the immersion order, in *Graph Theory and Combinatorics*, ed. by T. Warnow, B. Zhu. Lecture Notes in Computer Science, vol. 2697 (Springer, Berlin/New York), pp. 394–403
2. H. Adler, I. Adler, Interpreting nowhere dense graph classes as a classical notion of model theory. arXiv:1011.4016v1 [math.LO], (2010)
3. M. Aigner, G.M. Ziegler, *Proofs from the Book*, 2nd edn. (Springer, Berlin, 2004)
4. M. Ajtai, V. Chvátal, M.N. Newborn, E. Szemerédi, Crossing free subgraphs. Ann. Discret. Math. **12**, 9–12 (1982), Volume 60 of North-Holland Mathematical Studies, "Theory and practice of combinatorics"
5. M. Ajtai, Y. Gurevich, Monotone versus positive. J. ACM **34**, 1004–1015 (1987)
6. M. Ajtai, J. Komlós, E. Szemerédi, Topological complete subgraphs in random graphs. Studia Sci. Math. Hung. **14**, 293–297 (1979)
7. J. Alber, H. Bodlaender, H. Fernau, T. Kloks, R. Niedermeier, Fixed parameter algorithms for dominating set and related problems on planar graphs. Algorithmica **33**(4), 461–493 (2002)
8. J. Alber, H. Fan, M.R. Fellows, H. Fernau, R. Niedermeier, F. Rosamond, U. Stege, A refined search tree technique for dominating set on planar graphs. J. Comput. Syst. Sci. **71**(4), 385–405 (2005)
9. M.O. Albertson, G.G. Chappell, H.A. Kierstead, A. Kündgen, R. Ramamurthi, Coloring with no 2-colored P_4's. Electron. J. Comb. **11**(1), R26 (2004)
10. N. Alechina, Y. Gurevich, Syntax vs. semantics on finite structures, in *Structures in Logic and Computer Science, A Selection of Essays in Honor of Andrzej Ehrenfeucht*, ed. by J. Mycielsi, G. Rosenberg, A. Salomaa (Springer, Berlin/New York, 1997), pp. 14–33
11. N. Alon, Eigenvalues and expanders. Combinatorica **6**(2), 83–96 (1986)
12. ———, Subdivided graphs have linear ramsey numbers. J. Graph Theory **18**(4), 343–347 (1994)
13. N. Alon, F.R.K. Chung, Explicit construction of linear sized tolerant networks. Discret. Math. **72**, 15–19 (1989)
14. N. Alon, G. Ding, B. Oporowski, D. Vertigan, Partitioning into graphs with only small components. J. Comb. Theory Ser. B **87**(2), 231–243 (2003)
15. N. Alon, E. Fischer, M. Krivelevich, M. Szegedy, Efficient testing of large graphs. Combinatorica **20**, 451–476 (2000)
16. N. Alon, J. Grytczuk, Breaking the rhythm on graphs. Discret. Math. **308**, 1375–1380 (2008)

J. Nešetřil and P.O. de Mendez, *Sparsity*, Algorithms and Combinatorics 28, DOI 10.1007/978-3-642-27875-4, © Springer-Verlag Berlin Heidelberg 2012

17. N. Alon, J. Grytczuk, M. Hałuszczak, O. Riordan, Nonrepetitive colorings of graphs. Random Struct. Algorithm **21**(3–4), 336–346 (2002)
18. N. Alon, S. Gutner, Linear time algorithms for finding a dominating set of fixed size in degenerated graphs, in *COCOON*, ed. by G. Lin. LNCS, vol. 4598 (Springer, Berlin/New York, 2007), pp. 94–405
19. N. Alon, S. Hoory, N. Linial, The Moore bound for irregular graphs. Graph. Comb. **18**(1), 53–57 (2002)
20. N. Alon, M. Krivelevich, I. Newman, M. Szegedy, Regular languages are testable with a constant number of queries, in *Proceedings of the Fortieth Annual Symposium on Foundations of Computer Science* (IEEE Computer Society, Los Alamitos, 1999), pp. 656–666
21. N. Alon, T.H. Marshall, Homomorphisms of edge-colored graphs and Coxeter groups. J. Algebr. Comb. **8**, 5–13 (1998)
22. N. Alon, V.D. Milman, λ_1, isoperimetric inequalities for graphs, and superconcentrators. J. Comb. Theory Ser. B **38**(1), 73–88 (1985)
23. N. Alon, B. Mohar, D.P. Sanders, On acyclic colorings of graphs on surfaces. Isr. J. Math. **94**, 273–283 (1994)
24. N. Alon, P.D. Seymour, R. Thomas, A separator theorem for graphs with excluded minor and its applications, in *Proceedings of the 22nd Annual ACM Symposium on Theory of Computing* (Association for Computing Machinery, New York, 1990), pp. 293–299
25. ———, A separator theorem for nonplanar graphs. J. Am. Math. Soc. **3**, 801–808 (1990)
26. N. Alon, A. Shapira, A characterization of the (natural) graph properties testable with one-sided error, in *Proceedings of the 46th IEEE FOCS* (IEEE Computer Society, Los Alamitos, 2005), pp. 429–438
27. ———, Every monotone graph property is testable, in *Proceedings of the Thirty-Seventh Annual ACM Symposium on Theory of Computing (STOC)* (ACM Press, New York, 2005), pp. 128–137
28. ———, A characterization of the (natural) graph properties testable with one-sided error. SIAM J. Comput. **37**(6), 1703–1727 (2008)
29. N. Alon, R. Yuster, U. Zwick, Color-coding. J. ACM **42**(4), 844–856 (1995)
30. S. Arnborg, D.G. Corneil, A. Proskurowski, Complexity of finding embeddings in a k-tree. SIAM J. Algebr. Discret. Method **8**(2), 277–284 (1987)
31. S. Arnborg, B. Courcelle, A. Proskurowski, D. Seese, An algebraic theory of graph reduction, in *Proceedings of the Fourth Workshop on Graph Grammars and Their Applications to Computer Science*. Lecture Notes in Computer Science, vol. 532 (Springer, Berlin/New York, 1991), pp. 70–83
32. S. Arnborg, J. Lagergren, D. Seese, Easy problems for tree decomposable graphs. J. Algorithm **12**, 308–340 (1991)
33. S. Arnborg, A. Proskurowski, Linear time algorithms for NP-hard problems restricted to partial k-trees. Discret. Appl. Math. **23**(1), 11–24 (1989)
34. S. Arora, B. Barak, *Computational Complexity: A Modern Approach* (Cambridge University Press, Cambridge/New York, 2009)
35. S. Arora, C. Lund, R. Motwani, M. Sudan, M. Szegedy, Proof verification and the hardness of approximation problems. J. ACM **45**, 501–555 (1998)
36. S. Arora, S. Safra, Probabilistic checking of proofs: a new characterization of NP. J. ACM **45**, 70–122 (1998)
37. G. Atneosen, On the embeddability of compacta in N-books: intrinsic and extrinsic properties, PhD thesis, Michigan State University, 1968
38. A. Atserias, On digraph coloring problems and treewidth duality, in *20th IEEE Symposium on Logic in Computer Science (LICS)* (IEEE Computer Society, Los Alamitos, 2005), pp. 106–115
39. ———, On digraph coloring problems and treewidth duality. Eur. J. Comb. **29**(4), 796–820 (2008)

40. A. Atserias, A. Dawar, M. Grohe, Preservation under extensions on well-behaved finite structures, in *32nd International Colloquium on Automata, Languages and Programming (ICALP)*, ed. by Springer. Lecture Notes in Computer Science, vol. 3580 (Springer, Berlin/New York, 2005), pp. 1437–1449
41. A. Atserias, A. Dawar, P.G. Kolaitis, On preservation under homomorphisms and unions of conjunctive queries, in *Proceedings of the Twenty-Third ACM SIGMOD-SIGACT-SIGART Symposium on Principles of Database Systems* (ACM Press, New York, 2004), pp. 319–329
42. ———, On preservation under homomorphisms and unions of conjunctive queries. J. ACM **53**, 208–237 (2006)
43. T. Austin, T. Tao, On the testability and repair of hereditary hypergraph properties. arXiv:0801.2179v2 [math.CO], (2009)
44. C. Avart, V. Rödl, M. Schacht, Every monotone 3-graph property is testable. Electron. Notes Discret. Math. **22**, 539–542 (2005)
45. L. Babai, Automorphism group and category of cospectral graphs. Acta Math. Acad. Sci. Hung. **31**, 195–306 (1978)
46. L. Babai, A. Pultr, Endomorphism monoids and topological subgraphs of graphs. J. Comb. Theory Ser. B **28**(3), 278–283 (1980)
47. B.S. Baker, Approximation algorithms for NP-complete problems on planar graphs. J. Assoc. Comput. Mach. **41**(1), 153–180 (1994)
48. J. Bar-Ilan, G. Korstarz, D. Peleg, How to allocate network centers. J. Algorithm **15**(3), 385–415 (1993)
49. J. Barát, P.P. Varjú, On square-free vertex colorings of graphs. Studia Sci. Math. Hung. **44**(3), 411–422 (2007)
50. ———, On square-free edge colorings of graphs. Ars Comb. **87**, 377–383 (2008)
51. J. Barát, D.R. Wood, Notes on nonrepetitive graph colouring. Electron. J. Comb. **15**, R99 (2008)
52. L. Barrière, C. Huemer, D. Mitsche, D. Orden, On the Fiedler value of large planar graphs, in *Proceedings of the European Conference on Combinatorics, Graph Theory and Applications (EuroComb'11)*, Budapest, 2011
53. L. Barto, M. Kozik, Constraint satisfaction problems of bounded width, in *Annual IEEE Symposium on Foundations of Computer Science* (IEEE Computer Society, Los Alamitos, 2009), pp. 595–603
54. J. Beck, *Combinatorial Games: Tic-Tac-Toe Theory* (Cambridge University Press, Cambridge, 2008)
55. P. Bellenbaum, R. Diestel, Two short proofs concerning tree-decompositions. Comb. Probab. Comput. **11**, 541–547 (2002)
56. I. Benjamini, O. Schramm, Recurrence of distibutional limits of finite planar graphs. Electron. J. Probab. **6**(23), 13pp (2001)
57. I. Benjamini, O. Schramm, A. Shapira, Every minor-closed property of sparse graphs is testable, in *Proceedings of the 40th Annual ACM Symposium on Theory of Computing* (ACM Press, New York, 2008), pp. 393–402
58. C. Berge, Sur le couplage maximum d'un graphe. C. R. Acad. Sci. Paris Ser. I Math. **247**, 258–259 (1958)
59. ———, *Graphs and Hypergraphs*, 2nd edn. (North-Holland, Paris, 1973)
60. ———, *Graphes*, troisième edn. (Gauthier-Villars, Paris, 1983)
61. F.R. Bernhart, P.C. Kainen, The book thickness of a graph. J. Comb. Theory Ser. B **27**(3), 320–331 (1979)
62. N.L. Biggs, *Algebraic Graph Theory*. Cambridge Tracts in Mathematics, vol. 67 (Cambridge University Press, New York, 1974)
63. Y. Bilu, N. Linial, Lifts, discrepancy and nearly optimal spectral gap. Combinatorica **26**(5), 495–519 (2006)
64. R. Blankenship, Book embeddings of graphs, PhD thesis, Department of Mathematics, Louisiana State University, U.S.A., 2003

65. R. Blankenship, B. Oporowski, Drawing subdivisions of complete and complete bipartite graphs on books, Technical Report 1999-4, Department of Mathematics, Louisiana State University, U.S.A., 1999
66. M. Blum, M. Luby, R. Rubinfeld, Self-testing/correcting with applications to numerical problems. JCSS **47**, 549–595 (1993)
67. H.L. Bodlaender, Dynamic programming algorithms on graphs with bounded tree-width, in *Proceedings of the 15th International Colloquium on Automata, Languages and Programming*. Lecture Notes in Computer Science, vol. 317 (Springer, Berlin/New York ,1988), pp. 105–119
68. ———, *Polynomial Algorithms for Graph Isomorphism and Chromatic Index on Partial* k*-trees, SWAT 88*. Lecture Notes in Computer Science, vol. 318/1988 (Springer, Berlin/New York, 1988), pp. 223–232
69. ———, Polynomial algorithms for graph isomorphism and chromatic index on partial k-trees. J. Algorithm **11**(4), 631–643 (1990)
70. ———, A linear-time algorithm for finding tree-decompositions of small treewidth. SIAM J. Comput. **25**(6), 1305–1317 (1996)
71. ———, A partial k-arboretum of graphs with bounded treewidth. Theor. Comput. Sci. **209**(1–2), 1–45 (1998)
72. H.L. Bodlaender, J.S. Deogun, K. Jansen, T. Kloks, D. Kratsch, H. Müller, Zs. Tuza, *Rankings of Graphs*. Graph-Theoretic Concepts in Computer Science, Lecture Notes in Computer Science, vol. 903/1995 (Springer, Berlin/New York, 1995), pp. 292–304
73. H.L. Bodlaender, J.R. Gilbert, H. Hafsteinsson, T. Kloks, Approximating treewidth, pathwidth, frontsize, and shortest elimination tree. J. Algorithm **18**, 238–255 (1995)
74. T. Bohman, A. Frieze, T. Łuczak, O. Pikhurko, C. Smyth, J. Spencer, O. Verbitsky, First-order definability of trees and sparse random graphs. Comb. Probab. Comput. **16**(3), 375–400 (2007)
75. D. Bokal, G. Fijavž, B. Mohar, The minor crossing number. SIAM J. Discret. Math. **20**(2), 344–356 (2006)
76. B. Bollobás, *Random graphs* (Cambridge University Press, Cambridge, 2001)
77. B. Bollobás, A. Thomason, Proof of a conjecture of Mader, Erdős and Hajnal on topological complete subgraphs. Eur. J. Comb. **19**, 883–887 (1998)
78. C. Borgs, J. Chayes, L. Lovász, V.T. Sós, K. Vesztergombi, Counting graph homomorphisms, in *Topics in Discrete Mathematics*, ed. by M. Klazar, J. Kratochvíl, M. Loebl, J. Matoušek, R. Thomas, P. Valtr. Algorithms and Combinatorics, vol. 26 (Springer, Berlin/New York, 2006), pp. 315–371
79. C. Borgs, J.T. Chayes, L. Lovász, V.T. Sós, B. Szegedy, K. Vesztergombi, Graph limits and parameter testing, in *Proceedings of the 38th Annual ACM Symposium Principles of Distributed computing* (Springer, Berlin/Heidelberg, 2005), pp. 51–59
80. O.V. Borodin, On acyclic colorings of planar graphs. Discret. Math. **25**(3), 211–236 (1979)
81. J. Bőttcher, K.P. Pruessmann, A. Taraz, A. Wűrfl, Bandwidth, expansion, treewidth, separators, and universality for bounded degree graphs. Eur. J. Comb. **31**(5), 1217–1227 (2010)
82. J. Bourgain, N. Katz, T. Tao, A sum-product estimate in finite fields, and applications. Geom. Funct. Anal. **14**, 27–57 (2004)
83. B. Brešar, J. Grytczuk, S. Klavžar, S. Niwczyk, I. Peterin, Nonrepetitive colorings of trees. Discret. Math. **307**(2), 163–172 (2007)
84. B. Brešar, S. Klavžar, Square-free colorings of graphs. Ars Comb. **70**, 3–13 (2004)
85. H. Brezis, *Analyse Fonctionnelle: Théorie et Applications* (Dunod, Paris, 1999)
86. Y. Bu, N.W. Cranston, M. Montassier, A. Raspaud, W. Wang, Star-coloring of sparse graphs. J. Graph Theory **62**(3), 201–219 (2009)

87. S.A. Burr, P. Erdős, *On the Magnitude of Generalized Ramsey Numbers for Graphs.* Colloquia Mathematica Societatis Janos Bolyai 10 Infinite and Finite Sets, vol. 1 (North-Holland Pub. Co, Amsterdam, 1975), pp. 214–240
88. S.A. Burr, P. Erdős, and L Lovász, On graphs of ramsey type. Ars Comb. **1**, 167–190 (1976)
89. S. Ceri, G. Gottlob, L. Tanca, What you always wanted to know about datalog (and never dared to ask). IEEE Trans. Knowl. Data Eng. **1**, 146–166 (1989)
90. A.K. Chandra, P.M. Merlin, Optimal implementation of conjunctive queries in relational databases, in *Proceediongs of the 9th ACM Symposium on Theory of Computing* (ACM, New York, 1977), pp. 77–90
91. L.S. Chandran, A high girth graph construction. SIAM J. Discret. Math. **16**(3), 366–370 (2003)
92. J. Cheeger, *A lower bound for the smallest eigenvalue of the Laplacian.* Problems in analysis (Papers dedicated to Salomon Bochner, 1969), (Princeton University Press, Princeton, 1970), pp. 195–199
93. G.T. Chen, R.H. Schelp, Graphs with linearly bounded ramsey numbers. J. Comb. Theory Ser. B **57**(1), 138–149 (1993)
94. M. Chen, A. Raspaud, W. Wang, 8-star-choosability of a graph with maximum average degree less than 3. Discret. Math. Theor. Comp. Sci. **13**(3), 97–110 (2011)
95. G. Cherlin, S. Shelah, N. Shi, Universal graphs with forbidden subgraphs and algebraic closure. Adv. Appl. Math. **22**(4), 454–491 (1999)
96. M. Chrobak, D. Eppstein, Planar orientations with low out-degree and compaction of adjacency matrices. Theor. Comput. Sci. **86**, 243–266 (1991)
97. V. Chvátal, V. Rödl, E. Szemerédi, and W.T. Trotter Jr., The Ramsey number of a graph with bounded maximum degree. J. Comb. Theory Ser. B **34**(3), 239–243 (1983)
98. R.S. Cohen, Rank-non-increasing transformations on transition graphs. Inf. Control **20**(2), 93–113 (1972)
99. D. Coppersmith, S. Winograd, Matrix multiplication via arithmetic progressions. J. Symbol. Comput. **9**, 251–280 (1990)
100. R. Cori, *Un Code pour les Graphes Planaires et ses Applications*, vol. 27 (Société Mathématique de France, Paris, 1975)
101. R. Cori, A. Machì, Maps, hypermaps and their automorphisms. Expo. Math. **10**, 403–467 (1992)
102. B. Courcelle, Graph rewriting: an algebraic and logic approach, in *Handbook of Theoretical Computer Science*, vol. 2, ed. by J. van Leeuwen (Elsevier, Amsterdam, 1990), pp. 142–193
103. ———, The monadic second-order logic of graphs I: recognizable sets of finite graphs. Inf. Comput. **85**, 12–75 (1990)
104. B. Courcelle, J. Makowsky, U. Rotics, Linear time solvable optimization problems on graphs of bounded clique-width. Theory Comput. Syst. **33**(2), 125–150 (2000)
105. J. D. Currie, Pattern avoidance: themes and variations. Theor. Comput. Sci. **339**(1), 7–18 (2005)
106. S. Czerwiński, J. Grytczuk, Nonrepetitive colorings of graphs. Electron. Notes Discret. Math. **28**, 453–459 (2007)
107. A. Czumaj, A. Shapira, C. Sohler, Testing hereditary properties of nonexpanding bounded-degree graphs. SIAM J. Comput. **38**(6), 2499–2510 (2009)
108. B. Das, S. Datta, P. Nimbhorkar, Log-space algorithms for paths and matchings in k-trees, in *Proceedings of the 27th International Symposium on Theoretical Aspects of Computer Science (STACS 2010)*, ed. by J.-Y. Marion, T. Schwentick. Leibniz International Proceedings in Informatics (LIPIcs), vol. 5 (Schloss Dagstuhl–Leibniz-Zentrum fuer Informatik, Wadern, 2010), pp. 215–226

109. A. Dawar, Finite model theory on tame classes of structures, in *Mathematical Foundations of Computer Science 2007*, ed. by L. Kučera, A. Kučera. Lecture Notes in Computer Science, vol. 4708 (Springer, New York, 2007), pp. 2–12
110. ———, Homomorphism preservation on quasi-wide classes. J. Comput. Syst. Sci. **76**, 324–332 (2010)
111. A. Dawar, M. Grohe, S. Kreutzer, Locally excluding a minor, in *Proceedings of the 22nd IEEE Symposium on Logic in Computer Science* (IEEE Computer Society, Los Alamitos, 2007)
112. A. Dawar, S. Kreutzer, Domination problems in nowhere-dense classes, in *IARCS Annual Conference on Foundations of Software Technology and Theoretical Computer Science (FSTTCS 2009) (Dagstuhl, Germany)*, ed. by Ravi Kannan, K. Narayan Kumar. Leibniz International Proceedings in Informatics (LIPIcs), vol. 4 (Schloss Dagstuhl–Leibniz-Zentrum fuer Informatik, Wadern, 2009), pp. 157–168.
113. ———, Parametrized complexity of first-order logic, Technical Report 131, Electronic Colloquium on Computational Complexity, 2009
114. D. Dellamonica Jr.,V. Rödl, A note on thomassen's conjecture. J. Comb. Theory Ser. B **101**(6), 509–515 (2011)
115. E.D. Demaine, F. Fomin, M.T. Hajiaghayi, D.M. Thilikos, Fixed-parameter algorithms for (k, r)-center in planar graphs and map graphs. ACM Trans. Algorithm **1**(1), 33–46 (2005)
116. E.D. Demaine, M. Hajiaghayi, K.-I. Kawarabayashi, Algorithmic graph minor theory: decomposition, approximation, and coloring, in *Proceedings—Annual IEEE Symposium on Foundations of Computer Science, FOCS*, vol. 2005 (IEEE Computer Society, Los Alamitos, 2005), pp. 637–646
117. J.S. Deogun, T. Kloks, D. Kratsch, H. Muller, On vertex ranking for permutation and other graphs, in *Proceedings of the 11th Annual Symposium on Theoretical Aspects of Computer Science*, ed. by P. Enjalbert, E.W. Mayr, K.W. Wagner. Lecture Notes in Computer Science, vol. 775 (Springer, Berlin/New York, 1994), pp. 747–758
118. M. DeVos, G. Ding, B. Oporowski, D.P. Sanders, B. Reed, P.D. Seymour, D. Vertigan, Excluding any graph as a minor allows a low tree-width 2-coloring. J. Comb. Theory Ser. B **91**, 25–41 (2004)
119. M. DeVos, Z. Dvořák, J. Fox, J. McDonald, B. Mohar, D. Scheide, Minimum degree condition forcing complete graph immersion. arXiv:1101.2630, (2011)
120. M. DeVos, K. Kawarabayashi, B. Mohar, H. Okamura, Immersing small complete graphs. Ars Math. Contemp. **3**(2), 139–146 (2010)
121. R. Diestel, C. Rempel, Dense minors in graphs of large girth. Combinatorica **25**, 111–116 (2004)
122. G. Ding, Covering the edges with consecutive sets. J. Graph Theory **15**(5), 559–562 (1991)
123. ———, Subgraphs and well-quasi-ordering. J. Graph Theory **16**(5), 489–502 (1992)
124. T. Dinski, X. Zhu, A bound for the game chromatic number of graphs. Discret. Math. **196**(1–3), 109–115 (1999)
125. J. Dodziuk, Difference equations, isoperimetric inequality and transience of certain random walks. Trans. AMS **284**(2), 787–794 (1984)
126. R. G. Downey, M. R. Fellows, Parameterized computational feasibility, in Feasible Mathematics II, ed. by P. Clote, J. Remmel (Birkhäuser, Boston, 1995), pp. 219–244
127. ———, *Parameterized Complexity*, (Springer, New York, 1999)
128. W. Duckworth, N.C. Wormald, M. Zito, Maximum induced matchings of random cubic graphs. J. Comput. Appl. Math. **142**(1), 39–50 (2002)
129. V. Dujmović, P. Morin, D.R. Wood, Layout of graphs with bounded tree-width. SIAM J. Comput. **34**(3), 553–579 (2005)

130. V. Dujmović, A. Pór, D.R. Wood, Track layouts of graphs. Discret. Math. Theor. Comput. Sci. **6**(2), 497–522 (2004)
131. V. Dujmović, D.R. Wood, On linear layouts of graphs. Discret. Math. Theor. Comput. Sci. **6**(2), 339–358 (2004)
132. ———, Three-dimensional grid drawings with sub-quadratic volume, in *Proceedings of the 11th International Symposium on Graph Drawing (GD '03)*, ed. by G. Liotta. Lecture Notes in Computer Science, vol. 2912 (Springer, Berlin/New York, 2004), pp. 190–201
133. ———, Stacks, queues and tracks: layouts of graph subdivisions. Discret. Math. Theor. Comput. Sci. **7**, 155–202 (2005)
134. Z. Dvořák, Asymptotical structure of combinatorial objects, PhD thesis, Charles University, Faculty of Mathematics and Physics, 2007
135. ———, On forbidden subdivision characterizations of graph classes. Eur. J. Comb. **29**(5), 1321–1332 (2008)
136. ———, Constant-factor approximation of domination number in sparse graphs. arXiv:1110.5190v1 [math.CO], (2011)
137. Z. Dvořák, D. Kráľ, Algorithms for classes of graphs with bounded expansion. Lect. Notes Comput. Sci. **5911 LNCS**, 17–32 (2010)
138. Z. Dvořák, D. Kráľ, R. Thomas, Deciding first-order properties for sparse graphs, in *51st Annual IEEE Symposium on Foundations of Computer Science (FOCS 2010)*, (IEEE Computer Society, Los Alamitos, 2010), pp. 133–142
139. Z. Dvořak, S. Norine, Small graph classes and bounded expansion. J. Comb. Theory Ser. B **100**(2), 171–175 (2010)
140. N. Eaton, Ramsey numbers for sparse graphs. Discret. Math. **185**, 63–75 (1998)
141. H.-D. Ebbinghaus, J. Flum, *Finite Model Theory* (Springer, New York, 1996)
142. J. Edmonds, A combinatorial representation for polyhedral surfaces. Not. Am. Math. Soc. **7**, 643 (1960)
143. K. Edwards, The harmonious chromatic number of bounded degree graphs. J. Lond. Math. Soc. **55**(2), 435–447 (1997)
144. L.C. Eggan, Transition graphs and the star-height of regular events. Mich. Math. J. **10**(4), 385–397 (1963)
145. A. Ehrenfeucht, An application of games to the completeness problem for formalized theories. Fundam. Math. **49**, 129–141 (1961)
146. M. Elberfeld, A. Jakoby, T. Tantau, Logspace versions of the theorems of Bodlaender and Courcelle, in *Proceedings of the 51st Annual IEEE Symposium on Foundations of Computer Science (FOCS 2010)* (IEEE Computer Society, Los Alamitos, 2010), pp. 143–152
147. ———, Algorithmic meta theorems for circuit classes of constant and logarithmic depth. Electron. Colloq. Comput. Complex. **18**, 128 (2011)
148. G. Elek, The combinatorial cost. l'Enseignement Math. **53**, 225–236 (2007)
149. ———, L^2-spectral invariants and convergent sequences of finite graphs. J. Funct. Anal. **254**(10), 2667–2689 (2008)
150. ———, A regularity lemma for bounded degree graphs and its applications: parameter testing and infinite volume limits. arXiv:0711.2800, (2007)
151. G. Elek, G. Lippner, An analogue of the szemerédi regularity lemma for bounded degree graphs. arXiv:0809.2879v2 [math.CO], (2008)
152. Gy. Elekes, On the number of sums and products. Acta Arith. **81**, 365–367 (1997)
153. J. Ellis, H. Fan, M. Fellows, The dominating set problem is fixed parameter tractable for graphs of bounded genus. J. Algorithm **52**(2), 152–168 (2004)
154. E. Engeler, Äquivalenzklassen von n-tupeln. Zeit. Math. Log. Grund. der Math. **5**, 340–345 (1959)
155. H. Enomoto, M.S. Miyauchi, Embedding graphs into a three page book with $O(m \log m)$ crossings of edges over the spine. SIAM J. Discret. Math. **12**(3), 337–341 (1999)

156. H. Enomoto, M.S. Miyauchi, K. Ota, Lower bounds for the number of edge-crossings over the spine in a topological book embedding of a graph. Discret. Appl. Math. **92**(2–3), 149–155 (1999)
157. D. Eppstein, Subgraph isomorphism in planar graphs and related problems, in *Proceedings of the 6th Symposium Discrete Algorithms* (ACM and SIAM, New York, 1995), pp. 632–640
158. ———, Subgraph isomorphism in planar graphs and related problems. J. Graph Algorithm Appl. **3**(3), 1–27 (1999)
159. ———, Diameter and treewidth in minor-closed graph families. Algorithmica **27**, 275–291, (2000), Special issue on treewidth, graph minors, and algorithms
160. ———, Separating thickness from geometric thickness, in *Proceedings of the 10th International Symposium on Graph Drawing (GD '02)*, ed. by M.T. Goodrich, S.G. Kobourov. Lecture Notes in Computer Science, vol. 2528 (Springer, London, 2002), pp. 150–161
161. ———, Densities of minor-closed families. Electron. J. Comb. **17**, R136 (2010)
162. P. Erdős, On sequences of integers no one of which divides the product of two others and on some related problems. Mitteil. Forsch.-Inst. Math. Mech. Univ. Tomsk **2**, 74–82 (1938)
163. ———, Graph theory and probability. Canad. J. Math. **11**(1), 34–38 (1959)
164. ———, Problems and results in combinatorial analysis and graph theory, in *Proof Techniques in Graph Theory*, ed. by F. Harary (Academic Press, New York, 1969), pp. 27–35
165. ———, Problems and results in combinatorial analysis and graph theory. Discret. Math. **72**(1–3), 81–92 (1988)
166. P. Erdős, S. Fajtlowicz, On the conjecture of Hajós. Combinatorica **1**(2), 141–143 (1981)
167. P. Erdős, A. Hajnal, On chromatic number of graphs and set-systems. Acta Math. Acad. Sci. Hung. **17**, 61–99 (1966)
168. P. Erdős, R. Rado, Sets having a divisor property. Am. Math. Monthly **57**, 255–257 (1952)
169. ———, Intersection theorems for systems of sets. J. Lond. Math. Soc. **35**, 85–90 (1960)
170. P. Erdős, A. Rényi, The evolution of random graphs. Magyar Tud. Akad. Mat. Kutató Int. Közl **5**, 17–61 (1960)
171. R. Fagin, Generalized first-order spectra and polynomial-time recognizable sets, in *Complexity of Computation*, ed. by R. Karp. SIAM–AMS Proceedings, vol. 7 (American Mathematical Society, Providence, 1974), pp. 43–73
172. R. Faudree, M. Simonovits, On a class of degenerate extremal graph problems. Combinatorica **3**, 83–93 (1983)
173. R.J. Faudree, A. Gyárfás, R.H Schelp, Zs. Tuza, Induced matchings in bipartite graphs. Discret. Math. **78**(1–2), 83–87 (1989)
174. T. Feder, M.Y. Vardi, Monotone monadic SNP and constraint satisfaction, in *Proceedings of the 25th Annual ACM Symposium on Theory of Computing (STOC)* (ACM Press, New York, 1993), pp. 612–622
175. ———, The computational structure of monotone monadic SNP and constraint satisfaction: a study through datalog and group theory. SIAM J. Comput. **28**(1), 57–104 (1999) (English)
176. G. Fertin, A. Raspaud, B. Reed, On star coloring of graphs, in *Proceedings of the 27th International Workshop on Graph-Theoretic Concepts in Computer Science (WG '01)*. Lecture Notes in Computer Science, vol. 2204 (Springer, Berlin, 2001), pp. 140–153
177. M. Fiedler, Algebraic connectivity of graphs. Czechoslov. Math. J. **23**(2), 298–305 (1973)
178. J. Flum, M. Frick, M. Grohe, Query evaluation via tree-decompositions. J. ACM **49**(6), 716–752 (2002)

179. J. Flum, M. Grohe, Fixed-parameter tractability, definability, and model checking. SIAM J. Comput. **31**, 113–145 (2001)
180. J. Foniok, J. Nešetřil, C. Tardif, Generalised dualities and maximal finite antichains in the homomorphism order of relational structures. Eur. J. Comb. **29**(4), 881–899 (2008)
181. J. Foniok, C. Tardif, Adjoint functors and tree duality. Discret. Math. Theor. Comput. Sci. **11**(2), 97–110 (2009)
182. N. Fountoulakis, D. Kühn, D. Osthus, The order of the largest complete minor in a random graph. Random Struct. Algorithm **33**(2), 127–141 (2008)
183. J. Fox, B. Sudakov, Two remarks on the Burr–Erdős conjecture. Eur. J. Comb. **30**(7), 1630–1645 (2009)
184. R. Fraïssé, Sur une nouvelle classification des systèmes de relations. Comptes Rendus **230**, 1022–1024 (1950)
185. ———, Sur quelques classifications des systèmes de relations, PhD thesis, University of Paris, 1953, published in Publications Scientifiques de l'Université d'Alger, series A 1 (1954), 35–182., pp. 1022–1024
186. H. de Fraysseix, T. Matsumoto, P. Ossona de Mendez, P. Rosenstiehl, Regular orientations and graph drawing, in *Third Slovenian International Conference in Graph Theory*, University of Ljubljana, 1995, abstract, pp. 12–13
187. H. de Fraysseix, P. Ossona de Mendez, On regular orientations, in *Prague Midsummer Combinatorial Workshop* (Charles University, Department of Applied Mathematics, Praha, 1994), abstract, pp. 9–13
188. ———, Regular orientations, arboricity and augmentation, in *DIMACS International Workshop, Graph Drawing 94*. Lecture notes in Computer Science, vol. 894 (Springer, Berlin/New York, 1995), pp. 111–118
189. ———, On intersection graphs of Jordan arcs, Technical Report 146, CAMS, Paris, 1997
190. ———, Distributive lattices on planar graphs, in *Graph Algorithms and Applications*, ed. by T. Nishizeki, R. Tamassia, D. Wagner. Dagsthul-Seminar-Report, vol. 219 (Geschäftsstelle Schloss Dagstuhl, Saarbrücken, 1998), abstract, p. 30
191. ———, Intersection graphs of Jordan arcs, in *Contemporary Trends in Discrete Mathematics*. DIMACS Series in Discrete Mathematics and Theoretical Computer Science, vol. 49, DIMATIA-DIMACS, Štiřin 1997 Proceedings (American Mathematical Society, Providence, 1999), pp. 11–28
192. ———, Connectivity of planar graphs. J. Graph Algorithm Appl. **5**(5), 93–105 (2001)
193. ———, On topological aspects of orientations. Discret Math. **229**(1–3), 57–72 (2001)
194. ———, *PIGALE: Public Implementation of a Graph Algorithm Library and Editor*, Free Software (GPL licence), 2002
195. G. Liotta, R. Tamassia, I.G. Tollis (eds.), *Connectivity of Planar Graphs*. Graphs Algorithms and Applications, vol. 2 (World Scientific, Singapore, 2004)
196. ———, Stretching of Jordan arc contact systems, in *Graph Drawing*, ed. by G. Liotta. Lecture Notes in Computer Science, vol. 2912 (Springer, Berlin/New York, 2004), pp. 71–85
197. ———, Contact and intersection representations, in *Graph Drawing 2004*, ed. by J. Pach. Lecture Notes in Computer Science, vol. 3383 (Springer, Berlin/New York, 2005), pp. 217–227
198. ———, Barycentric systems and stretchability Discret. Appl. Math. **155**(9), 1079–1095 (2007)
199. ———, On representations by contact and intersection of segments. Algorithmica **47**(4), 453–463 (2007)
200. H. de Fraysseix, P. Ossona de Mendez, P. Rosenstiehl, On triangle contact graphs, in *Combinatorics, Geometry and Probability: A Tribute to Paul Erdős* (Cambridge University Press, Cambridge/New York, 1997), pp. 165–178

201. ———, Representation of planar hypergraphs by contacts of triangles, in *Proceedings of Graph Drawing 2007*. Lecture Notes in Computer Science, vol. 4875/2008 (Springer, Berlin/New York, 2008), pp. 125–136
202. M. Frick, M. Grohe, Deciding first-order properties of locally tree-decomposable structures. J. ACM **48**, 1148–1206 (2001)
203. ———, The complexity of first-order and monadic second-order logic revisited. Ann. Pure Appl. Log. **130**(1–3), 3–31 (2004), Papers presented at the 2002 IEEE Symposium on Logic in Computer Science (LICS)
204. J. Gagnon, Les graphes asymétriques minimaux de longueur induite 3, mémoire de maîtrise, Université de Montréal, 2006
205. H. Gaifman, On local and non-local properties, in *Proceedings of the Herbrand Symposium, Logic Colloquium '81* (North Holland Publ. Co, Amsterdam, 1982)
206. T. Galai, Transitiv orientierbare Graphen. Acta Math. Acad. Sci. Hung. **18**, 25–66 (1967)
207. Z. Galil, R. Kannan, E. Szemerédi, On 3-pushdown graphs with large separators. Combinatorica **9**(1), 9–19 (1989)
208. ———, On nontrivial separators for k-page graphs and simulations by nondeterministic one-tape Turing machines. J. Comput. Syst. Sci. **38**(1), 134–149 (1989)
209. T. Gallai, On directed graphs and circuits, in *Theory of Graphs (Proceedings of the Colloquium Tihany 1966)* (Academic Press, New York, 1968), pp. 115–118
210. G. Gallo, M.D. Grigoriadis, R.E. Tarjan, A fast parametric maximum flow algorithm and applications. SIAM J. Comput. **18**(1), 30–55 (1989)
211. A. Galluccio, L.A. Goddyn, P. Hell, High-girth graphs avoiding a minor are nearly bipartite. J. Comb. Theory Ser. B **83**(1), 1–14 (2001)
212. M.R. Garey, D.S. Johnson, *Computers and Intractability; a Guide to the Theory of NP-Completeness* (W. H. Freeman & Co., New York, 1990)
213. F. Gavril, J. Urrutia, An algorithm for fraternal orientation of graphs. Inf. Process. Lett. **41**, 271–274 (1992)
214. W. Gelade, Succinctness of regular expressions with interleaving, intersection and counting, in *MFCS '08 Proceedings of the 33rd international Symposium on Mathematical Foundations of Computer Science*. Lecture Notes in Computer Science, vol. 5162/2008 (Springer, Berlin, 2008), pp. 363–374
215. A.C. Giannopoulou, P. Hunter, D.M. Thilikos, LIFO-search: a min–max theorem and a searching game for cycle-rank and tree-depth. Discret. Appl. Math. submited (2011)
216. A.C. Giannopoulou, D.M. Thilikos, Obstructions for tree-depth, in *European Conference on Combinatorics, Graph Theory and Applications (EuroComb 2009)*. Electronic Notes in Discrete Mathematics, vol. 34(1), (Elsevier, Amsterdam, 2009), pp. 249–253
217. E.N. Gilbert, Random graphs. Ann. Math. Stat. **30**, 1141–1144 (1959)
218. J.R. Gilbert, J.P. Hutchinson, R.E. Tarjan, A separator theorem for graphs of bounded genus. J. Algorithm **5**(3), 375–390 (1984)
219. J.R. Gilbert, D.J. Rose, A. Edenbrandt, A separator theorem for chordal graphs. SIAM J. Algebr. Discret. Method **5**, 306–313 (1984)
220. O. Goldreich, *Computational Complexity: A Conceptual Perspective* (Cambridge University Press, Cambridge/New York, 2008)
221. O. Goldreich, S. Goldwasser, D. Ron, Property testing and its connection to learning and approximation. JACM **45**(4), 653–750 (1998)
222. O. Goldreich, D. Ron, Property testing in bounded-degree graphs. Algorithmica **32**, 302–343 (2002)
223. M.C. Golumbic, M. Lewenstein, New results on induced matchings. Discret. Appl. Math. **101**(1–3), 157–165 (2000)
224. E. Grädel, E. Rosen, On preservation theorems for two-variable logic. Math. Log. Q. **45**, 315–325 (1999)

225. R.L. Graham, V. Rödl, A. Ruciński, On graphs with linear Ramsey numbers. J. Graph Theory **35**, 176–192 (2000)
226. ______, On bipartite graphs with linear Ramsey numbers, Paul Erdös and his mathematics. Combinatorica **21**, 199–209 (2001)
227. R.L. Graham, B.L. Rothschild, J. Spencer, *Ramsey Theory* (Wiley, New-York, 1980)
228. M. Grohe, K. Kawarabayashi, D. Marx, P. Wollan, Finding topological subgraphs is fixed-parameter tractable, in *STOC '11 Proceedings of the 43rd Annual ACM symposium on Theory of computing* (ACM Press, New York, 2011), pp. 479–488
229. M. Grohe, S. Kreutzer, Methods for algorithmic meta theorems, in *Model Theoretic Methods in Finite Combinatorics, Contemporary Mathematics* (AMS, Providence, 2011), pp. 181–206
230. M. Grohe, D. Marx, On tree width, bramble size, and expansion. J. Comb. Theory Ser. B **99**(1), 218–228 (2009)
231. ______, Structure theorem and isomorphism test for graphs with excluded topological subgraphs. arXiv:1111.1109v1 [cs.DS], (2011)
232. M. Grohe, G. Turán, Learnability and definability in trees and similar structures. Theory Comput. Syst. **37**(1), 193–220 (2004)
233. B. Grünbaum, Acyclic colorings of planar graphs. Isr. J. Math. **14**, 390–408 (1973)
234. J. Grytczuk, Nonrepetitive colorings of graphs—a survey. Int. J. Math. Math. Sci. Art. ID 74639 (2007)
235. ______, Pattern avoidance on graphs. Discret. Math. **307**(11–12), 1341–1346 (2007)
236. ______, Thue type problems for graphs, points, and numbers. Discret. Math. **308**(19), 4419–4429 (2008)
237. B. Guenin, Packing T-Joins and edge colouring in planar graphs, Technical report CORR 2003-09, University of Waterloo, Department of Combinatorics and Optimization, 2003
238. Y. Gurevich, Toward logic tailored for computational complexity, in *Computation and Proof Theory*, ed. by M.M. Richter et al. Lecture Notes in Mathematics (Springer, Berlin/Heidelberg, 1984)
239. ______, On finite model theory (extended abstract), in *Feasible Mathematics*, ed. by S.R. Buss, P.J. Scott (Birkhaüser, Boston, 1990), pp. 211–219
240. A. Gyárfás, T. Jensen, M. Stiebitz, On graphs with strongly independent colour-classes. J. Graph Theory **46**, 1–14 (2004)
241. R. Halin, S-functions for graphs. J. Geom. **8**, 171–176 (1976)
242. P.L. Hammer, U.N. Peled, X. Sun, Difference graphs. Discret. Appl. Math. **28**(1), 35–44 (1990)
243. K. Hashiguchi, Algorithms for determining relative star height and star height. Inf. Comput. **78**(2), 124–169 (1988)
244. M. Hasse, Zur algebraischen Begründung der Graphentheorie I, Math. Nachr. **28**, 275–290 (1964/5)
245. P. Haxell, T. Szabó, G. Tardos, Bounded size components—partitions and transversals. J. Comb. Theory Ser. B **88**(2), 281–297 (2003)
246. L.S. Heath, F.T. Leighton, A.L. Rosenberg, Comparing queues and stacks as mechanisms for laying out graphs. SIAM J. Discret. Math. **5**(3), 398–412 (1992)
247. L.S. Heath, S.V. Pemmaraju, Stack and queue layouts of posets. SIAM J. Discret. Math. **10**(4), 599–625 (1997)
248. L.S. Heath, A.L. Rosenberg, Laying out graphs using queues. SIAM J. Comput. **21**(5), 927–958 (1992)
249. S. Hedetniemi, Homomorphisms of graphs and automata, Technical report 03105-44-T, University of Michigan, 1966

250. Z. Hedrlín, On universal partly ordered sets and classes. J. Algebr. **11**, 503–509 (1969)
251. L. Heffter, Über das Problem der Nachbargebiete. Math. Ann. **8**, 17–20 (1891)
252. P. Hell, J. Nešetřil, On the complexity of H-coloring. J. Comb. Theory Ser. B, **48**(1), 92–110 (1990)
253. ———, *Graphs and Homomorphisms*. Oxford Lecture Series in Mathematics and its Applications, vol. 28 (Oxford University Press, Oxford/New York, 2004)
254. ———, Colouring, constraint satisfaction, and complexity. Comput. Sci. Rev. **2**(3), 143–163 (2008)
255. P. Hell, J. Nešetřil, X. Zhu, Duality and polynomial testing of tree homomorphisms. Trans. Am. Math. Soc. **348**, 1281–1297 (1996)
256. G. Higman, Ordering by divisibility in abstract algebras. Proc. Lond. Math. Soc. **3**(2), 326–336 (1952)
257. W. Hodges, *Model Theory* (Cambridge University Press, Cambridge/New York, 1993)
258. ———, *A Shorter Model Theory* (Cambridge University Press, Cambridge, 1997)
259. S. Hoory, N. Linial, A. Wigderson, Expander graphs and their applications. Bull. Am. Math. Soc. **43**(4), 439–561 (2006)
260. J.E. Hopcroft, J.D. Ullman, *Introduction to Automata Theory, Languages and Computation* (Addison-Wesley, Reading, 1979)
261. G. Hotz, Ein Satz über Mittellinien. Arch. Math. **10**, 314–320 (1959)
262. ———, Arkadenfadendarstellung von Knoten und eine neue Darstellung der Knotengruppe. Abh. Math. Sem. Univ. Hamburg **24**, 132–148 (1960)
263. J. Hubička, J. Nešetřil, Finite paths are universal. Order **22**, 21–40 (2005)
264. ———, Universal partial order represented by means of oriented trees and other simple graphs. Eur. J. Comb. **26**(5), 765–778 (2005)
265. Y. Ishigami, Removal lemma for infinitely-many forbidden hypergraphs and property testing. arXiv:math/0612669 [math.CO], (2008)
266. L.O. James, R.G. Stanton, D.D. Cowan, Graph decomposition for undirected graphs, in *Proceedings of the Third Southeastern International Conference on Combinatorics, Graph Theory, and Computing (CGTC'72)*, Florida Atlantic University, 1972, pp. 281–290.
267. S. Janson, T. Łuczak, A. Ruciński, *Random Graphs* (Wiley, New York, 2000)
268. T. Jiang, Compact topological minors in graphs. J. Graph Theory (2010), published online.
269. I.A. Kanj, L. Perković, Improved parametrized algorithms for planar dominating set, in *MFCS '02 Proceedings of the 27th International Symposium on Mathematical Foundations of Computer Science*. Lecture Notes In Computer Science, (Springer, London, 2002), pp. 399–410
270. R. Kannan, Unraveling k-page graphs. Inf. Control **66**(1–2), 1–5 (1985)
271. K. Kawarabayashi, B. Reed, A separator theorem in minor-closed classes, in *51st Annual IEEE Symposium on Foundations of Computer Science (FOCS 2010)* (2010), pp. 153–162
272. H.A. Kierstead, A simple competitive graph coloring algorithm. J. Comb. Theory Ser. B **78**, 57–68 (2000)
273. H.A. Kierstead, W.T. Trotter, Planar graph coloring with an uncooperative partner. J. Graph Theory **18**(6), 569–584 (1994)
274. H.A. Kierstead, D. Yang, Orderings on graphs and game coloring number. Order **20**, 255–264 (2003)
275. D. Kirsten, Distance desert automata and the star height problem. RAIRO—Theor. Inf. Appl. **39**, 455–509 (2005)
276. D.J. Kleitman, The crossing number of $K_{5,n}$. J. Comb. Theory **9**, 315–323 (1970)

277. T. Kloks, *Treewidth: Computations and Approximations* (Springer, Berlin/New York, 1994)
278. D.E. Knuth, *The Art of Computer Programming*, vol. 1, 3rd edn. (Fundamental Algorithms), (Addison-Wesley, Reading, 1997)
279. P. Komárek, Good characterizations of graphs, PhD thesis, Charles University, Prague, 1988
280. J. Komlós, E. Szemerédi, Topological cliques in graphs. Comb. Probab. Comput. **3**, 247–256 (1994)
281. ———, Topological cliques in graphs II. Comb. Probab. Comput. **5**, 79–90 (1996)
282. S. Kopparty, B. Rossman, The homomorphism domination exponent. Eur. J. Comb. **32**(7), 1097–1114 (2011)
283. A. Kostochka, L. Pyber, Small topological complete subgraphs of "dense" graphs. Combinatorica **8**(1), 83–86 (1988)
284. A.V. Kostochka, On the minimum of the Hadwiger number for graphs with given average degree. Metody Diskretnogo Analiza **38**, 37–58 (1982), in Russian, English translation: *AMS Trans. (2)*, **132**(1986), 15–32
285. A.V. Kostochka, L.S. Melnikov, On bounds of the bisection width of cubic graphs, in *Fourth Czechoslovakian Symposium on Combinatorics, Graphs and Complexity*, ed. by J. Nesetril, M. Fiedler (Elsevier, Burlington, 1992), pp. 151–154
286. A.V. Kostochka, V. Rödl, On graphs with small Ramsey numbers. J. Graph Theory **37**, 198–204 (2001)
287. ———, On graphs with small Ramsey numbers II. Combinatorica **24**(3), 389–401 (2004)
288. A.V. Kostochka, B. Sudakov, On Ramsey numbers of sparse graphs. Comb. Probab. Comput. **12**, 627–641 (2003)
289. M. Kreidler, D. Seese, Monadic NP and graph minors, in *Computer Science Logic*, ed. by G. Gottlab, E. Grandjean, K. Seyr. Lecture Notes in Computer Science, vol. 1584 (Springer, Berlin/Heidelberg, 1999), pp. 126–141
290. K. Kremnizer, M. Szczesny, Feynman graphs, rooted trees, and Ringel-Hall algebras. Commun. Math. Phys. **289**(2), 561–577 (2009)
291. D. Kühn, D. Osthus, Topological minors in graphs of large girth. J. Comb. Theory Ser. B **86**(2), 364–380 (2002)
292. ———, Every graph of sufficiently large average degree contains a C_4-free subgraph of large average degree. Combinatorica **24**(1), 155–162 (2004)
293. G. Kun, M. Szegedy, A new line of attack on the dichotomy conjecture, in *Proceedings of the 41st Annual ACM Symposium on Theory of Computing* (Association for Computing Machinery, New York, 2009), pp. 725–734
294. A. Kündgen, M.J. Pelsmajer, Nonrepetitive colorings of graphs of bounded tree-width. Discret. Math. **308**(19), 4473–4478 (2008)
295. A. Kündgen, C. Timmons, Star coloring planar graphs from small lists. J. Graph Theory **63**(4), 324–337 (2010)
296. K. Kuratowski, Sur le problème des courbes gauches en topologie. Fundam. Math. **15**, 271–293 (1930)
297. B. Larose, C. Loten, C. Tardif, A characterisation of first-order constraint satisfaction problems. Log. Method Comput. Sci. **3**(4), 6, 22pp (2007) (electronic)
298. B. Larose, C. Tardif, Hedetniemi's conjecture and the retracts of a product of graphs. Combinatorica **20**(4), 531–544 (2000)
299. D. Lascar, *La théorie des modèles en peu de maux* (Cassini, Paris, 2009)
300. C. Lee, J. Lee, S.-i. Oum, Rank-width of random graphs. J. Graph Theory (2011). doi: 10.1002/jgt.20620 (in press)
301. F.T. Leighton, *Complexity Issues in VLSI* (MIT, Cambridge, 1983), ISBN 0262121042

302. Yusheng Li, C.C. Rousseau, L. Soltés, Ramsey linear families and generalized subdivided graphs. Discret. Math. **170**, 269–275 (1997)
303. L. Libkin, *Elements of Finite Model Theory* (Springer, Berlin/New York, 2004)
304. R. Lipton, R.E. Tarjan, A separator theorem for planar graphs. SIAM J. Appl. Math. **36**(2), 177–189 (1979)
305. R.J. Lipton, R.E. Tarjan, Applications of a planar separator theorem. SIAM J. Comput. **9**(3), 615–627 (1980)
306. J.W.H. Liu, The role of elimination trees in sparse factorization. SIAM J. Matrix Anal. Appl. **11**, 134–172 (1990)
307. M. Loebl, J. Nešetřil, Linearity and unprovability of set union problem, in *Proceedings of the Twentieth Annual ACM Symposium on Theory of Computing (STOC '88)* (ACM, New York, 1988), pp. 360–366
308. S. Lombardy, J. Sakarovitch, On the star height of rational languages, a new presentation for two old results, in *Words, Languages & Combinatorics III, Proceedings of the International Colloquium 2000*, Kyoto, ed. by M. Ito, T. Imaoka (World Scientific, River Edge, 2003), pp. 266–285
309. L. Lovász, Operations with structures. Acta Math. Hung. **18**, 321–328 (1967)
310. L Lovász, M. Freedman, A. Schrijver, Reflection positivity, rank connectivity, and homomorphisms in graphs. J. Am. Math. Soc. **20**, 37–51 (2007)
311. L. Lovász, J. Nešetřil, A. Pultr, On product dimension of graphs. J. Comb. Theory Ser. B **29**, 47–66 (1980)
312. L. Lovász, M.D. Plummer, *Matching Theory* (North-Holland, Amsterdam/New York, 1986)
313. L. Lovász, V.T. Sós, Generalized quasirandom graphs. J. Comb. Theory Ser. B **98**, 146–163 (2008)
314. L. Lovász, B. Szegedy, Limits of dense graph sequences. J. Comb. Theory Ser. B **96**, 933–957 (2006)
315. ———, Szemerédi lemma for the analyst. Geom. Func. Anal. **17**, 252–270 (2007)
316. A. Lubotzky, *Discrete Groups, Expanding Graphs and Invariant Measures*. Progress in Mathematics, vol. 125 (Brikhäuser, Basel/Boston, 1994)
317. A. Lubotzky, R. Phillips, P. Sarnak, Ramanujan graphs. Combinatorica **8**(3), 261–277 (1988)
318. T. Łuczak, B. Pittel, J.C. Wierman, The structure of a random graph at the point of the phase transition. Trans. AMS **341**, 721–748 (1994)
319. R.C. Lyndon, Properties preserved under homomorphism. Pac. J. Math. **9**, 129–142 (1959)
320. S. Mac Lane, I. Moerdijk, *Sheaves in Geometry and Logic: A First Introduction to Topos Theory*, Universitext (Springer-Verlag, New York, 1994)
321. F.R. Madelaine, Universal structures and the logic of forbidden patterns. Log. Method Comput. Sci. **5**(2), paper 13 (2009)
322. W. Mader, Homomorphiesätze für Graphen. Math. Ann. **178**, 154–168 (1968)
323. F. Manin, The complexity of nonrepetitive edge coloring of graphs. arXiv:0709.4497 [math.CO], (2007)
324. G.A. Margulis, Explicit group-theoretic constructions of combinatorial schemes and their applications in the construction of expanders and concentrators. Probl. Inf. Trans. **24**(1), 39–46 (1988)
325. D. Marker, *Model Theory: An Introduction*. Graduate Texts in Mathematics, vol. 217 (Springer, New York, 2001)
326. D. Marx, Can you beat treewidth?, in *48th Annual IEEE Symposium on Foundations of Computer Science (FOCS '07)* (IEEE Computer Society, Los Alamitos, 2007), pp. 169–179
327. D. Marx, M. Schaefer, The complexity of nonrepetitive coloring. Discret. Appl. Math. **157**, 13–18 (2009)
328. J. Matoušek, J. Nešetřil, *Invitation to Discrete Mathematics* (Oxford, Oxford University Press, 1998) (second printing 2008)

329. F. Mazoit, L. Lyaudet, S. Thomassé, Partitions versus sets: a case of duality. Eur. J. Comb. **31**(3), 681–687 (2010)
330. R.M. McConnell, J.P. Spinrad, Modular decomposition and transitive orientation. Discret. Math. **201**, 189–241 (1999)
331. R. McNaughton, The loop complexity of pure-group events. Inf. Control **11**(1–2), 167–176 (1967)
332. G.L. Miller, S.-H. Teng, W. Thurston, S.A. Vavasis, Geometric separators for finite-element meshes. SIAM J. Sci. Comput. **19**(2), 364–386 (1998)
333. G.L. Miller, W. Thurston, Separators in two and three dimensions, in *STOC '90: Proceedings of the Twenty-Second Annual ACM Symposium on Theory of Computing* (ACM, New York, 1990), pp. 300–309
334. M.S. Miyauchi, An $O(nm)$ algorithm for embedding graphs into a 3-page book. Trans. IEICE **E77-A**(3), 521–526 (1994)
335. ———, Embedding a graph into a $d+1$-page book with $\lceil m \log_d n \rceil$ edge crossings over the spine. IEICE Trans. Fundam. **E88-A**(5), 1136–1139 (2005)
336. J. Mycielski, Sur le coloriage des graphes. Colloq. Math. **3**, 161–162 (1955)
337. R. Naserasr, Homomorphisms and edge-coloring of planar graphs. J. Comb. Theory Ser. B **97**(3), 394–400 (2007)
338. R. Naserasr, Y. Nigussie, R. Škrekovski, Homomorphisms of triangle-free graphs without a K_5-minor. Discret. Math. **309**(18), 5789–5798 (2009), Combinatorics 2006, A Meeting in Celebration of Pavol Hell's 60th Birthday (May 1–5, 2006)
339. C. St J. A. Nash-Williams, On well-quasi-ordering trees, in *Theory of Graphs and Its Applications (Proceedings of the Symposium on Smolenice, 1963)* (Publishing House of the Czechoslovak Academy of Sciences, Prague, 1964), pp. 83–84
340. J. Nešetřil, *Ramsey Theory*, ed. by R.L. Graham, M. Grötschel, L. Lovász. Handbook of Combinatorics, vol. 2 (Elsevier, Amsterdam/New York, 1995), pp. 1331–1403
341. ———, Aspects of structural combinatorics. Taiwan. J. Math. **3**(4), 381–424 (1999)
342. ———, Homomorphism structure of classes of graphs. Comb. Probab. Comput. **8**, 177–184 (1999)
343. J. Nešetřil, J. Hubička, Homomorphism and embedding universal structures for restricted classes. (2012), to appear
344. J. Nešetřil, Y. Nigussie, Finite dualities and map critical graphs on a fixed surface. J. Comb. Theory Ser B **102**(1), 131–152 (2012)
345. J. Nešetřil, P. Ossona de Mendez, Colorings and homomorphisms of minor closed classes, in *Discrete & Computational Geometry: The Goodman-Pollack Festschrift*, ed. by B. Aronov, S. Basu, J. Pach, M. Sharir. Algorithms and Combinatorics, vol. 25 (Springer, Berlin/New York, 2003), pp. 651–664
346. ———, *Aspects algorithmiques des classes d'expansion bornée, 7e Journées Graphes et Algorithmes* (ed. by L. Esperet) (LaBRI – Université Bordeaux I), vol. 1372–05 (2005), pp. 55–58
347. ———, Cuts and bounds. Discret. Math. Struct. Comb.—Comb. Comput. Aspects Optim. Topol. Algebr. **302**(1–3), 211–224 (2005)
348. ———, The grad of a graph and classes with bounded expansion, in *7th International Colloquium on Graph Theory*, ed. by A. Raspaud, O. Delmas. Electronic Notes in Discrete Mathematics, vol. 22 (Elsevier, 2005) pp. 101–106.
349. ———, Folding. J. Comb. Theory Ser. B **96**(5), 730–739 (2006)
350. ———, Linear time low tree-width partitions and algorithmic consequences, in *STOC'06. Proceedings of the 38th Annual ACM Symposium on Theory of Computing* (ACM Press, 2006), pp. 391–400
351. ———, Low tree-depth partitions of classes with bounded expansion, in *Midsummer Combinatorial Workshop 2005 and DIMACS, DIMATIA, Rényi Workshop 2005*, Prague, ed. by J. Kára. KAM Series, vol. 2006-770, 2006, pp. 76–81

352. ______, Tree depth, subgraph coloring and homomorphism bounds. Eur. J. Comb. **27**(6), 1022–1041 (2006)
353. ______, Fraternal augmentations of graphs, coloration and minors, in *Proceedings of the Sixth Czech-Slovak International Symposium on Combinatorics, Graph Theory, Algorithms and Applications*, Elsevier. Electronic Notes in Discrete Mathematics, vol. 28 (2007), pp. 223–230.
354. ______, Grad and classes with bounded expansion I. Decompositions. Eur. J. Comb. **29**(3), 760–776 (2008)
355. ______, Grad and classes with bounded expansion II. Algorithmic aspects. Eur. J. Comb. **29**(3), 777–791 (2008)
356. ______, Counting homomorphisms to sparse graphs, in *European Conference on Combinatorics, Graph Theory and Applications (EuroComb 2009)*, ed. by J. Nešetřil, A. Raspaud, vol. 34 (Elsevier, Oxford, 2009), pp. 393–397
357. ______, How many F's are there in G? Eur. J. Comb. **32**(7), 1126–1141 (2011)
358. L. Lovasz, J. Nešetřil, P. Ossona de Mendez, A. Schrijver (eds.), Special issue on homomorphisms and limits. Eur. J. Comb. **32**(7), Elsevier, (2011)
359. J. Nešetřil, P. Ossona de Mendez, D.R. Wood, Characterizations and examples of graph classes with bounded expansion. Eur. J. Comb. **33**(3), 350–373 (2012)
360. J. Nešetřil, P. Ossona de Mendez, X. Zhu, Colouring edges with many colours in cycles, Technical Report 2011-1008, KAM-DIMATIA Series, 2011
361. J. Nešetřil, S. Poljak, Complexity of the subgraph problem. Comment. Math. Univ. Carol. **26.2**, 415–420 (1985)
362. J. Nešetřil, A. Pultr, On classes of relations and graphs determined by subobjects and factorobjects. Discret. Math. **22**, 287–300 (1978)
363. J. Nešetřil, A. Raspaud, Colored homomorphisms of colored mixed graphs. J. Comb. Theory Ser. B **80**(1), 147–155 (2000)
364. J. Nešetřil, V. Rödl, A short proof of the existence of highly chromatic hypergraphs without short cycles. J. Comb. Theory Ser. B **27**(2), 225–227 (1979)
365. ______, Chromatically optimal rigid graphs. J. Comb. Theory Ser. B **46**, 133–141 (1989)
366. J. Nešetřil, G. Sabidussi, Minimal asymmetric graphs of induced length 4. Graph Comb. **8**(4), 343–359 (1992)
367. J. Nešetřil, R. Šámal, Tension continuous maps-their structure and applications. Eur. J. Comb. **29**(4), 1025–1054 (2008)
368. J. Nešetřil, S. Shelah, On the order of countable graphs. Eur. J. Comb. **24**(6), 649–663 (2003)
369. J. Nešetřil, I. Švejdarová, Diameter of duals are linear. SIAM J. Discret. Math. **21**(2), 374–384 (2007)
370. J. Nešetřil, C. Tardif, Duality theorems for finite structures (characterizing gaps and good characterizations). J. Comb. Theory Ser. B **80**, 80–97 (2000)
371. J. Nešetřil, X. Zhu, On bounded treewidth duality of graphs. J. Graph Theory **23**(2), 151–162 (1996)
372. S. Norine, P. Seymour, R. Thomas, P. Wollan, Proper minor-closed families are small. J. Comb. Theory Ser. B **96**(5), 754–757 (2006)
373. R. Nowakowski, P. Winkler, Vertex-to-vertex pursuit in a graph. Discret. Math. **43**, 235–239 (1983)
374. P. Ossona de Mendez, On lattice structures induced by orientations, in *Proceedings of Graph Drawing '93* (CAMS, Paris, 1993), abstract, pp. 33–34
375. ______, *The Plane Bipolar Orientations' Lattice.* Sixth Franco-Japanese Days on Combinatorics and Optimization (École Normale Supérieure, Paris, 1993), abstract
376. ______, Orientations bipolaires, PhD thesis, École des Hautes Études en Sciences Sociales, Paris, 1994
377. J. Pach, G. Tóth, Graphs drawn with few crossings per edge. Combinatorica **17**(3), 427–439 (1997)

378. J. Pach, G. Tóth, Which crossing number is it anyway? J. Comb. Theory Ser. B **80**(2), 225–246 (2000)
379. C. Papadimitriou, *Computational Complexity* (Addison-Wesley Longman, Reading, 1994)
380. S.V. Pemmaraju, Exploring the powers of stacks and queues via graph layouts, PhD thesis, Virginia polytechnic institute and state university, U.S.A., 1992
381. G. Perarnau, O. Serra, On the tree-depth of random graphs, arXiv:1104.2132v1 [math.CO], (2011)
382. A. Pezarski, M. Zmarz, Non-repetitive 3-coloring of subdivided graphs. Electron. J. Combin. **16**(1), #N15 (2008)
383. M.S. Pinsker, On the complexity of a concentrator, in *7th International Teletraffic Conference*, International Advisory Council, 1973, pp. 318/1–318/4
384. B. Pittel, On a likely shape of the random Ferrers diagram, Adv. Appl. Math. **18**, 432–488 (1997)
385. J. Plehn, B. Voigt, Finding minimally weighted subgraphs, in *Proceedings of 16th International Workshop Graph-Theoretic Concepts in Computer Science*, ed. by Springer. Lecture Notes in Computer Science, vol. 484 (Springer, Berlin/Heidelberg, 1991), pp. 18–29
386. S. Plotkin, S. Rao, W.D. Smith, Shallow excluded minors and improved graph decomposition, in *5th Symposium on Discrete Algorithms* (SIAM, Philadelphia, 1994), pp. 462–470
387. K.-P. Podewski, M. Ziegler, Stable graphs. Fundam. Math. **100**, 101–107 (1978)
388. S. Poljak, Coloring digraphs by iterated antichains. Comment. Math. Univ. Carol. **32**(2), 209–212 (1991)
389. A. Pultr, V. Trnková, *Combinatorial Algebraic and Topological Representations of Groups, Semigroups and Categories* (North-Holland, Amsterdam/New York, 1980)
390. L. Pyber, V. Rödl, E. Szemerédi, Dense graphs without 3-regular subgraphs. J. Comb. Theory Ser. B **63**(1), 41–54 (1995)
391. A. Quillot, Jeux et pointes fixes sur les graphes, PhD thesis, Universite de Paris VI, 1978
392. S. Radziszowski, Small Ramsey numbers. Electron. J. Comb. **DS1**, (2004), dynamic survey
393. F.P. Ramsey, On a problem of formal logic. Proc. Lond. Math. Soc. Ser. 2 **30**, 264–286 (1930)
394. B. Reed, Tree width and tangles: a new connectivity measure and some applications, in *Surveys in Combinatorics*. London Mathematical Society Lecture Note Series, vol. 241 (Cambridge University Press, Cambridge/New York, 1997), pp. 87–162
395. B. Reed, M. Molloy, A bound on the strong chromatic index of a graph. J. Comb. Theory Ser. B **69**, 103–109 (1997)
396. G. Ringel, Das Geschlecht des vollstandingen paaren Graphen. Abh. Math. Sem. Univ. Hambg. **28**, 139–150 (1965)
397. N. Robertson, P.D. Seymour, Graph minors I. Excluding a forest. J. Comb. Theory Ser. B **35**(1), 39–61 (1983)
398. ———, Graph minors I–XX. J. Comb. Theory Ser. B (1983–2004)
399. ———, Graph minors III. Planar tree-width. J. Comb. Theory Ser. B **36**(1), 49–64 (1984)
400. ———, Graph minors II. Algorithmic aspects of tree-width. J. Algorithm. **7**(3), 309–322 (1986)
401. ———, Graph minors V. Excluding a planar graph. J. Comb. Theory Ser. B **41**(1), 92–114 (1986)
402. ———, Graph minors VI. Disjoint paths across a disc. J. Comb. Theory Ser. B **41**(1), 115–138 (1986)

403. ———, Graph minors VII. Disjoint paths on a surface. J. Comb. Theory Ser. B **45**(2), 212–254 (1988)
404. ———, Graph minors IV. Tree-width and well-quasi-ordering. J. Comb. Theory Ser. B **48**(2), 227–254 (1990)
405. ———, Graph minors IX. Disjoint crossed paths. J. Comb. Theory Ser. B **49**(1), 40–77 (1990)
406. ———, Graph minors VIII. A Kuratowski theorem for general surfaces. J. Comb. Theory Ser. B **48**(2), 255–288 (1990)
407. ———, Graph minors X. Obstructions to tree-decomposition. J. Comb. Theory Ser. B **52**(2), 153–190 (1991)
408. ———, Graph minors XI. Circuits on a surface. J. Comb. Theory Ser. B **60**(1), 72–106 (1994)
409. ———, Graph minors XII. Distance on a surface. J. Comb. Theory Ser. B **64**(2), 240–272 (1995)
410. ———, Graph minors XIII. The disjoint paths problem. J. Comb. Theory Ser. B **63**(1), 65–110 (1995)
411. ———, Graph minors XIV. Extending an embedding. J. Comb. Theory Ser. B **65**(1), 23–50 (1995)
412. ———, Graph minors XV. Giant steps. J. Comb. Theory Ser. B **68**(1), 112–148 (1996)
413. ———, Graph minors XVII. Taming a vortex. J. Comb. Theory Ser. B **77**(1), 162–210 (1999)
414. ———, Graph minors XVI. Excluding a non-planar graph. J. Comb. Theory Ser. B **89**(1), 43–76 (2003)
415. ———, Graph minors XVIII. Tree-decompositions and well-quasi-ordering. J. Comb. Theory Ser. B **89**(1), 77–108 (2003)
416. ———, Graph minors XIX. Well-quasi-ordering on a surface. J. Comb. Theory Ser. B **90**(2), 325–385 (2004)
417. ———, Graph minors XX. Wagner's conjecture. J. Comb. Theory Ser. B **92**(2), 325–357 (2004)
418. ———, Graph minors XXIII. Nash-Williams' immersion conjecture. J. Comb. Theory Ser. B **100**(2), 181–205 (2010)
419. N. Robertson, P.D. Seymour, R. Thomas, Hadwiger's conjecture for K_6-free graphs. Combinatorica **14**, 279–361 (1993)
420. V. Rödl, On the chromatic number of subgraphs of a given graph. Proc. Am. Math. Soc. **64**(2), 370–371 (1977)
421. V. Rödl, M. Schacht, Property testing in hypergraphs and the removable lemma, in *Proceedings of the Thirty-Ninth Annual ACM symposium on Theory of computing* (ACM Press, New York, 2007), pp. 488–495
422. ———, Regular partitions of hypergraphs: regularity lemmas. Comb. Probab. Comput. **16**, 833–885 (2007)
423. ———, Generalizations of the removal lemma. Combinatorica **29**(4), 467–501 (2009)
424. V. Rödl, R. Thomas, Arrangeability and clique subdivisions, in *The Mathematics of Paul Erdös*, ed. by R.L. Graham, J. Nešetřil (Springer, Berlin, 1991), pp. 236–239
425. B. Rossman, Homomorphism preservation theorems. J. ACM **55**(3), 1–53 (2008)
426. B. Roy, Nombre chromatique et plus longs chemins d'un graphe. Rev. Fr. Inform. Rech. Oper. **1**, 129–132 (1967)
427. R. Rubinfeld, M. Sudan, Robust characterization of polynomials with applications to program testing. SIAM J. Comput. **25**, 252–271 (1996)
428. C. Ryll-Nardzewski, On categoricity in power $\leq \aleph_0$. Bull. l'Acad. Pol. des Sci. **7**, 545–548 (1959)
429. W.J. Savitch, Relationships between nondeterministic and deterministic tape complexities. J. Comput. Syst. Sci. **4**(2), 177–192 (1970)

430. T.J. Schaefer, The complexity of satisfiability problems, in *STOC '78 Proceedings of the Tenth Annual ACM Symposium on Theory of Computing* (ACM Press, New York, 1978), pp. 216–226
431. P. Schaffer, Optimal node ranking of trees in linear time. Inf. Process. Lett. **33**, 91–96 (1989/90)
432. P. Scheffler, Linear-time algorithms for NP-complete problems restricted to partial k-trees. Preprint AdW d DDR, R-Math-03/87 (Karl-Weierstrass-Inst. für Math., Berlin, 1989)
433. D. Seese, Linear time computable problems and first-order descriptions. Math. Struct. Comput. Sci. **5**, 505–526 (1996)
434. P.D Seymour, R. Thomas, Graph searching and a min-max theorem for tree-width. J. Comb. Theory Ser. B **58**, 22–33 (1993)
435. S. Shelah, *Classification Theory and the Number of Non-isomorphic Models* (North-Holland, Amsterdam/New York, 1990)
436. L. Shi, Numbers and topologies: two aspects of Ramsey theory, PhD thesis, Mathematisch-Naturwissenschaftlichen Fakultät II, Humboldt-Universität zu Berlin, 2003
437. ———, Linear Ramsey numbers of sparse graphs. J. Graph Theory **50**(3), 175–185 (2005)
438. D.J. Skrien, A relationship between triangulated graphs, comparability graphs, proper interval graphs, proper circular arc graphs, and nested interval graphs. J. Graph Theory **6**(3), 309–316 (1982)
439. D.A. Spielman, S.-H. Teng, Spectral partitioning works: planar graphs and finite element meshes, in *37th Annual Symposium on Foundations of Computer Science (FOCS '96)* (IEEE Computer Society Press, Los Alamitos, 1996), pp. 96–106
440. ———, Spectral partitioning works: planar graphs and finite element meshes. Linear Algebra Appl. **421**(2–3), 284–305 (2007), Special Issue in honor of Miroslav Fiedler.
441. A. Steger, M.-l. Yu, On induced matchings. Discret. Math. **120**(1–3), 291–295 (1993)
442. I.A. Stewart, On the fixed-parameter tractability of parametrized model-checking problems. Inf. Process. Lett. **106**, 33–35 (2008)
443. L.J. Stockmeyer, The complexity of decision problems in automata theory, PhD thesis, Department of Electrical Engineering, MIT, 1974
444. L.J. Stockmeyer, V.V. Vazirani, NP-completeness of some generalizations of the maximum matching problem. Inf. Process. Lett. **15**(1), 14–19 (1982)
445. A. Stolboushkin, Finite monotone properties, in *Proceedings of 10th IEEE Symposium on Logic in Computer Science* (IEEE Computer Society Press, Los Alamitos, 1995), pp. 324–330.
446. A. Stolboushkin, M. Taitslin, Is first order contained in an initial segment of PTIME? in *Computer Science Logic*, ed. by L. Pacholski, J. Tiuryn. Lecture Notes in Computer Science, vol. 933 (Springer, Berlin/Heidelberg, 1995), pp. 242–248
447. L. Svenonius, $\aleph_0$–categoricity in first-order predicate calculus. Theoria **25**, 82–94 (1959)
448. B. Szegedy, Edge coloring models and reflection positivity. J. Am. Math. Soc. **20**, 969–988 (2007)
449. L.A. Székely, Crossing numbers and hard Erdős problems in discrete geometry. Comb. Probab. Comput. **6**(3), 353–358 (1997)
450. ———, A successful concept for measuring non-planarity of graphs: the crossing number. Discret. Math. **276**(1–3), 331–352 (2004)
451. E. Szemerédi, Regular partitions of graphs. Colloq. Int. CNRS **260**, 399–401 (1978)

452. W. Tait, A counterexample to a conjecture of Scott and Suppes. J. Symb. Log. **24**, 15–16 (1959)
453. T. Tao, The dichotomy between structure and randomness, arithmetic progression, and the primes, in *Proceedings of the International Congress of Mathematicians (Madrid 2006)*, ed. by European Mathematical Society, vol. 1 (European Mathematical Society, Zürich, 2007), pp. 581–608
454. C. Tardif, *Hedetniemi's Conjecture, 40 years Later*. Graph Theory Notes of New York LIV (New York Academy of Sciences, New York, 2008), pp. 46–57
455. A. Thomason, An extremal function for contractions of graphs. Math. Proc. Camb. Philos. Soc. **95**, 261–265 (1984)
456. ———, The extremal function for complete minors. J. Comb. Theory Ser. B **81**(2), 318–338 (2001)
457. C. Thomassen, Girth in graphs. J. Comb. Theory Ser. B **35**, 129–141 (1983)
458. ———, Five-coloring graphs on the torus. J. Comb. Theory Ser. B **62**(1), 11–33 (1994)
459. ———, A short list color proof of Grötzsch theorem, J. Comb. Theory Ser. B **88**, 189–192 (2003)
460. ———, Some remarks on Hajós' conjecture. J. Comb. Theory Ser. B **93**(1), 95–105 (2005)
461. A. Thue, Über unendliche Zeichenreihen. Norske Videnskabers Selskabs Skrifter Mat.-Nat. Kl. (Kristiana) **7**, 1–22 (1906)
462. W.T. Tutte, The factorization of linear graphs. J. Lond. Math. Soc. **22**, 107–111 (1947)
463. L. Valiant, A theory of the learnable. Commun. ACM **27**(11), 1134–1142 (1984)
464. V.N. Vapnik, A.Y. Chervonenkis, On the uniform convergence of relative sequences of events to their probabilities. Theory Probab. Appl. **16**, 264–280 (1971)
465. M. Vardi, The complexity of relational query languages, in *Proceedings of the 14th ACM Symposium on Theory of Computing* (ACM, New York, 1982), pp. 137–146
466. L.M. Vitaver, Determination of minimal colouring of vertices of a graph by means of boolean powers of incidence matrix. Doklady Akad. Nauk SSSR **147**, 758–759 (1962)
467. K. Wagner, Über eine Eigenschaft der ebenen Komplexe. Math. Ann. **114**, 570–590 (1937)
468. E. Welzl, Color families are dense, Theor. Comput. Sci. **17**, 29–41 (1982)
469. T.V. Wimer, Linear algorithms on k-terminal graphs, PhD thesis, Clemson University, 1987
470. D.R. Wood, Acyclic, star and oriented colourings of graph subdivisions. Discret. Math. Theor. Comput. Sci. **7**(1), 37–50 (2005)
471. N. Wormald, Models of random regular graphs, in *Surveys in Combinatorics* (Cambridge University Press, Cambridge/New York, 1999), pp. 239–298
472. M. Yannakakis, The complexity of the partial order dimension problem. SIAM J. Algebra. Discret. Method **3**, 351–358 (1982)
473. ———, Four pages are necessary and sufficient for planar graphs, in *18th Symposium of Theory of Computing* (ACM, New York, 1986), pp. 104–108
474. ———, Embedding planar graphs in four pages. J. Comput. Syst. Sci. **38**(1), 36–67 (1989)
475. X. Zhu, A survey on Hedetniemi's conjecture. Taiwan. J. Math. **2**(1), 1–24 (1998)
476. ———, Colouring graphs with bounded generalized colouring number. Discret. Math. **309**(18), 5562–5568 (2009)
477. ———, Fractional Hedetniemi's conjecture is true. Eur. J. Comb. **32**(7), 1168–1175 (2011)

Index

DOI 10.1007/978-3-642-27875-4,

MIX
Papier aus verantwortungsvollen Quellen
Paper from responsible sources
FSC® C105338

If you have any concerns about our products,
you can contact us on
ProductSafety@springernature.com

In case Publisher is established outside the EU,
the EU authorized representative is:
Springer Nature Customer Service Center GmbH
Europaplatz 3, 69115 Heidelberg, Germany

Printed by Libri Plureos GmbH
in Hamburg, Germany